Progress in Probability

Volume 22

Diffusion Processes and Related Problems in Analysis, Volume I

Diffusions in Analysis and Geometry

Mark A. Pinsky
Editor

1990

Springer Science+
Business Media, LLC

Mark A. Pinsky
Department of Mathematics
Northwestern University
Evanston, IL 60208-2730

Library of Congress Cataloging-in-Publication Data
Diffusion processes and related problems in analysis / [edited by]
Mark A. Pinsky.
p. cm. — (Progress in probability ; 22)
Papers presented at an international confernce held October
23-27, 1989 and sponsored by the Mathematics Department of
Northwestern University.
Includes bibliographical references.
Contents: v. 1. Diffusions in analysis and geometry
ISBN 978-1-4684-0566-8 ISBN 978-1-4684-0564-4 (eBook)
DOI 10.1007/978-1-4684-0564-4
1. Diffusion processes—Congresses. 2. Stochastic analysis–
–Congresses. I. Pinsky, Mark A., 1940– . II. Northwestern
University (Evanston, Ill.). Dept. of Mathematics. III. Series.
QA274.75.D54 1991
519.2—dc20 90-47081
CIP

Printed on acid-free paper.

ISBN 978-1-4684-0566-8

Camera-ready text prepared by the authors.

9 8 7 6 5 4 3 2 1

PREFACE

During the week of October 23–27, 1989, Northwestern University hosted an international conference on the theme "Diffusion Processes and Related Problems in Analysis." This was attended by 105 participants representing 14 different countries. The conference, which is part of the "Emphasis Year" program traditionally supported by the Mathematics Department, was additionally supported by grants from the National Science Foundation, the National Security Agency, the Institute for Mathematics and Applications, as well as by supplementary sources from Northwestern University.

The purpose of this meeting was to bring together workers in various parts of probability theory, mathematical physics, and partial differential equations. Previous efforts in this direction were represented by the 1987 AMS Summer Research Conference "Geometry of Random Motion" co-sponsored with Rick Durrett, the proceedings of which appeared as volume 73 in the AMS series "Contemporary Mathematics." The present effort is intended to extend beyond the strictly geometric theme and to include problems of large deviations, stochastic flows, and other areas of stochastic analysis in which diffusion processes play a leading role.

Admittedly the above definition of topics is somewhat "diffuse" in nature, leading to fears of a conference without a suitable focus. However, the organizers were delighted to find that, quite to the contrary, the meeting furnished a high level of scientific interchange and coherence. As much as we tried to feature the emerging younger figures, we were heartened to have the strong participation from many more recognized names. This interaction between people of different age groups and ethnic groups gave a healthy confirmation that our mathematical activities continue to provide "hands across the sea" as well as to unite otherwise disparate generations.

Many participants in 1989 conference had fond memories of the International Conference on Stochastic Analysis co-sponsored with Avner Freidman and held at Northwestern April 10–14, 1978 [published as *Stochastic Analysis* (Academic, 1978), ISBN 0-12-268380-3]. That meeting brought together such leading figures as A. Bensoussan, E. Dynkin, D. Elworthy, W. Fleming, R. Holley, N. Ikeda, K. Itô, P. Malliavin, S. Orey, G. Papanicolaou, E. Pardoux, D. Stroock, S. Varadhan, H. Tanaka, and S. Watanabe. At that time the power of probabilistic methods in analytical problems was just beginning to be appreciated. Hormander's theorem on hypoelliptic second-order differential operators was treated probabilistically for the first time by the "stochastic calculus of variations" developed by P. Malliavin and his followers. At this time it became especially clear that, in order to understand finite-dimensional problems ("downstairs"), it was well worth the effort to first understand the diffusion process at the infinite dimensional level ("upstairs"). The organization of the present volume is intended to reflect this order of priorities. The anticipated Volume II will continue to emphasize this viewpoint with the theme of "Stochastic Flows," a class of diffusion processes on the (infinite-dimensional) diffeomorphism group of a (finite-dimensional) manifold.

The present volume contains 30 papers, reflecting the above organization of the subject. We are indebted to the participants for providing their manuscripts promptly and to the many referees who were so helpful in providing their constructive comments and criticisms. The staff of the Mathematics Department is to be thanked for their assistance with the administration of the conference and subsequent arrangements. Mr. Edwin Beschler encouraged our efforts with the support of the Editorial Board of the series. We feel that our efforts will be successful if we have demonstrated the continuing richness of probability theory in its relations to analysis, geometry, and physics—alongside its known place in the spectrum of modern mathematics.

Mark A. Pinsky
Department of Mathematics
Northwestern University
Evanston, IL 60208

CONTENTS

Conference Participants
Evanston, Illinois, October 23–27, 1989

Robert Aebi (Zurich, CH)
Sergio Albeverio (Bochum, FRG)
Steven Altschuler (San Diego)
Ludwig Arnold (Bremen, FRG)
Dominique Bakry (Toulose)
Paolo Baldi (Catania, IT)
Rodrigo Banuelos (Lafayette)
Peter Baxendale (Los Angeles)
Denis Bell (Jacksonville)
Gerard BenArous (Paris, FRA)
Marc Berger (Atlanta)
Dante DeBlassie (College Station)
Alexandra Bellow (Evanston)
Douglas Blount (Salt Lake City)
Amit Bose (Ottawa, CAN)
Petra Boxler (Los Angeles)
Jim Brooks (Gainesville)
Rob Bradley (Evanston)
Gunnar Brosamler (Saarbrucken, FRG)
Wlodek Bryc (Cincinnati)
Krystoff Burdzy (Seattle)
Eric Carlen (Princeton)
Patrick Cattiaux (Etampes, FRA)
Isaac Chavel (New York)
David Chen (Evanston)
Mike Cranston (Rochester)
Hans Crauel (Bremen, FRG)
Donald Dawson (Ottawa, CAN)
Martin Day (Blacksburg)
Yves Derrienec (Brest, FRA)
Jean-Dominique Deuschel (Ithaca)
Nicolae Dinculeanu (Gainesville)
Tyrone Duncan (Lawrence)
Paul Dupuis (Amherst)
Rick Durrett (Ithaca)
Eugene Dynkin (Ithaca)
Alexandre Eisenberg (College Park)
David Elworthy (Coventry)
Stewart Ethier (Salt Lake City)
Edgar Feldman (New York)
Hans Föllmer (Bonn, FRG)
Hillel Furstenberg (Jerusalem, IS)

Victor Goodman (Bloomington)
Ho Ou (New Brunswick)
Elton Hsu (Evanston)
Saul Jacka (Coventry)
Leon Karp (New York)
Yuri Kifer (Jerusalem, IS)
Frank Knight (Urbana)
Tzong-Yow Lee (College Park)
John Lewis (Dublin, IRL)
Shenglin Lu (New York)
Paul Malliavin (Paris, FRA)
Peter March (Columbus)
Victor Mizel (Pittsburgh)
Salah Mohammed (Carbondale)
Tom Mountford (Los Angeles)
Carl Mueller (Rochester)
Masao Nagasawa (Zurich, CH)
S. Nakao (Osaka, JPN)
Shigeyoshi Ogawa (Kyoto, JPN)
Yukio Ogura (Saga, JPN)
Bernt Øksendal (Oslo, NRW)
George Papanicolaou (New York)
Etienne Pardoux (Marseille, FRA)
Mark Pinsky (Evanston)
Sidney Port (Los Angeles)
Philip Protter (Lafayette)
Joseph Pule (Dublin, IRL)
K. Ramachandran (Tampa)
Bruno Rémillard (Quebec, CAN)
Gareth Roberts (Nottingham, UK)
Michael Röckner (Edinburgh, UK)
Chris Rogers (Cambridge, UK)
Steven Rosenberg (Boston)
Boris Rozovskii (Charlotte)
Francesco Russo (Marseille, FRA)
Jamie San Martin (Lafayette)
Renming Song (Gainesville)
Seth Stafford (Ithaca)
Hiroshi Sugita (Saga, JPN)
Alain Sznitman (New York)
S. Takanobu (Nagoya, JPN)
John Taylor (Montreal, CAN)
Arkadii Tempelmann (Vilnius, USSR)
Aubrey Truman (Swansea, UK)
Naomasa Ueki (Osaka, JPN)

Michel Van den Berg (Edinburgh, UK)
Srinivasa Varadhan (New York)
Alexandre Wentsell (Moscow, USSR)
M. Vishik (Moscow, USSR)
Zoran Vondracek (Gainesville)
Joseph Watkins (Los Angeles)
Wojbor Woyczinski (Cleveland)
Volker Wihstutz (Charlotte)
Chuan-yi Xu (Evanston)
Z. Zhao (Columbia)
Weian Zheng (Shanghai, PRC)

Country codes: CH = Switzerland, FRG = West Germany, IT = Italy, FRA = France, CAN = Canada, IS = Israel, IRL = Ireland, JPN = Japan, NRW = Norway, PRC = Peoples Republic of China, UK = Great Britain, USSR = Soviet Union.

Part I
Infinite-dimensional Diffusion Processes

Martin Boundaries on Wiener Space

Hans Föllmer

0. Introduction

For Brownian motion on R^d, the Martin boundary associated to the integral representation of positive space-time harmonic functions is given by R^d. In this paper, our purpose is to look at the infinite-dimensional case and to investigate the Martin boundary for Brownian motion on $C[0,1]$.

In order to explain our approach, let us first review the probabilistic construction of the Martin boundary in the finite-dimensional case. Consider the class $\mathcal{H}$ of non-negative functions $h(x,t)$ on $R^d \times [0,\infty)$ normalized to $h(0,0) = 1$ which are space-time harmonic in the sense that

$$(0.1) \qquad P_s h(\cdot, t+s) = h(\cdot, t) \qquad (s,t \geq 0);$$

(P_s) denotes the semigroup of Brownian motion on R^d. The convex set $\mathcal{H}$ admits an integral representation of the form

$$(0.2) \qquad h(x,t) = \int_{R^d} K^y(x,t)\mu^h(dy)$$

in terms of the extreme points

$$(0.3) \qquad K^y(x,t) = \exp\left[(x,y) - \frac{t}{2}\,\|y\|^2\right] \qquad (y \in R^d),$$

and so R^d serves as the (space-time) Martin boundary of Brownian motion on R^d; cf., e.g., [8]. But (0.2) may also be viewed as an integral representation of the class of conditional Brownian motions induced by functions $h \in \mathcal{H}$. This observation goes back to J.L. Doob, and it is the key to the probabilistic construction of the Martin boundary. Let P on $\Omega = C([0,\infty), R^d)$ denote the distribution of Brownian motion on R^d starting in $0 \in R^d$, and define $X_t(\omega) = \omega(t)$. Any $h \in \mathcal{H}$ induces a martingale $h(X_t, t)$ $(t \geq 0)$ with respect to P, and a new probability measure P^h on Ω with local densities

$$\left.\frac{dP^h}{dP}\right|_{\mathcal{F}_t} = h(X_t, t) \qquad (t \geq 0) \tag{0.4}$$

on the σ-fields $\mathcal{F}_t = \sigma(X_s \cdot, s \leq t)$. P^h will be called the *h-process* associated to the space-time harmonic function h. Thus, we have identified $\mathcal{H}$ with the class of h-processes $\{P^h | h \in \mathcal{H}\}$. This class admits the following intrinsic characterization. For any $t > 0$, the conditional distribution of P^h with respect to the σ-field

$$\widehat{\mathcal{F}}_t = \sigma(X_s; s \geq t)$$

of future events after time t coincides with the conditional distribution $P[\cdot\,|\widehat{\mathcal{F}}_t](\omega)$ of the basic measure P. These conditional distributions can be defined explicitely in terms of a Brownian bridge from 0 to $X_t(\omega)$, in a consistent way and without the intervention of any $P-$ null sets. Thus, P^h belongs to the class

$$\mathcal{P} := \{Q |\; Q[\cdot\,|\widehat{\mathcal{F}}_t] = P[\cdot\,|\widehat{\mathcal{F}}_t] \quad (t > 0)\} \tag{0.5}$$

of all probability measures Q on Ω which are compatible with the local specification $P[\cdot\,|\widehat{\mathcal{F}}_t]$ $\;(t > 0)$ induced by P. But in the finite-dimensional case $d < \infty$, also the converse holds. In fact, we have $P[\cdot\,|\widehat{F}_u](\omega) \ll P$ on $\mathcal{F}_t$ whenever $0 < t < u$, and this implies

$$Q \ll P \;\text{ on } \mathcal{F}_t \qquad (Q \in \mathcal{P},\; t > 0). \tag{0.6}$$

It is then easy to show that the martingale of successive densities of a measure $Q \in \mathcal{P}$ with respect to P admits a version of the form (0.4) with some $h \in \mathcal{H}$, and so we have

$$\mathcal{P} = \{P^h |\, h \in \mathcal{H}\}. \tag{0.7}$$

In particular, the integral representation (0.2) in $\mathcal{H}$ translates into the integral representation

$$(0.8) \qquad Q = \int_{R^d} P^y \nu(dy)$$

of the convex set $\mathcal{P}$, where

$$(0.9) \qquad P^y = P^{K(\cdot,y)}$$

is the distribution of Brownian motion with constant drift $y \in R^d$.

From a probabilistic point of view, it is natural to prove directly the integral representation (0.8) for the class $\mathcal{P}$, using Dynkin's technique of sufficient statistics; cf [1,2]. The integral representation (0.2) for $\mathcal{H}$ then follows, due to (0.7) and (0.9). In infinite dimensions, however, the identification (0.9) of processes with functions breaks down. The reason is that we loose the local absolute continuity in (0.6). It now turns out that the probabilistic point of view is the more fundamental one. In fact, let P denote the distribution of Brownian motion with state space $B = \{x \in C[0,1] | \; x(0) = 0\}$. The integral representation in the class $\mathcal{P}$ defined with respect to P follows as in the finite-dimensional case, and the Martin boundary is given by B; this is shown in section 1. Extremal harmonic functions can be identified with those points in the Martin boundary which belong to the Cameron-Martin space H of all functions $y \in B$ which are absolutely continuous with derivative $\dot{y} \in L^2[0,1]$. Therefore, one might conjecture that an integral representation of space-time harmonic functions should only involve the Cameron-Martin space H. But an example in section 1 shows that this is not the case: We construct a space-time harmonic function whose representing measure ν is concentrated on the complement of H. Thus, B is the natural Martin boundary even if one is interested primarily in space-time harmonic functions rather than conditional Brownian motions.

In the finite-dimensional case, Itô 's formula

$$h(X_t,t) = \exp\left[\int_0^t \nabla \log h(X_s,s)ds - \frac{1}{2}\int_0^t \|\nabla log h(X_s,s)\|^2 ds\right]$$

together with (0.4) shows that P^h is given by a Girsanov transformation with drift $b(x,t) = \nabla \log h(x,t)$, and so the behavior of (X_t) under P^h is described by the stochastic differential equation

$$dX_t = dW_t + \nabla \log\, h(X_t, t)dt \tag{0.10}$$

where (W_t) is a Wiener process under P^h. In section 2 we show that an analogous description of h-processes can be given in infinite dimensions. This involves time reversal on Wiener space and a duality equation based on Malliavin calculus.

In section 3 we regard the infinite-dimensional Ornstein-Uhlenbeck process on B. Here again, the Martin boundary is given by B. In the recurrent reversible case, the Martin exit boundary can be identified with the Martin entrance boundary, and in this case the identification of the boundary has been obtained independently by M. Röckner [9]. But we can also take the transient Ornstein-Uhlenbeck process as our basic reference measure; the Martin boundary will remain the same. In the transient case, the process has a natural exit distribution given by a Wiener measure ν on B; in particular, we have $\nu(H) = 0$. This is another way of emphasizing that the Cameron-Martin space is not rich enough to serve as a Martin boundary in infinite dimensions.

The idea of constructing an infinite-dimensional Martin boundary in terms of the class $\mathcal{P}$ rather than in terms of space-time harmonic functions already appears in [5]. The example in section 1 was stimulated by a question of M. Röckner in Oberwolfach in July 89.

1. The Martin boundary of infinite-dimensional Brownian motion

Let P be the distribution of Brownian motion with state space

$$B := \{x \in C[0,1] | x(0) = 0\}.$$

Thus, P is a probability measure on the path space $\Omega := C([0,\infty), B)$ such that the coordinate process

$$X_t(\omega, \cdot) = \omega(t) \in B \qquad (t \geq 0)$$

has independent increments under P, and the distribution of $X_t - X_s$ on B is Wiener measure with variance $t - s$. Let H denote the Cameron-Martin space of all absolutely continuous functions $x \in B$ with derivative $\dot{x} \in L^2[0,1]$.

For our purposes, it will be convenient to introduce coordinates and to view Brownian motion on B as a countable collection of independent one-dimensional Brownian motions. This is made precise in the following

(1.1) **Remark.** Let $(\varphi_i)_{i \in I} \subseteq H$ denote the collection of Schauder functions on $[0,1]$. Then $(\dot{\varphi}_i)_{i \in I}$ is the orthonormal basis of Haar functions in $L^2[0,1]$, and each $x \in B$ can be represented in the form

$$x(\cdot) = \sum_i x_i \varphi_i(\cdot) \tag{1.2}$$

with coordinates $x_i = x(1)\dot{\varphi}_i(1) - \int_0^1 x\, d\dot{\varphi}_i$. Of course, x belongs to H if and only if $\sum_i x_i^2 < \infty$, and in that case we have

$$\dot{x}(\cdot) = \sum_i x_i \dot{\varphi}_i(\cdot)$$

with $x_i = \int_0^1 \dot{x}(\tau)\dot{\varphi}_i(\tau)d\tau$. (1.2) leads to the representation

$$X_t(\omega, \cdot) = \sum_{i \in I} X_t^i(\omega)\varphi_i(\cdot) \qquad (t \geq 0) \tag{1.3}$$

of Brownian motion on B in terms of a countable collection of independent one-dimensional Brownian motions $(X^i)_{i \in I}$. Conversely, (1.3) can be used as a *construction* of Brownian motion on B, in analogy to the Lévy-Ciesielski construction of one-dimensional Brownian motion in terms of independent normal random variables; cf., e.g., [7].

For $t > 0$ and for $y \in B$, let P_t^y denote the distribution on $C([0,t], B)$ of the Brownian bridge

$$X_s + (y - X_t) \cdot \frac{s}{t} \qquad (0 \leq s \leq t)$$

from $0 \in B$ to $y \in B$ at time t. In terms of these bridges, we can define an explicit version $\Pi_t(\omega, \cdot)$ of the conditional distribution of P with respect to the σ-field

$$\widehat{\mathcal{F}}_t := \sigma(X_s; s \geq t).$$

The collection of probability kernels $(\Pi_t)_{t>0}$ is a *local specification* in the sense of [3] and [2]. Let us now introduce the class $\mathcal{P}$ of all probability measures Q on the polish path space Ω which are compatible with this local specification. This means that, for each $t > 0$, the kernel Π_t is a version of the conditional distribution of Q with respect to $\widehat{\mathcal{F}}_t$. Now Dynkin's method allows us to construct a convenient conditional distribution Π_∞ for $\mathcal{P}$ with respect to the tail field

$$\widehat{\mathcal{F}} := \bigcap_t \widehat{\mathcal{F}}_t,$$

and this leads to an integral representation in $\mathcal{P}$:

(1.4) **Theorem.** *For any $Q \in \mathcal{P}$ there is exactly one probability measure ν on B such that*

$$Q = \int_B P^y \nu(dy) \tag{1.5}$$

where P^y is the distribution of $X_t + y \cdot t$ $(t \geq 0)$, i.e., of Brownian motion with drift $y \in B$. Conversely, any probability measure ν on B induces via (1.5) a measure $Q \in \mathcal{P}$.

Proof. Dynkin's general method leads to an integral representation (1.5) of $\mathcal{P}$ in terms of the set $\mathcal{P}_e$ of extreme points Q of $\mathcal{P}$; cf. [2]. We only have to identify $\mathcal{P}_e$ with B. Any $Q \in \mathcal{P}_e$ is of the form

$$(1.6) \qquad Q = \lim_n \Pi_{t_n}(\omega, \cdot)$$

for Q-almost all ω, where the limit is taken in the weak topology with respect to the polish space Ω; cf., e.g., [3]. In the coordinate picture (1.1), $\Pi_t(\omega,\cdot)$ resp. $P_{t_n}^{X_{t_n}(\omega)}$ corresponds to the product of one-dimensional Brownian bridges. (1.6) implies that Q corresponds to a product measure ΠQ^i, and that Q^i is the limit of one-dimensional Brownian bridges $P_{t_n}^{y_n^i}$. But the space-time Martin boundary of a one-dimensional Brownian motion is well known, and it follows that any such limit is a Brownian motion with some drift $y^i \in R^1$ given by

$$y^i = \lim_n \frac{y_n^i}{t_n}.$$

Thus, Q corresponds to a product of Brownian motions with drift y_i ($i \in I$). But since Q lives on Ω, these coordinates (y_i) must correspond to some point $y \in B$, in the sense of (1.2). Conversely, it is easy to see that any P^y with $y \in B$ is an extreme point of $\mathcal{P}$.

The integral representation (1.5) suggests to view B as the *Martin boundary* of Brownian motion on B. Let us now see how this boundary is related to the integral representation of space-time harmonic functions on B.

For any $y \in B$, the associated process P^y satisfies

$$(1.7) \qquad P^y \ll P \text{ on } \mathcal{F}_t \quad (t \geq 0),$$

if and only if the distribution μ_t^y of X_t under P^y is absolutely continuous with respect to the distribution μ_t of X_t under P, and this is equivalent to the condition that y belongs to the Cameron-Martin space H. In that case,

$$(1.8) \qquad \left.\frac{dP^y}{dP}\right|_{\mathcal{F}_t} = K^y(X_t, t),$$

where

$$K^y(x,t) = \frac{d\mu_t^y}{d\mu_t}(x) = \exp\left[\int_0^1 \dot{y}\, dx - \frac{1}{2}t\int_0^1 \dot{y}^2\, dt\right]$$

is defined for μ_t-almost all $x \in B$. It follows that any probability measure ν on H induces a measure $P^\nu = \int P^y \nu(dy)$ which is locally absolutely continuous with respect to P in the sense of (1.7), with

$$\frac{dP^\nu}{dP}\bigg|_{\mathcal{F}_t} = h(X_t, t) \quad (t \geq 0) \tag{1.9}$$

and

$$h(x,t) = \int_H K^y(x,t)\nu(dy). \tag{1.10}$$

Thus, we obtain a class of space-time harmonic functions h which can be represented as mixtures of the extremal functions $K^y (y \in H)$; for our present purpose, space-time harmonicity is simply defined in terms of the martingale property in (1.9). Let us now show that not every space-time harmonic function can be represented as a mixture (1.10) of extremal space-time harmonic functions:

(1.11) **Proposition.** *There exists a measure ν on B with $\nu(H) = 0$ such that P^ν is locally absolutely continuous with respect to P. Thus, the space-time harmonic function h given by*

$$h(X_t, t) = \frac{dP^\nu}{dP}\bigg|_{\mathcal{F}_t} \quad (t \geq 0)$$

does not admit an integral representation of the form (1.10).

Proof. Consider the product measure $\nu = \prod_i \nu_i$ on R^∞ with $\nu_i = N(0, \sigma_i^2)$. We assume that σ_i^2 converges to 0 slowly enough so that

$$\sum_i \sigma_i^4 < \infty \tag{1.12}$$

but

$$\sum_i \sigma_i^2 = \infty. \tag{1.13}$$

In particular, (1.12) implies that ν is concentrated on B, viewed as a subset of R^∞ via (1.2). The distribution ν_t of X_t under P^ν is of the form

$$\nu_t = \prod_i \int N(y_i t, t)\nu_i(dy_i) = \prod_i N(0, t(1+\sigma_i^2)).$$

We want to show that ν_t is absolutely continuous with respect to $\mu_t = \prod_i N(0,t)$; this implies that P^ν is absolutely continuous with respect to P on $\mathcal{F}_t$. To this end, we could use Kakutani's criterion. But the computation is simpler if we use instead relative entropy; this was suggested by Nina Gantert. Recall that the relative entropy between two probability measures μ_1 and μ_2 is defined as

$$H(\mu_1|\mu_2) = \int \log(\frac{d\mu_1}{d\mu_2})d\mu_1$$

if μ_1 is absolutely continuous with respect to μ_2, and $= \infty$ else. Since

$$H(N(0,\alpha)|N(0,\beta)) = \frac{1}{2}(\frac{\alpha}{\beta} - 1 - \log\frac{\alpha}{\beta}),$$

we obtain

$$\begin{aligned} H(\nu_t|\mu_t) &= \sum_i H(N(0,t(1+\sigma_i^2))|N(0,t)) \\ &= \frac{1}{2}\sum_i(\sigma_i^2 - \log(1+\sigma_i^2)) \le \sum_i \sigma_i^4 < \infty \end{aligned}$$

due to (1.12), and this implies absolute continuity of ν_t with respect to μ_t. In order to show $\nu(H) = 0$, it is enough to show that

$$\sum_i x_i^2 = \infty \quad \nu - a.s. \tag{1.14}$$

In fact,

$$\sum_{i=1}^n (x_i^2 - \sigma_i^2) \quad (n = 1, 2, ..)$$

is a martingale with respect to ν, bounded in $L^2(\nu)$ due to (1.12), hence ν-a.s. convergent to some finite limit. Thus (1.14) follows from (1.13).

2. The structure of h-processes

Consider a measure $Q \in \mathcal{P}$ which is locally absolutely continuous with respect to the basic measure P. Thus, Q is of the form (1.9) for some space-time harmonic function h. If Q has finite entropy with respect to P on each $\mathcal{F}_t$, i.e., if

$$\int \log \frac{dQ}{dP}\Big|_{\mathcal{F}_t} dQ = \int h(\cdot,t)\log h(\cdot,t) d\mu_t < \infty, \tag{2.1}$$

then the coordinate process is governed by a stochastic differential equation

$$dX = dW + b(X,t)dt \tag{2.2}$$

under P, where W is a Brownian motion on B, and b is an H-valued Markovian drift with finite energy

$$E_Q[\int_0^{t_0} \|b(\cdot,t)\|_H^2 dt] < \infty \tag{2.3}$$

for all $t_0 < \infty$; cf. [10]. Our aim is to obtain an explicit formula for the forward drift in terms of the function h. We start with the observation that the backward description is quite transparent: For any $Q \in \mathcal{P}$, the time-reversed process is an infinite-dimensional Brownian bridge converging to $0 \in B$. An infinite-dimensional analogue of the classical duality equation which couples the forward and the backward description then leads to the following identification of the forward drift, in analogy to the finite-dimensional case (0.10). In order to simplify the exposition we replace (2.1) by the stronger assumption

$$h(\cdot,t) \in \mathcal{L}^2(\mu_t) \qquad (t > 0), \tag{2.4}$$

and we use the notation of [12] for distributions on Wiener space.

(2.5) **Theorem.** *For almost all t, $h(\cdot,t)$ belongs to the Sobolev space $D_{2,1}$, and the drift is given by*

$$Dh(\cdot,t) = h(\cdot,t)b(\cdot,t). \tag{2.6}$$

Proof. Without loss of generality we restrict ourselves to the time interval $[0,1]$. Let $\widehat{Q}$ denote the time reversal of Q on $\Omega = C([0,1], B)$, i.e. the image of Q under the pathwise time reversal R on Ω given by $R\omega(t) = \omega(1-t)$. Under $\widehat{Q}$, the coordinate process is a Brownian bridge with drift

$$\hat{b}(x,t) = -\frac{x}{1-t}. \tag{2.7}$$

Now we use von Vintschger's infinite-dimensional version of the duality equation which relates forward and backward drift: For almost all $t \in [0,1]$, for smooth bounded F on B, and for $l \in H$,

$$\begin{aligned} \int (b(\cdot,t) + \hat{b}(\cdot,1-t), l) F d\nu_t &= -\int (DF, l) d\nu_t \\ &= -\int (DF, l) dh(\cdot,t) d\mu_t \end{aligned} \tag{2.8}$$

where ν_t resp. μ_t denotes the distribution of X_t under Q resp. P; cf. [10, p.52]. Integration by parts on the Wiener space (B, μ_t) leads to

$$\int (DF, l) h(\cdot,t) d\mu_t = -\int F(Dh(\cdot,t), l) d\mu_t + \frac{1}{t}\int F h(\cdot,t)(\cdot, l) d\mu_t. \tag{2.9}$$

This implies

$$\int (b(\cdot,t), l) F h(\cdot,t) d\mu_t = \int (Dh(\cdot,t), l) F d\mu_t,$$

hence (2.6). For a more careful discussion of the technical details in (2.8) and (2.9) we refer to [10] and [11].

3. The Martin boundary of the infinite-dimensional Ornstein-Uhlenbeck process

Let P be the distribution of Malliavin's Ornstein-Uhlenbeck process with state space B. Under P on $\Omega = C([0,\infty), B)$, the coordinate process satisfies the stochastic differential equation

$$dX = dW - \alpha X\,dt \tag{3.1}$$

with some $\alpha > 0$, where W is a Brownian motion with state space B. As in (1.3), the infinite-dimensional Ornstein-Uhlenbeck process can be constructed in terms of a countable collection of independent one-dimensional Ornstein-Uhlenbeck processes; cf. [6].

The construction of the space-time Martin boundary can be carried out in exact analogy to section 1. This leads to the identification of the Martin boundary with B, to the integral representation

$$Q = \int_B P^y \nu(dy) \qquad (Q \in \mathcal{P}) \tag{3.2}$$

of the class $\mathcal{P}$ defined in terms of the local specification of the Ornstein-Uhlenbeck process P, and to the following characterization of the extreme points $P^y (y \in B)$:

(3.3) **Theorem.** *For $y \in B$, the process P^y is the distribution of an Ornstein-Uhlenbeck process (3.1) around the curve $ye^{\alpha t} (t \geq 0)$.*

(3.4) **Remark.** This result, formulated in terms of the entrance rather than the exit boundary, was obtained independently by M. Röckner; cf. [9]. For the Ornstein-Uhlenbeck process, he also constructs in [9] a suitable version of our example in section 1.

The distribution P of the *recurrent* Ornstein-Uhlenbeck process (3.1) coincides with the extreme point P^0 associated with the constant function $0 \in B$. If ν is Wiener measure with variance $(2\alpha)^{-1}$, then the process

$$P^\nu = \int P^y \nu(dy) \tag{3.5}$$

is a ***transient*** Ornstein-Uhlenbeck process with stochastic differential equation

$$dX = dW + \alpha X\,dt. \tag{3.6}$$

In our discussion of the Martin boundary, we might as well take P^ν instead of P as our basic reference process; the boundary will remain the same. But for P^ν, the tail-field is generated by the ***exit point***

$$Y = \lim_{t\uparrow\infty} e^{-\alpha t} X_t, \tag{3.7}$$

and the ***exit distribution***, i.e., the distribution of Y under P^ν, is given by Wiener measure ν. In particular, the exit distribution is concentrated on the complement H^c of the Cameron-Martin space H. This shows, once again, that the Cameron-Martin space is not rich enough to serve as a Martin boundary.

References

[1] Dynkin, E.B.: Integral representation of excessive measures and excessive functions. Uspehi Mat. Nauk 27, Vol. 1, 43-90 (1972). English translation: Russian Math. Surveys 27, Vol. 1, 43-84 (1972).

[2] Dynkin, E.B.: Sufficient statistics and extreme points. Annals of Prob. 6, 705-730 (1978)

[3] Föllmer, H.: Phase transition and Martin boundary. Séminaire de Probabilités IX , Lecture Notes in Math. 465, 305-318, Springer (1975).

[4] Föllmer, H.: Time reversal on Wiener space. In: Proc. BiBos-Symposium "Stochastic processes - Math. and Physics". Lecture Notes in Math. 1158, 119-129, Springer (1986).

[5] Föllmer, H. and A. Wakolbinger: Time reversal of infinite-dimensional diffusions. Stochastic Processes and their Appl. 22, 59-77 (1986).

[6] Malliavin, P.: Stochastic calculus of variation and hypoelliptic operators. In: Proc. of the International Symposium on Stochastic Differential Equations Kyoto 1976, Tokyo (1978).

[7] Malliavin, P.: Calcul des variations stochastiques subordonné au processus de la chaleur. C. R. Acad. Sc. Paris, t. 295, 167-172 (1982).

[8] Robbins, H. and D. Siegmund: Statistical tests of power one and the integral representation of solutions of certain parabolic differential equations. Bull. Inst. Math. Acad. Sinica 1, 93-120 (1973).

[9] Röckner, M.: On the parabolic Martin boundary of the Ornstein-Uhlenbeck operator on Wiener space. Preprint.

[10] von Vintschger, R.S.: Zeitumkehr und invariante Masse für Diffusionen auf einem Wienerraum. Diss. ETH Zürich Nr. 8356 (1987)

[11] von Vintschger, R.S.: A duality equation for diffusions on an abstract Wiener space. In: Stachastic Analysis, Path Integration and Dynamics, ed. K.D. Elworthy and J.C. Zambrini, Pitman Research Notes in Mathematics Series 200, 215-223, Longman (1989)

[12] Yan, I.A.: Développement des distributions suivant les chaos de Wiener et applications àl'Analyse stochastique. Sém. Probabilités XXI, Lecture Notes Math. 1247, 27-33, Springer (1987).

Hans Föllmer
Institut für Angewandte Mathematik
Universität Bonn
Wegelerstr. 6
D 5300 Bonn

Hypoellipticity in Infinite Dimensions

Paul Malliavin

0. INTRODUCTION

Hörmander's theorem affirms the smoothness of the law of the diffusion associated with $\frac{1}{2}\sum A_k^2$ when the brackets of the vector fields $\{A_k\}$ generate the tangent space. This diffusion has for semi-classical approximation the horizontal control problem $\dot{\varphi} = \sum A_k \eta_k$. At this level the bracket phenomenon makes it possible to reach every point from a given starting point. In conclusion, in finite dimensions, the bracket phenomenon forces the spreading of the law and leads therefore to regularity.

In infinite dimensions we are interested in the *quasi-invariance* of measures under the infinitesimal action of suitable *vector fields*. For translation vector fields and gaussian measures the full answer is given by the Cameron-Martin theorem. The present development of analysis on the group of infinite dimensions [8] raises the natural question to construct measures on non-commutative groups having some good quasi-invariance property. If we choose a left-invariant semi-elliptic operator, then the law μ_t of the process at time t, starting at time zero from the identity, satisfies the semi-group property $\mu_t * \mu_{t'} = \mu_{t+t'}$. The quasi-invariance of μ_t leads directly to a hypoellipticity problem in infinite dimensions. Then bracket phenomena appear again and have the effect of *spreading the measure in the functional space*, which means to work now against the desired quasi-invariance.

To approach quasi-invariance we shall use the stochastic calculus of variations in a projected version, following closely the recent work of H. Airault [1]. The main difference between the infinite dimensional case and the finite dimensional case is that the tangent bundle on the target manifold *will be defined* by the stochastic calculus of variations. On this bundle a natural incommensurable Riemannian structure will appear. We shall get in this context a quasi-invariance theorem for a family of *non-constant* vector fields (constant meaning right invariant).

We have chosen to write this paper in the particular case of the loop group, but we shall not take advantage in this paper of the symmetries special to this case and our approach will be of a general nature. Better results can be obtained for the loop group taking advantage of its symmetries [6].

Contents

1. HYPOELLIPTIC OPERATORS ON LOOP GROUPS

We denote by G a compact, connected and simply connected Lie group. We shall write all the differential calculus on G as if G were a matrix group, *leaving to the reader its intrinsic transcription by the Maurer-Cartan form.* We denote by $\mathbf{L}(G)$ the loop group:

$$\mathbf{L}(G) = \text{ space of continuous maps of the circle } \mathbf{T} \text{ into } G.$$

We shall denote by $\mathcal{G}$ the Lie algebra of G, and by $\mathbf{L}(\mathcal{G})$ the Banach space of the loops on $\mathcal{G}$, that is the space of continuous maps of $\mathbf{T}$ into $\mathcal{G}$. Then $\mathbf{L}(\mathcal{G})$ looks formally like the Lie algebra of $\mathbf{L}(G)$. We choose on $\mathcal{G}$ an invariant euclidean metric. We shall denote by

$$\mathbf{L}_\alpha(\mathcal{G}) = \{u \in \mathbf{L}(\mathcal{G}) : \text{the } \alpha\text{-derivative of } u \text{ is in } L^2\}.$$

On $\mathbf{L}_\alpha(\mathcal{G})$ we define the Hilbertian structure

$$\|u\|_\alpha^2 = \sum_{n\in Z} \|\hat{u}(n)\|_{\mathcal{G}}^2 (1+n^2)^\alpha$$

where $\widehat{u}(n)$ denotes the nth-Fourier coefficient of u. Then, for $\alpha > \frac{1}{2}$ the canonical premeasure on $\mathbf{L}_\alpha(\mathcal{G})$ can be realized as a Borel measure on $\mathbf{L}\mathcal{G}$. This means, in classical language, that the following random series

$$u(\theta) = G_0 + \sum_{n>0} \frac{1}{(1+n^2)^{\alpha/2}} (G_n \cos n\theta + G'_n \sin n\theta)$$

where G_k, G'_k are independent normal variables on $\mathcal{G}$, converges almost surely in $\mathbf{L}(\mathcal{G})$.

Denote by b_k, b'_k independent Brownian motions on $\mathcal{G}$, starting at 0 at time zero. Then we define a *$L(\mathcal{G})$-valued continuous process* $t \to x(.,t)$ by the random series

$$x(\theta,t) = b_0(t) + \sum_{n>0} \frac{1}{(1+n^2)^{\alpha/2}}(b_n(t)\cos n\theta + b'_n(t)\sin n\theta). \tag{1}$$

We shall denote by X the corresponding probability space.

We shall denote by ε_s an orthonormal basis of $\mathcal{G}$. Then

$$b_k(t) = \sum_{s=1}^{d} \varepsilon_s b_k^s(t).$$

We shall define $e_{k,s}, e'_{k,s} \in \mathbf{L}(\mathcal{G})$ by

$$e_{k,s} = \varepsilon_s \quad \frac{\cos k\theta}{(1+k^2)^{\alpha/2}}$$
$$e'_{k,s} = \varepsilon_s \quad \frac{\sin k\theta}{(1+k^2)^{\alpha/2}}$$

We shall denote by $\mathcal{L}_e$ the right derivative following $e \in \mathbf{L}(\mathcal{G})$ defined by

$$(\mathcal{L}_e u)(\gamma) = \{\frac{d}{d\tau}u(\gamma\exp(\tau e))\}_{\tau=0}.$$

Then the purpose of this paper is to study the following operator on $\mathbf{L}(G)$:

$$\Delta^\alpha = \frac{1}{2}\sum_{k,s}(\mathcal{L}^2_{e_{k,s}} + \mathcal{L}^2_{e'_{k,s}}).$$

finite subset S of $\mathbf{T}$ we denote by $\mathcal{E}_S$ the corresponding evaluation map:

$$\mathcal{E}_S : \mathbf{L}(G) \to G^S.$$

oth function v on G^S we shall denote $\widetilde{v} = v \circ \mathcal{E}_S$. The family of ions $\widetilde{v}$, where v and S vary, will be called the vector space of *cylindrical functions.*

We shall consider the following family of Itô's stochastic differential equations in t, where θ is a fixed parameter:

$$\begin{cases} d_t g_x(\theta,t) = g_x(\theta,t)(d_t(x(\theta,t) - a\,dt)) \\ g_x(\theta,0) = e \qquad (e = \text{identity element of } G) \end{cases} \tag{2}$$

with

$$a = c_\alpha \sum_{s=1}^{d} \varepsilon_s^2 \text{ and } c_\alpha = 1 + 2\sum_{n>0} \frac{1}{(1+n^2)^\alpha}.$$

PROPOSITION 1.1. *It is possible to find a version of the solution of (2) such that the map*

$$t \to g_x(\,.\,,t)$$

defines a $\mathbf{L}(G)$ *- valued continuous process.*

The projection of this process by an evaluation map is a left-invariant diffusion on G^S.

For every cylindrical function $\widetilde{u}$ *we have*

$$\lim_{\varepsilon\downarrow 0} \varepsilon^{-1} E^{\mathcal{N}_t}\{\widetilde{u}(g_x(\,.\,,t+\varepsilon) - \widetilde{u}(g_x(\,.\,,t)))\} = (\Delta^\alpha \widetilde{u})(g_x(\,.\,,t)).$$

Before proving this proposition it will be convenient to introduce some notations. We want to realize the underlying probability space X on a fixed space independent of α. This space will be the space of the *white noise on the half cylinder* $\mathbf{T}\times\mathbf{R}^+$. We shall denote this space by $\widetilde{Y}$. To every measurable set A on X is associated a gaussian variable on $\widetilde{Y}$ with variance equal to the Lebesgue measure of A. We denote by $Y = \widetilde{Y}\otimes\mathcal{G}$ the probability space of the $\mathcal{G}$-valued white noise. The probability measure of Y has to be realized as a Borel measure on the Sobolev space of order $-1-\varepsilon$. The Cameron-Martin space of Y is $L^2(\mathbf{T}\times R^+,\mathcal{G})$. We shall consider the kernel

$$P_\alpha(\theta) = 1 + 2\sum_{n>0} \frac{\cos n\theta}{(1+n^2)^{\alpha/2}}.$$

We have for instance

$$P_2(\theta) = \text{ const } (e^{-\theta} + e^{\theta-2\pi}) \text{ for } \theta \in]0,2\pi[\,.$$

Then we have

1.2. LEMMA. *Define*

$$u_y(\theta,t) = \int_{\mathbf{T}}\int_{[0,t]} P_\alpha(\theta-\theta')\mathbf{1}_{t'<t}y(d\theta'\otimes dt).$$

Then u_y is isomorphic in law with x defined in (1). Therefore the Cameron-Martin space associated with X is

$$H_{1,\alpha}(R \times \mathbf{T}; \mathcal{G}) = H_1([0, +\infty[\,; H_\alpha(\mathbf{T})) \otimes \mathcal{G}$$

where $H_\alpha(\mathbf{T})$ is the Sobolev space of functions having their α-derivative in $L^2(\mathbf{T})$. The dual norm $H_{-\alpha}(\mathbf{T})$ has the following expression

$$\|f\|_{H_{-\alpha}(T)} = (P_{2\alpha} * f | f)_{L^2(\mathbf{T})}.$$

We remark that $H_\alpha(\mathbf{T})$, under our permanent assumption $\alpha > \frac{1}{2}$, is an algebra of continuous functions. Then, given a finite subset S of $\mathbf{T}$, we define J_S as the ideal of functions of $H_\alpha(\mathbf{T})$ vanishing on S. Then we denote $H_\alpha(S) = H_\alpha(\mathbf{T})/J_S$ and $H_\alpha(S)$ is a Hilbert space of dimension $|S|$. Then $H_\alpha(S)$ is the dual space of $H_{-\alpha}(S)$, the space of measures on $\mathbf{T}$, with support on S, the norm being inherited from the imbeding in $H_{-\alpha}(\mathbf{T})$. We have then

1.3. LEMMA. *The map*

$$t \to \{x(\theta, t)\}_{\theta \in S}$$

is the Brownian motion on the euclidean space $H_\alpha(S) \otimes \mathcal{G}$.

We shall now prove 1.1. By Kolmogoroff's method it is sufficient to estimate

$$E(\{d(g_x(\theta, t) g_x^{-1}(\theta', t))\}^r).$$

Then if we consider $S = (\theta, \theta')$ then the covariance matrix defining $H_{-\alpha}(S)$ is

$$\begin{pmatrix} P_{2\alpha}(0) & P_{2\alpha}(\theta - \theta') \\ P_{2\alpha}(\theta' - \theta) & P_{2\alpha}(0) \end{pmatrix}$$

Denote

$$p(\varphi) = 2(P_{2\alpha}(0) - P_{2\alpha}(\varphi)).$$

We have

1.4. LEMMA. *Denote by δ the Brownian motion on G. Then*

$$g(\theta, t) g^{-1}(\theta', t) \text{ and}$$
$$\delta(p(\theta - \theta')t)$$

have the same law.

PROOF: Denote

$$v(t) = g(\theta, t) g^{-1}(\theta', t).$$

Then, in Stratonovitch notation, we have

$$dv(t) = g_x(\theta, t) \circ [d_t x(\theta, t) - d_t x(\theta', t)] \circ g^{-1}(\theta', t)$$
$$dv(t) = v(t) Ad(g^{-1}(\theta', t))[d_t x(\theta, t) - d_t x(\theta', t)].$$

We remark $x(\theta, t) - x(\theta, t')$ is, after the time change $t \mapsto tp(\theta - \theta')$, a Brownian motion. We remark also that $Ad(g)$ is an orthogonal transformation of $\mathcal{G}$. Therefore the map

$$B \rightarrow \int_0^t Ad(g^{-1}(\theta', t)) dB(t)$$

is an isomorphism of the probability space of the $\mathcal{G}$-valued Brownian motion B. These two remarks prove 1.4.

1.5. PROOF OF PROPOSITION 1.1: Consider the Riemannian distance d on G induced by the metric of $\mathcal{G}$. Define

$$Y(t) = d^2(g_x(\theta, t), g_x(\theta', t)), \quad \varphi = \theta - \theta'.$$

Then $Y(t)$ is a semi-martingale and by 1.3. there exists an abstract Brownian motion b such that

$$dY(t) = (p(\varphi))^{\frac{1}{2}} c_1 db + p(\varphi) c_2 dt$$

where c_1 and c_2 are bounded by absolute constants. Therefore, by [5], there exists a constant $c_3(r)$ such

$$E((\sup_{0<t<t_0} Y(t))^r) \leq c_3(r) [p(\varphi)]^{r/2}.$$

As $p(\varphi)$ satisfies a Hölder contition at $\varphi = 0$, the Kolmogoroff criterion gives the result.

1.6. STOCHASTIC CALCULUS OF VARIATIONS. *We shall fix θ_0 and we denote*

$$\Phi_{\theta_0}(x)(t) = g_x(\theta_0, t).$$

We want to express the derivative of $\Phi_\theta(x)$. Choose $h \in H_{1,\alpha}(\mathcal{G})$. Then

LEMMA.

$$\left\{\frac{d}{d\varepsilon} g_{x+\varepsilon h}(t)\right\}_{\varepsilon=0} = \left(\int_0^t Ad(g_x(\theta_0,\xi))\dot{h}(\xi,\theta_0)d\xi\right)\Phi_{\theta_0}(x)(t).$$

PROOF: Denote

$$k(t) = \{\frac{\partial}{\partial \varepsilon} g_{x+\varepsilon h}(t)\}_{\varepsilon=0} \quad g_x^{-1}(t).$$

Now differentiating in ε the S.D.E

$$d_t[g_{x+\varepsilon h}] = g_{x+\varepsilon h}[du_x - a\,dt + \varepsilon \dot{h}\,dt]$$

we obtain

$$d_t(kg) = kg[dx - a\,dt] + g\dot{h}\,dt$$
$$(dk)g + kdg = kg[dx - a\,dt] + g\dot{h}\,dt$$
$$dk(t) = Ad(g_x(t))\dot{h}(t)dt.$$

∎

1.7. THE REGULARITY GAP. *Take for instance $\alpha = 1$. Then*

$$h(\xi,.) \in H_1(\mathbf{T};\mathcal{G}) \qquad \text{and}$$
$$Ad(g_x(\xi,.)) \in H_{1/2-\varepsilon}(\mathbf{T}; GL(\mathcal{G})) \qquad \text{for all } \varepsilon > 0$$

Therefore

$$\int_0^1 Ad(g_x(.,\xi))h(\xi,.)d\xi \in H_{\frac{1}{2}-\varepsilon}(\mathbf{T};\mathcal{G})$$

The natural space ([6]) would be $H_1(\mathbf{T}; GL(\mathcal{G}))$. The loss of regularity corresponds to the bracket phenomenon.

2. RIEMANNIAN METRIC ON L(G)

To simplify the notation we will denote by $\gamma \in \mathbf{L}(G)$, $t \to \gamma_x(t)$ the process $t \to g_x(\theta,t)$, and we will forget as much as possible the variable θ. We denote

$$\Phi(x,t) = \gamma_x(t).$$

We denote by $\rho_{\gamma,t}(dx)$ the conditional probability under the conditioning $\Phi(x,t) = \gamma$.

We shall consider the Hilbert space

$$\mathcal{H}_{\gamma,t} = L^2(X, \rho_{\gamma,t}; H_{1,\alpha}([0,t] \times \mathbf{T}; \mathcal{G})).$$

where

$$H_{1,\alpha} = \{f; \frac{\partial}{\partial t}(\frac{\partial}{\partial \theta})^\alpha f \in L^2\}$$

We denote

$$\Phi'(x).h = \int_0^t Ad(\gamma_x(\xi)).\dot{h}(\xi)d\xi.$$

We define a map

$$A_\theta : \mathcal{H}_{\gamma,t} \to \mathcal{G} \qquad \text{by}$$

$$A_\theta(z) = \int_X [\Phi'(x).z(x)](\theta)\rho_{\gamma,t}(dx).$$

We shall prove

2.1. LEMMA. *A_θ is a continuous linear map of $\mathcal{H}_{\gamma,t}$ into $\mathcal{G}$.*

PROOF: As $Ad(g)$ is an orthogonal transformation we have

$$\|Ad(g_x(\theta,\xi))\dot{h}(\xi,\theta)\|_{\mathcal{G}} = \|\dot{h}(\xi,\theta)\|_{\mathcal{G}}$$

Therefore

$$\|A_\theta(z)\|_{\mathcal{G}} \leq \int_{X\times[0,t]} \|\dot{z}(x,\theta)\|_{\mathcal{G}}\rho_{\gamma,t}(dx)d\xi.$$

There exists a constant c such that

$$\|z(x,\theta)\|_{\mathcal{G}} \leq c\|z(x)\|_{H_\alpha(\mathcal{G})}.$$

Then

2.1.1. $$\|A_\theta(z)\|_{\mathcal{G}} \leq c\|z\|_{L^2}.$$

2.2. DEFINITION. *Denote*

$$T_{\gamma,t}(\mathbf{L}) = \mathcal{H}_{\gamma,t}/M$$

where $M = \cap_{\theta\in\mathbf{T}} A_\theta^{-1}(0)$.

Then M is a closed subspace of $\mathcal{H}_{\gamma,t}$. Therefore $T_{\gamma,t}(\mathbf{L})$ is a Hilbert space. We shall call $T_{\gamma,t}(\mathbf{L})$ the tangent space at γ at time t. The collection of $\{T_{\gamma,t}(\mathbf{L})\}$ will be called the Riemannian tangent bundle on $\mathbf{L}$ at time t.

2.3. REALIZATION OF THE TANGENT SPACE IN $\mathbf{L}(\mathcal{G})$. *All the functions $Ad(g_x(\theta,t))z(x)(\theta,t)$ being continuous in θ, we could take as universal Banach space the space $C(\mathbf{T};\mathcal{G})$ of continuous functions with values in $\mathcal{G}$. Then*

$$\Phi'(x).z(x) \in C(\mathbf{T};\mathcal{G}).$$

Then, if $z \in \mathcal{H}_{\gamma,t}$, the following integral converges by 2.1.1 in $\mathbf{L}(\mathcal{G})$

$$\delta(z) = \int_X \Phi'(x)z(x)\rho_{\gamma,t}(dx)$$

and the mapping $z \to \delta(z)$ defines a realization of $T_{y,t}(\mathbf{L})$ as a subspace of $\mathbf{L}(\mathcal{G})$.

2.5. NORM ON THE COTANGENT SPACE. *We remark, as $H_\alpha(\mathbf{T}) \subset C(\mathbf{T})(\alpha > \frac{1}{2})$ then $H_{-\alpha}(\mathbf{T})$ contains the space of Radon measures on $\mathbf{T}$ that we shall denote $M(\mathbf{T})$.*

We shall identify $\mathcal{G}^*$ with $\mathcal{G}$. Then $M(\mathbf{T}) \otimes \mathcal{G} = M(\mathbf{T},\mathcal{G})$ is contained in the cotangent spaces whatsoever (γ,t).

If $\sigma \in M(\mathbf{T};\mathcal{G})$ we define

$$\|\sigma\|^*_{\gamma,t} = \sup_{\|z\|_{\mathcal{H}_{\gamma,t}} \leq 1} \langle \delta(z), \sigma \rangle.$$

2.5.1. PROPOSITION.

$$(\|\sigma\|^*_{\gamma,t})^2 = \iint_{T^2} (P_{2\alpha}(\theta-\theta')K_{\gamma,t}(\theta,\theta')\sigma(d\theta)|\sigma(d\theta'))_{\mathcal{G}}$$

where

$$K_{\gamma,t}(\theta,\theta') = \int_X \int_0^t Ad(g_x(\theta',\xi)g_x^{-1}(\theta,\xi))d\xi\, \rho_{\gamma,t}(dx).$$

PROOF:

$$\begin{aligned}\langle \delta(z), \sigma \rangle &= \int_{X\times[0,t]} \langle Ad(\gamma_x(\xi))z(x,\xi), \sigma\rangle \rho_{\gamma,t}(dx)\,dt \\ &= \int_{X\times[0,t]} \langle z(x,\xi), Ad(\gamma_x^{-1}(\xi)\sigma\rangle \rho_{\gamma,t}(dx)\,dt.\end{aligned}$$

$$\begin{aligned}(\|\sigma\|^*_{\gamma,t})^2 &= \int_{X\times[0,t]} \|Ad(\gamma_x^{-1}(\xi))\sigma\|^2_{H_{-\alpha}(\mathbf{T})} d\xi \\ &= \int_{X\times[0,t]\times\mathbf{T}^2} P_{2\alpha}(\theta-\theta')(Ad(g_x^{-1}(\xi,\theta))\sigma(d\theta)|Ad(g_x^{-1}(\xi,\theta')\sigma(d\theta')).\end{aligned}$$

3. DIVERGENCES

We shall say that a differential form λ on $\mathbf{L}(G)$ has a *divergence at the level t* if there exist an integrable function $\delta_t\lambda := \delta$ such that for every cylinder function $\widetilde{u}$

$$\int_{\mathbf{L}(G)} (d\widetilde{u}|\lambda)_{*\gamma,t}\,\mu_t(d\gamma) = \int_{\mathbf{L}(G)} (\widetilde{u}\,\delta_t\lambda)(\gamma)\mu_t(d\gamma)$$

3.1. THEOREM. *Let $\sigma \in M(\mathbf{T};\mathcal{G}^*)$, λ_σ the differential form associated; then $\delta(\lambda_\sigma)$ exists and $\delta(\lambda_\sigma) \in L^2$.*

PROOF: Consider the stochastic integral

$$J(x) = \int_0^t \langle Ad(\gamma_x^{-1}(\xi))dx(\xi), \sigma\rangle.$$

We want to prove

$$\delta\lambda_\sigma(\gamma_0) = E(J(x)|\gamma_x(t) = \gamma_0).$$

Therefore we have to compute

$$E(\widetilde{u}(\gamma_x(t)(\delta\lambda_\sigma)(\gamma_x(t))) = E(\widetilde{u}(\gamma_x(t))J(x)).$$

Using the Itô-Haussman-Bismut-Ocone [2] [7] representation we have

$$u(\gamma_x(t)) = \int_0^t E^{\mathcal{N}_\xi}((\partial^{\text{left}}_{Ad(\gamma_x(\xi))e_k}\widetilde{u})(\gamma_x(\xi))dx^k(\xi)$$

Therefore

$$\begin{aligned} E(\widetilde{u}(\gamma_x(t))J(x)) &= E\left(\int_0^t \sum_k (\partial^{\text{left}}_{Ad(\gamma_x(\xi)e_k)}\widetilde{u})(\gamma_x(t))\right)(\langle Ad(\gamma_x(\xi))e_k, \sigma\rangle)\,d\xi \\ &= E\left(\int_o^t (Ad\gamma_x^{-1}(\xi)d\widetilde{u}(\gamma(t))|Ad(\gamma_x^{-1}(\xi)\sigma)d\xi = E(E^{\gamma_x(t)=\gamma_0}\ldots\right) \\ &= \int_X (d\widetilde{u}|\sigma)_{*\gamma,t}d\mu_t. \end{aligned}$$

3.2. ADMISSIBLE VECTOR FIELDS. *Given a differential form λ_σ there exists a unique vector field $Z_{\gamma,t}$ such that*

$$(d\widetilde{u}|\lambda_\sigma)_{*\gamma,t} = \langle Z^\sigma_{\gamma,t}, d\widetilde{u}\rangle.$$

Then we have by translation of the previous paragraph.

3.2.1. PROPOSITION. *The divergence of $Z^{\sigma}_{\gamma,t}$ exists and is in L^2.*

We get in this way a lot of vector fields which are vectors of derivation for the meausre μ_t. Of course these vector fields are not "natural" because they are non-constant (which means, right invariant). To deal with constant vector fields it will be good to solve the following two preliminary problems:

(i) To get a lower bound of the $\| \|^*_{\gamma,t}$ independent of γ.

(ii) To get good estimates of the Christofel symbol.

3.2.2. Remark. In the usual stochastic calculus of variations a key step is the inversion of the *covariance matrix*. In the point of view of H. AIRAULT [1] the covariance matrix is replaced by an *averaged covariance* matrix which has therefore *more positivity*.

Then (i) corresponds in the functional analysis case to this problem of invertibility. In [4], the invertibility problem has been raised for the usual covariance.

3.2.3. Remark. It is quite possible that for a *generic* hypoelliptic operator, which is a small perturbation of a *left* invariant but itself no longer left invariant, then a right invariant vector field will not be admissible (see [3] for such examples).

4. INCOMMENSURABILITY OF THE RIEMANNIAN METRIC

THEOREM. *We shall take in this paragraph $\alpha = 1$. We say that γ is equivalent to γ' if the two metrics $\| \|^*_{\gamma,t}$ and $\| \|^*_{\gamma',t}$ defined on $M(\mathbf{T};\mathcal{G})$ are equivalent. Then the equivalence classes are of measure zero.*

PROOF: We shall prove the contrapositive of this statement. Then there will exist a Borel set A, with $\mu_t(A) > 0$ and a constant c such that

$$\| \|_{\bullet,\gamma} \leq c\| \|_{\bullet,\gamma'} \text{ for all } \gamma, \gamma' \in A. \tag{4.1}$$

4.2. LEMMA. *There exists a constant c such that*

$$\int (\|\sigma\|^*_{\gamma,t})^2 d\mu_t(\gamma) \leq c\|\sigma\|^2_{H_{-1}}(T;\mathcal{G})$$

for all $\sigma \in M(\mathbf{T};\mathcal{G})$.

PROOF: To simplify the computation, we choose on $\mathcal{G}$ a metric such that

$$\sum \varepsilon_s^2 = -\text{ Identity}.$$

Then $g^{-1}(\theta', t)g(\theta, t)$ has the same law as $\delta(p(\varphi)t)$ where $\delta(t)$ satisfies the Itô stochastic differential equation

$$d\delta = \delta\left(\sum \varepsilon_s db^s - I dt\right).$$

Define

$$M(t) = E(Ad(\delta(t))) \quad , \quad S(t) = E(M(t)).$$

Then

$$\frac{dS}{dt} = -S(t) \text{ and } S(t) = e^{-t}.$$

Therefore

$$E((\|\sigma\|^*_{\gamma,t})^2) = \iint_{T^2} (f(\theta-\theta')P_2(\theta-\theta')\sigma(d\theta)|\sigma(d\theta'))_{\mathcal{G}}$$

where

$$f(\varphi) = \frac{1 - e^{-(P_2(0)-P_2(\varphi))}}{P_2(0) - P_2(\varphi)}.$$

Define

$$c_n = \int_{\mathbf{T}} f(\varphi)\exp(-in\varphi)d\varphi.$$

Then it results from Beurling's theory of negative definite functions that

$$\exp(-\lambda(P(0)) - P(\varphi)) \text{ is a positive definite function.}$$

Therefore $c_n \geq 0$ for all n and

$$E((\|\sigma\|^*_{\gamma,t})^2) = \sum c_n \frac{1}{p^2+1}\|\hat{\sigma}(n+p)\|^2_{\mathcal{G}}.$$

We have to prove

(i)
$$\sum_{p+n=r} c_n \frac{1}{p^2+1} = 0\left(\frac{1}{r^2}\right).$$

We compute in the sense of distributions the second derivative of f; we get

$$f'' = \text{ Dirac measure at } 0 + \text{ a bounded function.}$$

Therefore

$$c_n = \frac{1}{n^2} + o\left(\frac{1}{n^2}\right)$$

which implies (i).

4.3. **Proof of the theorem:** We shall integrate 4.1. in $\gamma' \in A$, leaving γ fixed. We get

$$(\|\sigma\|_{\gamma,t})^2 \leq cE(1_A(\gamma')(\|\sigma\|^*_{\gamma',t})^2 \leq c\|\sigma\|^2_{H_{-1}(\mathbf{T},\mathcal{G})}$$

for all $\sigma \in M(\mathbf{T},\mathcal{G})$ and $\gamma \in A$.

Consider now the eigenfunction $\varphi^S_{\gamma,t}$ of $K_{\gamma,t}$ in $L^2(\mathbf{T})$. Then

$$K_{\gamma,t}(\theta,\theta') = \sum \lambda^S_{\gamma,t}\, \varphi^S_{\Gamma,t}(\theta)\, \varphi^S_{\gamma,t}(\theta')$$

and

$$(\|\sigma\|^*_{\gamma,t})^2 = \sum \lambda^S_{\gamma,t} \|\varphi^S \sigma\|^2_{H_{-1}}.$$

According to the regularity gap $K_{\gamma,t}(\theta,\theta')$ is not Hölder $\frac{1}{2}+\varepsilon$ in θ therefore it is the same for $\varphi^S_{\gamma,t}(\theta)$. We shall get

$$\|\varphi^1_{\gamma,t}\sigma\|_{H_{-1}} \leq c\|\sigma\|_{H_{-1}}$$

which is absurd.

5. TOWARDS CHRISTOFEL SYMBOL

The previous paragraph has shown that the Riemannian metric on $\mathbf{L}(G)$ is globally very irregular. We want now to prove a statement of regularity in the reverse direction.

5.1. **Theorem.** *Denote by ∇ the gradient associated to the Laplacian Δ^α. Fix $\sigma \in M(\mathbf{T};\mathcal{G})$ and consider*

$$f(\gamma,t) = (\|\sigma\|^*_{\gamma,t})^2.$$

Then

$$\int_{t_0}^{t_1} \int_X |\nabla f|^2 d\mu_t \otimes dt < \infty.$$

5.2. **Lemma.**

$$(\frac{\partial}{\partial t} + \Delta^\alpha) f(\gamma,t) \geq 0.$$

Proof: We have

$$(\frac{\partial}{\partial t} + \Delta^\alpha) K_{\gamma,t}(\theta,\theta') = Ad(\gamma^{-1}(\theta')\gamma(\theta)).$$

Therefore

$$(\frac{\partial}{\partial t} + \Delta^\alpha) f(\gamma, t) = \|Ad(\gamma)\sigma\|^2_{H_{-\alpha}} \geqslant 0$$

5.3. PROOF OF THE THEOREM: We have

$$(\frac{\partial}{\partial t} + \Delta^\alpha) f^2 \geq \|\nabla f\|^2.$$

Therefore if $t_0 < t$,

$$\int_X f^2(\gamma, t_1)\mu_{t_1}(d\gamma) - f^2(\gamma, t_0)\mu(d\gamma) \geq \int_{t_0}^{t_1} \|\nabla f\|^2 d\mu_t \, dt.$$

The finiteness of the left hand side results from the inequality

$$f(\gamma, t) \leq \|\sigma\|^2_M$$

where $\| \, \|_M$ denotes the norm on Radon measures.

6. EXPRESSION OF THE PROJECTED ORNSTEIN-UHLENBECK PROCESS

We choose an orthonormal basis e^*_k of $H_{-\alpha}(\mathbf{T}; \mathcal{G})$ which is composed of measures. We denote

$$a_{k,s}(\gamma, t) = (e^*_k | e^*_s)_{*\gamma,t}.$$

We could choose this basis such that

$$\{e^*_k\} \quad k \in [1, d2^n] \text{ is a basis of } H_{-\alpha}(S_n)$$

where S_n is the set of 2^n-th roots of unity. We shall denote

$$E^{S_n}(a_{k,s}) = a_{k,s,n}$$

where E^{S_n} denotes the conditional expectation following the evaluation $\mathcal{E}_{S_n}$. We shall denote by $a_n^{k,s}$ the inverse matrix associated to $a_{k,s,n}$, $k, s \in [1, 2^n d]$. Then we denote

$$J_k(x) = \int_0^t \langle Ad(g_x(\xi)) dx(\xi), e_k \rangle$$
$$p^k = E^\Phi(J_k(x))$$
$$p_n^k = E^{S_n}(p_k).$$

6.1. PROPOSITION. *Fix t and consider the map $\Psi_S(x) : x \to \mathcal{E}_{S_n}(g_x(t))$. Then the projected Ornstein-Uhlenbeck generator in the sense of [1] on X is the following operator on G^{S_n}.*

$$\frac{1}{2} a_n^{k,s} \partial_{e_k}^{\text{left}} \partial_{e_s}^{\text{left}} - \frac{1}{2} p_n^k \partial_{e_i}^{\text{left}}$$

where e_i is the dual basis in $H_\alpha(\mathrm{T}; \mathcal{G})$ of the e_i^*. Similarly the heat equation on X has for projection the following operator

$$\frac{1}{2} a_n^{k,s} \partial_{e_k}^{\text{left}} \partial_{e_s}^{\text{left}}.$$

The proof can be made by a direct computation.

REFERENCES

[1] H. Airault, *Projection of the infinitesimal generator of a diffusion*, J. of Functional Analysis 85 (1989), 353–391.

[2] J.-M. Bismut, *Martingales, Malliavin Calculus and hypoellipticity under general condition*, Z. Wahr. 56 (1981), 469–505.

[3] V.I. Borgachev, *Some results on differentiable measures*, Translated 1986, Matematik Sbornik **127** 169 (1985), 335–349.

[4] Yu. L. Dalestkii, *Stochastic differential geometry*, Russian Math. Surveys 38 (1983), 87–111.

[5] N. Ikeda and S. Watanabe *A comparison theorem for solutions of stochastic differential equations*, Osaka, J. Math **14** (1977), 285–300.

[6] M.-P. Malliavin et P. Malliavin, *Quasi-invariant integration on Loop groups*, J. of Funct. Analysis (September 1990).

[7] D. Ocone, *Malliavin's calculus and stochastic integral representations of functionals of diffusion process*, Stochastics 12 (1984), 161–186.

[8] A. Pressley and G. Segal, *Loop groups*, Oxford (1986).

Paul Malliavin
10, rue Saint Louis en l'Isle
75004 - PARIS

On Interacting Ornstein–Uhlenbeck Processes

Weian Zheng

§0. Introduction

A system of interacting Ornstein-Uhlenbeck processes moving under an external potential h is governed by the equations of motion

$$
(1)\qquad \begin{aligned}
dX_i(t) &= V_i(t)dt \\
dV_i(t) &= -\sum_{j\neq i} m^{-1}\nabla g(X_i(t) - X_j(t))dt - m^{-1}\nabla h(X_i(t))dt \\
&\quad - \frac{\sigma^2\beta}{2m}V_i(t)dt + \sigma dW_i(t).
\end{aligned}
$$

(i,j=1,2,...), where $V_i(t)$ is the velocity of the i-th particle at time t, $X_i(t)$ is its position in R^d, m is the mass of a single particle, and σ, β are constants. To simplify

Project supported by National Natural Science Foundation of China and partly supported by SERC Grant Number GR/D/31928 of Britain

notation, we will always suppose $\sigma = \beta = m = 1$ in the following discussion. The particles interact on each other by a pair-force $-\nabla g$, and are driven by

$$-\nabla h(X_i(t)) - \frac{1}{2}V_i(t) + \frac{dW_i(t)}{dt}. \tag{2}$$

There are already many discussions about the existence of interacting Wiener processes under the condition that the interacting potential g is smooth or not too far from being smooth (see Lang[4], Guo and Papanicolau[5], and Fritz [3]etc.) Rost[6]discussed the interacting Wiener processes of hard spheres in R^1.

In this paper, we prove an existence theorem for (1) under the condition that the particles are hard spheres in R^d. If the pair-force is strong enough, then the "hard sphere"condition is not necessary. That is different from those authors who technically assume the pair-potential is not too strong.

§I. Some Results About the Weak Convergence

Denote by $C([0,1] \longrightarrow R^d)$ $(d = 1, 2, ..., \infty)$ the space of continuous R^d-valued functions on the unit interval $[0, 1]$ with the metric

$$\rho(x, y) = \sum_{k=1}^{d} 2^{-k}[\sup_s |x_k(s) - y_k(s)|(1 + \sup_s |x_k(s) - y_k(s)|)^{-1}]$$

where $x = (x_k(s)), y = (y_k(s))(k = 1, ..., d; s \in [0, 1])$ are elements of $C([0, 1] \longrightarrow R^d)$. Under the above metric ρ, $C([0, 1] \longrightarrow R^d)$. is separable and complete.

LEMMA 1. Suppose $X^{(n)}(t) = (X_1^{(n)}(t), X_2^{(n)}(t), ...)$ $(t \in [0,1],\ n = 1, 2, ..., \infty)$ is a sequence of stochastic processes (which may not be defined on the same probability space). Then the laws of $\{X^{(n)}\}_{n=1,2,...}$ converge on $C([0,1] \to R^\infty)$ to that of $X^{(\infty)}$ if and only if $(X_1^{(n)}, X_2^{(n)}, ..., X_d^{(n)})$ converge weakly to $(X_1^{(\infty)}, X_2^{(\infty)}, ..., X_d^{(\infty)})$ on $C([0,1] \to R^\infty)$ for all finite d.

Proof. It is sufficient to prove the "if" case. Denote by U the class of all open balls in $C([0,1] \to R^\infty)$ (d=1,2,...), and denote by U' the class of all the finite intersection of elements in U'. Now let U'' be the class of all the $P^{(\infty)}$–continuity set in U'. We denote by $S(x, \varepsilon)$ the balls in R^∞ with x as its center and ε as its radius, we can easily verify that: (i) U'' is closed under finite intersections, and (ii) for every $x \in R^\infty$ and $\varepsilon > 0$ there is $A \in U''$ with $x \in A^\circ \subseteq A \subseteq S(x, \varepsilon)$. Since R^∞ is separable and from the hypothesis of the lemma, $P^{(n)}(A) \to P^{(\infty)}(A)$ $(\forall A \in U'')$, where $P^{(n)}$ is the probability on $C([0,1] \to R^d)$ by $X^{(n)}$, we can apply a result of Billingsley ([1], Corollary 1, p.14) to the class U'' satisfying (i) and (ii).

From the above lemma, we can easily get

THEOREM 1. Suppose $X^{(n)}(t) = (X_1^{(n)}(t), X_2^{(n)}(t), ...)$ $(t \in [0,1],\ n = 1, 2, ...)$ is a sequence of stochastic processes. If all component processes $\{X_i^{(n)}(t)\}_n$ are tight, then the laws of $\{X^{(n)}(t)\}_n$ are tight as well.

Proof. By the diagonal method, we can find a subsequence $X^{(n')}$ such that each $X_i^{(n')}$ converges to some $X^{(\infty)_i}$. Then we can easily verify that $X^{(n')}$ converges to $X^{(\infty)}$ by using Lemma 1.

§II. Finite Particles Case

(3) We assume

i) the function g in (1) is symmetric, i.e. $g(x) = g(y)$ if $|x| = |y|$;

ii) g has compact support;

iii) g is continuously differentiable on $R^d \backslash \{x \in R^d,\ |x| \leq r\}$, where $r \geq 0$ is the radius of particles;

iv) $r > 0$, and $|g(x)|(|x| - r)^2 \longrightarrow +\infty$ when $|x| \searrow r$;

We will also consider the case where $r = 0$. In that case we need the strengthened hypothesis:

iv') $g(x)|x|^{d+1} \longrightarrow +\infty$ when $|x| \to 0$,

or

iv'') $g(x) \geq 0$, and $|x|^2 g(x) \longrightarrow +\infty$ when $|x| \searrow r = 0$.

(4) For the external potential h, we assume

v) h is twice differentiable on its domain $A \in R^d$;

vi) h is bounded from below on A, and $h(x) \to \infty$ when $x \to \partial A$.

THEOREM 2. Suppose that there are n particles and their positions at time 0 are $x_1, ..., x_n$ where $\min_{i \neq j} |x_i - x_j| > r$ and $x_i \in A$ $(i = 1, ..., n)$. And suppose that $v_1, ..., v_n \in R^d$. If (3) and (4) hold, then the stochastic differential equation (1) has a unique solution under the initial condition

$$V_i(0) = v_i, \qquad X_i(0) = x_i \qquad (i = 1, ..., n). \tag{5}$$

Proof. i) Define the compact set for each integer k,

$$K_k = \{(y_1, ..., y_n) \in R^{nd}; \min_{i \neq j} |y_i - y_j| \geq \frac{1}{k} + r, \ \inf_{x \notin A} \min_i |x - y_i| \geq \frac{1}{k}\}.$$

By the theory of stochastic differential equations, we know that there are a stopping time T and an $(\mathcal{F}_t)$–semimartingale $X(t) \in R^{nd}$ satisfying (5) and (1) before T, where $(\mathcal{F}_t)$ is the natural filtration of $\{W_i(t),\ i = 1, ..., n\}$. Denote $T_k = \inf\{t,\ X(t) \notin K_k\}$. Then $T_k \nearrow T$. We are going to show $T_k \nearrow \infty$, which gives us the existence of a global solution of (1). This method has already been used in the theory of stochastic mechanics.

ii) By Ito's formula and the antisymmetry of ∇g,

$$\sum_i h(X_i(t)) + \frac{1}{2}\sum_{i=1}^{n}\sum_{j \neq i} g(X_i(t) - X_j(t)) - \sum_i h(X_i(0))$$
$$- \frac{1}{2}\sum_{i=1}^{n}\sum_{j \neq i} g(X_i(0) - X_j(0))$$
$$= \int_0^t \sum_{i=1}^{n}\sum_{j \neq i} \nabla g(X_i(s) - X_j(s))V_i(s)ds + \int_0^t \sum_{i=1}^{n} \nabla h(X_i(s))V_i(s)ds.$$

And we have also

$$\frac{1}{2}\sum_i |V_i(t)|^2 - \frac{1}{2}\sum_i |V_i(0)|^2$$
$$= - \int_0^t \sum_{i=1}^{n}\sum_{j \neq i} \nabla g(X_i(s) - X_j(s))V_i(s)ds - \int_0^t \sum_i \nabla h(X_i(s))V_i(s)ds$$
$$- \frac{1}{2}\int_0^t \sum_i |V_i(s)|^2 ds + \int_0^t \sum_i V_i(s)dW_i(s) + \frac{1}{2}tnd\ .$$

Combining the above two formulas, we get

$$
\begin{aligned}
&\sum_i h(X_i(t)) + \frac{1}{2}\sum_{i=1}^{n}\sum_{j\neq i} g(X_i(t) - X_j(t)) + \frac{1}{2}\sum_i |V_i(t)|^2 \\
&+\frac{1}{2}\sum_i \int_0^t |V_i(s)|^2 ds \\
(6)\qquad &= \sum_i h(X_i(0)) + \frac{1}{2}\sum_{i=1}^{n}\sum_{j\neq i} g(X_i(0) - X_j(0)) + \frac{1}{2}\sum_i |V_i(0)|^2 \\
&+\sum_i \int_0^t V_i(s)dW_i(s) + \frac{1}{2}tnd
\end{aligned}
$$

Therefore

$$
\begin{aligned}
&\sum_i E[h(X_i(T_k \wedge t))] + \frac{1}{2}\sum_{i=1}^{n}\sum_{j\neq i} E[g(X_i(T_k \wedge t) - X_j(T_k \wedge t))] \\
(7)\qquad &+\frac{1}{2}\sum_i E[|V_i(T_k \wedge t)|^2] + \frac{1}{2}E[\sum_i \int_0^{T_k \wedge t} |V_i(s)|^2 ds] \\
&\leq \sum_i h(x_i) + \frac{1}{2}\sum_{i=1}^{n}\sum_{j\neq i} g(x_i - x_j) + \frac{1}{2}\sum_i |v_i|^2 + \frac{1}{2}tnd
\end{aligned}
$$

Define $A = \{\omega;\ lim_k(t \wedge T_k(\omega)) < t\}$. If $P[A] > 0$, the left-hand side of (7) tends to $+\infty$ when $k \to +\infty$. But the right-hand side is bounded. Thus $P[A] = 0$. Therefore $T_k \nearrow +\infty$ *a.s.*

Under the hypothesis in (3) (need only that one of the conditions *iv*), *iv'*) or *iv''*) holds.) we have the following estimate:

LEMMA 2. There is a constant $C > -\infty$ such that

$$\sum_{i=1}^{n}\sum_{j\neq i} g(x_i - x_j) \geq nC\,, \qquad \forall x_i \in R^d, n = 1,2,...$$

Proof. It is sufficient to prove the lemma under $iv')$. Furthermore we need in this proof only a weaker condition as follows:

$$(8) \qquad g(x)|x|^d \longrightarrow +\infty \qquad when\ |x| \to 0.$$

Suppose that $b > 0$ is a constant such that $\sup_{|x|\geq b} |g(x)| = 0$. Denote $a = \inf_x g(x)$. Then, under Condition (8) there is $k > 0$ such that $|a|(2kdb+1)^d \leq \frac{1}{4}g(\frac{1}{k})$ and $g(x) \geq g(\frac{1}{k})$ when $|x| \leq \frac{1}{k}$. We divide the whole space R^d into identical cubes such that the side length of each cube is $(kd)^{-1}$. The particles $\{x_i\}$ are in these blocks. Define q as the maximum number of particles in one block. We denote by A any block which has q particles. Then we get

$$\sum_{x_i,x_j\in A} g(x_i - x_j) \geq q(q-1)g(\frac{1}{k}),$$

because the distance between any pair of particles in A is less than $\frac{1}{k}$. Since the support of g is of radius b, there are at most $(2kdb+1)^d$ blocks within a distance less than b to A. Thus,

$$\sum_{x_i\in A,\, x_j\notin A} g(x_i - x_j) \geq aq^2(2kdb+1)^d\,.$$

From the above two inequalities, we get

$$\sum_{x_i, x_j \in A} g(x_i - x_j) + \sum_{x_i \in A, x_j \notin A} g(x_i - x_j) \geq 0 .$$

If $q \geq 2$, from the last inequality, we get

$$\sum_{i,j} g(x_i - x_j) \geq \sum_{x_i,\ x_j \notin A} g(x_i - x_j) .$$

So the particles outside of A form a new set for which the sum $\sum_{i,j} g(x_i - x_j)$ is smaller or equal to the previous one. We repeat the above procedure if there are more than one particles in any block in the new set .... In the end we get a set of particles such that there is at most one particle in each block. Therefore for any fixed particle x_i, there are at most $(2kdb+1)^d$ particles lying within a distance less than b to it in the final set. Denote $C = a(2kdb+1)^d$. We get this lemma.

THEOREM 3. For any fixed t, there is a constant C such that the following estimates hold:

$$\text{(9)} \quad E\{\sum_i \int_0^t |V_i(s)|^2 ds\} \leq 2\sum_i h(x_i) + \sum_{i=1}^{n} \sum_{j \neq i} g(x_i - x_j) + \sum_i |v_i|^2 + nC ,$$

$$\text{(10)} \quad E\{\sup_{s<t}[\sum_i h(X_i(s))]\} \leq 5\sum_i h(x_i) + \frac{5}{2}\sum_{i=1}^{n} \sum_{j \neq i} g(x_i - x_j) + \frac{5}{2}\sum_i |v_i|^2 + nC ,$$

$$E\{\sup_{s<t}[\sum_{i=1}^{n}\sum_{j\neq i}g(X_i(s)-X_j(s))]\}$$
$$\leq 10\sum_i h(x_i)+5\sum_{i=1}^{n}\sum_{j\neq i}g(x_i-x_j)+5\sum_i|v_i|^2+nC, \tag{11}$$

and

$$E\{\sup_{s<t}[\sum_i|V_i(s)|^2]\}\leq 10\sum_i h(x_i)+5\sum_{i=1}^{n}\sum_{j\neq i}g(x_i-x_j)+5\sum_i|v_i|^2+nC. \tag{12}$$

Proof. From (7) and Lemma 2, we get easily (9). We are going to prove (10) from (6).

$$\sum_i h(X_i(t))\leq\sum_i h(x_i)+\frac{1}{2}\sum_{i=1}^{n}\sum_{j\neq i}g(x_i-x_j)+\frac{1}{2}\sum_i|v_i|^2+\sum_i\int_0^t V_i(s)dW_i(s)$$
$$+\frac{1}{2}tnd-\frac{1}{2}\sum_{i=1}^{n}\sum_{j\neq i}g(X_i(t)-X_j(t)).$$

From an inequality in martingale theory,

$$E\{\sup_{s<t}|\sum_i\int_0^s V_i(u)dW_i(u)|\}\leq E^{\frac{1}{2}}[(\sup_{s<t}|\sum_i\int_0^s V_i(u)dW_i(u)|)^2]$$
$$\leq 2\{\sup_{s<t}E^{\frac{1}{2}}[|\sum_i\int_0^s V_i(u)dW_i(u)|^2]\}=2\{\sup_{s<t}E^{\frac{1}{2}}[(\sum_i\int_0^s|V_i(u)|^2du]\}$$
$$\leq 2\{E[\sum_i\int_0^t|V_i(u)|^2du]+1\}.$$

Therefor we get (10). The proof of the other inequalities are similar.

§III. Infinite Particles Case

We will use weak convergence techniques to prove the existence of the infinite particle system (1). Our argument depends on a groupwise symmetry assumption on the initial data. We introduce

DEFINITION. Given two positive integers $p < \infty$ and $K \leq \infty$ (K may take $+\infty$ as its value). Let x_i and v_i ($i = 1, ..., pK$) be the initial positions and initial velocities respectively such that

$$\inf_{i \neq j} |x_i - x_j| > 0.$$

We say they are **groupwise symmetrically positioned** if we can renumber them as $x_{i,j}$ and $v_{i,j}$ ($i = 1, ..., K$; $j = 1, 2, ..., p$) in a way such that the following condition is satisfied: for any fixed $1 \leq j \leq p$, any pair of integers α and β, we can find two new Cartesian coordinate systems Ξ and Ξ', with $x_{\alpha,j}$ and $x_{\beta,j}$ as their origins respectively, through translation and rotation of the original one such that the set $\{x_{i,j}, v_{i,j}\}$ under Ξ-representation are just the same as that under Ξ'-representation (the only change is that of the order). When $K = \infty$, we will assume in addition that there is a sequence of groupwise symmetrically positioned finite sets

$$\Phi_n = \{x_{i,j}^{(n)}, v_{i,j}^{(n)}; i = 1, ..., n, j = 1, ..., p.\}$$

such that

$$\lim_n x_{i,j}^{(n)} = x_{i,j}, \qquad \lim_n v_{i,j}^{(n)} = v_{i,j}. \tag{13}$$

REMARK 1. For example, when $d = 2$, let's consider the case where the positions are

$$\{\ldots, (-2,0), (-1,0), (0,0), (1,0), (2,0)\ldots\}$$

and the velocities are equal to the constant vector $(1,0)$. They are groupwise symmetrically positioned with $j = 1$. We can chose Φ_n as follows: for $1 \le i \le n$, $x_{i,1}^{(n)}$ are distributed with equal distance between the neighboring points on the circle which is centered at $(0, (2\pi)^{-1}n)$ and tangent to the x-axis at $(0,0)$; $v_{i,1}^{(n)}$ is defined as the unit tangent vector to that circle at $x_{i,1}^{(n)}$. The Φ_n are groupwise symmetrically positioned finite sets and (13) holds if we number the points in the set properly.

We say that a function $f(x)$ is fit for the groupwise symmetrically positioned $\{x_{i,j}, v_{i,j}\}$ if its value is invariant under the above coordinate system changes.

LEMMA 3. Suppose that (3), (4) and the groupwise symmetry assumption hold. Suppose also that there is a constant b such that $0 < b < +\infty$ and $x_i \in R^{d-1} \times (-b, b)$ $(i = 1, 2, \ldots)$. If the external potential function h is defined on a region $A \subseteq R^d$ such that

$$h(x)|dis(x, \partial A)|^2 \to \infty \tag{14}$$

when $x \to \partial A$ (where $dis(x, \partial A)$ is the distance between x and ∂A), and h is fit for $\{(x_i, v_i)\}$ then (1) has a solution $\{X_i(t), V_i(t)\}$ with the initial distribution

$$(15) \qquad V_i(0) = v_i, \qquad X_i(0) = x_i \qquad (i = 1, 2, ...).$$

Proof. We need only one of the hypotheses (3)iv), iv') or iv'') to hold. To simplify our description, we will assume that (3)iv') holds and we only consider the two dimensional ($d = 2$) case. The proofs in the other cases are similar. Also to simplify our description, we suppose that all the initial positions are on the Y-axis and $p = 1$. According to our hypothesis, there are groupwise symmetrically positioned finite sets Φ_n such that (13) holds (see Remark 1). We replace h by a sequence $\{h_n)\}_n$ which are fit for Φ_n respectively and $h_n(.) \longrightarrow h(.)$.

Now consider the n particles in Φ_n :

$$(16) \qquad \begin{aligned} &dX_i^{(n)}(t) = V_i^{(n)}dt, \\ &dV_i^{(n)}(t) = -\sum_{j \neq i} \nabla g(X_i^{(n)}(t) - X_j^{(n)}(t))dt - \nabla h_n(X_i^{(n)}(t))dt \\ &-\frac{1}{2}V_i^{(n)}(t)dt + dW_i^{(n)}(t) \end{aligned}$$

with the initial values

$$(17) \qquad X_i^{(n)}(0) = x_i^{(n)}, \qquad V_i^{(n)}(0) = v_i^{(n)},$$

(i,j=1,...,n). We have discussed its existence in Theorem 2. From the symmetry hypothesis and (9), we get

$$(18). \; E[\int_0^1 |V_i^{(n)}|^2 dt] \leq n^{-1}(2\sum_i^n h_n(x_i) + \sum_{j \neq i} g(x_i - x_j) + \sum_i^n |v_i|^2 + nC) \leq \underline{C} < \infty$$

where $\underline{C}$ is some constant independent of n. From Lemma 2 of [7], we deduce that for any fixed i, $\{X_i^{(n)}\}_n$ is tight on $C([0,1] \to R^d)$. Therefore from Theorem 1 we deduce that $\{X^{(n)}\}_n$ is tight on $C([0,1] \to R^\infty)$.. Let P be any limiting law. We are going to show that under P, the coordinate processes $\{X_i^{(n)}(t)\}$ satisfy (1). Since $C([0,1] \to R^\infty)$ is a Polish space, we can apply the well-known Skorohod's theorem (see Billingsley [2], Theorem 3.3) to $X_\cdot$. Thus we can realize all our processes on a same probability space $(\Omega, \mathcal{F}, P)$ such that $X_i^{(n)}$ converges to X_i a.s. under the topology of uniform convergence.

ii) We need the following proposition:

Proposition. Let $f \in C_0^1(R^d)$. Given $\varepsilon > 0$, there is $k < \infty$ such that

$$\sum_{i=k}^{\infty} E[\sup_{t \le 1} |f(X_i^{(n)}(t))|] < \varepsilon, \qquad (n = 1, ..., \infty). \tag{19}$$

Proof. We introduce for any n a family of functions $f_{n,i}$ as follows: when $n = \infty$, we set $f_{n,i}(x) = f(x + x_i - x_1)$. When $n < \infty$, the particles are distributed along a circle denoted by C_n such that the distances between the neighboring particles are equal (see Remark 1). Suppose that the arc between $x_1^{(n)}$ and $x_i^{(n)}$ is equal to $u(i)\pi$, then we rotate f $(-u(i)\pi)$ around the center of C_n to get $f_{n,i}$. Thus we can easily verify that $f(X_i^{(n)}(t))$ has the same law as $f_{n,i}(X_1^{(n)}(t))$. Therefore (19) is equivalent to

$$\sum_{i=k}^{\infty} E[\sup_{t \le 1} |f_{n,i}(X_1^{(n)}(t))|] < \varepsilon, \qquad (\forall n). \tag{20}$$

Since $X_1^{(n)}$ has the velocity $V_1^{(n)}$,

$$E[\sup_{t\leq 1}|f_{n,i}(X_1^{(n)}(t))|] \leq E[\int_0^1 |\nabla f_{n,i}(X_{n,1}(t))||V_1^{(n)}(t)|dt].$$

From the Cauchy inequality,

$$E[\int_0^1 (\sum_{i=k}^{\infty}|\nabla f_{n,i}(X_1^{(n)}(t))||V_1^{(n)}(t)|)dt]$$
$$\leq E^{\frac{1}{2}}[\int_0^1 (\sum_{i=k}^{\infty}|\nabla f_{n,i}(X_1^{(n)}(t))|^2dt]E^{\frac{1}{2}}[\int_0^1 |V_1^{(n)}(t)|^2dt].$$

Therefore we get

$$E[\sup_{t\leq 1}|f_{n,i}(X_1^{(n)}(t))|] \leq E^{\frac{1}{2}}[\int_0^1 (\sum_{i=k}^{\infty}|\nabla f_{n,i}(X_1^{(n)}(t))|^2dt]E^{\frac{1}{2}}[\int_0^1 |V_1^{(n)}(t)|^2dt].$$

From the discussion in i), we know that $E^{\frac{1}{2}}[\int_0^1 |V_1^{(n)}(t)|^2dt]$ is uniformly bounded in n. Thus we need only to show the first factor will become sufficiently small when k becomes large. Nevertheless

$$\sum_{i=k}^{\infty}|\nabla f_{n,i}(.)| \tag{21}$$

is uniformly bounded in n and converge to 0 when $k \nearrow \infty$, and we can always choose a $k(Q)$ for any bounded set $Q \subset R^d$ such that

$$\sum_{i=k(Q)}^{\infty} |\nabla f_{n,i}(x)| = 0 \qquad (\forall x \in Q, \forall n).$$

Therefore,

$$E^{\frac{1}{2}}\Big[\int_0^1 \Big(\sum_{i=k(Q)}^{\infty} |\nabla f_{n,i}(X_{n,1}(t))|\Big)^2 dt\Big] \tag{22}$$
$$\leq C(P[\omega;\ \exists t \in [0,1]\ s.t.\ X_1^{(n)}(t,\omega) \notin Q])^{\frac{1}{2}}.$$

where C is an upper bound of (21). We take, in particular, $Q = \{x; |x - x_1^{(n)}| < q\}$, then

$$P[\omega;\ \exists t \in [0,1]\ s.t.\ X_1^{(n)}(t,\omega) \notin Q] \leq P[\omega;\ (\int_0^1 |V_1^{(n)}(t,\omega)|^2 dt)^{\frac{1}{2}}] \geq q]$$
$$\leq q^{-2} E[\int_0^1 |V_1^{(n)}|^2 dt] .$$

When q is sufficiently large, the right-hand side of (22) will be arbitrarily small and therefore we get (21).

iii) We are going to show the following inequalities:

$$\inf_{0\leq s\leq 1} |X_i(s) - X_j(s)| > 0 \quad a.s. \quad (i \neq j); \tag{23}$$

$$\inf_{0\leq s\leq 1} [dis(X_i(s), \partial A)] > 0 \quad a.s. \quad (\forall i). \tag{24}$$

We prove only (23). The proof of (24) is similar. We have

$$d|X_i - X_j|(t) = \sum_{m=1}^{d} (X_{i,m}(t) - X_{j,m}(t))|X_i(t) - X_j(t)|^{-1}(V_{i,m}(t) - V_{j,m}(t))dt. \tag{25}$$

where $X_{i,m}(t)$ is the m-th component of $X_i(t)$. Fix some $\underline{r}$ and define a new function on R_+ :

$$G(x) = \begin{cases} 0, & \text{if } x \geq \underline{r}; \\ \int_x^{\underline{r}} |g(s)|^{\frac{1}{2}} ds, & \text{otherwise.} \end{cases} \tag{26}$$

Since $|g(s)|s^2 \to +\infty$ when $s \to +0$, we get $G(0) = +\infty$. But from (25) and (26) we get

$$\begin{aligned} &G(|X_i(t) - X_j(t)|) - G(|X_i(0) - X_j(0)|) \\ &= -\int_0^t |g(X_i(s) - X_j(s)|^{\frac{1}{2}} |X_i(s) - X_j(s)|^{-1} \\ &\quad \sum_m (X_{i,m}(s) - X_{j,m}(s))(V_{i,m}(s) - V_{j,m}(s)) ds. \end{aligned} \tag{27}$$

If we take $f(.) = g^-(.)$ in (24) and use (11), we can easily verify

$$E[g^-(X_i^{(n)}(s) - X_j^{(n)}(s))] < C, \qquad E[g^+(X_i^{(n)}(s) - X_j^{(n)}(s))] < C$$

where C is some constant. So that from the Fatou lemma we get

$$E|g(X_i(s) - X_j(s)| < 2C \ . \tag{28}$$

From (27),

$$\begin{aligned} &E[\sup_{t \leq 1} |G(|X_i(t) - X_j(t)|) - G(|X_i(0) - X_j(0)|)|] \\ &\leq E^{\frac{1}{2}} \int_0^1 |g(X_i(t) - X_j(t))| dt \, E^{\frac{1}{2}} \int_0^1 |V_i(s) - V_j(s)|^2 ds. \end{aligned}$$

Thus, using (18) and (28), we deduce

$$E\{\sup_{t\leq 1}[G(|X_i(t) - X_j(t)|)]\} < \infty$$

which is equivalent to (23).

iv) By (23), (24) and the uniform convergence of $X_i^{(n)}(.)$ to $X_i(.)$, we have

$$\int_0^t \nabla g(X_i^{(n)}(s) - X_j^{(n)}(s))ds \longrightarrow \int_0^t \nabla g(X_i(s) - X_j(s))ds \qquad a.s. \tag{29}$$

and

$$\int_0^t h_p(X_i^{(n)}(s))ds \longrightarrow \int_0^t \nabla h(X_i(s))ds \qquad a.s. \tag{30}$$

By (19), we know that for sufficiently large k, the probability of the following events

$$\{\omega;\ \varepsilon < \sup_{t\leq 1}|\sum_{j=1,j\neq i}^{k}\sum_{0}^{t}\nabla g(X_i^{(n)}(s,\omega) - X_j^{(n)}(s,\omega))ds$$

$$- \sum_{j=1,j\neq i}^{\infty}\sum_{0}^{t}\nabla g(X_i^{(n)}(s,\omega) - X_j^{(n)}(s,\omega))ds|\}$$

will be arbitrarily small ($n = 1, ..., +\infty$). Therefore we deduce from (29) that

(31)

$$(Pr - \lim) \sum_{(j,j\neq i)} \int_0^t \nabla g(X_{n,i}(s) - X_{n,j}(s))ds = \sum_{(j,j\neq i)} \int_0^t \nabla g(X_i(s) - X_j(s))ds\ .$$

where $(Pr - \lim)$ means limit in probability. Similarly, we can also deduce that $\int_0^t h_n(X_i^{(n)}(s))ds$ converges to $\int_0^t h(X_i(s))ds$. Now, we have

$$V_i^{(n)}(t) - V_i^{(n)}(0) = -\sum_{j\neq i}\sum_{0}^{t}\nabla g(X_i^{(n)}(s) - X_j^{(n)}(s))ds$$

$$-\sum_{0}^{t}\nabla h_n(X_i^{(n)}(s))ds - \frac{1}{2}X_i^{(n)}(t) + \frac{1}{2}X_i^{(n)}(0) + W_i^{(n)}(t)\ .$$

From the convergence of each term, we get the convergence of $V_i^{(n)}(t)$ to some $V_i(t)$. We can easily verify $X_i(t)$ and $V_i(t)$ satisfy (1). Therefore we get the lemma.

The following theorem get rid of the additional assumption of Lemma 3.

THEOREM 4. Let $h = 0$. Suppose that (3) holds and x_i and v_i $(i = 1, ..., pK)$ are groupwise symmetrically positioned. Then (1) has a solution $\{X_i(t), V_i(t)\}$ with the initial condition (15).

Proof. We define for $x \in R^d$, $y \in R$ and $k = 1, 2, ...$, the following functions:

$$h_k(x, y) = \begin{cases} (\frac{1}{k} - |y|^2)^{-3}, & \text{if } |y| < k^{-\frac{1}{2}}; \\ +\infty, & \text{otherwise.} \end{cases}$$

Then for fixed k, $h_k(x, y)$ is a function defined on R^{d+1} and satisfies the hupothesis of Lemma 3. We also define a new initial condition on R^{d+1} :

$$X_i^{(k)}(t) = (x_i, 0), \qquad V_i^{(k)}(0) = (v_i, 0). \tag{32}$$

Using Lemma 3, we get a sequence of processes $\{X_i^{(k)}(t), V_i^{(k)}(t)\}$ which satisfies (1) for $h(.) = h_k(.)$ and (31). Since the right-hand sides of the estimates in (9)-(12) do not depend on the choice of k, we can repeat the proof of Lemma 3 to deduce that the laws of $\{X_i^{(k)}(t), V_i^{(k)}(t)\}$ are tight. Denote $\{X_i(t), V_i(t)\}$ as one of the limiting process. From the fact that

$$P[X_i^{(k)}(t) \in R^d \times [-k^{-\frac{1}{2}}, k^{\frac{1}{2}}], \forall t \in [0,1]] = 1 ,$$

we get

$$P[X_i(t) \in R^d \times \{0\},\ \forall t \in [0,1]] = 1\,.$$

Therefore we can restrict $X_i(t)$ on R^d and the theorem is proved.

We can also get rid of the condition $h = 0$ by the Girsanov theorem as follows:

THEOREM 5. Let $h \in C_0^1(R^d)$. Suppose that (3) and (4) hold. And suppose that the x_i and v_i ($i = 1, \ldots, pK$) are groupwise symmetrically positioned. If the particles are hard-sphered (i.e. (3)iv) holds), then (1) has a solution with the initial condition (15).

Proof. Let $\{X_i(t),\ V_i(t)\}$ be the solution of (1) when h=0. Define a new process by

$$Z_t = exp\{-\int_0^t \sum_i \nabla h(X_i(s))dW_i(s) - \frac{1}{2}\int_0^t \sum_i |\nabla h(X_i(s))|^2 ds\}.$$

Since there is only a bounded number of particles in the support of $h(.)$, we deduce that

$$\int_0^t \sum_i |\nabla h(X_i(s))|^2 ds < C\,,$$

where C is some constant. Denote by P the probability law of $\{X_i(t),\ V_i(t)\}$, and by $(\mathcal{F}_t)$ the natural filtration. Then Z_t is an $(\mathcal{F}_t)$-martingale. Let $dP^* = Z_1 dP$. From the Girsanov theorem we know that under the new law P^*, $\{X_i(t),\ V_i(t)\}$ are just the processes asserted by the Theorem.

REFERENCES

[1]Billingsley, P. Convergence of Probability Measures, John Wiley & Sons, Inc., New York, 1968;

[2]Billingsley, P. Weak convergence of probability measures, CBMS Regional Conference Series in Math., n.6, Amer. Math. Soc.;

[3]Fritz, J. Gradient dynamics of infinite point systems, Annals of Probability, 1987, Vol.15, No.2;

[4]Lang, R. Unendlich-dimensionale Wienerprozesse mit Wechselwirkung I, Zeit. Wahrsch. verw. Gebiete 1977, 39;

[5]Guo, M.Z. and Papanicolau, G. Bulk diffusion for interacting Brownian particles, Progress in Physics — Statistical Physics and Dynamical Systems, (S.Szasz ed.) Birkhäuser, Boston, 1985;

[6]Rost, H. Diffusion de sphères dans la droite réelle: comportement macroscopique et équilibre local, Séminaire de Probabilit'es XVIII, Lecture Notes in Mathematics 1059, 1984;

[7]Zheng, W. Tightness results for laws of diffusion processes: application to stochastic mechanics, Ann. Inst. Henri Poincare, Vol.21, n.2, 1985.

Department of Math. Statistics

East China Normal University, Shanghai, China

Part II
Finite-dimensional Diffusion Processes

Some Phenomena of the Characteristic Boundary Exit Problem

Martin V. Day

§1: Introduction

The exit problem concerns the asymptotic behavior of the first exit position $x^\epsilon(\tau_D^\epsilon)$ of a diffusion

$$dx^\epsilon(t) = \mathbf{b}(x^\epsilon(t))dt + \epsilon^{1/2}\sigma(x^\epsilon(t))dw(t); \quad x^\epsilon(0) = x_0 \tag{1.1}$$

from a region D. Assume D is attracted to a unique stable critical point at the origin by the deterministic system

$$\dot{x}^0(t) = \mathbf{b}(x^0(t)). \tag{1.2}$$

Let

$$\mu^\epsilon(x_0, A) = P_{x_0}[x^\epsilon(\tau_{\partial D}^\epsilon) \in A], \quad A \subseteq \partial D,$$

be the exit distribution on ∂D. ($\tau_{\partial D}^\epsilon$ is defined in (1.4) below.) Our goal is to study the asymptotic ($\epsilon \downarrow 0$) behavior of $\mu^\epsilon(x_0, dy)$.

The work of Wentzell and Freidlin [11] on this problem assumes that (1.2) enters D nontangentially:

$$\langle \mathbf{b}(y), \mathbf{n}(y) \rangle < 0, \quad \text{all } y \in \partial D,$$

where $\mathbf{n}(y)$ is the unit outward normal. We are concerned here with the case of characteristic boundary:

$$\langle \mathbf{b}(y), \mathbf{n}(y) \rangle = 0, \quad \text{all } y \in \partial D.$$

Several papers have addressed aspects of the characteristic boundary problem ([3, 4, 5, 12] for instance), but it is not as clearly understood as the nontangential case (see [6] and [11]). The usual large deviations results continue to hold in the characteristic boundary case, [4], but are not as satisfying as in the nontangential

[1]This resreach was supported in part by the National Science Foundation under grant DMS-8420755

case. In particular it is more common for the quasipotential function $V(y)$ to achieve its minimum over ∂D on a large subset of ∂D. For instance if ∂D is a periodic orbit of (1.2) then V will be constant over ∂D. The question we will address here is what can be said about the weak limiting behavior of μ^ϵ as $\epsilon \downarrow 0$.

In this paper we will consider only a simplified 2-dimensional version of the problem which allows some explicit calculations to be employed. We discover several types of behavior which have not been anticipated in earlier work, showing that the problem is more complicated than previously thought. For instance if ∂D is a single periodic orbit of (1.2) then the exit measures can "cycle" around ∂D as periodic functions of $\log(\epsilon^{-1/2})$, failing to converge. In fact such nonconvergence would appear to be typical for ∂D periodic; only for very special examples will the exit measures converge. This will be discussed in §4. When ∂D contains isolated rest points of (1.2), both stable and unstable, then one would expect the exit measure to concentrate on the stable ones. We will see that this can occur in different ways, with the measures for $\epsilon > 0$ either being centered about the rest points or converging to them from the sides ("skewing" as it called in [3]). However we will also see that it is possible for the exit measure to concentrate at the unstable rest points. Moreover it is possible for the limiting exit measure to have a positive density between critical points. These are discussed in §5.

We take the region D to be the unit disk,

$$D = \{x \in \mathbb{R}^2 : \ \|x\| < 1\}.$$

Thus ∂D is the unit circle. We will use the coordinates (ρ, θ) where θ is the usual polar angle and $\rho = 1 - \|x\|$ is the distance from the boundary. For (1.1) we assume a system which in the annulus $0 \le \rho < 1/2$ can be expressed as

$$\begin{aligned} d\rho^\epsilon(t) &= B\rho^\epsilon(t)dt + \epsilon^{1/2}d\beta_1(t) \\ d\theta^\epsilon(t) &= b(\theta^\epsilon(t))dt + \epsilon^{1/2}d\beta_2(t). \end{aligned} \tag{1.3}$$

Here $B > 0$ is a constant (fixed throughout) and $b(\theta)$ is some smooth 2π–periodic function. The $\beta_i(t)$ are independent scalar Brownian motions. In the center disk of radius $1/2$,

$$C = \{(\rho, \theta) : \ \rho > \frac{1}{2}\},$$

we assume that the definitions are extended in some way, producing a nondegenerate diffusion x^ϵ defined up to the exit time

$$\tau_D^\epsilon = \inf\{t > 0 : \ \rho^\epsilon(t) = 0\}, \tag{1.4}$$

and so that the corresponding deterministic system (1.2) does attract all of D to 0.

Notice that (1.3) is *not* strictly speaking of the form (1.1). Indeed, if (1.3) is rewritten in cartesian coordinates we get an $\mathcal{O}(\epsilon)$ contribution to the drift term;

$\mathbf{b}(x^\epsilon(t)) + \epsilon \mathbf{b}_1(x^\epsilon(t))$ instead of the fixed $\mathbf{b}(x^\epsilon(t))$ in (1.1). It turns out that such an $\mathcal{O}(\epsilon)$ perturbation of the drift does not alter the main features of our calculations below. The technical justification of this assertion will be given in a forthcoming more comprehensive treatment of (1.1). Our purpose here is to use the simple structure of (1.3) to exhibit some qualitative features of the characteristic boundary exit problem.

In [3] Bobrovsky and Zeitouni also carry out some related calculations for a system like (1.3). Their effort is focused on the skewing phenomenon alone. Our approach sharpens their estimate on the ratio of eigenvalues for which this occurs from $\frac{-\lambda}{B} < 2/3$ ($\frac{g}{k}$ in their Theorem 3.1) to $\frac{-\lambda}{B} < 1$, as well as (we think) providing a clearer intuitive explanation of the phenomena.

§2: Conditioning on Exit

Consider the disk G of radius $3/4$,

$$G = \{\rho > \frac{1}{4}\}$$

which is intermediate between the unit disk D and the center disk C already mentioned: $C \subseteq G \subseteq D$. For $x_0 \in C$ consider the exit distribution from G:

$$\nu^\epsilon(x_0, F) = P_{x_0}[x^\epsilon(\tau^\epsilon_{\partial G}) \in F], \quad F \subseteq \partial G$$

where

$$\tau^\epsilon_{\partial G} = \inf\{t > 0 : \rho^\epsilon(t) = \frac{1}{4}\}.$$

In addition for $x_0 \in D \backslash C$ define the first contact time for C:

$$\tau^\epsilon_{\partial C} = \inf\{t > 0 : \rho^\epsilon(t) = \frac{1}{2}\}.$$

We can give the following heuristic description of how the exit to ∂D occurs, for asymptotically small ϵ. Starting at x_0 the process first falls into the vicinity of 0 where it looses it dependence on the initial point x_0. Hence μ^ϵ (and ν^ϵ) are nearly independent of x_0:

$$\mu^\epsilon(x_0, dy) \simeq \mu^\epsilon(dy) \qquad \nu^\epsilon(x_0, dy) \simeq \nu^\epsilon(dy).$$

(See (2.2) below.) After a long time the process makes an excursion away from 0 and reaches ∂G for the first time at time $\tau^\epsilon_{\partial G}$, with distribution ν^ϵ. Next with a very small probability

$$\gamma^\epsilon = P_y[\tau^\epsilon_{\partial D} < \tau^\epsilon_{\partial C}], \quad y \in \partial G,$$

the process proceeds the rest of the way to ∂D without re-entering C. On the other hand, with probability $1 - \gamma^\epsilon$, after $\tau^\epsilon_{\partial G}$ the process falls back into C, without reaching ∂D, and back to the vicinity of 0 and this all begins anew. Thus we see a sequence of essentially independent excursions from 0 to ∂G, each landing with distribution ν^ϵ and followed by an attempt to proceed from the current position on ∂G to ∂D, the attempt succeeding with probability γ^ϵ. If the attempt fails, we go on to the next in the sequence of excursions from 0. Eventually one of the attempts to proceed from $y \in \partial G$ to ∂D succeeds, with $x^\epsilon(\tau^\epsilon_{\partial D})$ landing at a point on ∂D given by the *conditional* distribution

$$Q^\epsilon(y, A) = P_y[x^\epsilon(\tau^\epsilon_{\partial D}) \in A \mid \tau^\epsilon_{\partial C} < \tau^\epsilon_{\partial D}].$$

The upshot of this intuitive description is that

$$\mu^\epsilon(A) = Q^\epsilon(\nu^\epsilon, A), \tag{2.1}$$

the latter being shorthand for $\int_{\partial G} Q^\epsilon(y, A)\nu^\epsilon(dy)$.

We can make (2.1) precise if μ^ϵ and ν^ϵ are defined as averages (using the same notational convention)

$$\mu^\epsilon(A) = \mu^\epsilon(\pi^\epsilon, A) \qquad \nu^\epsilon(F) = \nu^\epsilon(\pi^\epsilon, F)$$

where π^ϵ is a certain probability measure on ∂C. (Choose π^ϵ so that, for all measurable $H \subseteq \partial C$, $\int P_y[x^\epsilon(\tau^\epsilon_{\partial C}) \in H \mid \tau^\epsilon_{\partial C} < \tau^\epsilon_{\partial D}]\nu^\epsilon(\pi^\epsilon, dy) = \pi^\epsilon(H)$.) A conditioning argument can now be given to justify (2.1) rigorously. It is important for (2.1) that γ^ϵ is independent of $y \in \partial G$, a fact which follows from our choice of circular domains and the separation of the ρ^ϵ and θ^ϵ equations in (1.3). The basic exponential leveling result [7] (see also [8]) says that

(2.2) $|\mu^\epsilon(x, A) - \mu^\epsilon(A)| \to 0$ uniformly over all $x_0 \in \bar{C}$, and measurable $A \subseteq \partial D$,

and likewise for ν^ϵ. Thus the weak limiting behavior of $\mu^\epsilon(x_0, dy)$ is the same as for $\mu^\epsilon(dy)$.

Notice that the exit problem for G is a nontangential one, so all the results for that better understood case apply to ν^ϵ. Moreover since the dynamics of the process in C can be chosen in different ways we can arrange for ν^ϵ to behave in various ways while retaining (1.3) in the annulus $D\backslash C$. In particular given any $y^* \in \partial G$ we can arrange for $\nu^\epsilon \Rightarrow \delta_{y^*}$ by adjusting the dynamics in C so that y^* is the unique minimizer of the quasipotential $V(y)$ over $y \in \partial G$.

The special structure that we have assumed in (1.3) makes the evaluation of $Q^\epsilon(y_0, A)$ especially simple. In (ρ, θ) coordinates $y_0 \in \partial G$ corresponds to $\rho_0 = 1/4$ and some θ_0. If we let $\theta^\epsilon(\theta_0; t)$ denote the solution the θ^ϵ equation of (2.1) with $\theta^\epsilon(0) = \theta_0$, then since $A \subseteq \partial D$ is naturally identified with a set of θ values, $Q^\epsilon(y_0, dx)$ is the distribution of $\theta^\epsilon(\theta_0; \tau^\epsilon_{\partial D})$ conditional on $\rho^\epsilon(0) = 1/4$ and $\tau^\epsilon_{\partial D} < \tau^\epsilon_{\partial C}$. Since the $\theta^\epsilon(\cdot)$ and $\rho^\epsilon(\cdot)$ in (1.3) are completely independent, the conditioning affects only the distribution of the random time $\tau^\epsilon_{\partial D}$, not the distribution of the $\theta^\epsilon(t)$ paths. We will use the notation τ^ϵ_* in place of $\tau^\epsilon_{\partial D}$ when it is considered subject to this conditional. Thus τ^ϵ_* is a random variable independent of $\theta^\epsilon(\theta_0;\cdot)$ and $Q^\epsilon(x_0, dx)$ is the distribution of $\theta^\epsilon(\theta_0;\tau^\epsilon_*)$. When θ_0 is distributed according to ν^ϵ it is natural to write $\theta^\epsilon(\nu^\epsilon;t)$ for the corresponding process. With these conventions (2.1) says that μ^ϵ is the distribution of $\theta^\epsilon(\nu^\epsilon;\tau^\epsilon_*)$. We will write this as

$$\mu^\epsilon =^{\text{distn.}} \theta^\epsilon(\nu^\epsilon;\tau^\epsilon_*).$$

Our plan then is as follows.

1. Determine asymptotically the distribution of the conditioned exit time τ^ϵ_*. This is carried out in §3. We will see that $\tau^\epsilon_* = \mathcal{O}(B^{-1}\log(\epsilon^{-1/2}))$.

2. Analyze the distribution of $\theta^\epsilon(t)$ on the time scale determined by the τ^ϵ_* asymptotics. This will be done in §4 and 5.

With adequate understanding of these issues we will be able to describe various possible behaviors of μ^ϵ in terms of the boundary dynamics

$$\dot\theta^0 = b(\theta^0); \quad \theta^0(0) = \theta_0$$

on the unit circle, and $\nu^0 = \lim \nu^\epsilon$ as determined by the interior dynamics.

§3: Asymptotics of the Conditional Exit Time

This section is devoted to the first of the two goals just described: asymptotic analysis of the conditioned exit time τ^ϵ_*. This depends only on the radial part of the process, $\rho^\epsilon(t)$. We can study τ^ϵ_* by means of some explicit calculations. Our conclusions are stated as Proposition 3.3 at the end of the section.

First make the change of variables

$$y(t) = B^{1/2}\epsilon^{-1/2}\rho^\epsilon(t); \quad s = Bt.$$

It follows that

$$dy(s) = y(s)ds + d\bar\beta(s); \quad y(0) = y_0 \tag{3.1}$$

where $d\bar\beta(s) = -\sqrt{B}d\beta_1(t)$ is a new Brownian motion and, since $\rho^\epsilon(0) = 1/4$, $y_0 = B^{1/2}\epsilon^{-1/2}/4$. In other words

$$\rho^\epsilon(t) = \epsilon^{1/2}B^{-1/2}y(Bt) \tag{3.2}$$

where $y(s)$ is the (repelling) Ornstein – Uhlenbeck process (3.1). The transition density for $y(s)$ is given explicitly by

$$p(x;s,y) = [2\pi\sigma^2(s)]^{-1/2}e^{-(y-e^s x)^2/2\sigma^2(s)}$$

where

$$\sigma^2(s) = \frac{1}{2}(e^{2s}-1).$$

For any a let η_a denote the first time $y(\cdot) = a$:

$$\eta_a = \inf\{s > 0 : \ y(s) = a\}.$$

Note that $\eta_a = +\infty$ with positive probability if $0 \leq a < y_0$. Consider the (sub)probability distribution of η_0:

$$F_\infty(y_0, s) = P_{y_0}[\eta_0 \leq s].$$

Using a reflection argument,

$$\begin{aligned} F_\infty(y_0,s) &= 2P_{y_0}[y(s) \leq 0] \\ &= 2\int_{-\infty}^{0} p(y_0;s,z)dz \\ &= 2\Phi(y_0 r(s)), \text{ for } y_0 \geq 0 \end{aligned} \tag{3.3}$$

where

$$r(s) = \frac{e^s}{\sigma(s)} = \sqrt{2}(1-e^{-2s})^{-1/2} \tag{3.4}$$

and $\Phi(\cdot)$ is the tail of the standard normal distribution:

$$\Phi(x) = \int_x^\infty g(u)du, \quad g(u) = \frac{1}{\sqrt{2\pi}}e^{-u^2/2}.$$

Therefore if G_∞ is the *conditional* distribution of η_0 given $\eta_0 < \infty$,

$$G_\infty(y_0,s) = P_{y_0}[\eta_0 \leq s | \ \eta_0 < \infty],$$

then

$$G_\infty(y_0,s) = cF_\infty(y_0,s), \quad \text{where } c^{-1} = F_\infty(y_0,\infty).$$

We will need to know the asymptotic behavior of $G_\infty(y_0, ds)$ as $y_0 \to +\infty$. The following proposition provides this.

PROPOSITION 3.1. *For any bounded $\psi(\cdot) \in C(\mathbb{R})$,*

$$\left| \int_0^\infty \psi(s - \log(y_0))G_\infty(y_0,ds) \ - \ \int_{-\infty}^\infty \psi(s)q(s)ds \right| \to 0$$

as $y_0 \to +\infty$, where $q(s)$ is the probability density

$$q(s) = 2e^{-(2s+e^{-2s})}.$$

PROOF: According to what we have already said, it suffices to show that, as $y_0 \to +\infty$,

$$cF_\infty(y_0, \log(y_0) + s) \to e^{-e^{-2s}},$$

since the latter is the distribution function of the density q. This will follow from (3.3) and the fact that

$$\Phi(x) \sim x^{-1} g(x); \tag{3.6}$$

see (3.9) below. First since $r(+\infty) = \sqrt{2}$ we have

$$c^{-1} = F_\infty(y_0, \infty) = 2\Phi(y_0\sqrt{2}) \sim 2(\sqrt{2}y_0)^{-1} g(\sqrt{2}y_0) = \frac{1}{y_0\sqrt{\pi}} e^{-y_0^2},$$

or

$$c \sim \sqrt{\pi} y_0 e^{y_0^2}.$$

Next check that

$$y_0 r(\log(y_0) + s) = \sqrt{2} y_0^2 (y_0^2 - e^{-2s})^{-1/2}.$$

Therefore using (3.6) again,

$$\begin{aligned} F_\infty(y_0, \log(y_0) + s) &= 2\Phi(y_0 r(\log(y_0) + s) \\ &\sim \frac{1}{\sqrt{\pi}} y_0^{-2} (y_0^2 - e^{-2s})^{-1/2} \exp\left(-(\sqrt{2} y_0^2 (y_0^2 - e^{-2s})^{-1/2})^2/2\right). \end{aligned}$$

Thus

$$G_\infty(y_0, \log(y_0) + s) = cF_\infty(y_0, \log(y_0) + s) \sim \sqrt{\frac{y_0^2 - e^{-2s}}{y_0^2}} \exp\left(\frac{-y_0^2}{y_0^2 - e^{-2s}} e^{-2s}\right).$$

Since, for any fixed s,

$$\frac{y_0^2}{y_0^2 - e^{-2s}} \to 1, \text{ as } y_0 \to \infty$$

it follows that

$$G_\infty(y_0, \log(y_0) + s) \to e^{-e^{-2s}}$$

as claimed.

Next for any $0 < y_0 < a$ consider the distributions of η_0 restricted to and conditional on $\eta_0 < \eta_a$, respectively:

$$\begin{aligned} F_a(y_0, s) &= P_{y_0}[\eta_0 \le s;\ \eta_0 < \eta_a] \\ G_a(y_0, s) &= P_{y_0}[\eta_0 \le s \mid \eta_0 < \eta_a] = \frac{F_a(y_0, s)}{F_a(y_0, \infty)}. \end{aligned}$$

By virtue of (3.2) τ_*^ϵ has distribution

$$\begin{aligned} P[\tau_{\partial D}^\epsilon \le t \mid \tau_{\partial D}^\epsilon < \tau_{\partial C}^\epsilon] &= P_{y(0)=B^{1/2}\epsilon^{-1/2}/4}[\eta_0 \le Bt \mid \eta_0 < \eta_{\frac{1}{2}B^{1/2}\epsilon^{-1/2}}] \\ &= G_{\frac{1}{2}B^{1/2}\epsilon^{-1/2}}\left(B^{1/2}\epsilon^{-1/2}/4, Bt\right). \end{aligned} \tag{3.7}$$

The next proposition shows that the parametric dependence of G on $a = \frac{1}{2}B^{1/2}\epsilon^{-1/2}$ does indeed behave as the notation suggests as $a \to \infty$.

Proposition 3.2.

$$|G_a(y_0,s) - G_\infty(y_0,s)| \to 0$$

uniformly in s as $a^2 - y_0^2 \to +\infty$, with $y_0 \geq 1$.

Proof: It is sufficient to show that

$$\frac{F_a(y_0,s)}{F_\infty(y_0,s)} \to 1$$

uniformly in s and $y_0 \geq 1$ as $a^2 - y_0^2 \to +\infty$. Now

$$F_\infty(y_0,s) = P_{y_0}[\eta_0 \leq s] = F_a(y_0,s) + P_{y_0}[\eta_0 \leq s;\ \eta_a \leq \eta_0]$$

and

$$\begin{aligned} P_{y_0}[\eta_0 \leq s;\ \eta_a \leq \eta_0] &\leq P_{y_0}[\eta_a \leq \eta_0 \leq \eta_a + s] \\ &= P_{y_0}[\eta_a \leq \eta_0] P_a[\eta_0 \leq s] \\ &\leq P_a[\eta_0 \leq s] = F_\infty(a,s). \end{aligned}$$

Thus we have the following inequalities:

$$F_a(y_0,s) \leq F_\infty(y_0,s) \leq F_a(y_0,s) + F_\infty(a,s)$$

$$\frac{F_a(y_0,s)}{F_\infty(y_0,s)} \leq 1 \leq \frac{F_a(y_0,s)}{F_\infty(y_0,s)} + \frac{F_\infty(a,s)}{F_\infty(y_0,s)}$$

or

$$0 \leq 1 - \frac{F_a(y_0,s)}{F_\infty(y_0,s)} \leq \frac{F_\infty(a,s)}{F_\infty(y_0,s)}. \tag{3.8}$$

This inequality will allow us to prove the proposition by working with F_∞ alone, for which we have an explicit expression. From (3.3) and (3.4), $F_\infty(a,s) = 2\Phi(a \cdot r(s))$. Using the classical bounds ([10] pg.175)

$$(x^{-1} - x^{-3})g(x) \leq \Phi(x) \leq x^{-1}g(x) \tag{3.9}$$

we get

$$\frac{F_\infty(a,s)}{F_\infty(y_0,s)} \leq \frac{(ar(s))^{-1}}{(y_0r(s))^{-1} - (y_0r(s))^{-3}} \cdot \frac{g(ar(s))}{g(y_0r(s))}.$$

Now

$$\frac{(ar)^{-1}}{(y_0r)^{-1} - (y_0r)^{-3}} = \frac{y_0}{a}\left[\frac{(y_0r)^2}{(y_0r)^2 - 1}\right].$$

The expression in brackets decreases in $y_0r > 1$. Since $r(s) \geq \sqrt{2}$ it follows that for $y_0 \geq 1$

$$\frac{(ar)^{-1}}{(y_0r)^{-1} - (y_0r)^{-3}} \leq 2\frac{y_0}{a}$$

and therefore

$$\frac{F_\infty(a,s)}{F_\infty(y_0,s)} \leq 2\frac{y_0}{a}\frac{g(ar(s))}{g(y_0r(s))} = 2\frac{y_0}{a}e^{-\frac{1}{2}r(s)^2(a^2-y_0^2)}$$
$$\leq 2\frac{y_0}{a}e^{-(a^2-y_0^2)}.$$

Since this last expression $\to 0$ as $a^2 - y_0^2 \to +\infty$, the proposition follows from (3.8).

∎

We can now put the pieces together to exhibit the asymptotics of τ_*^ϵ. Assign the parametric values

$$a = B^{1/2}\epsilon^{-1/2}/2 \quad \text{and} \quad y_0 = B^{1/2}\epsilon^{-1/2}/4.$$

Then

$$a^2 - y_0^2 = B\epsilon^{-1}\frac{3}{16} \to \infty.$$

Therefore for any bounded continuous function ψ we have, starting with (3.7),

$$\begin{aligned}
E[\psi(\tau_*^\epsilon - B^{-1}\log\epsilon^{-1/2})] &= \int_0^\infty \psi(B^{-1}(s - \log\epsilon^{-1/2}))G_a(y_0, ds)\\
&\sim \int_0^\infty \psi(B^{-1}(s - \log\epsilon^{-1/2}))G_\infty(y_0, ds)\\
&= \int_0^\infty \psi(B^{-1}(s - \log y_0 + \log(B^{1/2}/4)))G_\infty(y_0, ds)\\
&\sim \int_{-\infty}^\infty \psi(B^{-1}(s + \log(B^{1/2}/4)))q(s)ds\\
&= \int_{-\infty}^\infty \psi(t)Bq(Bt - \log(B^{1/2}/4))dt.
\end{aligned}$$

This proves the following, which is the whole point of this section.

Proposition 3.3. *As $\epsilon \downarrow 0$,*

$$\left| E[\psi(\tau_*^\epsilon - B^{-1}\log\epsilon^{-1/2})] - \int_{-\infty}^\infty \psi(t)Bq(Bt - \log(B^{1/2}/4))dt \right| \to 0$$

for every bounded $\psi \in C(\mathbb{R})$, where $q(s) = 2e^{-(2s+e^{-2s})}$.

The interpretation for our representation of μ^ϵ in (2.2) is that

$$\tau_*^\epsilon \sim T^\epsilon + C$$

where

$$T^\epsilon = B^{-1}\log\epsilon^{-1/2}$$

and C is a random variable (independent of the angular process $\theta^\epsilon(\cdot)$) with density given by $Bq(Bt - \log(B^{1/2}/4))$. The implication for (2.2) is that

$$\mu^\epsilon \sim^{\text{distn.}} \theta^\epsilon(\nu^\epsilon, T^\epsilon + C).$$

It is interesting that the distribution of the conditioned exit time τ_*^ϵ neither focuses nor disperses as $\epsilon \downarrow 0$, rather the asymptotic effect is merely translation by T^ϵ while the law (that of $\mathcal{C}$) remains fixed. The Fourier series calculation of (4.3) below is the only place where the particular distribution of $\mathcal{C}$ is relevant. Otherwise our conclusions will be based on estimates for $\theta^\epsilon(T^\epsilon + c)$ which hold uniformly over c in compacts.

§4: Periodic Boundaries

We now turn to the analysis of $\theta^\epsilon(T^\epsilon + \mathcal{C})$. In this section we consider the case in which the unit circle ∂D is a periodic orbit. In terms of the deterministic angular dynamics

$$\dot{\theta}^0 = b(\theta^0); \quad \theta^0(0) = \theta_0, \tag{4.1}$$

this means that $|b(\theta)| > 0$ for all θ. We will assume that α is a positive constant for which

$$\alpha^{-1} \leq b(\theta) \leq \alpha.$$

The basic result is that $\theta^\epsilon(\cdot) \sim \theta^0(\cdot)$ holds on time scales $\mathcal{O}(\log(\epsilon^{-1/2}))$ in length.

Proposition 4.1. *$|\theta^\epsilon(T^\epsilon + c) - \theta^0(T^\epsilon + c)| \to 0$ in probability, uniformly over $\theta^\epsilon(0) = \theta_0 \in \mathbb{R}$ and c in compacts.*

Proof: Define

$$\phi(\theta) = \int_0^\theta b(u)^{-1} du.$$

Think of this as a reparameterization of the θ–axis. In this parameterization we have

$$\frac{d}{dt}\phi(\theta^0) = b(\theta^0)^{-1} \cdot \dot{\theta}^0 \equiv 1$$

and

$$d\phi(\theta^\epsilon) = [\frac{\epsilon}{2}\phi''(\theta^\epsilon) + 1]dt + \epsilon^{1/2} b(\theta^\epsilon)^{-1} d\beta_2(t).$$

Thus $\phi(\theta^0(t)) = \phi(\theta_0) + t$ and

$$\phi(\theta^\epsilon(t)) - \phi(\theta^0(t)) = \frac{\epsilon}{2}\int_0^t \phi''(\theta^\epsilon(s))ds + \epsilon^{1/2}\int_0^t b(\theta^\epsilon(s))^{-1} d\beta_2(s). \tag{4.2}$$

$K = \|\phi''\|_u$ is finite by periodicity, and so

$$\left|\frac{\epsilon}{2}\int_0^{T^\epsilon + c} \phi''(\theta^\epsilon(s))ds\right| \leq \epsilon K(T^\epsilon + c)/2 \to 0.$$

The stochastic integral in (4.2) has mean 0 and variance given by

$$\epsilon E[\int_0^{T^\epsilon + c} |b(\theta^\epsilon(s))^{-1}|^2 ds] \leq \epsilon\alpha^2 \cdot (T^\epsilon + c) \to 0.$$

Both of the preceding convergences are uniform over c in any compact set. It follows that

$$|\phi(\theta^\epsilon(T^\epsilon + c)) - \phi(\theta^0(T^\epsilon + c))| \to 0 \quad \text{in probability,}$$

again with uniformity for c in compacts. Finally, since

$$|\phi(b) - \phi(a)| \leq \alpha^{-1}|b - a| \quad \text{or} \quad |b - a| \leq \alpha|\phi(b) - \phi(a)|,$$

the convergence claimed in the proposition follows. ∎

The following corollary expresses the implications for θ^ϵ.

COROLLARY 4.2. *If the unit circle is a periodic orbit of (4.1), then for any continuous function $\phi(\cdot)$ on the unit circle*

$$E[\phi(\theta^\epsilon(\theta_0; \tau_*^\epsilon))] \sim E[\phi(\theta^0(\theta_0; T^\epsilon + C))],$$

uniformly over $x_0 = (1/4, \theta_0) \in \partial G$.

To consider the consequences for the asymptotic behavior of μ^ϵ it is convenient to work in t-space rather than θ-space. To this end fix a reference point θ^* and let Θ be the map from $\mathbb{R}$ onto the unit circle given by the solution of (4.1) with $\theta^0(0) = \theta^*$; i.e. $\Theta(t) = \theta^0(\theta^*; t)$. Suppose $T > 0$ is the fundamental period of (4.1). Let I be the "T-circle" obtained by identifying all points in $\mathbb{R}$ that differ by integral multiples of T. Then we can consider Θ as acting I. Thus I is identified with the unit circle through the map Θ. We will let S_t denote the shift maps on I,

$$S_t(u) = u + t \quad \text{(with } T\text{-periodic identification).}$$

Through Θ this corresponds to evolution by t in θ^0.

Every measure on ∂D or ∂G has a corresponding image on I. Let $\tilde{\mu}^\epsilon$ denote the image of μ^ϵ on I, $\tilde{\mu}^\epsilon = \mu^\epsilon \circ \Theta$ (i.e. $\tilde{\mu}^\epsilon(A) = \mu^\epsilon(\Theta(A))$ for $A \subseteq I$). We will assume that the exit distributions from G converge,

$$\nu^\epsilon \Rightarrow \nu^0, \quad \text{as } \epsilon \to 0.$$

Let $\tilde{\nu} = \nu^0 \circ \Theta$ be the image of ν^0 on I. The above corollary together with (2.3) tells us how to construct $\tilde{\mu}^\epsilon$ from $\tilde{\nu}$ asymptotically. First convolve $\tilde{\nu}$ with σ, where σ is the distribution of C on I (after periodic identification). Then shift the convolution $\tilde{\nu} * \sigma$ around I by T^ϵ:

$$\tilde{\mu}^\epsilon \sim (\tilde{\nu} * \sigma) \circ S_{-T^\epsilon}.$$

First of all we see that $\tilde{\mu}^\epsilon$, and hence μ^ϵ, converges (weakly) if and only if $\tilde{\nu} * \sigma$ is the uniform distribution on I, in which case $\lim \tilde{\mu}^\epsilon$ is also uniform. In terms of Fourier series (T-periodic) the uniformity of $\tilde{\nu} * \sigma$ is equivalent to

$$\widehat{\tilde{\nu} * \sigma}(n) = \hat{\tilde{\nu}}(n) \cdot \hat{\sigma}(n) = 0 \quad \text{for all } n \neq 0.$$

However $\hat{\sigma}(n)$ can be computed explicitly using the density q of Proposition 3.3:

$$\begin{aligned}\hat{\sigma}(n) &= E[e^{i\frac{2\pi}{T}n\mathcal{C}}] \\ &= z_n \int_{-\infty}^{\infty} e^{i\frac{2\pi}{BT}ns} q(s)ds, \text{ where } z_n = e^{in\frac{2\pi}{BT}\log(B^{1/2}/4)} \\ &= z_n \Gamma(1 - i\frac{\pi}{BT}n), \end{aligned} \tag{4.3}$$

according to [1] (26.1.30). But then by [1] (6.1.32),

$$|\hat{\sigma}(n)|^2 = \frac{\frac{\pi^2}{BT}n}{\sinh(\frac{\pi^2}{BT}n)}$$

which is $\neq 0$ for $n \neq 0$. Therefore $\tilde{\nu} * \sigma$ is uniform if and only if $\tilde{\nu}$ is itself uniform!

Notice that the uniformity of $\tilde{\nu}$ and $\lim \tilde{\mu}^\epsilon$ on I is equivalent to ν^0 and $\lim \mu^\epsilon$ being the (unique) invariant probability measure for (1.4) on the circle (∂G and ∂D respectively), i.e. having density proportional to $|b(\theta)|^{-1}$. As we have already said, the known theory of the noncharacteristic boundary exit problem for G applies to $\nu^0 = \lim \nu^\epsilon$. Those results make it clear that $\tilde{\nu}$ will be uniform (and hence μ^ϵ will converge) only in exceptional cases. Among other things this would require that the quasipotential V be constant over ∂G. V is determined by the dynamics inside G, including the part inside the center disk C which we have left unspecified. Thus it would seem that only in examples with a lot of angular symmetry and special structure will $\tilde{\nu}$ be uniform and μ^ϵ converge.

The more typical situation is that $\tilde{\nu}$ and $\tilde{\nu} * \sigma$ are non-uniform over I. $\tilde{\mu}^\epsilon$ will then consist (asymptotically) of $\tilde{\nu} * \sigma$ rotating or "cycling" around I periodically with respect to $T^\epsilon = B^{-1} \log \epsilon^{-1/2}$. μ^ϵ is the image of $\tilde{\mu}^\epsilon$ through Θ back on ∂D. If λ is the corresponding image of $\tilde{\nu} * \sigma$, $\lambda = (\mu * \sigma) \circ \Theta^{-1}$, then λ is not invariant for (4.1). μ^ϵ is (asymptotically) the distribution of $\theta^0(T^\epsilon)$ where the initial point θ_0 is random with distribution λ. This is expressed by (4.4) in the following summary of our findings.

THEOREM 4.1. *If ∂D is periodic for (1.2), then μ^ϵ converges if and only if ν^ϵ converges to the probability measure with density proportional to $|b(\theta)|^{-1}$ on ∂G, in which case μ^ϵ converges to the same probability measure, but on ∂D. Otherwise μ^ϵ cycles around ∂D without converging, being given by*

$$\mu^\epsilon \sim^{\text{distn.}} \theta^0(\lambda; B^{-1} \log \epsilon^{-1/2}) \tag{4.4}$$

where λ is a measure on ∂D which is not invariant for (4.1).

In some sense the conditioning argument of §2 has made the interior of D repelling instead of attracting. In this light the above result bears some resemblance to Eizenberg's example [9] of nonconvergent exit measures for a repelling critical point. There are differences however. In Eizenberg's example the nonconvergence arises from behavior in a neighborhood of the critical point. In our case it is a boundary phenomenon.

In closing this section notice that Theorem 4.1 above *is* consistent with the usual large deviations principle for the characteristic boundary exit problem, established in [4]. Indeed, if ∂D is periodic then the quasipotential V is constant over ∂D. Therefore the usual large deviations principle reduces to the assertion that

$$\epsilon \log \mu^\epsilon(A) \to 0 \tag{4.5}$$

for any measurable $A \subseteq \partial D$ with nonempty interior. But notice that σ, and hence $\tilde{\nu} * \sigma$, has a positive density on the unit circle. Therefore $\underline{\lim}\, \mu^\epsilon(A) > 0$, which implies (4.5).

§5: Boundary Critical Points

Now we consider the case in which the boundary dynamics (4.1) have isolated critical points. Again, μ^ϵ is given by the limiting distribution of $\theta^\epsilon(\theta_0; T^\epsilon + C)$ with θ_0 distributed according to ν^ϵ. For the most part $\lim \mu^\epsilon$ will depend only on $\nu^0 = \lim \nu^\epsilon$, which we again assume to exist, and the behavior of $\theta^\epsilon(\theta_0; T^\epsilon + c)$ for fixed θ_0 and c. However there are some cases whose analysis requires more care and effort than we are ready to give here. Instead of trying to state a comprehensive theorem, we will settle for a discussion of some phenomena which are readily accessible using the approach we have developed.

Consider a section of ∂D between two critical points: $\theta_a \le \theta \le \theta_r$ where θ_a is attracting and θ_r is repelling, having "tangential" eigenvalues

$$\lambda_a = b'(\theta_a) \le 0 \quad \text{and} \quad \lambda_r = b'(\theta_r) \ge 0.$$

We want to consider the contribution of $\theta^\epsilon(\theta_0; \tau_*^\epsilon + C)$ to μ^ϵ for the different θ_0 in this range.

First consider the attracting point $\theta = \theta_a$. Large deviations theory tells us that it takes an $\mathcal{O}(e^{\delta/\epsilon})$ amount of time, for some $\delta > 0$, for $\theta^\epsilon(t)$ to leave the attracting well of θ_a with any significant probability. Since $\tau_*^\epsilon = \mathcal{O}(B^{-1} \log \epsilon^{-1/2}) \ll e^{\delta/\epsilon}$ it follows that $\theta^\epsilon(\tau_*^\epsilon) \to \theta_a$ almost surely. This means that θ_a yields a

$$\text{contribution of } \nu^0(\{\theta_a\})\delta_{\theta_a} \text{ to } \mu^0 = \lim \mu^\epsilon.$$

The situation for $\theta_a < \theta_0 < \theta_r$ is similar. $\theta^\epsilon(\theta_0; t)$ follows $\theta^0(\theta; t)$ into the vicinity of θ_a and does not come out again until $\mathcal{O}(e^{\delta/\epsilon}) \gg \tau_*^\epsilon$. So we get an additional

$$\text{contribution of } \nu^0((\theta_a, \theta_r))\delta_{\theta_a} \text{ to } \mu^0 = \lim \mu^\epsilon. \tag{5.1}$$

However the nature of this convergence depends on whether or not $\theta^\epsilon(\theta_0;t)$ has actually reached θ_a or not by the time τ_*^ϵ when we stop it. If so the above contribution to μ^0 will be roughly centered about θ_a; if not the contribution will be supported entirely on one side of θ_a, converging to it as $\epsilon \downarrow 0$. The following proposition tells us which case applies. The basic fact is that $\theta^\epsilon(\theta_0;t)$ takes $\mathcal{O}(-\lambda_a^{-1}\log\epsilon^{-1/2})$ to reach θ_a for the first time.

PROPOSITION 5.1. *Suppose θ_a is an isolated attracting critical point of (4.1) with $\lambda_a = \frac{\partial}{\partial\theta}b(\theta_a) \le 0$, $\lambda_a \ne -B$. Then for all $\theta_0 \ne \theta_a$ which are attracted to θ_a by (4.1),*

$$P_{\theta_0}[\theta^\epsilon(s) = \theta_a \text{ for some } s \le \tau_*^\epsilon] \to \begin{cases} 0 & \text{if } -B < \lambda_a \le 0 \\ 1 & \text{if } \lambda_a < -B \end{cases}.$$

We will omit the details of proof. The approach is to compare $\theta^\epsilon(t)$ to its linearization at θ_a, for which explicit calculations are possible. Also note that we have omitted the critical case of $\lambda_a = -B$, in which we would expect a mixture of the two effects.

We see then that there is a qualitative difference in the way convergence to the contribution (5.1) occurs, depending on the relative sizes of λ_a and B. If $\lambda_a < -B$ then the contribution to μ^ϵ of $\theta^\epsilon(\theta,\tau_*^\epsilon)$ for $\theta_a < \theta < \theta_r$ is roughly centered around θ_a. However if $-B < \lambda_a$ then this contribution to μ^ϵ is supported all on the θ_r side of θ_a, converging to it as $\epsilon \downarrow 0$. This is the "skewing" effect first observed by Bobrovsky and Schuss [2] and discussed by Bobrovsky and Zeitouni [3].

Now consider the contribution of the repelling point θ_r to μ^ϵ. First we note that $\nu^0(\{\theta_r\})$ can be positive if θ_r is an isolated minimum of the quasipotential $V(\theta)$ over ∂G, as determined by the dynamics inside G. There are several things to consider here. Although the deterministic process $\theta^0(\theta_r;t)$ remains at θ_r for all t, for $\theta_0 \ne \theta_r$ however $\theta^0(\theta_0;t)$ diverges from θ_r. Thus $\theta^\epsilon(\theta_r,t)$ will leave the vicinity of θ_r, but as above, there is a question of whether or not this occurs in the $\mathcal{O}(B^{-1}\log\epsilon^{-1/2})$ time before τ_*^ϵ. Also (2.2) says μ^ϵ is the distribution of $\theta^\epsilon(\theta,\tau_*^\epsilon)$ where θ is distributed according to ν^ϵ. Is the ϵ-dependence of ν^ϵ significant for $\theta^\epsilon(\nu^\epsilon;\tau_*^\epsilon)$, or can we simply consider $\theta^\epsilon(\theta_r;\tau_*^\epsilon)$ instead? The following proposition give us at least partial answers to these questions.

PROPOSITION 5.2. *Suppose θ_r is an isolated critical point of (4.1) with eigenvalue $\lambda_r = \frac{\partial}{\partial\theta}b(\theta_r) \ge 0$, $\lambda_r \ne B$. Consider any $\delta > 0$.*

i) If $0 \le \lambda_r < B$ then

$$P_{\theta_0}[|\theta^\epsilon(t) - \theta_r| < \delta \text{ for all } t \le \tau_*^\epsilon] \to 1$$

uniformly over $|\theta_0 - \theta_r| \le \epsilon^\gamma$ for any $\gamma > \lambda_r/2B$.

ii) If $B < \lambda_r$ then

$$P_{\theta_0}[|\theta^\epsilon(t) - \theta_r| < \delta \text{ for all } t \le \tau_*^\epsilon] \to 0$$

uniformly in θ_0.

Again we omit the details of proof, but will give at least a heuristic justification. Consider linearizing the θ^ϵ equation about θ_r so that $\theta^\epsilon(t) \approx \theta_r + y^\epsilon(t)$ where

$$dy^\epsilon(t) = \lambda_r y^\epsilon(t)dt + \epsilon^{1/2}d\beta(t); \quad y^\epsilon(0) = \theta_0 - \theta_r.$$

$y^\epsilon(t)$ is a Gaussian random variable with

$$\text{mean } = e^{\lambda_r t}y^\epsilon(0) \quad \text{and variance} = \epsilon\frac{e^{2\lambda_r t} - 1}{2\lambda_r}.$$

Thus with $t = B^{-1}\log\epsilon^{-1/2} + c$ the variance is

$$\epsilon(\epsilon^{\lambda_r/B}e^{2\lambda_r c} - 1)/2\lambda_r \sim \epsilon^{1-\frac{\lambda_r}{B}} \cdot 2e^{2\lambda_r c}/\lambda_r$$
$$\to \begin{cases} 0 & \text{if } \lambda_r < B \\ +\infty & \text{if } \lambda_r > B \end{cases}.$$

And the mean is

$$y_0\epsilon^{-\frac{\lambda_r}{2B}}e^{\lambda_r c}.$$

Thus if ν^ϵ is concentrating θ_0 at θ_r faster than ϵ^γ with some $\gamma > \lambda_r/2B$ then it is legitimate to replace the ν^ϵ-distributed θ_0 by θ_r itself. The actual proof of Proposition 5.2 uses a comparison argument to relate θ^ϵ to the linearization.

Returning to our discussion of how the ν^0 mass of θ_r shows up in μ^ϵ, consider the case in which $\lambda_r > B$. The implication of part ii) of the proposition is that $\theta^\epsilon(t)$ leaves a given neighborhood of θ_r before $t = \tau_*^\epsilon$. Thus the mass $\nu^0(\{\theta_r\})$ is split into two parts which concentrate at θ_a and the next attracting critical point $\theta_a' > \theta_r$ on the other side of θ_r. (θ_a could coincide with θ_a' if there is only one attracting critical point.) We are not prepared to say what the proportions of this split are. One might expect half the mass to go to either side of θ_i^*, but this appears to be very sensitive to how the appropriate part of ν^ϵ focuses at the critical point θ_i^*.

Moving to the case in which $\lambda_r < B$, the proposition implies that if the appropriate part of ν^ϵ focuses θ_0 at θ_r fast enough, then $\theta^\epsilon(t)$ will not be able to move away from the critical point before τ_*^ϵ. Thus depending on the behavior of ν^ϵ, it is possible for the mass $\nu^0(\{\theta_r\})$ to remain concentrated at θ_r in μ^ϵ as $\epsilon \downarrow 0$. It is easy to give a sufficient condition for this to occur. As we said above, for ν^0 to give positive mass to θ_r requires in particular that the quasipotential function V have a minimum over ∂G at θ_r. Assume the following:

ASSUMPTION 5.1. *V has only a finite number of minima over ∂G and at each of them*

$$\partial^2 V/\partial\theta^2 > 0.$$

It follows then from known results about the noncharacteristic boundary exit problem from G (see [6]) that, near θ_r,

$$\nu^\epsilon(d\theta) \sim \epsilon^{-1/2}R_0(\theta)e^{-V(\theta)/\epsilon}d\theta$$

where R_0 is a positive continuous function. The Laplace asymptotics for such a density imply that

$$\nu^\epsilon((\theta_r - \epsilon^\gamma, \theta_r + \epsilon^\gamma)) \to \nu^0(\{\theta_r\})$$

for any $\gamma < 1/2$. If $\lambda_r < B$ then we can pick $\lambda_r/2B < \gamma < 1/2$. This means that ν^ϵ is focusing the part of its mass that converges to θ_r within the set of θ_0 in which part i) of Proposition 5.2 applies. Hence under Assumption (5.1) it does follow that μ^ϵ retains a mass of $\nu^0(\{\theta_r\})$ concentrated at θ_r in the limit. In particular it can happen that $\mu^0 = \lim \mu^\epsilon$ give positive probability to the repelling critical points on ∂D!

Although we neglected the critical case of $\lambda_a = -B$ for attracting critical points, the critical case of $\lambda_r = B$ for repelling points is especially interesting. For the linearized process above, the variance of $y^\epsilon(T^\epsilon + c)$ approaches a positive finite value. This suggests that the limiting distribution of $\theta^\epsilon(\tau_*^\epsilon)$ might have a positive density on an interval surrounding the critical point. It turns out that this can in fact happen in some examples, which is remarkable since then $\lim \mu^\epsilon$ would not be supported on the critical points alone. This means that there exist counter- examples to the conjecture made in §6 of [5], which in the present context would have predicted that $\mu^0 = \lim \mu^\epsilon$ is an invariant probability measure for the deterministic process (1.2) on the boundary. Thus the conjecture of [5] is false.

It may be that there are yet a few more surprises waiting in the cases we have not covered or in the general case of (1.1). However limited the approach and equations (1.3) of this paper may be, we have uncovered some possible phenomena that have not been mentioned in any of the literature on the characteristic boundary exit problem up to now.

REFERENCES

[1] M. Abramowitz and I. A. Stegun, *Handbook of Mathematical Functions*, Dover, New York, 1972.

[2] B.-Z. Bobrovsky and Z. Schuss, A singular perturbation method for the computation of the mean first passage time in a nonlinear filter, SIAM J. Appl. Math. 42 (1982), pp.174-187.

[3] B.-Z. Bobrovsky and O. Zeitouni, Some results on the problem of exit from a domain, preprint.

[4] M. V. Day, Large deviations results for the exit problem with characteristic boundary, J. Math. Anal. and Appl. (to appear).

[5] M. V. Day, Boundary local time and small parameter exit problems with characteristic boundaries, SIAM J. Math. Anal. 20 (1989), pp. 222-248.

[6] M. V. Day, Recent progress on the small parameter exit problem, Stochastics 20 (1987), pp. 121-150.

[7] M. V. Day, Exponential leveling for stochastically perturbed dynamical systems, SIAM J. Math. Anal. 13 (1982), pp. 532-540.

[8] A. Eizenberg, The exponential leveling and the Ventcel-Freidlin "minimal action" function, preprint.

[9] A. Eizenberg, The exit distributions for small random perturbations of dynamical systems with a repulsive type stationary point, Stochastics 12 (1984), pp.251-275.

[10] W. Feller, *An Introduction to Probability Theory and its Applications* v.1, J. Wiley, New York, 1966.

[11] M. I. Freidlin and A. D. Wentzell, *Random Perturbations of Dynamical Systems*, Springer Verlag, New York, 1984.

[12] B. J. Matkowsky, Z. Schuss and C. Tier, Diffusion across characteristic boundaries with critical points, SIAM J. Appl. Math. 43 (1983), pp. 673-695.

Martin Day
Department of Mathematics
Virginia Tech (VPI & SU)
Blacksburg, VA 24061-0123

Harmonic Measure for Random Genetic Drift

S.N. Ethier* and R.C. Griffiths

An explicit formula is found for the Lebesgue density of the harmonic measure associated with the finite-dimensional diffusion process that models pure random genetic drift. Also, the corresponding Dirichlet problem is discussed.

1. Introduction

The diffusion process in the n-dimensional simplex

$$\Delta_n = \{x = (x_1, \ldots, x_n)\colon x_1 \geq 0, \ldots, x_n \geq 0,\ x_1 + \cdots + x_n \leq 1\} \tag{1}$$

with generator

$$L = \frac{1}{2} \sum_{i,j=1}^{n} x_i(\delta_{ij} - x_j) \frac{\partial^2}{\partial x_i\, \partial x_j} \tag{2}$$

models pure random genetic drift in a population with $n+1$ alleles, $A_0, A_1, \ldots, A_n$. When the process is in state $x = (x_1, \ldots, x_n) \in \Delta_n$, the allele frequencies are $x_0, x_1, \ldots, x_n$, respectively, where

$$x_0 = 1 - x_1 - \cdots - x_n. \tag{3}$$

For each $x \in \Delta_n$ let $P_x \in C([0,\infty), \Delta_n)$ be the unique solution of the martingale problem for L acting on $C^2(\Delta_n)$ starting at x (see [2]); then $\{P_x,\ x \in \Delta_n\}$ is a Feller continuous, strong Markov family. Let $\{X(t),\ t \geq 0\}$ denote the canonical coordinate process on $C([0,\infty),\Delta_n)$ (i.e., $X(t)(\omega) = \omega(t)$), and put $X(\infty) \equiv (0,\ldots,0) \in \Delta_n$. Denote by $\partial\Delta_n$ the boundary of Δ_n and define the hitting time $\tau\colon C([0,\infty),\Delta_n) \mapsto [0,\infty]$ of $\partial\Delta_n$ by

$$\tau = \inf\{t \geq 0\colon X(t) \in \partial\Delta_n\}, \tag{4}$$

where $\inf \emptyset = \infty$. It is easily seen that $E^{P_x}[\tau] < \infty$ for all $x \in \Delta_n$.

Our aim here is to find the P_x-distribution of $X(\tau)$ for each $x \in \Delta_n$. This is the harmonic measure on $\partial\Delta_n$ associated with the degenerate elliptic operator L.

*Research supported in part by NSF grant DMS-8902991.

In the population genetics context, it is the joint distribution of the frequencies of alleles $A_1, \ldots, A_n$ at the time of the first loss of an allele.

The case $n = 1$ is trivial, so let us assume hereafter that $n \geq 2$. In Section 2 we show for each x in Δ_n°, the interior of Δ_n, that the P_x-distribution of $X(\tau)$ is absolutely continuous with respect to $(n-1)$-dimensional Lebesgue measure on $\partial\Delta_n$, and we evaluate its density explicitly. In Section 3 we discuss existence and uniqueness of solutions of the Dirichlet problem associated with L.

Some additional related results will appear elsewhere.

2. Harmonic measure

Define

$$H_i = \{x \in \Delta_n \colon x_i = 0\} \tag{5}$$

and

$$H_i^\circ = \{x \in \Delta_n \colon x_i = 0,\ x_j > 0 \text{ for all } j \in \{0, 1, \ldots, n\} - \{i\}\} \tag{6}$$

for $i = 0, 1, \ldots, n$; here and below x_0 is given by (3). Let

$$S = \partial\Delta_n - (H_0^\circ \cup \cdots \cup H_n^\circ). \tag{7}$$

The following lemma is a special case of Proposition 4.4 of [3].

Lemma 1. *For each $x \in \Delta_n^\circ$, $P_x\{X(\tau) \in S\} = 0$.*

This result says that the harmonic measure is concentrated on the union of the interiors H_i° of the faces H_i of the simplex Δ_n. In the population genetics context, it says that with probability 1 the first two alleles lost are not lost at exactly the same time.

For $i = 1, \ldots, n$, define $\pi_i \colon \Delta_{n-1} \mapsto H_i$ by $\pi_i(x) = (x_1, \ldots, x_{i-1}, 0, x_i, \ldots, x_{n-1})$, and define $\pi_0 \colon \Delta_{n-1} \mapsto H_0$ by $\pi_0(x) = (1 - x_1 - \cdots - x_{n-1}, x_1, \ldots, x_{n-1})$. Let λ_{n-1} denote $(n-1)$-dimensional Lebesgue measure on Δ_{n-1}, and define λ to be the Borel measure on $\partial\Delta_n$ whose restriction to H_i is equal to $\lambda_{n-1}\pi_i^{-1}$ ($i = 0, 1, \ldots, n$); λ is what we mean by $(n-1)$-dimensional Lebesgue measure on $\partial\Delta_n$.

Let us define $h_0 \colon \Delta_n^\circ \times H_0^\circ \mapsto (0, \infty)$ by

$$\begin{aligned} h_0(x, y) = {} & \sum_{k=n+1}^{\infty} (1 - x_0)^{k-1} x_0 \\ & \cdot \sum_{\alpha \in \mathbf{N}^n \colon |\alpha| = k-1} \binom{k-1}{\alpha_1, \ldots, \alpha_n} \left(\frac{x_1}{1 - x_0}\right)^{\alpha_1} \cdots \left(\frac{x_n}{1 - x_0}\right)^{\alpha_n} \\ & \cdot \frac{\Gamma(\alpha_1 + \cdots + \alpha_n)}{\Gamma(\alpha_1) \cdots \Gamma(\alpha_n)} y_1^{\alpha_1 - 1} \cdots y_n^{\alpha_n - 1}, \end{aligned} \tag{8}$$

where $|\alpha| = \alpha_1 + \cdots + \alpha_n$ for $\alpha \in \mathbf{N}^n$. In terms of the geometric, multinomial, and Dirichlet densities, (8) can be expressed more concisely as

$$h_0(x,y) = \sum_{k=n+1}^{\infty} \operatorname{geom}(k; x_0) \cdot \sum_{\alpha \in \mathbf{N}^n : |\alpha| = k-1} \operatorname{mult}(\alpha; k-1, (1-x_0)^{-1}x) \operatorname{Dir}(y; \alpha). \tag{9}$$

For $i = 0, 1, \ldots, n$, we define $\rho_i \colon \Delta_n \to \Delta_n$ by $\rho_i(x) = (x_0, \ldots, \check{x}_i, \ldots, x_n)$, where the ˇ notation signifies a deleted component, and we note that ρ_i maps H_i° onto H_0°. Of course, ρ_0 is the identity map on Δ_n. Finally, the function $h \colon \Delta_n^\circ \times \partial\Delta_n \mapsto (0, \infty)$ is defined on $\Delta_n^\circ \times (\partial\Delta_n - S)$ by

$$h(x,y) = h_0(\rho_i(x), \rho_i(y)) \quad \text{if} \quad y \in H_i^\circ, \qquad i = 0, 1, \ldots, n, \tag{10}$$

and on $\Delta_n^\circ \times S$ by continuity. We can now state our main result.

Theorem 1. *For each $x \in \Delta_n^\circ$, the P_x-distribution of $X(\tau)$ on $\partial\Delta_n$ is absolutely continuous with respect to λ and has density $h(x, \cdot)$.*

Proof. Given f in $B(\partial\Delta_n)$, the space of bounded Borel functions on $\partial\Delta_n$, define $u \colon \Delta_n^\circ \mapsto \mathbf{R}$ by

$$u(x) = \int_{\partial\Delta_n} f(y)\, h(x,y)\, \lambda(dy). \tag{11}$$

For $N \geq n+1$ define $h_0^N \colon \Delta_n^\circ \times H_0^\circ \mapsto (0, \infty)$ as in (8) or (9) but with $\sum_{k=n+1}^N$ in place of $\sum_{k=n+1}^\infty$; $h^N \colon \Delta_n^\circ \times \partial\Delta_n \mapsto (0,\infty)$ as in (10) but with h_0^N in place of h_0; and $u^N \colon \Delta_n^\circ \mapsto \mathbf{R}$ as in (11) but with h^N in place of h.

Using the identity

$$\operatorname{Dir}(y; \alpha) = (k-2) \cdots (k-n) \operatorname{mult}(\alpha - 1; k-n-1, y) \tag{12}$$

in (9), where $\alpha - 1 = (\alpha_1 - 1, \ldots, \alpha_n - 1)$, we find that

$$\begin{aligned} h_0(x,y) &\leq \sum_{k=n+1}^{\infty} (k-2) \cdots (k-n) \operatorname{geom}(k; x_0) \\ &= (n-1)!\,(1-x_0)^n / x_0^{n-1}, \qquad (x,y) \in \Delta_n^\circ \times H_0^\circ, \end{aligned} \tag{13}$$

hence

$$h(x,y) \leq (n-1)! \max_{0 \leq i \leq n} (1-x_i)^n / x_i^{n-1}, \qquad (x,y) \in \Delta_n^\circ \times \partial\Delta_n. \tag{14}$$

Similar estimates show that $\{h^N\}$ converges to h uniformly on compact subsets of $\Delta_n^\circ \times \partial\Delta_n$ (implying that h is continuous), hence $\{u^N\}$ converges to u uniformly on compact subsets of Δ_n°. In fact, $\{u^N\}$ converges boundedly as well, since

$$
\begin{aligned}
(15)\qquad |u^N(x)| &\le \|f\| \int_{\partial\Delta_n} h^N(x,y)\,\lambda(dy) \\
&= \|f\| \sum_{i=0}^{n} \int_{\Delta_{n-1}^\circ} h_0^N(\rho_i(x),\pi_0(y))\,\lambda_{n-1}(dy) \\
&= \|f\| \sum_{i=0}^{n} \sum_{k=n+1}^{N} \operatorname{geom}(k;x_i) \\
&\qquad \cdot \sum_{\alpha\in\mathbf{N}^n:|\alpha|=k-1} \operatorname{mult}(\alpha;k-1,(1-x_i)^{-1}\rho_i(x)) \\
&\le (n+1)\,\|f\|, \qquad x\in\Delta_n^\circ.
\end{aligned}
$$

For $\alpha \in \mathbf{N}^n$ define $f_\alpha \in C^2(\Delta_n)$ by $f_\alpha(x) = x_0 x_1^{\alpha_1}\cdots x_n^{\alpha_n}$ and check that

$$
(16)\qquad Lf_\alpha = \sum_{i:\alpha_i\ge 2} \binom{\alpha_i}{2} f_{\alpha-e_i} - \binom{|\alpha|+1}{2} f_\alpha,
$$

where $(e_i)_j = \delta_{ij}$ for $i,j = 1,\ldots,n$. For $\beta \in (\mathbf{Z}_+)^n$ and $y \in H_0^\circ$ let $\beta! = \beta_1!\cdots\beta_n!$ and $y^\beta = y_1^{\beta_1}\cdots y_n^{\beta_n}$, and recall that $\alpha - 1 = (\alpha_1 - 1,\ldots,\alpha_n - 1)$ for $\alpha \in \mathbf{N}^n$. Noting that h_0^N can be extended to $\Delta_n \times H_0^\circ$ in such a way that

$$
(17)\qquad h_0^N(\cdot,y) = \sum_{k=n+1}^{N} \sum_{\alpha\in\mathbf{N}^n:|\alpha|=k-1} \frac{(k-1)!\,(k-2)!}{\alpha!\,(\alpha-1)!}\, y^{\alpha-1} f_\alpha \in C^2(\Delta_n)
$$

for each $y \in H_0^\circ$, (16) implies that

$$
\begin{aligned}
(18)\qquad L[h_0^N(\cdot,y)] &= \sum_{k=n+1}^{N-1} \Bigg\{ - \sum_{\alpha\in\mathbf{N}^n:|\alpha|=k-1} \frac{(k-1)!\,(k-2)!}{\alpha!\,(\alpha-1)!}\, y^{\alpha-1} \binom{k}{2} f_\alpha \\
&\qquad + \sum_{\alpha\in\mathbf{N}^n:|\alpha|=k} \frac{k!\,(k-1)!}{\alpha!\,(\alpha-1)!}\, y^{\alpha-1} \sum_{i:\alpha_i\ge 2} \binom{\alpha_i}{2} f_{\alpha-e_i} \Bigg\} \\
&\qquad - \sum_{\alpha\in\mathbf{N}^n:|\alpha|=N-1} \frac{(N-1)!\,(N-2)!}{\alpha!\,(\alpha-1)!}\, y^{\alpha-1} \binom{N}{2} f_\alpha \\
&= - \sum_{\alpha\in\mathbf{N}^n:|\alpha|=N-1} \frac{(N-1)!\,(N-2)!}{\alpha!\,(\alpha-1)!}\, y^{\alpha-1} \binom{N}{2} f_\alpha
\end{aligned}
$$

for each $y \in H_0^\circ$; here we use the observation that the expression within braces is 0 for each k. Arguing as in (13), this gives

$$
(19)\qquad |L[h_0^N(\cdot,y)](x)| \le \frac{1}{2} N(N-1)\cdots(N-n)(1-x_0)^{N-1} x_0
$$

for all $(x,y) \in \Delta_n \times H_0^\circ$. The corresponding extension of u^N to Δ_n given by

$$
(20)\qquad u^N(x) = \sum_{i=0}^{n} \int_{\Delta_{n-1}^\circ} f(\pi_i(y))\, h_0^N(\rho_i(x),\pi_0(y))\,\lambda_{n-1}(dy), \qquad x \in \Delta_n,
$$

belongs to $C^2(\Delta_n)$ and (since $L(f_\alpha \circ \rho_i) = (Lf_\alpha) \circ \rho_i$)

$$(21) \qquad Lu^N(x) = \sum_{i=0}^{n} \int_{\Delta_{n-1}^\circ} f(\pi_i(y))\, L[h_0^N(\cdot, \pi_0(y))](\rho_i(x))\, \lambda_{n-1}(dy)$$

for all $x \in \Delta_n$. By (19),

$$(22) \qquad |Lu^N(x)| \le \frac{1}{2} N(N-1)\cdots(N-n)\, \|f\| \sum_{i=0}^{n} (1-x_i)^{N-1} x_i/(n-1)!$$

for all $x \in \Delta_n$, implying that $Lu^N \to 0$ uniformly on compact subsets of Δ_n°.

For $\delta > 0$ define τ_δ: $C([0,\infty), \Delta_n) \mapsto [0,\infty]$ by

$$(23) \qquad \tau_\delta = \inf\{t \ge 0\colon \min_{0 \le i \le n} X_i(t) \le \delta\},$$

where $\inf \emptyset = \infty$. Applying the optional stopping theorem to the P_x-martingale $u^N(X(t)) - \int_0^t Lu^N(X(s))\, ds$ and the stopping time τ_δ and then letting $N \to \infty$, we therefore have

$$(24) \qquad E^{P_x}[u(X(\tau_\delta))] = u(x), \qquad x \in \Delta_n^\circ,\ \delta > 0;$$

here we use the previously noted fact that $E^{P_x}[\tau] < \infty$. So far, f has been assumed only to belong to $B(\partial\Delta_n)$. Let us now assume also that f is continuous at each point of $\partial\Delta_n - S$. Suppose for the moment we can show that

$$(25) \qquad \lim_{x \to x^\circ} u(x) = f(x^\circ), \qquad x^\circ \in \partial\Delta_n - S.$$

Then, letting $\delta \to 0$,

$$(26) \qquad \begin{aligned} E^{P_x}[u(X(\tau_\delta))] &= E^{P_x}[u(X(\tau_\delta)); X(\tau) \in \partial\Delta_n - S] \\ &\to E^{P_x}[f(X(\tau)); X(\tau) \in \partial\Delta_n - S] \\ &= E^{P_x}[f(X(\tau))], \qquad x \in \Delta_n^\circ, \end{aligned}$$

where the two equalities use Lemma 1, and the limit depends on the fact that $X(\tau_\delta) \to X(\tau)$ on $\{\tau < \infty\}$, on (25), and on the boundedness of u (see (15)). By (11), (24), and (26),

$$(27) \qquad E^{P_x}[f(X(\tau))] = \int_{\partial\Delta_n} f(y)\, h(x,y)\, \lambda(dy), \qquad x \in \Delta_n^\circ.$$

Since in particular (27) holds for all $f \in C(\partial\Delta_n)$, it must hold for all $f \in B(\partial\Delta_n)$, and the theorem follows.

It remains only to verify (25) under the assumptions that $f \in B(\partial\Delta_n)$ and f is continuous at each point of $\partial\Delta_n - S$. Write

(28)
$$\begin{aligned} u(x) &= \sum_{j=0}^{n} \int_{\Delta_{n-1}^\circ} f(\pi_j(y))\, h_0(\rho_j(x), \pi_0(y))\, \lambda_{n-1}(dy) \\ &= \sum_{j=0}^{n} \biggl\{ \sum_{k=n+1}^{\infty} \operatorname{geom}(k; x_j) \\ &\qquad \cdot \sum_{\alpha\in\mathbf{N}^n:|\alpha|=k-1} \operatorname{mult}(\alpha; k-1, (1-x_j)^{-1}\rho_j(x)) \\ &\qquad \cdot \int_{\Delta_{n-1}^\circ} f(\pi_j(y))\, \operatorname{Dir}(\pi_0(y); \alpha)\, \lambda_{n-1}(dy) \biggr\} \\ &\equiv \sum_{j=0}^{n} u_j(x), \qquad x \in \Delta_n^\circ. \end{aligned}$$

Fix $i \in \{0, 1, \ldots, n\}$ and $x^\circ \in H_i^\circ$. If $j \in \{0, 1, \ldots, n\} - \{i\}$, then

(29)
$$\begin{aligned} |u_j(x)| &\le \sum_{k=n+1}^{\infty} \operatorname{geom}(k; x_j) \\ &\qquad \sum_{\alpha\in\mathbf{N}^n:|\alpha|=k-1} \operatorname{mult}(\alpha; k-1, (1-x_j)^{-1}\rho_j(x))\, \|f\| \\ &\le \sum_{k=n+1}^{\infty} (1-x_j)^{k-1} x_j \,(1-(1-(1-x_j)^{-1}x_i)^{k-1})\, \|f\| \end{aligned}$$

for all $x \in \Delta_n^\circ$, so $u_j(x) \to 0$ as $x \to x^\circ$. On the other hand, for $N \ge n+1$ and $\varepsilon > 0$,

(30)
$$\begin{aligned} |u_i(x) - f(x^\circ)| &\le \sum_{k=n+1}^{N} \operatorname{geom}(k; x_i)\, \|f\| \\ &+ \sum_{k=N+1}^{\infty} \operatorname{geom}(k; x_i) \sum_{\alpha\in A^c} \operatorname{mult}(\alpha; k-1, (1-x_i)^{-1}\rho_i(x))\, \|f\| \\ &+ \sum_{k=N+1}^{\infty} \operatorname{geom}(k; x_i) \sum_{\alpha\in A} \operatorname{mult}(\alpha; k-1, (1-x_i)^{-1}\rho_i(x)) \\ &\qquad \cdot \biggl| \int_{\Delta_{n-1}^\circ} f(\pi_i(y))\, \operatorname{Dir}(\pi_0(y); \alpha)\, \lambda_{n-1}(dy) \\ &\qquad\qquad - f(\pi_i((k-1)^{-1}(\alpha_2, \ldots, \alpha_n))) \biggr| \\ &+ \sum_{k=N+1}^{\infty} \operatorname{geom}(k; x_i) \sum_{\alpha\in A} \operatorname{mult}(\alpha; k-1, (1-x_i)^{-1}\rho_i(x)) \\ &\qquad \cdot |f(\pi_i((k-1)^{-1}(\alpha_2, \ldots, \alpha_n))) - f(x^\circ)| \\ &+ \biggl| \sum_{k=N+1}^{\infty} \operatorname{geom}(k; x_i) \sum_{\alpha\in A} \operatorname{mult}(\alpha; k-1, (1-x_i)^{-1}\rho_i(x)) - 1 \biggr|\, \|f\| \end{aligned}$$

for all $x \in \Delta_n^\circ$, where $A = \{\alpha \in \mathbf{N}^n\colon |\alpha| = k-1,\ |(k-1)^{-1}\alpha - (1-x_i)^{-1}\rho_i(x)| \le \varepsilon\}$ and $A^c = \{\alpha \in \mathbf{N}^n\colon |\alpha| = k-1,\ \alpha \notin A\}$. The second term on the right side of (30) can be estimated using Chebyshev's inequality. For the third term we use the fact that if K is a bounded open convex set in $\mathbf{R}^d$, Y is a K-valued random variable, g is bounded and continuous on K, and $\delta > 0$, then

$$|E[g(Y)] - g(E[Y])| \le w_g(\delta; E[Y]) + 2\delta^{-2}\,\|g\| \sum_{j=1}^{d} \operatorname{Var}(Y_j), \tag{31}$$

where $w_g(\delta; y)$ denotes the modulus of continuity of g at $y \in K$: $w_g(\delta; y) = \sup_{z \in K: |z-y| \le \delta} |g(z) - g(y)|$. We also use

$$w_g(\delta; y) \le w_g(\delta + \eta; y^\circ) + w_g(\eta; y^\circ) \le 2w_g(\delta + \eta; y^\circ) \quad \text{if} \quad |y - y^\circ| \le \eta. \tag{32}$$

The fourth term can be estimated in terms of the modulus of continuity of f at x°. The fifth term is bounded by $(1 - (1 - x_i)^N)\,\|f\|$ plus the second term plus a term involving a sum over all $\alpha \in (\mathbf{Z}_+)^n - \mathbf{N}^n$ with $|\alpha| = k - 1$. Putting all this together, (30) implies that

$$\begin{aligned} |u_i(x) - f(x^\circ)| \le{} & ((1 - x_i)^n - (1 - x_i)^N)\,\|f\| + 2nN^{-1}\varepsilon^{-2}\,\|f\| \\ & + 2w_f(\varepsilon\sqrt{n} + \varepsilon + |(1 - x_i)^{-1}x - x^\circ|; x^\circ) \\ & + 2(N+1)^{-1}\varepsilon^{-2}\,\|f\| \\ & + w_f(\varepsilon + |(1 - x_i)^{-1}x - x^\circ|; x^\circ) \\ & + (1 - (1 - x_i)^N)\,\|f\| \\ & + \sum_{j \in \{0,\ldots,n\} - \{i\}} \frac{x_i}{x_i + x_j}\,\|f\|, \qquad x \in \Delta_n^\circ. \end{aligned} \tag{33}$$

Letting $x \to x^\circ$, $N \to \infty$, and $\varepsilon \to 0$ in that order, we see that $u_i(x) \to f(x^\circ)$ and hence that (25) holds. This completes the proof.

The proof of (25) would have been slightly easier if we had assumed that $f \in C(\partial\Delta_n)$, and in fact this is all we needed to get (27) for all $f \in B(\partial\Delta_n)$. The conclusion that (25) holds under the weaker assumption on f is needed in the next section.

3. Dirichlet problem

The results of the preceding section readily give a useful existence and uniqueness theorem for the Dirichlet problem associated with the degenerate elliptic operator L.

Theorem 2. *Let f be a bounded Borel function on $\partial\Delta_n$ that is continuous at each point of $\partial\Delta_n - S$. Then*

$$u(x) \equiv E^{P_x}[f(X(\tau))] = \int_{\partial\Delta_n} f(y)\,h(x,y)\,\lambda(dy) \tag{34}$$

defines the unique bounded $C^2(\Delta_n^\circ)$ solution u of the differential equation

$$Lu = 0 \quad \text{on} \quad \Delta_n^\circ \tag{35}$$

subject to the boundary condition (25).

Proof. The second equality in (34) is just (27).

In the proof of Theorem 1 we saw that $u^N \to u$ and $Lu^N \to 0$ uniformly on compact subsets of Δ_n°. Similar calculations imply that the first- and second-order partial derivatives of u^N converge uniformly on compact subsets of Δ_n°, and it follows that $u \in C^2(\Delta_n^\circ)$ and $Lu = 0$ on Δ_n°. The boundary condition (25) has already been proved under the stated conditions on f.

Now let u be any bounded $C^2(\Delta_n^\circ)$ function satisfying (35) and (25). Apply the optional stopping theorem to get (24), note that (26) holds (here we use the assumed boundedness of u), and conclude that the first equality in (34) holds. This implies the uniqueness assertion and completes the proof.

Example 1. Littler [4] discovered that in the case $n = 2$,

$$u(x) \equiv x_1 x_2 \Big(\frac{1}{1-x_1} + \frac{1}{1-x_2}\Big) \tag{36}$$

satisfies $Lu = 0$ on Δ_2°, $u = 1$ on H_0°, and $u = 0$ on $H_1^\circ \cup H_2^\circ$, and he quoted Chapter 13 of [1] to conclude that $P_x\{X(\tau) \in H_0^\circ\} = u(x)$ for all $x \in \Delta_2^\circ$. But the results of that chapter do not immediately apply because of the discontinuous boundary values of u. Theorem 2 above, however, justifies the stated conclusion. (See Corollary 4.1 of [3] for a slightly different approach.)

Example 2. We can generalize the result of the previous example by applying Theorem 1. Indeed, for arbitrary $n \geq 2$ and $x \in \Delta_n^\circ$,

$$\begin{aligned}
&P_x\{X(\tau) \in H_0^\circ\} \\
&= \int_{H_0^\circ} h(x,y)\,\lambda(dy) \\
&= \int_{\Delta_{n-1}^\circ} h_0(x, \pi_0(y))\,\lambda_{n-1}(dy) \\
&= \sum_{k=n+1}^{\infty} \operatorname{geom}(k; x_0) \sum_{\alpha \in \mathbf{N}^n : |\alpha| = k-1} \operatorname{mult}(\alpha; k-1, (1-x_0)^{-1} x) \\
&= \sum_{k=n+1}^{\infty} (1-x_0)^{k-1} x_0 \\
&\qquad \cdot \Big\{1 - \sum_{m=1}^{n-1} (-1)^{m-1} \sum_{I \subset \{1,\dots,n\} : |I| = m} \Big(1 - (1-x_0)^{-1} \sum_{j \in I} x_j\Big)^{k-1}\Big\} \\
&= x_0 \sum_{m=0}^{n-1} (-1)^m \sum_{I \subset \{1,\dots,n\} : |I| = m} \Big(1 - x_0 - \sum_{j \in I} x_j\Big)^n \Big/ \Big(x_0 + \sum_{j \in I} x_j\Big),
\end{aligned} \tag{37}$$

where empty sums are 0. The special case in which $n = 2$ is consistent with Example 1. Denoting the right side of (37) by $u_0(x)$, it is easy to see that, for $i = 0, 1, \ldots, n$ and $x \in \Delta_n^\circ$, we have $P_x\{X(\tau) \in H_i^\circ\} = u_0(\rho_i(x))$ and therefore the conditional P_x-distribution of $\pi_i^{-1}(X(\tau))$, given that $X(\tau) \in H_i^\circ$, has $(n-1)$-dimensional Lebesgue density $h_0(\rho_i(x), \pi_0(\cdot))/u_0(\rho_i(x))$ on Δ_{n-1}°.

REFERENCES

[1] DYNKIN, E. B., *Markov Processes* **2**. Springer, Berlin, 1965.

[2] ETHIER, S. N., A class of degenerate diffusion processes occurring in population genetics. *Comm. Pure Appl. Math.* **29** (1976), 483–493.

[3] ETHIER, S. N., Limit theorems for absorption times of genetic models. *Ann. Probability* **7** (1979), 622–638.

[4] LITTLER, R. A., Loss of variability at one locus in a finite population. *Math. Biosciences* **25** (1975), 151–163.

S. N. ETHIER
DEPARTMENT OF MATHEMATICS
UNIVERSITY OF UTAH
SALT LAKE CITY, UTAH 84112, USA

R. C. GRIFFITHS
DEPARTMENT OF MATHEMATICS
MONASH UNIVERSITY
CLAYTON, VICTORIA 3168, AUSTRALIA

Boundary Hitting Approximations for Markov Processes

Gareth Roberts

1. Introduction

In this paper, we consider the problem of approximating the distribution of first hitting time of time-dependent boundaries, by certain classes of one-dimensional Markov processes. Many of the techniques to be used here, were originally derived for diffusion processes in Roberts (1989, a), using techniques involving deriving stochastic inequalities for conditioned diffusions. However, due to the need for such hitting-time approximations in statistics, particularly in sequential analysis and change point analysis, we extend the techniques of Roberts (1989, a), to cover a larger class of processes, called strongly stochastically monotone processes. This class of processes includes all time-inhomogeneous diffusions, and it will be seen that it is the natural class of processes to consider when searching for inequalities for conditioned Markov processes.

We find that the class of strongly stochastically monotone processes contains many interesting discrete-time Markov processes, including partial sums of certain independent and identically distributed (i.i.d.) random variables, which is of particular use for sequential analysis applications. This problem of approximating time-dependent boundary hitting times for sums of i.i.d. random variables is complicated by the need to estimate 'overshoot' distributions, see for example Siegmund (1985) and Woodroofe (1982).

An alternative approach is to use a diffusion approximation, in this case Brownian motion with constant drift, and to approximate the hitting-time distribution for the diffusion. Although this approach avoids making overshoot corrections, the diffusion approximation is often crude leading to inaccurate results.

Many techniques exist for approximating boundary hitting times for diffusions. These include the method of images (see Lévy, 1965), the method of weighted likelihoods (Robbins and Siegmund, 1970), and the tangent approximation (Strassen, 1967). These are summarized in the excellent book by Lerche (1986).

Here we shall adopt a different approach, achieving estimates for hazard functions of hitting times directly from the distribution of the diffusion at time t conditioned to remain below the boundary until time t. These methods allow us to prove asymptotic results for the hitting-time distribution which are illustrated by examples of Brownian motion hitting-time approximations. These results will be expressed as stochastic inequalities, and will appear in section 2.

In section 3, we improve and generalize the inequalities for conditioned processes to apply to the larger class of strongly stochastically monotone processes. In this way we can avoid the need to make diffusion approximations or overshoot corrections when dealing with partial sum hitting times.

All the results in this paper are one dimensional, and the techniques used offer no obvious generalizations. Indeed the function-analytic preliminaries of subsection 2.1 do not even generalize to higher dimensions. The justification for developing methods for one-dimensional problems only lies in the statistical applications mentioned earlier.

2. Conditioned Diffusions

2.1. *Functional-Analytic Preliminaries*

Let X be a time-homogeneous non-degenerate one-dimensional diffusion with generator

$$G = \frac{1}{2}\sigma^2(x)\frac{d^2}{dx^2} + \xi(x)\frac{d}{dx},$$

where σ and ξ are such that X has bounded scale and speed measures on bounded intervals, and σ and ξ are sufficiently nice to allow X to have a unique weak solution which is almost surely finite.

Let $\tau = \inf\{t \geq 0; x_t \notin (a,b)\}$, for $a < b$, be a time-homogeneous exit time for X and define

$$\phi(t,x) = \mathbf{P}[\tau > t \mid X_0 = x]. \tag{2.1}$$

Then

$$\frac{\partial \phi}{\partial t} = G\phi \tag{2.2}$$

and $\phi(0, x) = 1$, $x \in (a, b)$.

If a and b are finite, then the relevant boundary conditions, $\phi(t, a) = \phi(t, b) = 0$, are sufficient to solve (2.2). Furthermore, the space

$$V = \{C^2 \text{ functions } f \text{ such that } f(a) = f(b) = 0\} \tag{2.3}$$

is a pre-Hilbert space with integral norms, and if we choose the inner product

$$\langle f, g \rangle = \int f(x) g(x) \rho(x) dx \tag{2.4}$$

where $\rho = dm/ds$, the derivative of the speed measure with respect to the scale measure for X, then G is actually self-adjoint. The eigenfunctions of G, $\{e_i,\ i = 1, 2, \ldots\}$, are therefore orthogonal under $\langle \cdot, \cdot \rangle$ and span V, so that eigenfunction expansions can be given explicitly for ϕ:

$$\phi(t, x) = \sum_{i=1}^{\infty} e^{\lambda_i t} e_i(x) \langle e_i, 1 \rangle.$$

The eigenvalues $\{\lambda_i, i = 1, 2, \ldots\}$ are all real and negative (by the maximum principle for parabolic operators, and the non-degeneracy of X), and furthermore the spectrum of G is discrete since $[a, b]$ is finite and again X is non-degenerate. Finally, the finiteness of $[a, b]$ also ensure that we can work in the nice L^2 space even though this is an unnatural space in which to express probabilities.

So it is clear that the hitting time τ has asymptotic exponential decay with rate λ_1.

Suppose that we are interested in the behaviour of $\mu_t = [X_t \mid \tau > t]$. (Note that we shall use $[\cdot]$ to denote the distribution of $\cdot$, or the probability measure it induces.) It is clear that $\theta(t, x) = \mathbf{P}[X_t \le y,\ \tau > t \mid X_0 = x]$ for some $y \in [a, b]$, satisfies (2.1) with initial conditions $\theta(0, x) = I[x \le y]$, and so,

$$\mathbf{P}[x_t \le y \mid \tau > t, X_0 = x] = \frac{\sum_{i=0}^{\infty} e^{\lambda_i t} e_i(x) \langle \phi_i, I[x \le y] \rangle}{\sum_{i=0}^{\infty} e^{\lambda_i t} e_i(x) \langle e_i, 1 \rangle}$$

and that asymptotically,

$$\mathbf{P}[X_t \le y \mid \tau > t] \simeq \frac{\langle e_1, I[x \le y] \rangle}{\langle e_1, 1 \rangle}.$$

The asymptotic density is hence given by,

$$f_{[X_t \mid \tau > t]}(y) \simeq \frac{e_1(y) \rho(y)}{\langle e_1, 1 \rangle},$$

and we shall denote this distribution, $\lim_{t\to\infty}[X_t \mid \tau > t]$ by δ_∞.

Note that in the case when a or b is infinite, δ_∞ may or may not exist. In the examples that follow, it suffices that in the case where X is an Ornstein–Uhlenbeck process, δ_∞ exists since the discreteness of the spectrum is assured by Molachanov's theorem, see Roberts (1989, a) for details.

The generalization of the above theory to higher dimensions is restricted to the case where the drift ξ is the gradient of some potential, when we can essentially generalize the notions of speed and scale and thus provide a density with respect to which the diffusion generator is self-adjoint, see Pinsky (1985).

For a more detailed presentation of these function-analytic preliminaries, see Jacka and Roberts (1987).

2.2. *Stochastic Inequalities*

In this subsection, we are concerned with the following time-inhomogeneous problem. Let X be a time-inhomogeneous one-dimensional diffusion, and let τ be a hitting time defined by

$$\tau_f = \inf\{t \geq 0; X_t \geq f(t)\}$$

by some continuous function f, and such that $\mathbf{P}[\tau > s] > 0\ \forall s \geq 0$. We wish to estimate $\gamma_t = [X_t \mid \tau_f > t]$.

We shall use the usual stochastic order relation: if N_1 and N_2 are random variables, then $[N_1] \overset{\text{st}}{\leq} [N_2]$ if and only if

$$\mathbf{P}[N_1 > x] \leq \mathbf{P}[N_2 > x], \quad \forall x \in \mathbf{R}.$$

In order to manipulate such conditional distributions effectively, we need the following comparison results.

Lemma 2.1. *Denote by γ_t^i, the distribution of $[X_t \mid \tau_f > t,\ X_0 \sim \gamma^i]$ for starting distributions γ^i, $i = 1, 2$. If γ_1 and γ_2 are distributions such that $\gamma^1 \overset{\text{st}}{\leq} \gamma^2$, then $\gamma_t^1 \overset{\text{st}}{\leq} \gamma_t^2$.*

Lemma 2.2. *Let X^1 and X^2 be diffusions satisfying*

$$dX_t^i = \sigma(t, X_t^i)dB_t + \xi^i(t, X_t^i)dt \tag{2.5}$$

for σ and ξ sufficiently smooth to allow weak comparison theorems to hold (see for example Roberts (1987)*). Also let*

$$\tau_f^i = \inf\{t \geq 0; X_t^i \geq f(t)\}.$$

If

$$\xi^1(t,x) \le \xi^2(t,x), \quad \forall (t,x) \in [0,\infty) \times \mathbf{R}, \tag{2.6}$$

then

$$X_t^1 \overset{\text{st}}{\le} X_t^2, \quad \forall t \in [0,\infty).$$

The proofs of Lemmas 2.1 and 2.2 can be found in Roberts (1989, a) and rely on the observation that a process $X \mid \tau > t$ is also a diffusion. We intend to use these results to approximate γ_t, and consequently the hazard function of the distribution of the hitting time at time t. Natural candidates for estimates for the γ_t's are the time-homogeneous quasi-stationary distributions discussed in subsection 2.1. We therefore also require the following comparison results for stationary distributions.

Lemma 2.3. *Suppose f and g are functional boundaries such that $f(t) \le g(t)$ $\forall t \ge 0$, and $\mathbf{P}[\tau_f < t] < 1$ for our process X. Then*

$$[X_t \mid \tau_f > t] \overset{\text{st}}{\le} [X_t \mid \tau_g > t].$$

For diffusion processes, Lemma 2.3 can be proved many ways, including coupling and h-transforms. It is instructive to consider such proofs with a view to generalizing to a larger class of processes. See Roberts (1989a,b) for proofs of Lemma 2–3 under varying regularity conditions.

Lemma 2.4. *Let X be a time-homogeneous diffusion, and let τ_a be the time-homogeneous hitting time:*

$$\tau_a = \inf\{t \ge 0; X_t \ge a\}.$$

Suppose also that $\delta_\infty^a = \lim_{t\to\infty} [X_t \mid \tau > t]$ exists, at least for two values of a, $a_1 < a_2$ say. Then,

$$\delta_\infty^{a_1} \overset{\text{st}}{\le} \delta_\infty^{a_2}.$$

Proof. This follows by taking the limit as $t \to \infty$ of the inequality in Lemma 2.3.

Lemma 2.5. *In the time-homogeneous setup of the previous lemma, suppose $\mu_0 \overset{\text{st}}{\le} \delta_\infty^a$. Then*

$$\mu_t = [X_t \mid \tau > t, X_0 \sim \mu_0] \overset{\text{st}}{\le} \delta_\infty^a.$$

Lemma 2.6. *Suppose X^1 and X^2 are time-homogeneous diffusions satisfying the* S.D.E.'s 2.5 *and constraint* 2.6, *such that* $\delta_\infty^{\xi_i} = \lim_{t\to\infty}[X_t^i \mid \tau_a > t]$ *exists* $(i = 1, 2)$. *Then*

$$\delta_\infty^{\xi_1} \overset{\text{st}}{\leq} \delta_\infty^{\xi_2}.$$

Proof. This follows by taking to the limit the inequality in Lemma 2.2.

We are now ready to state the main results of this subsection.

Theorem 2.7. *Let X be a time-homogeneous diffusion, and let f be a non-decreasing functional boundary such that* $\mathbf{P}[\tau_f > t] > 0$, $\forall t \geq 0$. *Suppose that the time-homogeneous quasi-stationary distributions,* $\delta_\infty^{f(t)} = \lim_{s\to\infty}[X_s \mid \tau_{f(t)} > s]$ *exists at least for t in some range* $[0, T]$, *say. Then,*

$$[X_t \mid \tau_f > t, X_0 \sim \delta_\infty^{f(0)}] \overset{\text{st}}{\leq} \delta_\infty^{f(t)}. \tag{2.7}$$

Proof.

$$[X_t \mid \tau_f > t, X_0 \sim \delta_\infty^{f(0)}] \overset{\text{st}}{\leq} [X_t \mid \tau_{f(t)} > t, X_0 \sim \delta_\infty^{f(0)}] \overset{\text{st}}{\leq} \delta_\infty^{f(t)}$$

since the first inequality follows directly from Lemma 2.3, and the second from Lemmas 2.4 and 2.5.

Theorem 2.8. *Let X satisfy,*

$$dX_t = \sigma(X_t)dB_t + \xi(t, X_t)dt,$$

where $\xi(\cdot, x)$ is a non-decreasing function for all $x \in \mathbf{R}$. Let $\delta_\infty^{\xi(t,\cdot)}$ exist (adopting the notation of Lemma 2.6) *for each of the time-inhomogeneous diffusions,*

$$dX_s^t = \sigma(X_s^t)dB_s + \xi(t, X_s)ds$$

$t \in [0, T]$ *say, and the hitting time,*

$$\tau_a^t = \inf\{t \geq 0; X_t^t \geq a\}, \quad t \in [0, \tau], \text{ say.}$$

Then,

$$[X_t \mid \tau_a > t, X_0 \sim \delta_\infty^{\xi(0,\cdot)}] \overset{\text{st}}{\leq} \delta_\infty^{\xi(t,\cdot)}. \tag{2.8}$$

Proof. By Lemmas 2.2, 2.5 and 2.6,

$$[X_t \mid \tau_a > t, X_0 \sim \delta_\infty^{\xi(0,\cdot)}] \overset{\text{st}}{\leq} [X_t^t \mid \tau_a^t > t, X_0 \sim \delta_\infty^{\xi(0,\cdot)}] \overset{\text{st}}{\leq} \delta_\infty^{\xi(t,0)}.$$

Of course, Theorems 2.7 and 2.8 have analogies in the cases where the boundary is non-increasing and the drift is non-increasing, respectively. Also it is instructive to note that in the cases where the boundary is constant and the drift is constant, respectively, we trivially have equality in equations 2.7 and 2.8, and in fact intuitively, the 'closer' the problem is to being time homogeneous, in some sense, the tighter the bounds in 2.7 and 2.8. This suggests that the search for powerful inequalities for time-inhomogeneous problems involves searching for transformations that will convert the problem to a less time-dependent one.

2.3. *Examples Involving Brownian Motion*

In this section, we give examples of applications of Theorems 2.7 and 2.8, in the case where the Markov process is simple Brownian motion, $\{B_t, t \geq 1\}$.

Theorem 2.9. *Suppose f is a simple lower case boundary, that is, f is differentiable and $f(t)f'(t)$ is asymptotically non-increasing to 0. Then if $\tau_f = \inf\{t \geq 1;\ |B_t| \geq f(t)\}$,*

$$\mathsf{P}[\tau_f > t] = h(t)\exp\left(-\frac{\beta(t)\pi^2}{8a^2}\right),$$

where β is the solution of

$$\beta'(t) = \frac{1}{f^2(t)}, \quad \text{with } \beta(1) = 0,$$

and $h(t)$, the sub-exponential term, satisfies the bounds

$$\exp\left(k_1 \int_0^{\beta(t)} r(s)\,ds\right) \leq h(t) \leq k_2 t^p,$$

where k_1, k_2 and p are positive constants and $r(t) = f(\alpha(t))f'(\alpha(t))$, where $\alpha(\cdot)$ is the inverse of $\beta(\cdot)$.

Proof. The lower bound follows from making the transformation $Z_t = e^{-t/2}B_{\exp\{t\}}$ and applying a Theorem 2.7 type result, and the upper bound follows from transforming $Z_t = B_{\alpha(t)}/f(\alpha(t))$ and using a Theorem 2.8 type result. The details appear in Roberts (1989, a).

Definition. Define the function $m(\cdot)$ to be the inverse function of $z(\cdot)$, where $z(c)$ is the smallest positive zero of $M(-c, \frac{1}{2}, \frac{1}{2}x^2)$, where M is the confluent hypergeometric function, (see Abramowitz and Stegun (1972)).

Theorem 2.10. *Suppose f is an upper case boundary, that is, a boundary f such that $f(t)/t^{\frac{1}{2}} \uparrow \infty$, satisfying certain smoothness conditions, (see conditions* (B1)–(B4) p. 39 *of Roberts* (1989, a)), *or alternatively, suppose f is an approximate square-root boundary ($f(t) = a(t)t^{\frac{1}{2}}$, where $a(t) \to a(\infty) > 0$) satisfying other smoothness conditions (see* p. 25 *of Roberts* (1989 ,a)). *Then,*

$$\mathsf{P}[\tau_f > t] = k(1 + o(1)) \exp\left(- \int_1^t \frac{m(a(s))}{s} ds \right)$$

for some positive constant k, as $t \to \infty$.

Proof. See the proofs of Theorems 3.2 and 5.2 in Roberts (1989, a).

Note that by letting $t = \infty$, and giving an asymptotic expansion for μ, we can prove the Kolmogorov–Erdös–Feller–Petrowski theorem as a corollary to Theorem 2.10.

Definition. Let $-n(\cdot)$ be the inverse function of $z(\cdot)$ where $z(\lambda)$ is the smallest zero of the equation

$$D_{2\lambda}(-x) = 0,$$

where D is the Whitaker function or parabolic cylinder function (see Abramowitz and Stegun (1972)).

Theorem 2.11. *Let $\tau_f = \inf\{t \geq 1,\ B_t \geq f(t)\}$ for some nicely behaved upper case boundaries f (for details of regularity conditions, we again refer to Roberts* (1989, a)) *then*

$$\mathsf{P}[\tau > t] = k(1 + o(1)) \exp\left(- \int_1^t \frac{n(a(s))}{s} ds \right),$$

where k is a constant and $a(s) = f(s)s^{-\frac{1}{2}}$.

Proof. This is Theorem 6.3.2 of Roberts (1989, a). It is proved by the usual manipulation techniques and needs certain regularity conditions on n.

Other examples of applications for Brownian motion are also given in Roberts (1989, a).

3. Inequalities for Conditional Markov Processes

It is clear that the methodology introduced in Theorems 2.7 and 2.8 is powerful, as demonstrated in subsection 2.3. However, if we are interested in approximating conditioned Markov processes, then stochastic inequalities are not necessarily the right inequalities to be using. For instance, stochastic inequalities are not in general preserved by conditioning, so why should Lemmas 2.1 to 2.5 hold in general for Markov processes?

The power behind the techniques in Section 2 is largely derived from Lemmas 2.2 and 2.3, both inequalities which depend crucially on an ordering of the state space. It would seem natural therefore, as a first attempt to generalizing these results to larger classes of Markov processes, to consider the class of stochastically monotone processes. Stochastically monotone processes were introduced by Daley (1968) and are processes which preserve order, in the sense that, $\mathbf{P}[X_t \in y \mid X_0 = x]$ is non-increasing as a function of x for all y. However, because stochastic inequalities are not preserved by conditioning, it is easy to construct counter-examples to the conjecture that stochastically monotone processes satisfy a Lemma 2.3 type inequality. (See Roberts, 1989, b.)

In fact the natural class of processes to consider for generalizations of Lemmas 2.2 and 2.3 is a subclass of stochastically monotone processes, which we will call strongly stochastically monotone processes.

Definition. (a) $\mu_1 \overset{\text{sst}}{\leq} \mu_2$, and we say that μ_1 is strongly stochastically less than or equal to μ_2 if and only if the Radon–Nikodym derivative

$$\frac{\mu_2(dx)}{\mu_1(dx)}$$

is non-decreasing in x on $\text{supp}(\mu_2) \cup \text{supp}(\mu_1)$. Note that we make no assumption of continuity between measures, and $\mu_2(dx)/\mu_1(dx)$ is allowed to take values 0 or ∞.

(b) A process, X, is strongly stochastically monotone if

$$\forall t,\ y_1 < y_2, \quad [X_t \mid X_0 = y_1] \overset{\text{sst}}{\leq} [X_t \mid X_0 = y_2].$$

We shall always assume that we are considering the cadlag version of X.

Note that all time-inhomogeneous diffusions are strongly stochastically monotone. This can be shown by a simple coupling argument.

The results for this section can now be stated.

Theorem 3.1. *Let X be a strongly stochastically monotone process and let f and g be functional boundaries such that $f(t) \le g(t)\ \forall t \ge 0$, and $\mathbf{P}[\tau_f > t] > 0$. Then*

$$[X_t \mid \tau_f > t] \overset{\text{sst}}{\le} [X_t \mid \tau_g > t].$$

Theorem 3.2. *Let X^1 and X^2 be strongly stochastically monotone processes such that $[X_t^1] \overset{\text{sst}}{\le} [X_t^2]\ \forall t \ge 0$. Also suppose that f is a functional boundary such that $\mathbf{P}[\tau_f(X^1) > t] > 0$, where $\tau_f(X^i)$ is the hitting time of f for the process X^i. Then,*

$$[X_t^1 \mid \tau_f(X^1) > t] \overset{\text{sst}}{\le} [X_t^2 \mid \tau_f(X^2) > t].$$

The proof of Theorem 3.1 relies on an FKG type inequality for a backwards induction argument in discrete time, together with a limiting argument to show that the effect of conditioning to remain below a discretized boundary converges to the effect of conditioning to remain below the continuous boundary. Details can be found in Roberts (1989, b). Theorem 3.2 is proved in an identical way to Theorem 3.1.

REFERENCES

Abramowitz, M. and Stegun, I. A. (1972). *Handbook of mathematical functions.* Dover, New York.

Daley, D. J. (1968). Stochastically monotone Markov chains. *Z. Wahrscheinlichkeitsth. Verw Geb.*, **10**, 305–317.

Jacka, S. D. and Roberts, G. O. (1987). Conditional diffusions; their infinitesimal generators and limit laws. *Warwick University Statistics Department Research report* no. **127.**

Levy, P. (1965). *Processus Stochastiques et Mouvement Brouwnien.* Gauthier-Villars, Paris.

Lerche, H. R. (1986). Boundary crossings of Brownian motion. *Lecture notes in Statistics,* **40**, Springer-Verlag, Berlin-Heidelberg-New York-Tokyo.

Pinsky, R. G. (1985). On the convergence of diffusion processes conditioned to remain in a bounded region for large time to limiting positive recurrent diffusion processes. *Ann. Prob.*, **13**, 2, 363–378.

Robbins, H. and Siegmund, D. (1970). Boundary crossing probabilities for the Wiener process and partial sums. *Ann. Math. Statist.* **41**, 1410–1429.

Roberts, G. O. (1987). A weak comparison theorem for SDE's. *Warwick University Statistics Department Research report* no. **146**.

Roberts, G. O. (1989, (a)). Asymptotic approximations for Brownian motion boundary hitting times. Submitted to *Annals of Probability.*

Roberts, G. O. (1989, (b)). A comparison theorem for conditional Markov processes. Submitted to *J. Applied Probability.*

Siegmund, D. (1985). *Sequential Analysis.* Springer-Verlag, Heidelberg.

Strassen, V. (1967). Almost sure behaviour of sums of independent random variables and martingales. *Proc. Fifth Berkeley Symp. Math. Statist. Probab.,* Univ. of Cal. Press, Vol. III, Part I, 315–343.

Woodroofe, M. (1982). Nonlinear Renewal Theory in Sequential Analysis, *Regional Conference series in applied mathematics of SIAM,* **39**, Philadelphia.

Department of Mathematics
University of Nottingham
University Park
Nottingham NG7 2RD
United Kingdom

Brownian Motion in a Wedge with Variable Skew Reflection: II

L.C.G. Rogers

1. Introduction. Let $D = \{z \equiv re^{i\theta} \in \mathbb{C} : \quad r > 0, \quad 0 < \theta < \xi\}$ be a wedge in the complex plane, and suppose given on ∂D a vector field pointing into D, which will specify the direction of reflection of Brownian motion in D when it hits the boundary. There are several natural questions which one can ask about this process, in particular;

(1.i) Can one construct the process, at least until it first reaches 0?

(1.ii) Does the process approach 0 ?

(1.iii) If so, will the process reach 0 in finite time ?

(1.iv) If so, can the process be extended beyond its first hit on 0 so as to have continuous paths, and to spend no time in 0 ?

(1.v) If so, is such an extension unique ?

In the case where the directions of reflection are constant on each side of the wedge, these questions were completely answered in [7] (see also [5] for an excursion-theoretic derivation of the results.) The case of variable direction of reflection is considerably more difficult, and has been begun in [6], to which this paper is a sequel.

In order to state the main results of this paper, we review the notation and results of the earlier paper [6]. The construction (1.i) is sufficiently straightforward as to need no comment here. For convenience, we may assume that the directions of reflection on ∂D are normal outside a neighbourhood of 0, since the answers to (1.ii-v) are determined by what happens in a neighbourhood of 0. The first step in [6] is to transform D by the map $z \mapsto -1/z^{\pi/\xi}$, which takes D to $\mathbf{H}$, mapping 0 to $+\infty$, and preserving the directions of reflection on the boundary; thus (1.ii) is equivalent to the question, 'Is the skew-reflecting Brownian motion in $\mathbf{H}$ transient ?' (*Aside*: the 'skew-reflecting Brownian mapping theorem', to which we have just appealed , is an 'obvious fact', but, thanks to Chris Burdzy's insistence, I have set down a proper statement and proof of the result in the appendix!)

The reflection vector field on $\partial\mathbf{H}$ is specified by a function $\theta : \mathbf{R} \to (-\pi/2, \pi/2)$, the reflection at $x \in \mathbf{R}$ being in a direction making angle $\theta(x)$ in a clockwise sense with the inward-pointing normal. In [6], the function θ was taken to be C^1 with bounded derivatives, satisfying $|\tan\theta(x)| \leq A(1+|x|)$ for some A. These conditions are unnecessarily restrictive - local Hölder α for some $\alpha \in (0,1]$ would suffice, and even weaker conditions work in some sense, as we shall see.

Define now the analytic function $\psi : \mathbf{H} \to \mathbf{C}$ by

$$\psi(z) = \exp\left[\int_{-\infty}^{\infty} \frac{\theta(x)dx}{\pi}\left\{\frac{1}{x-z} - \frac{x}{1+x^2}\right\}\right]. \tag{2}$$

We shall insist that ψ can be extended continuously to $\overline{\mathbf{H}}$ (a local Hölder condition on θ will ensure this). *The essential property of ψ is that its argument at $x \in \mathbf{R}$ is $\theta(x)$.* To see why this is relevant, suppose that $h : \overline{\mathbf{H}} \to \mathbf{R}$ is C^2. Then a simple Itô-formula calculation (carried out in [6]) shows that, if Z is the skew reflecting Brownian motion, then $h(Z_t)$ is a local martingale if and only if

$$\triangle h = 0 \quad \text{in} \quad \mathbf{H} \tag{3.i}$$

$$\tan\theta(x)\frac{\partial h}{\partial x}(x) + \frac{\partial h}{\partial y}(x) = 0 \quad \forall x \in \mathbf{R} \tag{3.ii}$$

If now we take g to be the conjugate function to h, so that $f \equiv g + ih$ is analytic in $\mathbf{H}$, the boundary condition (3.ii) can be restated as

$$\mathrm{Re}\big(f'(x)/\psi(x)\big) = 0 \quad \text{for all} \quad x \in \mathbf{R}. \tag{4}$$

The key to the study of this problem appears to be the construction of suitable analytic f satisfying the boundary condition (4).

To state the main results of this paper, we need some simple facts about analytic functions $\phi : \mathbf{H} \to \mathbf{H}$. This class of functions is called the class of *Pick* functions by Donoghue [2], from whose book we quote the following facts.

Every Pick function ϕ has the representation

(5)
$$\phi(z) = \int_{-\infty}^{\infty} \frac{\mu(dx)}{\pi}\left[\frac{1}{x-z} - \frac{x}{1+x^2}\right] + c_1 z + c_2,$$

where c_1, c_2 are real constants, $c_1 \geq 0$, and μ is a measure satisfying the integrability condition

$$\int \frac{\mu(dx)}{1+x^2} < \infty.$$

If $\mathrm{Im}\phi$ can be extended continuously to $\overline{\mathsf{H}}$, then we may take $\mu(dx) = \mathrm{Im}\phi(x)dx$. The constant c_1 is identified as

(6)
$$c_1 = \lim_{b\uparrow\infty} \phi(ib)/ib.$$

Notice that $i\psi$ is a Pick function, so has a representation (5).

THEOREM 1. *The skew-reflecting Brownian motion may with positive probability never hit* $(-\infty, 0)$ *if and only if*

(7.i)
$$\int_1^{\infty} \left(\frac{\pi}{2} - \theta(x)\right)\frac{dx}{x} < \infty.$$

If this condition holds, then the process is transient, and the original skew-reflecting Brownian motion in the wedge D reaches 0 *in finite time almost surely.*

If condition (7.i) holds, then a one-sided escape to ∞ down R^+ is possible; the analogous condition for a one-sided escape to ∞ down R^- to be possible is of course

(7.ii)
$$\int_{-\infty}^{-1} \left(\theta(x) + \frac{\pi}{2}\right)\frac{dx}{|x|} < \infty.$$

The proof of this result is in §2. Having decided when a one-sided escape is possible, and what happens in that case, we turn our attention to the remaining cases.

THEOREM 2. *If* $c_1 > 0$, *then the skew-reflecting Brownian motion in* H *is recurrent. If* $c_1 = 0$, *then the skew-reflecting Brownian motion in* H *is recurrent if and only if*

(8)
$$\int_{-\infty}^{\infty} \frac{\mathrm{Im}\, i\psi(x)}{1+|x|}dx = +\infty.$$

This result is proved in §3. The constant c_1 is the constant appearing in the Pick function representation (5) of $i\psi$.

Theorem 2 improves on Theorem 2 in [6], where the function $\Psi(z) \equiv \int_0^z \psi(\omega)d\omega$ was used to map $\mathbf{H}$ to the domain $\Psi(\mathbf{H})$; the process was transient if $\Psi(\mathbf{H})$ was contained in a wedge of angle $< \pi$, and recurrent if $\Psi(\mathbf{H})$ contains a non-vertical half-plane. Evidently, there is a gap between these conditions, and Burdzy and Marshall [1] have refined the complex mapping techniques of [6] to obtain a result equivalent to Theorem 2. We prove this equivalence in the appendix. In fact, the skew-reflecting Brownian motion in $\mathbf{H}$ can with advantage be defined via a time change of the vertically reflecting Brownian motion in $\Psi(\mathbf{H})$, mapped back under Ψ^{-1}. As Burdzy remarks, this allows us to define the skew-reflecting Brownian motion in cases where the direction of reflection is much less regular; an account of this work (with D. Marshall) is in preparation.

Let us make a few remarks on the methodology used here to prove Theorem 2. We shall be making use of the same classical techniques of complex analysis which Dynkin [3] used (see also Malyutov [4]), with the aim of constructing a Green's function for the process. Let $G(\cdot,\cdot)$ be the Green's function for the process (assumed for the moment to be transient). Now it is clear that we may decompose

$$G(z,z_0) = G^\theta(z,z_0) + h(z,z_0), \tag{9}$$

where G^θ is the Green's function for Brownian motion in $\mathbf{H}$ killed on first hitting $\mathbf{R}$,

$$\begin{aligned} G^\theta(z,z_0) &= \frac{1}{\pi}\log\left|\frac{z-\bar{z}_0}{z-z_0}\right| \\ &= \operatorname{Im}\frac{i}{\pi}\{\log(z-\bar{z}_0) - \log(z-z_0)\}, \end{aligned} \tag{10}$$

and where formally

$$h(z,z_0) = E^z\left[\int_\tau^\infty \delta_{z_0}(Z_t)dt\right],$$

where $\tau \equiv \inf\{u : Z_u \in \mathbf{R}\}$. It is easy to see that $h(\cdot, z_0)$ is harmonic, and non-negative. If we fix z_0 for the time being, and drop it from the notation, we may take the conjugate function g to h to form $f = g + ih$, and then

$$G(z) = \operatorname{Im}\left(\frac{i}{\pi}\{\log(z-\bar{z}_0) - \log(z-z_0)\} + f(z)\right) \tag{11}$$

Since G must satisfy the boundary condition (3.ii), we have from (4) that

$$\operatorname{Im}\frac{1}{i\psi(x)}\left\{\frac{i}{\pi}\left(\frac{1}{x-\bar{z}_0} - \frac{1}{x-z_0}\right) + f'(x)\right\}$$

$$(12) \qquad = \mathrm{Im}\frac{1}{i\psi(x)}\left\{\frac{2y_0}{\pi|x-z_0|^2} + f'(x)\right\} = 0 \quad \forall x \in \mathbf{R},$$

where $z_0 \equiv x_0 + iy_0 \in \mathsf{H}$.

While it is not clear that a Green's function with sufficient regularity properties to justify the above steps should exist, we shall instead use the condition (12) to build a candidate Green's function, and use the properties of the function constructed to decide recurrence or transience.

It is slightly disappointing that the clean complex mapping techniques of [6] must give way to the analysis of various integral expressions, but if you want an integral test for transience, you must expect to deal with integrals ....! Nonetheless, those complex mapping techniques serve us well in the proof of Theorem 1.

In the final section of this paper, we turn our attention to the question (1.iii) – will the process in the wedge reach 0 in finite time? We are unable to give a complete answer to this question, but the following result shows that the question is an interesting one.

THEOREM 3. *If the angle ξ of the wedge is greater than $\pi/2$ and if the skew-reflecting Brownian motion Z can approach 0, then Z will reach* 0 *in finite time.*

If the angle ξ of the wedge is less than $\pi/2$, then it is possible for Z to approach 0 *but never reach* 0 *in finite time.*

We then proceed to analyse the Green's function in more detail, to obtain a criterion for the mean passage time to 0 to be finite. Recall that the wedge D has been mapped to H by the function $z \mapsto -1/z^{\pi/\xi}$, and that the Green's function defined by (9) is defined for skew-reflecting Brownian motion ζ in H, which is the time-change of the image of skew-reflecting Brownian motion Z in D. Thus to go back from skew-reflecting Brownian motion ζ in H to D, we must firstly change time by the additive functional $\int_0^t (\xi/\pi)^2 |\zeta_s|^{-2-2\xi/\pi} ds$, and then map $\zeta \mapsto (-1/\zeta)^{\xi/\pi}$. Thus Z will reach 0 in finite time if and only if $\int_0^\infty |\zeta_s|^{-2-2\xi/\pi} I_{\{|\zeta_s|\geq 1\}} ds < \infty$. The final result gives a necessary and sufficient condition for the mean of this random variable to be finite.

THEOREM 4. *Assume that neither of* (7.i) , (7.ii) *holds, and that Z approaches* 0. *For $\xi < \pi/2$, the mean time spent by Z in $\{z : |z| \leq 1\}$ before reaching* 0 *is finite if and only if*

$$\int_{-\infty}^{\infty} dx \int_{-\infty}^{\infty} dv \frac{\mathrm{Im}(i\psi(x))\mathrm{Im}(-1/i\psi(v))}{1+|v|^{1+2\xi/\pi}} \frac{\tan^{-1}(x) - \tan^{-1}(v)}{x-v} < \infty.$$

Acknowledgement. It is a pleasure to thank Chris Burdzy and Don Marshall for numerous stimulating exchanges on these and other related questions.

2. Escape down a side. Let us define the Pick functions

$$\phi_+(z) \equiv \exp\left[\int_0^\infty \frac{\theta(x)+\pi/2}{\pi}dx\left(\frac{1}{x-z}-\frac{x}{1+x^2}\right)\right],$$

$$\phi_-(z) \equiv \exp\left[\int_\infty^0 \frac{\theta(x)+\pi/2}{\pi}dx\left(\frac{1}{x-z}-\frac{x}{1+x^2}\right)\right],$$

so that $\phi_+\phi_- = i\psi$, and ϕ_+ (respectively, ϕ_-) is real positive on $\mathbf{R}^-$ (respectively $\mathbf{R}^+$). We emulate the complex mapping methods of [6] by defining the analytic function

$$\Phi_+(z) \equiv \int_i^z \phi_+(\omega)d\omega.$$

This is 1-1 on $\mathbf{H}$ (see Lemma 1 of [6]). The effect of using ϕ_+ is to consider the original skew-reflection problem *but replacing θ in $\mathbf{R}^-$ by the constant value $-\pi/2$.* Thus Φ_+ sends $\mathbf{R}^-$ to a horizontal half line; indeed, $\phi_+(\omega)$ is integrable near zero, and for $t<0$, $\partial\Phi_+/\partial t = \phi_+(t)$ is real positive. Thus the region $\Phi_+(\mathbf{H})$ looks like:

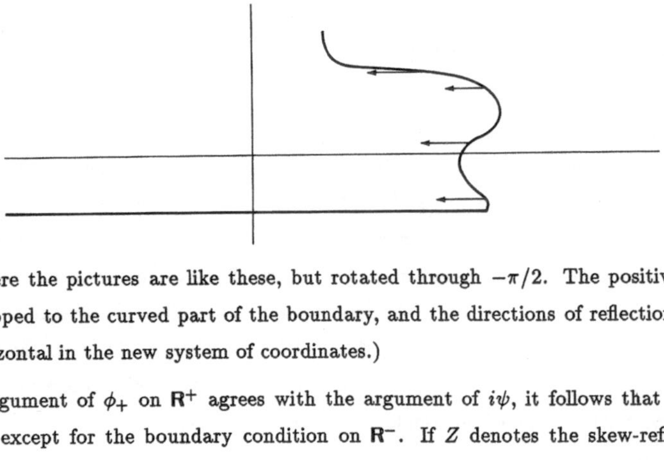

(cf. [6], where the pictures are like these, but rotated through $-\pi/2$. The positive half line has mapped to the curved part of the boundary, and the directions of reflection have become horizontal in the new system of coordinates.)

Since the argument of ϕ_+ on $\mathbf{R}^+$ agrees with the argument of $i\psi$, it follows that $\mathrm{Im}\Phi_+$ satisfies (3) except for the boundary condition on $\mathbf{R}^-$. If Z denotes the skew-reflecting Brownian motion in $\mathbf{H}$, and if $\tau \equiv \inf\{u : Z_u \in \mathbf{R}^-\}$, then certainly the process $\mathrm{Im}\Phi_+(Z_{t\wedge\tau})$ is a continuous local martingale, and is bounded below, so is almost surely convergent. If $\mathrm{Im}\Phi_+$ were bounded above, then the local martingale is bounded, so is closed on the right by its terminal value $\mathrm{Im}\Phi_+(Z_\tau)$. This implies that $P(\tau<\infty)<1$. On the other hand, if $\mathrm{Im}\Phi_+$

were not bounded above, then the continuous local martingale $\mathrm{Im}\Phi_+(Z_{t\wedge\tau})$ converges, to some value at least $\alpha \equiv \inf \mathrm{Im}\Phi_+(z)$. If it converged to some value strictly greater than α, then, since Z keeps on hitting $\mathbf{R}$, we would conclude that Z converged to some point of $\mathbf{R}$, which is impossible since $\mathrm{Im}(Z)$ is a reflecting Brownian motion on $\mathbf{R}^+$. To summarize, then,

$$(13) \qquad P(\tau < \infty) = 1 \quad \textit{if and only if } \mathrm{Im}\Phi_- \textit{ is unbounded below.}$$

The aim is therefore to prove that (7.i) is equivalent to the condition that $\mathrm{Im}\Phi_+$ is bounded above. We do this by showing that each is equivalent to the condition $\lim_{b\to\infty} ib\phi_+(ib)$ *exists and is finite.* From the definition of ϕ_+, we have

$$\begin{aligned}\log\phi_+(ib) + \log(ib) &= i\pi + \int_0^\infty \frac{dx}{\pi}\left[\frac{1}{x-ib} - \frac{x}{1+x^2}\right]\left(\theta(x) - \frac{\pi}{2}\right)\\ &= i\pi + \int_0^\infty \left(\theta(x) - \frac{\pi}{2}\right)\frac{dx}{\pi}\left[\frac{ib}{x^2+b^2} + \frac{x}{x^2+b^2} - \frac{x}{1+x^2}\right],\end{aligned}$$

from which the equivalence of (7.i) and the existence of $\lim ib\phi_+(ib)$ follows easily.

Now since ϕ_+ is a Pick function, since ϕ_+ is locally Hölder continuous away from 0, and since the singularity of ϕ_+ at 0 is at worst $|z|^{-\gamma}$, where $\gamma = (\theta(0) + \pi/2)/\pi < 1$, ϕ_+ has the Pick function representation (5)

$$(14) \qquad \phi_+(z) = \int \frac{\mathrm{Im}\phi_+(x)}{\pi}\left[\frac{1}{x-z} - \frac{x}{1+x^2}\right]dx + c_1 z + c_2.$$

It is a simple matter to deduce from this that $\lim ib\phi_+(ib)$ exists if and only if

$$(15) \qquad \mathrm{Im}\phi_+ \in L^1 \quad \textit{and} \quad \phi_+(z) = \int \mathrm{Im}\phi_+(x)\frac{dx}{\pi(x-z)}.$$

Using the representation (14), we can obtain an expression for $\Phi_+(a+ib) \equiv \int_i^{a+ib} \phi_+(\omega)d\omega$; a few calculations reduce the imaginary part to

$$\int_0^\infty \mathrm{Im}\phi_+(t)\frac{dt}{\pi}\left[\tan^{-1}\left(\frac{t}{b}\right) - \tan^{-1}\left(\frac{t-a}{b}\right) - \frac{(b-1)t}{1+t^2}\right] + 2c_1ab + (b-1)c_2.$$

If this remains bounded as $a + ib$ moves through $\mathbf{H}$, we see by setting $b = 1$ and letting $a \to \infty$ that $c_1 = 0$ and $\mathrm{Im}\phi_+ \in L^1$. Then holding $a = 0$ and letting $b \to \infty$, we see that boundedness of $\mathrm{Im}\Phi_+$ implies that the third term in the integral must exactly balance the term $(b-1)c_2$; thus if $\mathrm{Im}\Phi_+$ is bounded, then (15) must hold. The converse is easy to see using the above representation of $\mathrm{Im}\Phi_+(a+ib)$.

Since $\mathrm{Im}\Phi_+$ is bounded below, we have $\mathrm{Im}\Phi_+$ *is bounded if and only if* (7.i) *holds.*

In either of the situations (7), then, it is possible for the skew-reflecting Brownian motion to make a one-sided escape to infinity. Any such escape must be very rapid indeed, for consider the process $\log Z_t$, which is a skew-reflecting Brownian motion in the strip $\{z : 0 \le \mathrm{Im}(z) \le \pi\}$, run with clock $\int_0^t |Z_s|^{-2} ds$. Now if Z makes a one-sided escape to infinity (along $\mathbf{R}^+$, say) it has to be that $\mathrm{Im}\log Z_t$ converges to zero (because $\mathrm{Im}\log Z_t$ is the time change of a reflecting Brownian motion in [0,1]) and so the clock must be almost surely convergent:

$$\int_0^\infty |Z_s|^{-2} ds < \infty \quad \text{a.s.}$$

However, the map $z \mapsto (1/z)^\eta$, $\eta \equiv \xi/\pi$ takes $\mathbf{H}$ back to skew-reflecting Brownian motion in D with the clock $\int_0^t |Z_s|^{-2-2\eta} ds$. Because of (16) and the transience of Z, this ensures that $\int_0^\infty |Z_s|^{-2-2\eta} ds < \infty$ a.s., and so skew-reflecting Brownian motion in D *reaches* 0 *in finite time whenever it reaches* 0 *down one side only.*

For later use, we note that

(16) *if either of the conditions* (7) *holds, then* $\displaystyle\int \frac{\mathrm{Im}\, i\psi(x)}{1+|x|} dx < \infty$, *and* $c_1 = 0$.

Indeed, assuming (7.i), from (15) we see that $t\phi_+(-t)$ remains bounded as $t \to +\infty$, and so the integral over $(-\infty, 0)$ is finite (since ϕ_- is a Pick function, it is integrable with respect to the Cauchy density.) For the integral over $(0,\infty)$, we observe that from the representation

$$\phi_-(a) = \int_{-\infty}^0 \mathrm{Im}\phi_-(x) \frac{dx}{\pi} \{\frac{1}{x-a} - \frac{x}{1+x^2}\} + c_1 a + c_2$$

that $\phi_-(a)$ is $O(a)$ as $a \to \infty$. Since $\mathrm{Im}\phi_+$ is integrable, the conclusion follows . To see that c_1 must be zero, recall from (6) that

$$c_1 = \lim_{b\to\infty} i\psi(ib)/ib = \lim_{b\to\infty} -ib\phi_+(ib)\phi_-(ib)/b^2 = 0,$$

since $ib\phi_+(ib)$ converges.

3. Proof of Theorem 2. Let us fix $z_0 \equiv x_0 + iy_0 \in \mathbf{H}$. We shall construct some analytic $f : \mathbf{H} \to \mathbf{C}$ which is C^1 on $\overline{\mathbf{H}}$, and satisfying

(17) $$\mathrm{Im}\left[\frac{f'(x)}{i\psi(x)}\right] = \mathrm{Im}\left[\frac{-1}{i\psi(x)}\right] \frac{2y_0}{\pi|x-z_0|^2} \qquad \text{for } x \in \mathbf{R},$$

which is simply the condition (12) rephrased. If we now define $\tilde{\psi}(z) \equiv \psi(y_0 z + x_0)$, $\tilde{f}(z) \equiv \frac{\pi}{2} f(y_0 z + x_0)$, then (17) becomes

$$\text{(18)} \qquad \operatorname{Im}\left[\frac{\tilde{f}'(x)}{i\tilde{\psi}(x)}\right] = \operatorname{Im}\left[\frac{-1}{i\tilde{\psi}(x)}\right]\frac{1}{1+x^2} \quad \text{for} \quad x \in \mathbf{R},$$

effectively simplifying (17) to the case $z_0 = i$. Notice that $-1/i\tilde{\psi}$ is a Pick function so for some real $c_1' \geq 0, c_2'$, we may express

$$\text{(19)} \qquad \frac{-1}{i\tilde{\psi}(z)} = \int \frac{\gamma(x)dx}{\pi}\left\{\frac{1}{x-z} - \frac{x}{1+x^2}\right\} + c_1' z + c_2',$$

where $\gamma(x) \equiv \operatorname{Im}(-1/i\tilde{\psi}(x))$. The first step is to form the analytic extension into $\mathbf{H}$ of the right-hand side of (18). Since $-1/i\tilde{\psi}$ is a Pick function, it follows that $\gamma(x)/(1+x^2)$ is integrable, so the analytic extension may be given as

$$\text{(20)} \qquad \rho(z) \equiv \int \frac{\gamma(x)dx}{\pi(1+x^2)(x-z)}.$$

Now observe that

$$\begin{aligned}(1+z^2)\rho(z) &= \int \frac{\gamma(x)dx}{\pi}\left(\frac{1}{x-z} - \frac{x}{1+x^2}\right) - z\int \frac{\gamma(x)dx}{\pi(1+x^2)} \\ &= \left\{-\frac{1}{i\tilde{\psi}(z)} - c_1' z - c_2'\right\} - z\left\{\operatorname{Im}\left(-\frac{1}{i\tilde{\psi}(i)}\right) - c_1'\right\},\end{aligned}$$

in view of the representation (19);

$$= -\frac{1}{i\tilde{\psi}(z)} - c_2' - z\operatorname{Im}\left(-\frac{1}{i\tilde{\psi}(i)}\right).$$

If we now write $-1/i\tilde{\psi}(i) \equiv \alpha + i\beta$, then $c_2' = \alpha$ (from (19)), and we have

$$\begin{aligned}(1+z^2)\rho(z) &= -\frac{1}{i\tilde{\psi}(z)} - \alpha - z\beta \\ &= \frac{1}{i\tilde{\psi}(i)} - \frac{1}{i\tilde{\psi}(z)} - \beta(z-i).\end{aligned}$$

In view of (18), this leads us to define $\tilde{f}$ to within an additive constant by

$$\text{(20)} \qquad \begin{aligned}\tilde{f}'(z) &= i\tilde{\psi}(z)\rho(z) \\ &= \left(\frac{i\tilde{\psi}(z)}{i\tilde{\psi}(i)} - 1\right)/(1+z^2) - \frac{\beta i\tilde{\psi}(z)}{z+i}.\end{aligned}$$

To understand better what $\tilde{f}$ must be, we integrate the two terms of $\tilde{f}'$ separately.

$$\int^z \frac{d\omega}{1+\omega^2}\left(\frac{i\tilde{\psi}(\omega)}{i\tilde{\psi}(i)} - 1\right) = \frac{1}{i\tilde{\psi}(i)}\int^z \frac{d\omega}{1+\omega^2}\left\{\int \frac{\mathrm{Im}\, i\tilde{\psi}(x)}{\pi}dx\left(\frac{1}{x-\omega} - \frac{1}{x-i}\right) + c_1(\omega - i)\right\}$$

$$= \frac{1}{i\tilde{\psi}(i)}\left[\int \frac{\mathrm{Im}\, i\tilde{\psi}(x)dx}{\pi(1+x^2)}\int^z d\omega\left(\frac{1}{\omega+i} - \frac{1}{\omega - x}\right) + c_1 \log(z+i)\right]$$

$$= \frac{1}{i\tilde{\psi}(i)}\left[\int \frac{\mathrm{Im}\, i\tilde{\psi}(x)}{\pi(1+x^2)}dx\left\{\log(z+i) - \log(z-x)\right\} + c_1\log(z+i)\right],$$

(21)

and integrating the second term yields

$$-\beta\int^z \frac{d\omega}{\omega+i}\left\{\int \frac{\mathrm{Im}\, i\tilde{\psi}(x)}{\pi}dx\left(\frac{1}{x-\omega} - \frac{x}{1+x^2}\right) + c_1\omega + c_2\right\}$$

$$= -\beta\int \frac{\mathrm{Im}\, i\tilde{\psi}(x)}{\pi(1+x^2)}dx\left\{(x-i)\big(\log(z+i) - \log(z-x)\big) - x\log(z+i)\right\}$$

$$-\beta c_1 z - \beta(c_2 - ic_1)\log(z+i). \tag{22}$$

Using the fact $-1/i\tilde{\psi}(i) = \alpha + i\beta$, adding (21) and (22) gives us that to within an additive constant

$$\tilde{f}(z) = \int \frac{\mathrm{Im}\, i\tilde{\psi}(x)}{\pi(1+x^2)}dx\left\{(\alpha+\beta x)\log(z-x) - \alpha\log(z+i)\right\}$$
$$- c_1\alpha\log(z+i) - \beta c_1 z - \beta c_2\log(z+i).$$

This simplifies when we notice that

$$\int \frac{\mathrm{Im}\, i\tilde{\psi}(x)}{\pi(1+x^2)}dx = \mathrm{Im}(i\tilde{\psi}(i)) - c_1,$$

removing the terms in $\log(z+i)$ and leaving

$$\tilde{f}(z) = \int \frac{\mathrm{Im}\, i\tilde{\psi}(x)}{\pi(1+x^2)}(\alpha+\beta x)\log(z-x)dx - \beta c_1 z. \tag{23}$$

Since it is only the imaginary part which will interest us, we can write $z = a + ib$ and deduce that

$$\mathrm{Im}\big(\tilde{f}(z) - \tilde{f}(i)\big) = \int \frac{\mathrm{Im}\, i\tilde{\psi}(x)}{\pi(1+x^2)}(\alpha+\beta x)\left\{\tan^{-1}\left(\frac{x-a}{b}\right) - \tan^{-1}(x)\right\}dx - \beta c_1(b-1). \tag{24}$$

Theorem 2 will follow from an analysis of the properties of $\mathrm{Im}\tilde{f}$ as given by (24).

The argument which gave (12) shows that if we add the function $\frac{1}{2}\log|z+i| - \frac{1}{2}\log|z-i|$ to $\mathrm{Im}\tilde{f}$, then we obtain a function harmonic in $\mathsf{H}\backslash\{i\}$, continuous in $\overline{\mathsf{H}}$, and satisfying the boundary condition (3.ii) on $\mathbf{R}$.

Case 1: $c_1 = 0, \quad \int \mathrm{Im}\, i\psi(x)(1+|x|)^{-1}dx < \infty.$

In this case, it is evident from (24) that $\mathrm{Im}\tilde{f}$ is bounded, and so the function $h^*(z) \equiv \mathrm{Im}\tilde{f}(z) + \frac{1}{2}\log|z+i| - \frac{1}{2}\log|z-i|$ is harmonic in $\mathsf{H}\backslash\{i\}$, satisfies the boundary condition (3.ii) on $\mathbf{R}$ and is bounded below, hence $h^*(Z_t)$ is a local martingale bounded below, hence $h^*(Z_t)$ is convergent almost surely, forcing $|Z_t| \to \infty$ a.s..

Case 2: $c_1 > 0$ *or* $\int \mathrm{Im}\, i\psi(x)(1+|x|)^{-1}dx = +\infty.$

Define the function

$$K(x) \equiv \int_0^x \frac{t\,\mathrm{Im}\, i\tilde{\psi}(t)}{\pi(1+t^2)}dt.$$

Clearly, K is non-negative, increasing in $(0,\infty)$, and decreasing in $(-\infty,0)$. Moreover it is easy to see that $K(x)/|x| \to 0$ as $|x| \to \infty$.

Now consider the coefficient of β in the right-hand side of (24):

$$\begin{aligned} &\int \frac{x\,\mathrm{Im}\, i\tilde{\psi}(x)}{\pi(1+x^2)}dx\{\tan^{-1}(\frac{x-a}{b}) - \tan^{-1}(x)\} \\ &= -\int K(x)\Big\{\frac{b}{b^2+(x-a)^2} - \frac{1}{1+x^2}\Big\}dx \\ &\leq \int K(x)\frac{dx}{1+x^2}, \end{aligned}$$

which is finite, because

$$\int \frac{K(x)}{x^2}dx = \int \frac{\mathrm{Im}\, i\tilde{\psi}(t)}{\pi(1+t^2)}dt$$

is finite, $i\tilde{\psi}$ being a Pick function. Thus

$$\begin{aligned} \mathrm{Im}(\tilde{f}(z) - \tilde{f}(i)) = \beta \int K(x)\{\frac{1}{1+x^2} - \frac{b}{b^2+(x-a)^2}\}dx \\ - \beta c_1(b-1) + 0(1) \end{aligned}$$

is a harmonic function in H, *bounded above*. Moreover, it is easy to see that $\mathrm{Im}\tilde{f}(ib) \to -\infty$ as $b \uparrow \infty$, either because $c_1 > 0$, or else because c_1 is zero and the integral $\int \mathrm{Im}\, i\psi(x)(1+|x|)^{-1}dx$ is divergent. This last implies that K is unbounded, and monotonicity of K in each half of $\mathbf{R}$ gives the desired conclusion.

We now consider as before the function $h^*(z) \equiv \mathrm{Im}\tilde{f}(z) + \frac{1}{2}\log|z+i| - \frac{1}{2}\log|z-i|$. As before, $h^*(Z_t)$ is a continuous local martingale, and so a.s. either converges or oscillates between $+\infty$ and $-\infty$. But away from some neighbourhood of i, h^* is bounded above, and tends to $-\infty$ along the ray $i\mathbf{R}^+$. If the process Z were transient, since the argument of Z

oscillates back and forth between 0 and π (see (16)), it must be that Z keeps crossing $i\mathbf{R}^+$ at an arbitrarily large distance from 0. Thus $h^*(Z_t)$ cannot be bounded below. Hence $h^*(Z_t)$ cannot be bounded above, which can only be happening if Z keeps entering a neighbourhood of i, implying that Z is recurrent, a contradiction.

4. Reaching the vertex of the wedge in finite time. The first objective of this section is to prove Theorem 3. Let Z be skew-reflecting Brownian motion in the wedge D, and let $g(z) \equiv z^{\pi/\xi}$ be the obvious analytic map taking D to $\mathbf{H}$, fixing 0. By the skew-reflecting Brownian mapping theorem, $g(Z)$ is a time-change of skew-reflecting Brownian motion in $\mathbf{H}$. We could therefore alternatively start with a skew-reflecting Brownian motion ζ in $\mathbf{H}$, time-change it by the additive functional $A_t \equiv \int_0^t (\xi/\pi)^2 |\zeta_s|^{2(\xi/\pi)-2} ds$, and map back to D by g^{-1} to obtain a skew-reflecting Brownian motion in D. The skew-reflecting Brownian motion in D will reach 0 in finite time if and only if

$$\int_0^\tau |\zeta_s|^{2(\xi/\pi)-2} ds \quad < \infty$$

where $\tau = \inf\{t : \zeta_t = 0\}$. But if $\xi \in (\frac{\pi}{2}, \pi]$, then $\beta \equiv 2(\xi/\pi) - 2 \in (-1, 0]$, and so

$$\int_0^\tau |\zeta_s|^\beta ds \le \int_0^\tau (\text{Im } \zeta_s)^\beta ds \quad < \infty,$$

since Im ζ_s is just a reflecting Brownian motion on $\mathbf{R}^+$ and the additive functional $\int_0^t (\text{Im } \zeta_s)^\beta ds = \int_0^\infty L(t,x) x^\beta dx$ is evidently finite-valued (where $\{L(t,x) : t \ge 0, x \ge 0\}$ is the jointly continuous local time of Im ζ). For $\xi > \pi$, $\beta = 2(\xi/\pi) - 2 > 0$, to the finiteness of A_t is trivial. All that remains is to notice that τ must be finite a.s. in the case where 0 can be approached - because if ζ were to approach zero but never reach it, then Im ζ_t would tend to 0 as $t \to \infty$, which is *not* a property of reflecting Brownian motion on $\mathbf{R}^+$!

To compete the proof of Theorem 3, we must build an example of a wedge D with opening $\zeta < \pi/2$ where Z will approach but never reach 0. The example is best understood by firstly transforming D by log, taking D to $S \equiv \{(x + iy) : 0 \le y \le \xi\}$, and then specifying the directions of reflection on ∂S.

The strip S is going to be split into an infinite sequence of 'boxes' by the reals $\beta_0 = 0 > \alpha_0 > \beta_1 > \alpha_1 > \cdots$; in the boxes $\{(x + iy) : \alpha_n < x \le \beta_n, \quad 0 \le y \le \xi\}$ the reflection at the upper and lower edges will be normal, and in the boxes $\{(x + iy) : \beta_n <$

$x \le \alpha_{n-1}, 0 \le y \le \xi\}$, the reflection at the upper and lower edges will be degenerate, pushing to the left.

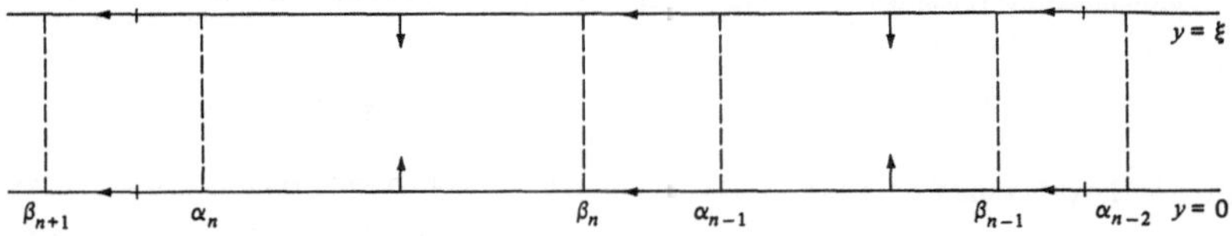

(It may seem improper to allow a tangential reflection. An example could doubtless be constructed where the reflection was extremely and increasingly close to tangential between β_n and α_{n-1}, and also varying smoothly, but this is a pointless embellishment; the real point is that the qualitative behaviour described is possible.) We refer to boxes of the first kind as 'pools', and boxes of the second kind as 'valves'. We envisage that the lengths of the boxes are going to infinity, but that $\beta_n - \alpha_n \gg \alpha_{n-1} - \beta_n$; a pool is *much* longer than the valve immediately to the right. The name 'valve' is very appropriate, because if one considers Brownian motion in a valve, as soon as it touches one of the horizontal sides it is immediately swept to the left-hand end. Thus it is very easy for Brownian motion to pass through a valve from right to left, but almost impossible for it to pass through from left to right. The purpose of 'pools' is to provide a place for Brownian motion to swim around and spend time, so that the total time for the skew-reflecting Brownian motion in D to reach 0 will accumulate to $+\infty$. This suggests that one makes the length $\beta_n - \alpha_n$ of the pool very large; but not so large that the probability that the process escapes through the valve to the right before it escapes through the valve to the left becomes too big. It is not clear that these conflicting aims can be achieved simultaneously; but, if $\xi < \pi/2$, we shall see that they can.

We begin with the analysis of valves. We shall consider a valve of length $l \equiv \alpha_{n-1} - \beta_n$, and estimate the rate of excursions from the left-hand end of the valve which reach the right-hand end. For simplicity, let us suppose that the left-hand end of the box is on the imaginary axis. It is easy to see that for $0 \le y \le \xi$, $\quad x > 0$ small

$$P^{x+iy} \text{ (reach right-hand end of valve before any other edge)}$$

$$\le P^{x+i\xi/2} \text{ (reach right-hand end of valve before any other edge),}$$

and it is not too hard to persuade oneself that this last quantity varies proportional to $e^{-\pi l/\xi}\sinh(\pi x/\xi)$, by considering the harmonic function $h(x,y) = \sinh(\pi x/\xi)\sin(\pi y/\xi)$ which vanishes on three sides of the valve. An application of the optional sampling theorem gives

$$E^{x+i\xi/2}(\sin(\pi Y_\tau/\xi); \quad X_\tau = l) = \sinh(\pi x/\xi)/\sinh(\pi l/\xi)$$

where τ is the exit time from the valve. This argument is not conclusive, but a firm proof can be given proceeding via another route. One firstly transforms $\{x+iy : x \geq 0, \ 0 \leq y \leq \xi\}$ to $\overline{\mathbf{H}}$ and then observes what has happened to the edge Re $z = l$. This has been mapped to a somewhat complicated curve in $\mathbf{H}$, but this curve is contained between two concentric circular arcs with centre on $\mathbf{R}$. The probability of leaving the valve firstly through the right-hand side is at most the probability that Brownian motion in $\mathbf{H}$ hits the outer of the two circular arcs before it hits $\mathbf{R}$, and this is a straightforward matter to compute. (The Brownian motion starts at a point exterior to both circular arcs). We now deduce an upper bound for the measure of excursions starting at the left-hand end of the valve and reaching the right end; we divide by $2x$ (the measure of Brownian excursions from $0 \in \mathbf{R}$ which get above x) and let $x \downarrow 0$. The conclusion is that the measure of excursions from the left end of the box which get through to the right end is at most

$$\frac{25\pi}{2\xi}e^{-\pi l/\xi} \equiv ce^{-\pi l/\xi}. \tag{25}$$

The other part of the analysis of valves is concerned with showing that if one starts at the right end of the valve, the measure of excursions which get through to the left end is bounded below. For simplicity, suppose the right end is on the imaginary axis. The ingredients of the argument are similar; the worst possible case is when one starts on the line $y = \xi/2$, and now the probability of leaving the valve anywhere but the right end is at least the probability of leaving the semi-infinite strip $\{x+iy : x \geq 0, \ 0 \leq y \leq \xi\}$ anywhere but the right end. This is easier to deal with than the other case, and leads to a lower bound for the measure of excursions from the right end of the valve which get through to the left end; it must be at least $1/\xi$.

Thus if we define

$$p_n \equiv \sup_{0\leq y\leq \xi} P^{\beta_n+iy}(X \quad \text{reaches } \beta_{n-1} \text{ before } \beta_{n+1}),$$
$$p'_n \equiv \sup_{0\leq y\leq \xi} P^{\alpha_n+iy}(X \quad \text{reaches } \beta_n \text{ before } \beta_{n+1}),$$
$$s_{n-1} \equiv \beta_{n-1} - \alpha_{n-1} \quad , \quad l \equiv \alpha_{n-1} - \beta_n$$

(where $Z_t \equiv X_t + iY_t$ is the skew-reflecting Brownian motion in S), we can estimate

$$p'_n \le \frac{1}{2s_n}\left(\frac{1}{2s_n}+\frac{1}{\xi}\right)^{-1} = \xi(\xi+2s_n)^{-1} \tag{26}$$

because the rate of excursions of X from $x = \alpha_n$ which get to $x = \beta_n$ before returning is $1/(2s_n)$, and the rate of excursions of X from $x = \alpha_n$ which escape through to $x = \beta_{n+1}$ before returning to $x = \alpha_n$ is, as we have just seen, at least $\pi/2\xi$.

By considering the situation when X starts at β_n, and decomposing at the times of reaching the ends of boxes, one estimates similarly

$$p_n \le \frac{ce^{-\pi l/\xi}(p'_{n-1}+(1-p'_n)p_n)+\frac{1}{2s_n}p'_np_n}{ce^{-\pi l/\xi}+\frac{1}{2s_n}}. \tag{27}$$

Rearrangment, and use of the estimate (26), lead finally to the estimate

$$p_n \le \left(1+\frac{\xi+2s_{n-1}}{\xi+2s_n}\,\frac{e^{\pi l/\xi}}{c\xi}\right)^{-1} \tag{28}$$

We shall aim to choose the α_n, β_n in such a way that always $p_n \le (2+\delta)^{-1} < \frac{1}{2}$, because then if one looks at the sequence of valves which Z visits, this sequence goes almost surely to $-\infty$, by comparison with simple random walk. The estimate (28) ensures that this will happen whenever

$$\frac{\xi+2s_{n-1}}{\xi+2s_n}\,\frac{e^{\pi(\alpha_{n-1}-\beta_n)/\xi}}{c\xi} \ge 1+\delta \qquad \text{for all } n. \tag{29}$$

We turn now to the analysis of pools, which is considerably simpler. The pools are the places where the process accumulates 'time' but 'time' measured by the additive functional $\int_0^t \exp(2X_s)ds$, which is the clock used to transform the time-scale into the time-scale for Brownian motion in D. If we take the nth pool, abbreviating α_n, β_n to α, β, we shall obtain

$$E^{\beta+iy}\left[\exp(-\tfrac{1}{2}\int_0^{\tau_\alpha} e^{2X_s}I_{(\alpha,\beta]}(X_s)ds)\right]$$

where $\tau_\alpha \equiv \inf\{t : X_t = \alpha\}$. This is a simple problem in one-dimensional diffusion theory; it is not hard to prove that

$$E^{\beta+iy}\left[\exp(-\tfrac{1}{2}\int_0^{\tau_\alpha} e^{2X_s}I(\alpha,\beta](X_s)ds)\right]$$

$$= \frac{I_0'(e^\beta)K_0(e^\beta)-K_0'(e^\beta)I_0(e^\beta)}{I_0'(e^\beta)K_0(e^\alpha)-K_0'(e^\beta)I_0(e^\alpha)}. \tag{30}$$

The numerator of (30) is $e^{-\beta}$. Using the inequalities $I_0'(x) \geq x/2$, $K_0(x) \geq -\gamma_1 \log x$, $-K_0'(x) \geq (1-\gamma_2 x)/x$, which, for appropriate constants γ_1, γ_2, are valid for all $x \in (0,1]$, we deduce that (30) is at most

$$(1 - \gamma_2 e^{\beta} - \tfrac{1}{2}\gamma_1 \alpha e^{2\beta})^{-1}. \tag{31}$$

We shall pick $\beta_0 = 0$ and insist that for all n, $\alpha_n = -e^{-\beta_n}$. This will ensure the upper bound

$$E^{\beta_n + iy} \exp(-\tfrac{1}{2} \int_0^{\tau(\alpha_n)} e^{2X_s} I_{\alpha_n, \beta_n]}(X_s) ds) \leq (1 + \gamma_1/3)^{-1} \tag{32}$$

for all large enough n. Now the total 'time' taken to reach 0, $\int_0^\infty \exp(2X_s)ds$, is at least the sum of the contributions between first hitting β_n and first hitting α_n. Thus

$$\begin{aligned} E \exp(-\tfrac{1}{2} \int_0^\infty e^{2X_s} ds) &\leq \prod_{n=0}^{\infty} E^{\beta_n + i\xi/2} [\exp(-\tfrac{1}{2} \int_0^{\tau(\alpha_n)} e^{2X_s} I_{(\alpha_n, \beta_n]}(X_s) ds)] \\ &= 0 \end{aligned}$$

using (32). This relationship between α_n, β_n will ensure that a.s. the 'time' taken to reach 0 is infinite, but can we pick the β_n in such a way that (29) holds?

If we have determined $\alpha_{n-1}, \beta_{n-1}, \ldots, \alpha_0, \beta_0$, we determine β_n so as to satisfy (29); we want

$$\frac{e^{\pi(\alpha_{n-1} - \beta_n)/\xi}}{\xi + 2(\beta_n - \alpha_n)} \geq \frac{c\xi(1+\delta)}{\xi + 2(\beta_{n-1} - \alpha_{n-1})},$$

but, since $\alpha_n = -e^{-2\beta_n}$, this is the same as

$$\frac{e^{\pi(\alpha_{n-1} - \beta_n)/\xi}}{\xi + 2(\beta_n + e^{-2\beta_n})} \geq \frac{c\xi(1+\delta)}{\xi + 2(\beta_{n-1} - \alpha_{n-1})}$$

and since $\pi/\xi > 2$ by assumption, we can always achieve this inequality by letting β_n go far enough to the left.

The proof of Theorem 3 is complete.

The last item to deal with now is the proof of Theorem 4. For this, we return to the analysis of §3, and study the Green's function more closely, assuming that one-sided escape to ∞ is impossible (i.e that neither of (7.i), (7.ii) holds) and that the process is transient, that is

$$c_1 = 0, \quad \int \frac{\operatorname{Im} i\psi(x)}{1+|x|} dx < \infty. \tag{33}$$

Now consider the expression (24) for $\text{Im}\,\tilde{f}$, which simplifies here to

$$\text{Im}\,\tilde{f}(z) = \int \frac{\text{Im}\,i\tilde{\psi}(x)}{\pi(1+x^2)}(\alpha+\beta x)\tan^{-1}\left(\frac{x-a}{b}\right)dx + \quad \text{const.,} \tag{34}$$

in view of (33). If we consider this function for $a \in \mathbf{R}$, the limit of $\text{Im}(\tilde{f}(a) - \tilde{f}(-a))$ as $a \to \infty$ is

$$\begin{aligned} -\pi \int \frac{\text{Im}\,i\tilde{\psi}(x)}{\pi(1+x^2)}(\alpha+\beta x)dx &= -\pi\{\alpha\ \text{Im}\,i\tilde{\psi}(i) + \beta \int \frac{\text{Im}\,i\tilde{\psi}(x)x\,dx}{\pi(1+x^2)}\} \\ &= -\pi\{\alpha\ \text{Im}\,i\tilde{\psi}(i) + \beta\ \text{Re}\,(i\tilde{\psi}(i) - c_2)\}, \end{aligned} \tag{35}$$

where we refer to the Pick function representation (5) of $i\tilde{\psi}$;

$$i\tilde{\psi}(z) = \int \frac{\text{Im}\,i\tilde{\psi}(x)dx}{\pi(x-z)} + c_2.$$

Now $\alpha + i\beta = -1/i\tilde{\psi}(i)$ by definition, so (35) simplifies to $\beta\pi c_2$, and $\beta > 0$. But recall that one-sided escape to ∞ is ruled out, so that the process Z must keep on visiting $(0,\infty)$ and $(-\infty, 0)$ arbitrarily far from 0. Moreover, $h^*(z) = \text{Im}\,\tilde{f}(z) + \frac{1}{2}\log|z+i| - \frac{1}{2}\log|z-i|$ is harmonic in $\mathsf{H}\backslash\{i\}$, satisfies the boundary condition on $\mathbf{R}$ and is bounded below, so $h^*(Z_t)$ converges almost surely. But this cannot happen if $\lim_{a\to\infty} \text{Im}\,(\tilde{f}(a) - \tilde{f}(-a)) \neq 0$; so the conclusion is that c_2 *must be zero*, and

$$i\tilde{\psi}(z) = \int \frac{\text{Im}\,i\tilde{\psi}(x)}{\pi(x-z)}dx. \tag{36}$$

We are now going to return to (17) and derive another expression for the difference $h(z,z_0) \equiv G(z,z_0) - G^\theta(z,z_0)$ of the Green's function and the Green's function for the killed process. Holding $z_0 \equiv x_0 + iy_0$ fixed for now, we shall build f to satisfy (17) by defining

$$\begin{aligned} f'(z) &\equiv i\psi(z) \int \frac{\text{Im}\,(-1/i\psi(x))}{\pi(x-z)} \frac{2y_0}{\pi|x-z_0|^2}dx \\ &= \int \frac{\text{Im}\,i\psi(v)dv}{\pi(v-z)} \int \frac{\text{Im}\,(-1/i\psi(x))}{\pi(x-z)} \frac{2y_0}{\pi|x-z_0|^2}dx \end{aligned} \tag{37}$$

so that

$$f(z) - f(i) = 2y_0 \int\int \frac{\text{Im}\,i\psi(v)\text{Im}\,(-1/i\psi(x))}{\pi^3|x-z_0|^2(v-x)}\Big[\log(\omega - v) - \log(\omega - x)\Big]_i^z dv\,dx$$

and hence

$$\begin{aligned} \text{Im}\,(f(z) - f(i)) = \int\int \frac{\text{Im}\,i\psi(v)\text{Im}\,(-1/i\psi(x))2y_0}{\pi^3|x-z_0|^2(v-x)}dx\,dv\{&\tan^{-1}\left(\frac{v-a}{b}\right) - \tan^{-1}\left(\frac{x-a}{b}\right) \\ &- \tan^{-1}(v) + \tan^{-1}(x)\}. \end{aligned}$$

The arbitrary constant in the definition of f is determined in such a way that

$$\text{(38)} \qquad \operatorname{Im} f(z) = \int\int \operatorname{Im} i\psi(v) \operatorname{Im}(-1/i\psi(x)) \frac{2y_0 dv\, dx}{\pi^3|x-z_0|^2} \frac{\tan^{-1}(\frac{v-a}{b}) - \tan^{-1}(\frac{x-a}{b})}{v-x},$$

the integral being absolutely convergent in view of the integrability condition (33), the integrability of $(1+x^2)^{-1}\operatorname{Im}(-1/i\psi(x))$, and the elementary estimate;

$$\text{(39)} \qquad 0 \le \frac{\tan^{-1}(v) - \tan^{-1}(x)}{v-x} \le c.(1+|x|+|v|)^{-1}.$$

Now, comparing the definitions (20) and (37) of $\tilde{f}$ and f, we see that $\tilde{f}' = \frac{\pi}{2} y_0 f'(y_0 z + x_0)$, implying that $\tilde{f}(z) - \frac{\pi}{2} f(y_0 z + x_0)$ is constant. It follows from (34), (38) that

$$\int \frac{\operatorname{Im} i\tilde{\psi}(x)}{\pi(1+x^2)}(\alpha + \beta x)\frac{2}{\pi}\tan^{-1}(\frac{x-a}{b})\,dx$$

$$\text{(40)} \qquad = \int\int \operatorname{Im} i\psi(v) \operatorname{Im}(-1/i\psi(x)) \frac{2y_0 dv\, dx}{\pi^3|x-z_0|^2} \frac{\tan^{-1}(\frac{v-a}{b}) - \tan^{-1}(\frac{x-a}{b})}{v-x},$$

the two sides differing by a constant, which is seen to be 0 if we let $b \to \infty$, while $a = 0$. The right-hand side is evidently non-negative, the left-hand side is evidently a bounded harmonic function, Now, as we have seen, $\operatorname{Im} \tilde{f}(z) + \frac{\pi}{2} G^\vartheta(z, z_0)$ is harmonic, satisfies the boundary condition, and is bounded below; from (40) therefore

$$G(z,z_0) - G^\vartheta(z,z_0) = \int\int \operatorname{Im} i\psi(v) \operatorname{Im}(-1/i\psi(x)) \frac{2y_0\, dv\, dx}{\pi^3|x-z_0|^2} \frac{\tan^{-1}(\frac{v-a}{b}) - \tan^{-1}(\frac{x-a}{b})}{v-x}.$$

(41)

If we specialize to $z = i$, we have that

$$G(i,z_0) - G^\vartheta(i,z_0) = \int\int \frac{2y_0 \operatorname{Im} i\psi(v) \operatorname{Im}(-1/i\psi(x))}{\pi^3|x-z_0|^2} dv\, dx \frac{\tan^{-1}(v) - \tan^{-1}(x)}{v-x},$$

and hence $E^i[\int_\tau^\infty \phi(z_s)ds] < \infty$ if and only if

$$\int\int \operatorname{Im} i\psi(v) \operatorname{Im}(-1/i\psi(x)) dv\, dx \frac{\tan^{-1}(v) - \tan^{-1}(x)}{v-x} \int\int \frac{y_0}{|x-z_0|^2}\phi(z_0)dx_0 dy_0 < \infty.$$

(42)

The case of major interest to us is the case where $\phi(z) = |z|^{-2-2\xi/\pi} I_{(|z|\ge 1)}$, as was explained in the Introduction; the finiteness of the integral (42) in this case will guarantee that the process in the wedge will reach the vertex in finite time. It is straightforward to calculate for $v > 0, \eta > 0$, that

$$\int_{|z|\geq 1} |z|^{-2-\eta} \operatorname{Im}(\frac{1}{v-z})dx\, dy$$
$$= 2v^{-1-\eta} \int_{1/v}^{\infty} (\log|1+y| - \log|1-y|)y^{-1-\eta}\, dy,$$

and as $v \to \infty$ this behaves like $(1 + v^{1+\eta})^{-1}$ for $\eta < 1$, and like $(1 + v^2)^{-1}$ for $\eta > 1$. The case we are particularly concerned with corresponds to $\eta = 2\xi/\pi$ (<1 by assumption), and so the finiteness condition (42) becomes the condition stated in Theorem 4, concluding the proof.

Appendix: (i) A skew-reflecting Brownian mapping theorem

Let D_1, D_2 be two open simply-connected domains in $\mathbf{C}$, with closures $\overline{D}_1, \overline{D}_2$, and suppose that $f : \overline{D}_1 \to \overline{D}_2$ is a C^2 diffeomorphism, whose restriction to D_1 is analytic. Let v be a vector field defined on ∂D_1, everywhere inward-pointing and non-vanishing, and let $\tilde{v}$ be the corresponding vector field on ∂D_2, defined by $\tilde{v}(f(z)) = f'(z)v(z), z \in \partial D_1$. Suppose that Z is a skew-reflecting Brownian motion in $\overline{D}_1$ with the vector field v specifying the directions of reflection;

$$Z_t = Z_0 + W_t + \int_0^t v(Z_s)dL_s$$

where W is a standard Brownian motion in $\mathbf{C}$, L is a continuous increasing process satisfying $\int_0^t I_{D_1}(Z_s)dL_s \equiv 0$, and Z spends no time on $\partial D_1(\int_0^t I_{\partial D_1}(Z_s)ds = 0)$.

Then $f(Z_t) \equiv \tilde{Z}_t$ is a skew-reflecting Brownian motion in $\overline{D}_2$, run with the clock $\int_0^t |f'(Z_s)|^2 ds$.

Proof. Applying Itô's formula to $f(Z_t)$, we obtain

$$d\tilde{Z}_t \equiv df(Z_t) = f'(Z_t)dW_t + \tilde{v}(\tilde{Z}_t)dL_t.$$

Thus if $A_t \equiv \int_0^t |f'(Z_s)|^2 ds$, (τ_t) is the inverse to A, we deduce that

$$\zeta_t \equiv \tilde{Z}_{\tau_t} = \tilde{Z}_0 + \beta_t + \int_0^t \tilde{v}(\zeta_s)d\lambda_s,$$

where $\lambda_t \equiv L_{\tau_t}$, and β is a complex Brownian motion. Moreover, $\int_0^t I_{D_2}(\zeta_s)d\lambda_s = \int_0^{\tau_t} I_{D_1}(Z_s)dL_s = 0$, and $\int_0^t I_{\partial D_2}(\zeta_s)ds = \int_0^{\tau_t} I_{\partial D_1}(Z_u)|f'(Z_u)|^2 du = 0$, which establishes that ζ is skew-reflecting Brownian motion in $\overline{D}_2$.

(ii) Equivalence of the conditions of Burdzy & Marshall and Theorem 2

Burdzy and Marshall work with a Brownian motion for which the singular point is 0 not ∞, so we must transform the problem we have been working with so that ∞ becomes 0; this is easily achieved by the mapping $z \mapsto -1/z$, which converts ψ to $\tilde{\psi}$ defined by $\tilde{\psi}(z) = \psi(-1/z)$. The criterion of [1] for the Brownian motion to reach 0 with positive probability is that

$$\int_0^1 \frac{dt}{t} \mathrm{Im} i\tilde{\psi}(it) < \infty$$

Now $c_1 = \lim_{b\to\infty} i\psi(ib)/ib = \lim_{t\downarrow 0} -it(-i\tilde{\psi}(it))$, so in the case where $c_1 > 0$, we have that

$$\int_0^1 \frac{dt}{t} \mathrm{Im} i\tilde{\psi}(it) \sim \int_0^1 \frac{dt}{t}\frac{c_1}{t} = \infty,$$

so that when $c_1 > 0$, the [1] criterion for hitting the point fails, as it must. When $c_1 = 0$, the Pick function $i\tilde{\psi}$ has a representing measure which has no mass at zero, and is simply $\mathrm{Im} i\tilde{\psi}(x)dx$. Let the constants in the representation (5) of $i\tilde{\psi}$ be denoted $\tilde{c}_1$, $\tilde{c}_2$, so that

$$\begin{aligned}\int_0^1 \frac{dt}{t} \mathrm{Im} i\tilde{\psi}(it) &= \int_0^1 \frac{dt}{t} \left(\int_{-\infty}^{\infty} \frac{\mathrm{Im} i\tilde{\psi}(x)}{\pi} \frac{tdx}{x^2+t^2} + \tilde{c}_1 t\right) \\ &= \tilde{c}_1 + \int_{-\infty}^{\infty} \frac{\mathrm{Im} i\tilde{\psi}(x)}{\pi} \frac{dx}{x} \tan^{-1}(1/x) \\ &\sim \tilde{c}_1 + \int_{-\infty}^{\infty} \frac{\mathrm{Im} i\tilde{\psi}(x)}{\pi} \frac{dx}{|x|(1+|x|)}\end{aligned}$$

which is divergent if and only if (8) holds.

REFERENCES

[1] BURDZY, K., and MARSHALL, D. Hitting a boundary point with reflected Brownian motion. Preprint, 1990.

[2] DONOGHUE, W.F., *Monotone matrix functions and analytic continuation*, Springer, Berlin, 1974.

[3] DYNKIN, E.B., Martin boundaries and non-negative solutions of a boundary value problem with a directional derivative. *Uspeki Mat. Nauk*, **19**, 3-50, 1964.

[4] MALYUTOV, M.B., Brownian motion with reflection and the inclined derivative problem. *Soviet Math. Doklady*, **5**, 822-825, 1964.

[5] ROGERS, L.C.G., A guided tour through excursions. *Bull. London Math. Soc.*, **21**, 305-341, 1989.

[6] ROGERS, L.C.G., Brownian motion in a wedge with variable skew reflection. To appear in *Trans. Amer. Math. Soc.*, 1990.

[7] VARADHAN, S.R.S., and WILLIAMS, R.J., Brownian motion in a wedge with oblique reflection. *Comm. Pure Appl. Math.* **38**, 405-443, 1985.

L. C. G. Rogers
Statistical Laboratory
University of Cambridge
16 Mill Lane
Cambridge CB2 1SB
Great Britain

A Generalised Arc-Sine Law and Nelson's Stochastic Mechanics of One-Dimensional Time-Homogeneous Diffusions

Aubrey Truman and David Williams

Abstract

A generalised arc-sine law is established for one-dimensional time-homogeneous diffusions by considering the quantum Hamiltonian for the corresponding Nelson stochastic mechanics. Moreover our methods also give elementary proofs of some variants of Lévy's formulae by establishing some new relations with an associated Dirichlet quantum mechanical Hamiltonian.

Introduction

In this paper we present some new results for one-dimensional time-homogeneous diffusions, including the diffusions which arise in Nelson's stochastic mechanics from stationary states. One of these results is a generalised arc-sine law for sufficiently well-behaved time-homogeneous diffusions X on the real line satisfying

$$dX(t) = b(X(t))dt + dB(t) \quad , \tag{0}$$

B being a $BM(\mathbb{R})$ process, the boundary being inaccessible for X.

Let $\tau_x(a) = \inf\{s>0 : X(s) = a \,|\, X(o) = x\}$ and set $L^{\pm}(t) = \mathrm{Leb}\{s\,\varepsilon[o,t] : X(s) \gtrless a\}$. Then for the time-homogeneous one-dimensional diffusions considered here we show that for any constants $\alpha, \lambda_+, \lambda_- > 0$

$$\int_0^\infty e^{-\alpha t}\, \mathbb{E}_a\{e^{-\lambda_+ L^+(t) - \lambda_- L^-(t)}\}dt = \frac{\mathrm{Disc}\,|_a (\alpha+\lambda)^{-1} \frac{d}{dx}\, \mathbb{E}\{e^{-(\alpha+\lambda)\tau_x(a)}\}}{\mathrm{Disc}\,|_a \frac{d}{dx}\, \mathbb{E}\{e^{-(\alpha+\lambda)\tau_x(a)}\}} , \tag{1}$$

where $\lambda = \lambda(x) = \lambda_{\pm}$ for $x \gtrless a$ and $\mathrm{Disc}\,|_a f(x) = f(a+) - f(a-)$.

For Brownian motion for instance $\mathbb{E}\{e^{-\lambda\tau_x(a)}\} = e^{-\sqrt{2\lambda}\,|x-a|}$, for each $\lambda > 0$, and setting $\lambda_+ = 0$, $\lambda_- = \lambda$ in Eq.(1), gives

$$\int_0^\infty e^{-\alpha t}\, \mathbb{E}_a\{e^{-\lambda L^-(t)}\}dt = \frac{1}{[(\alpha+\lambda)\alpha]^{1/2}} = \int_0^\infty \frac{e^{-\alpha t}}{\pi} \left\{\int_0^t \frac{e^{-\lambda\theta}d\theta}{[\theta(t-\theta)]^{1/2}}\right\} dt, \tag{2}$$

so that in this case

$$\mathbf{P}_0 \,[L^-(t) \le \theta] = \int_0^{\theta/t} \frac{ds}{\pi[(1-s)s]^{1/2}} = \frac{2}{\pi} \text{ arc-sine} \left(\left(\frac{\theta}{t}\right)^{1/2}\right) , \tag{3}$$

for $0 \le \theta \le t$, the usual arc-sine law. (See Karatzas and Shreve Ref.(1)).

The above follows from a special case of the Batchelor-Truman formula (see Ref.(2)) and some variants of some little known formulae of Lévy established here. Let $L^a(t)$ be the local time at a upto time t, $L^a(t) = \int_0^t \delta(X(s) - a)ds$, a being a regular point for the diffusion X and set $\gamma^a(t) = \inf\{u > 0 : L^a(u) > t\}$.

We shall show that

$$\mathbb{E}_a \left\{ e^{-\lambda L^{\pm}(\gamma^a(t))} \right\} = \exp \left\{ -t \int_0^{\infty} (1-e^{-\lambda s})\, d\nu^{\pm}(s) \right\} , \tag{4}$$

where the Poisson-Lévy excursion measures are given by

$$d\nu^{\pm}(s) = -d_s \int_{x \gtrless a} dx\, \rho_0(x)(\rho_0(a))^{-1}\, \mathbf{P}\,(\tau_x(a)\, \varepsilon\, ds) , \tag{5}$$

ρ_0 being the invariant density for the diffusion X and $\rho_0(a) \neq 0$ by assumption. To prove our results we show that X can be regarded as a ground-state Nelson diffusion process corresponding to the potential $V(x) = 2^{-1}(b'(x) + b^2(x))$. A simple study of a Dirichlet Hamiltonian associated with the potential V leads to our results. We have not striven to establish the results in maximum generality. Rather we have restricted the class of diffusions being considered to simplify our arguments and to focus attention on the main ideas. We hope to publish a paper in the near future extending our results to a wider class of diffusions.

Introduction to Nelson's Stochastic Mechanics

Let $\Psi = \Psi(x,t)$ for $(x,t)\ \varepsilon\ \mathbb{R} \times \mathbb{R}^+$ be a classical solution of the Schrödinger equation for a unit mass quantum particle Q moving in one dimension in the force field $-V'(x)$,

$$i\, \Psi^* \frac{\partial \Psi}{\partial t} = -2^{-1}\, \Psi^* \frac{\partial^2 \Psi}{\partial x^2} + V|\Psi|^2 , \tag{6}$$

Ψ^* being the complex conjugate of Ψ.

Setting $\Psi = e^{R+iS}$ for real-valued R and S, equating real and imaginary parts of the last equation gives

$$\frac{\partial R}{\partial t} = -\left(\frac{\partial R}{\partial x}\frac{\partial S}{\partial x} + 2^{-1}\frac{\partial^2 S}{dx^2}\right) , \tag{7}$$

and

$$\frac{\partial S}{\partial t} = 2^{-1}\left(\left(\frac{\partial R}{\partial x}\right)^2 - \left(\frac{\partial S}{\partial x}\right)^2 + \frac{\partial^2 R}{dx^2}\right) - V \ . \tag{8}$$

The first of these equations expresses conservation of particle number in that

$$\mathbf{P}\,(Q \varepsilon A \text{ at time } t) = \int_A |\Psi(x,t)|^2\, dx = \int_A \rho(x,t)\, dx \ ,$$

where $\rho = |\Psi|^2 = e^{2R}$ is the quantum mechanical particle density and Eq.(7) can be rewritten

$$\frac{\partial \rho}{\partial t} = \frac{\partial}{\partial x}\left(-\rho\frac{\partial S}{\partial x}\right) = \frac{\partial}{\partial x}\left(2^{-1}\frac{\partial \rho}{\partial x} - b\rho\right) \ , \tag{9}$$

for $b = b(x,t)$ given by $b = \frac{\partial}{\partial x}(R+S)$.

The last equation is just the forward Kolmogorov equation for the diffusion

$$dX(t) = b(X(t),t)\, dt + dB(t) \ , \tag{10}$$

where the forward drift $b = \frac{\partial}{\partial x}(R+S)$, B being a BM ($\mathbb{R}$) process with

$$\mathbb{E}\,(B(t)\,B(s)) = \min\,(s,t) \ . \tag{11}$$

Following Nelson, define the mean forward and backward time derivatives $D\pm$ by

$$D\pm f(X(t),t) = \lim_{h\downarrow 0} \mathbb{E}\left\{\frac{f(X(t\pm h),t\pm h) - f(X(t),t)}{\pm h}\,\middle|\, X(t)\right\} \ , \tag{12}$$

so that

$$D_+ X(t) = b(X(t),t) = \frac{\partial}{\partial x}(S+R)\,(X(t),t) \ . \tag{13}$$

Using Itô's formula, Nelson showed that for sufficiently well-behaved f and g, we have

$$D_+ f(X(t),t) = \left(\frac{\partial f}{\partial t} + b\frac{\partial f}{\partial x} + 2^{-1}\frac{\partial^2 f}{\partial x^2}\right)(X(t),t)$$

and the product rule

$$\frac{d}{dt}\,\mathbb{E}\{f\,g\} = \mathbb{E}\{(D_+f)\,g + f(D_-g)\} \ .$$

Nelson deduced that for a diffusion with density ρ

$$D_-X(t) = \left(b - \frac{\partial}{\partial x}\ln\rho\right)(X(t),t) \overset{\text{def}}{=} b_*(X(t),t) \ , \tag{14}$$

giving in our case

$$D_-X(t) = (S-R)(X(t),t) \tag{15}$$

and

$$D_- f(X(t),t) = \left(\frac{\partial f}{\partial t} + b_* \frac{\partial f}{\partial x} - 2^{-1}\frac{\partial^2 f}{\partial x^2}\right)(X(t),t) \quad . \tag{16}$$

It follows that

$$\begin{aligned} 2^{-1}(D_+D_- + D_-D_+)\,X(t) &= 2^{-1}\left\{D_+\frac{\partial}{\partial x}(S-R) + D_-\frac{\partial}{\partial x}(S+R)\right\} \\ &= \frac{\partial}{\partial x}\left\{\frac{\partial S}{\partial t} - 2^{-1}\left(\left(\frac{\partial R}{\partial x}\right)^2 - \left(\frac{\partial S}{\partial x}\right)^2 + \frac{\partial^2 R}{dx^2}\right)\right\}(X(t),t) \\ &= -V'(X(t)) \ , \end{aligned} \tag{17}$$

by virtue of Eq.(8) . Therefore, we see that the Schrödinger equation embodies a dynamical principle for the diffusion X, a stochastic version of Newton's 2nd Law of Motion. This is called the Nelson-Newton law. (See Refs. (3) and (4)).

Now specialize to stationary states $\Psi_E(x,t) = f_E(x)\, e^{-iEt}$, where f_E is the ground-state eigenfunction of the Hamiltonian $H = -2^{-1}\frac{d^2}{dx^2} + V(x)$, corresponding to the eigenvalue $E = \inf \operatorname{spec}(H)$,

$$(H-E)\, f_E = 0 \ , \tag{18}$$

$f_E \in L^2(\mathbb{R})$, with $\|f_E\|_{L^2} = 1$. We assume that V is continuous and for simplicity that $\lim_{x\to\pm\infty} V(x)$ exists. Then H is the ordinary differential operator

$H = -2^{-1}\frac{d^2}{dx^2} + V(x)$ limit point at $\pm\infty$, generating the semigroup exp(-tH) given by a Feynman-Kac formula (see e.g. p.150 in Ref.(5) and Ref.(6)). Moreover, $f_E \in C^2(\mathbb{R})$ with $f_E > 0$ and f_E'/f_E bounded. Therefore, in this case

$$dX(t) = \frac{d}{dX(t)} \ln f_E(X(t))\, dt + dB(t) = b(X(t))\, dt + dB(t)) \ , \tag{19}$$

B being a BM($\mathbb{R}$) process above, b being a bounded drift i.e. X is a one-dimensional time-homogeneous process with generator,

$$L_x = 2^{-1}\frac{d^2}{dx^2} + \frac{d}{dx}\ln f_E(x)\frac{d}{dx} = -f_E^{-1}(x)\,(H-E)\,f_E(x) \ , \tag{20}$$

the last identity following from Eq.(18). Since the transition density $p_t(x,y)\,dy = \mathbf{P}(X(t) \in dy \mid X(o) = x)$ satisfies

$$\frac{\partial p_t}{\partial t}(x,y) = L_y^*\, p_t(x,y) \ , \tag{21}$$

with

$$\lim_{t\downarrow 0} p_t(x,y) = \delta(x-y) \ , \tag{22}$$

and $L_y^* = -f_E(y)\left(-2^{-1}\frac{d^2}{dy^2} + V(y) - E\right) f_E^{-1}(y)$, we expect that

$$p_t(x,y) = f_E^{-1}(x) \exp\left\{-t(H-E)\right\}(x,y) f_E(y) \quad , \tag{23}$$

with invariant density ρ_0 given by

$$\rho_0(x) = f_E^2(x) \quad . \tag{24}$$

Now let

$$\tau_x(a) = \inf\{s > 0 : X(s) = a \,|\, X(0) = x\} \quad . \tag{25}$$

Then, for any intermediate point a between x and y, for the diffusing particle to have gone from x to y in time t the first hitting time of a must be less than t. Therefore, because the diffusion starts afresh at each time u,

$$p_t(x,y) = \int_0^t \mathbf{P}(\tau_x(a) \in du)\, p_{t-u}(a,y) \quad . \tag{26}$$

Taking Laplace transforms, since the continuity of V ensures that $p_s(x,y)$ is continuous in x and y, letting $y \to a$, we obtain for each $\lambda > 0$

$$\mathbb{E}\left\{e^{-\lambda\tau_x(a)}\right\} = \tilde{p}_\lambda(x,a) / \tilde{p}_\lambda(a,a) \quad , \tag{27}$$

where $\tilde{p}_\lambda(x,a) = \int_0^\infty e^{-\lambda s} p_s(x,a)\, ds$. Therefore, if Eq.(23) is valid, for the ground-state Nelson diffusion the first hitting time $\tau_x(a)$ will satisfy

$$\mathbb{E}\left\{e^{-\lambda\tau_x(a)}\right\} = \frac{f_E^{-1}(x)(H+\lambda-E)^{-1}(x,a) f_E(a)}{(H+\lambda-E)^{-1}(a,a)} \quad , \tag{28}$$

for each $\lambda > 0$, $(H+\lambda-E)^{-1}(x,a)$ being the resolvent kernel given by $\int_0^\infty e^{-\lambda t} \exp(-t(H-E))(x,a)\, dt$. We now prove Eq.(23).

Let f be any bounded measurable function. Then by the Girsanov-Cameron-Martin theorem, because $b = f_E'/f_E$ is bounded (see e.g. Ref(1) or (8)),

$$\mathbb{E}_x\left\{f(X(t))\right\} = \mathbb{E}\left\{\exp\left\{-\tfrac{1}{2}\int_0^t b^2(x+B(s))\, ds + \int_0^t b(x+B(s))\, dB(s)\right\} f(x+B(t))\right\}. \tag{29}$$

Writing $d \ln f_E = (f_E'/f_E)\, dB(s) + 2^{-1}\left[(f_E f_E'' - f_E'^2)/f_E^2\right] ds$ and using

$$-2^{-1}\, f_E'' + (V - E)\, f_E = 0 \ , \tag{30}$$

we obtain

$$\mathbb{E}_x\left\{ f(X(t)) \right\} = e^{Et} f_E^{-1}(x)\, \mathbb{E}\left\{ \exp\left\{ -\int_0^t V(x+B(s))\, ds \right\} (f_E f)(x+B(t)) \right\} . \tag{31}$$

Therefore,

$$\mathbb{E}_x\left\{ f(X(t)) \right\} = e^{Et} f_E^{-1}(x)\, \mathbb{E}\left\{ \mathbb{E}\left\{ \exp\left(-\int_0^t V(x+B(s))\, ds \right) \Big| \, B(t) \right\} (f_E f)(x+B(t)) \right\} . \tag{32}$$

But $B(s) = \alpha_t(s) + \frac{s}{t} B(t)$, for $s \le t$, $\alpha_t(s)$ being the Brownian bridge on $[0,t]$ and, since $\mathbf{P}(x+B(t) \,\varepsilon\, dy) = (2\pi t)^{-1/2} \exp(-(x-y)^2/2t)\, dy$, we arrive at

$$\mathbb{E}_x\left\{ f(X(t)) \right\} = f_E^{-1}(x) \int_{-\infty}^{\infty} dy\, e^{Et}\, \mathbb{E}\left\{ \exp\left(-\int_0^t V\left(x + \tfrac{s}{t}(y-x) + \alpha_t(s)\right) ds \right) \right\}$$
$$(2\pi t)^{-1/2}\, e^{-\frac{(x-y)^2}{2t}}\, (f_E f)(y) \tag{33}$$

Therefore, since from Ref(6) for continuous bounded V,

$$\exp\left\{ -t(H-E) \right\}(x,y) = \mathbb{E}\left\{ \exp\left(-\int_0^t V\left(x + \tfrac{s}{t}(y-x) + t^{1/2} \alpha_1(\tfrac{s}{t})\right) ds \right\}$$
$$(2\pi t)^{-1/2}\, e^{-\frac{(x-y)^2}{2t} + Et} \ . \tag{34}$$

This proves the desired identity.

We summarise the above in:-

Proposition

Let the real valued potential $V \,\varepsilon\, C(\mathbb{R})$ be such that $\lim_{x \to \pm\infty} V(x)$ exists. Set the quantum mechanical Hamiltonian $H = -2^{-1} \frac{d^2}{dx^2} + V(x)$. If $p_t(x,y)$ is the transition density corresponding to the ground-state Nelson diffusion for the eigenfunction f_E with $(H - E) f_E = 0$, with eigenvalue $E = \inf \mathrm{spec}\,(H)$, then for $t > 0$,

$$p_t(x,y) = f_E^{-1}(x) \exp\left\{ -t(H-E) \right\}(x,y)\, f_E(y) \ , \tag{35}$$

with invariant measure $f_E^2(x)\, dx$. Also, for $\tau_x(a)$ the first hitting time of a starting from x, for each $\lambda > 0$,

$$\mathbb{E}\{e^{-\lambda\tau_x(a)}\} = \frac{f_E^{-1}(x)(H+\lambda-E)^{-1}(x,a)f_E(a)}{(H+\lambda-E)^{-1}(a,a)} \quad . \tag{36}$$

We see that the Hamiltonian H has an important role in giving the distribution of first hitting times. In the next section we identify the Hamiltonian for a one-dimensional time-homogeneous diffusion.

Stochastic Dynamics of One-Dimensional Time-Homogeneous Diffusions

Let X be the solution of the time-homogeneous diffusion in one-dimension

$$dX(t) = b(X(t))\,dt + dB(t) \quad , \tag{37}$$

where we assume that b is bounded and $b \in C^1(\mathbb{R})$, B being a BM ($\mathbb{R}$) process as above. Also, assume that the corresponding invariant density $\rho_0 \in L^1(\mathbb{R})$, where

$$\rho_0(x) = \exp\left\{2\int^x b(u)\,du\right\} / \int_{-\infty}^{\infty} dy \exp\left\{2\int^y b(u)\,du\right\} \tag{38}$$

after normalisation.

Then it is easy to check that $f_0(x) = \rho_0^{1/2}(x)$ satisfies

$$-2^{-1} f_0''(x) + V(x)\, f_0(x) = 0 \quad , \tag{39}$$

where the potential function $V(x) = 2^{-1}(b^2(x) + b'(x))$. Also, in the invariant density state

$$b_*(x) = b(x) - \frac{\partial}{\partial x} \ln \rho_0(x) = -b(x) \tag{40}$$

and so

$$2^{-1}(D_+D_- + D_-D_+)X(t) = 2^{-1}(D_- - D_+)b = -\,2^{-1}(b'' + 2b'b)(X(t)) = -V'(X(t)), \tag{41}$$

as expected.

From the above we expect that the formally symmetric Hamiltonian

$$H_0 = -2^{-1}\frac{d^2}{dx^2} + 2^{-1}(b^2(x) + b'(x)) \tag{42}$$

determines the distribution of first hitting times. Firstly the formulae in Eqs.(35) and (36) must be valid because of the operator identity

$$2^{-1}\frac{d^2}{dx^2} + b(x)\frac{d}{dx} = -f_0^{-1}(x)\left\{-2^{-1}\frac{d^2}{dx^2} + V(x)\right\} f_0(x) \quad . \tag{43}$$

The last equation follows from Eq.(39). Imitating the proof of the last proposition using Eq.(39) instead of Eq.(30) proves the analogue of Eqs.(35) and (36). We aim for a slightly different result here.

Proposition

Set $H_0 = -2^{-1}\frac{d^2}{dx^2} + 2^{-1}(b'(x) + b^2(x))$, $b \varepsilon C^1(\mathbb{R})$ being bounded, and define the Dirichlet Hamiltonians $H_\pm(a)$ by

$$H_\pm(a) = \lim_{\lambda\uparrow\infty}(H_0 + \lambda\,\chi_\mp(x)) \quad , \tag{44}$$

where $\chi_\mp(x)$ are the characteristic functions of $\{x : x \lesseqgtr a\}$ for each fixed $a \varepsilon \mathbb{R}$, respectively.

Then for the above diffusion for $x \gtreqless a$

$$\mathbb{P}(\tau_x(a) > t) = f_0^{-1}(x)\{\exp(-t\,H_\pm(a))\,f_0(x)\} \quad , \tag{45}$$

respectively, where by hypothesis $f_0 \varepsilon L^2(\mathbb{R})$ is given by $f_0 = \rho_0^{1/2}$, ρ_0 being the invariant density, $\rho_0 \varepsilon L^1(\mathbb{R})$.

Proof

For definiteness sake assume that $x > a$. Since $\int_0^t \chi_-(X_x(s))\,ds > 0$, if $X_x(s_0) < a$ for some $s_0 \varepsilon (o,t)$,

$$\mathbb{P}(\tau_x(a) > t) = \lim_{\lambda\uparrow\infty} \mathbb{E}\left\{\exp\left(-\lambda\int_0^t \chi_-(X_x(s))\,ds\right)\right\} \quad . \tag{46}$$

The Girsanov-Cameron-Martin formula then gives for bounded b (see e.g. Ref.(11))

$$\mathbb{P}(\tau_x(a) > t) = \lim_{\lambda\uparrow\infty} \mathbb{E}\left\{\exp\left\{\int_0^t b(B_x(s))dB(s) - 2^{-1}\int_0^t b^2(B_x(s))ds - \lambda\int_0^t \chi_-(B_x(s))ds\right\}\right\} \quad , \tag{47}$$

with $B_x(s) = x + B(s)$, B being a $BM(\mathbb{R})$ process starting at 0. From Itô's formula

$$\ln f_0(B_x(t)) - \ln f_0(x) = \int_0^t b(B_x(s))dB(s) + 2^{-1}\int_0^t \left(\frac{f_0'}{f_0}\right)'(B_x(s))ds \tag{48}$$

$$= \int_0^t b(B_x(s))dB(s) - 2^{-1}\int_0^t b^2(B_x(s))ds + 2^{-1}\int_0^t \frac{f_0''(B_x(s))}{f_0(B_x(s))}\,ds \tag{49}$$

and the result follows from the Feynman-Kac formula and Eq.(39).

Definition

Define the mean first hitting time of a, $\bar{\tau}(a)$, by

$$\mathbb{P}(\bar{\tau}(a)\ \varepsilon\ ds) = \int \rho_0(x)\, \mathbb{P}(\tau_x(a)\ \varepsilon\ ds)\, dx \overset{\text{def}}{=} H_a(s)\, ds \ , \tag{50}$$

ρ_0 being the invariant measure. Also, define $H_a^{\pm}(\cdot)$ by

$$\int_{x \gtrless a} \rho_0(x)\, \mathbb{P}(\tau_x(a)\ \varepsilon\ ds)\, dx = H_a^{\pm}(s)\, ds \ . \tag{51}$$

Then it follows that:

Corollary

$$H_a(s) = H_a^{+}(s) + H_a^{-}(s) \tag{52}$$

where

$$H_a^{\pm}(s) = \big(f_0\, , H\pm(a)\exp(-\,s\ H\pm(a))\, f_0\big)_{L^2} \ , \quad s > 0 \ , \tag{53}$$

$H\pm(a)$ being the Dirichlet Hamiltonians defined above, f_0 the square root of the invariant density.

Remark

The above results apply to ground-state Nelson diffusions with energy E in continuous potentials V if we make the transcriptions

$H_0 = -\,2^{-1}\dfrac{d^2}{dx^2} + V(x) - E$, $f_0 = f_E$, the corresponding ground-state eigenfunction.

Lemma

Because $\rho\ \varepsilon\ L^1(\mathbb{R})$, $\{-\infty,\infty\}$ is an inaccessible boundary for the diffusion X.

Proof

We need to prove that, for the scale function S, $\int^y dS(x)$ diverges as $y \to \pm\infty$ i.e. $\int^y \rho_0^{-1}(x)dx$ diverges as $y \to \pm\infty$. Observe that

$$\int^y 1.dx = \int^y f_0(x)\, f_0^{-1}(x)dx \le \left\{\int^y \rho_0(x)dx\right\}^{1/2} \left\{\int^y \rho_0^{-1}(x)dx\right\}^{1/2} \le \left\{\int^y \rho_0^{-1}(x)dx\right\}^{1/2} \tag{54}$$

and $\int^y 1\, dx$ diverges as $y \to \infty$. //

Set $L^{\pm}(t) = \mathrm{Leb}\{s \varepsilon (0,t) : X(s) \gtrless a\}$. We conclude this section with a special case of the Batchelor-Truman formula.

Proposition

For the above one-dimensional time-homogeneous diffusion, with inaccessible boundary, for each $\alpha, \lambda > 0$,

$$\int_0^{\infty} e^{-\alpha t}\, \mathbb{E}_a(e^{-\lambda L^-(t)}) = \frac{\bar{\mathbb{E}}_-(e^{-(\alpha+\lambda)\tau(a)}) + \bar{\mathbb{E}}_+(e^{-\alpha\tau(a)})}{(\alpha+\lambda)\, \bar{\mathbb{E}}_-(e^{-(\alpha+\lambda)\tau(a)}) + \alpha\, \bar{\mathbb{E}}_+(e^{-\alpha\tau(a)})} , \tag{55}$$

where $\bar{\mathbb{E}}_{\pm}(\cdot) = \int_{x \gtrless a} dx\, \rho_0(x)\, \mathbb{E}_x\,(\cdot)$.

Proof

Set $u(x) = \int_0^{\infty} e^{-\alpha t} \mathbb{E}_x\{e^{-\lambda L^-(t)}\}dt = \int_0^{\infty} e^{-\alpha t}\, \mathbb{E}_x\{e^{-\lambda \int_0^t \chi_-(X(s))ds}\}dt$, where χ_- is the characteristic function of $\{x : x < a\}$. Applying the Girsanov-Cameron-Martin theorem,

$$u(x) = f_0^{-1}(x) \int_0^{\infty} e^{-\alpha t}\, \mathbb{E}_x\{e^{-\int_0^t (V+\lambda\chi_-)(B(s))ds}\, f_0(B(t))\}dt , \tag{56}$$

for each $\alpha, \lambda > 0$, V being $\frac{1}{2}(b' + b^2)$.

From Kac's theorem, because $f_0 \varepsilon L^2$, $u(x)$ is the piecewise C^2 solution of

$$(L_x - (\lambda\, \chi_-(x) + \alpha))\, u(x) = -1 , \tag{57}$$

with $u(a+) = u(a-)$ and $u'(a+) = u'(a-)$ (see Ref.(3)). However, because of the inaccessible boundary, the only solutions of

$$L_x h(x) = \gamma h(x) \qquad (\gamma > 0) \tag{58}$$

bounded in a neighbourhood of $x = \pm\infty$ are $\mathbb{E}\{e^{-\gamma\tau_x(a)}\}$. Writing

$$u(x) = \frac{1}{(\alpha+\lambda)} + A\, \mathbb{E}\{e^{-(\alpha+\lambda)\tau_x(a)}\} , \qquad x < a , \tag{59}$$

$$u(x) = \frac{1}{\alpha} + B\, \mathbb{E}\{e^{-\alpha\tau_x(a)}\} , \qquad x > a , \tag{60}$$

and satisfying the above boundary condition gives

$$u(a) = \frac{\left[(\alpha+\lambda)^{-1} \frac{d}{dx}\Big|_{x=a-} \mathbb{E}\{e^{-(\alpha+\lambda)\tau_x(a)}\} - \alpha^{-1} \frac{d}{dx}\Big|_{x=a+} \mathbb{E}\{e^{-\alpha\tau_x(a)}\}\right]}{\left[\frac{d}{dx}\Big|_{x=a-} \mathbb{E}\{e^{-(\alpha+\lambda)\tau_x(a)}\} - \frac{d}{dx}\Big|_{x=a+} \mathbb{E}\{e^{-\alpha\tau_x(a)}\}\right]} \tag{61}$$

The desired result follows because $h(x) = \mathbb{E}\{e^{-\gamma\tau_x(a)}\}$ satisfies

$2^{-1}\frac{d}{dM(x)}\frac{d}{dS(x)}h(x) = \gamma h(x)$, with $\frac{dh(x)}{dS(x)} = 0$ at $x = \pm\infty$, S being the scale function, M the speed measure, with $dS(x) = \rho_0^{-1}(x)dx$, $dM(x) = \rho_0(x)dx$. //

Excursions in Stochastic Mechanics

Let a be a regular point for the diffusion X, so that

$$\mathbb{E}\{\tau_a(a) = 0\} = 1 \ . \tag{62}$$

We shall consider excursions for the process X from the point a.

Local Time L^a (See Rogers and Williams Ref(8)).

The local time at a upto time t, $L^a(t)$, is defined through Trotter's theorem and the identity

$$\int_0^t f(X(s))ds = \int_{-\infty}^{\infty} f(a)\, L^a(t)da \ , \tag{63}$$

for all bounded measurable functions f. Formally then

$$L^a(t) = \int_0^t \delta(X(s) - a)ds \ . \tag{64}$$

The points of increase of $L^a(\cdot)$ are therefore the zeros of $(X(\cdot) - a)$.
Firstly observe that

$$\mathbb{E}\left\{\text{Leb}\,\{t : X(t) = a\}\right\} = 0 \ , \tag{65}$$

where Leb denotes Lebesgue measure. For

$$\mathbb{E}\left\{\text{Leb}\,\{t : X(t) = a\}\right\} = \mathbb{E}\left\{\int_0^\infty \chi_{\{a\}}(X(t))dt\right\} = \int_0^\infty \mathbb{P}(X(t) = a)dt = 0 \ . \tag{66}$$

Therefore, Leb $\{t : X(t) = a\} = 0$, almost surely. (See Rogers Ref(9))
Secondly, denoting by p_t the transition density for the process X,

$$\mathbb{E}_b(L^a(t)) = \mathbb{E}_b\left\{\int_0^t \delta(X(s) - a)ds\right\} = \int_0^t p_s(b,a)ds \ . \tag{67}$$

We now consider how long one has to wait before the local time at a equals t.

<u>Subordinator γ^a</u>

Define γ^a, the right inverse of L^a, by

$$\gamma^a(t) = \inf\{u > 0 : L^a(u) > t\} \quad , \tag{68}$$

so that $L^a(\gamma^a(t)) = t$. Then each $\gamma^a(t)$ is a stopping time with $X(\gamma^a(t)) = a$.

Moreover, $\gamma^a(t)$ is a subordinator in the sense that

$$\gamma^a(t+s)(w) - \gamma^a(t)(w) = \gamma^a(s)\left((\theta_{\gamma^a(t)})(w)\right) \quad , \tag{69}$$

where $\theta_{\gamma^a(t)}$ is the usual shift

$$\theta_{\gamma^a(t)}(w)(s) = w(s+\gamma^a(t)) - w(\gamma^a(t)) \quad . \tag{70}$$

The importance of the subordination property is that it implies the existence of a constant $\Psi(\lambda)$ with

$$\mathbb{E}_a(e^{-\lambda\gamma^a(t)}) = \exp(-t\Psi(\lambda)) \quad , \quad \text{each } \lambda > 0 \quad . \tag{71}$$

To see this observe that

$$\mathbb{E}_a(e^{-\lambda\gamma^a(t+s)}) = \mathbb{E}_a\left(e^{-\lambda\gamma^a(t)}\,\mathbb{E}(e^{-\lambda(\gamma^a(t+s) - \gamma^a(t))} \mid \mathcal{F}_{\gamma^a(t)})\right) \quad . \tag{72}$$

By the strong Markov property

$$\mathbb{E}(e^{-\lambda(\gamma^a(t+s) - \gamma^a(t))} \mid \mathcal{F}_{\gamma^a(t)}) = \mathbb{E}_a(e^{-\lambda\gamma^a(s)}) \quad , \tag{73}$$

so

$$\mathbb{E}_a(e^{-\lambda\gamma^a(t+s)}) = \mathbb{E}_a(e^{-\lambda\gamma^a(t)})\,\mathbb{E}_a(e^{-\lambda\gamma^a(s)}) \quad , \tag{74}$$

proving the existence of $\Psi(\lambda)$. We now calculate $\Psi(\lambda)$.

Using the inverse property of γ^a, $\int_0^\infty e^{-\lambda\gamma^a(t)}\,dt = \int_0^\infty e^{-\lambda s}\,dL^a(s)$, giving

$$\Psi(\lambda)^{-1} = \int_0^\infty \mathbb{E}_a(e^{-\lambda\gamma^a(t)})dt = \mathbb{E}_a\left\{\int_0^\infty e^{-\lambda s}\,dL^a(s)\right\} = \int_0^\infty e^{-\lambda s}\,d(\mathbb{E}_a(L^a(s))$$

$$= \tilde{p}_\lambda(a,a) \quad , \tag{75}$$

~ being Laplace tranform. Hence, we obtain

$$\mathbb{E}_a\left\{e^{-\lambda\gamma^a(t)}\right\} = \exp\left\{\frac{-t}{\tilde{p}_\lambda(a,a)}\right\} \quad . \tag{76}$$

Lemma

Assume that the diffusion X has an invariant measure $\rho_0\ (\geq 0)$, with $\int\rho_0(x)p_s(x,y)dx = \rho_0(y)$, each $s>0$ (i.e. $L_x^*\rho(x) = 0$, L_x^* being the L^2 - adjoint of the generator $L_x = 2^{-1}\frac{d^2}{dx^2} + b(x)\frac{d}{dx}$) .

Then for each $t,\ \lambda>0$

$$\mathbb{E}_a\left\{e^{-\lambda\gamma^a(t)}\right\} = \exp\left\{-t\lambda\frac{\mathbb{E}(e^{-\lambda\tau(a)})}{\rho_0(a)}\right\}, \tag{77}$$

with $\mathbb{E}\left\{e^{-\lambda\tau(a)}\right\} = \int\rho_0(x)\,\mathbb{E}(e^{-\lambda\tau_x(a)})dx = \int_0^\infty e^{-\lambda s}H_a(s)ds$, H_a being given by above.

Proof

Recall that from Eq.(27)

$$\mathbb{E}\{e^{-\lambda\tau_x(a)}\} = \int_0^\infty e^{-\lambda s}p_s(x,a)ds\,/\,\tilde{p}_\lambda(a,a) . \tag{78}$$

Multiplying both sides by $\rho_0(x)$ and integrating by Fubini's theorem

$$\mathbb{E}(e^{-\lambda\tau(a)}) = \int_0^\infty e^{-\lambda s}ds\int_{-\infty}^\infty \rho_0(x)\,p_s(x,a)dx\,/\,\tilde{p}_\lambda(a,a) = \lambda^{-1}\rho_0(a)\,/\,\tilde{p}_\lambda(a,a) . \tag{79}$$

//

Corollary

If $\rho_0\,\varepsilon\,L^1(\mathbb{R})$, then $\mathbb{P}(\gamma^a(t)<\infty) = \mathbb{P}(L^a(\infty)>t) = 1$.

Proof

$$\lim_{\lambda\downarrow 0}\mathbb{E}(e^{-\lambda\tau(a)}) = \int\rho_0(x)\,\mathbb{P}(\tau_x(a)<\infty)dx \leq \|\rho_0\|_{L^1} < \infty . \ //$$

We also have:

Proposition

For the above diffusion X for any regular point a, for each $t,\ \lambda>0$,

$$\mathbb{E}_a\left\{e^{-\lambda\gamma^a(t)}\right\} = \exp\left\{-t\int_0^\infty(1-e^{-\lambda s})\frac{(-dH_a(s))}{\rho_0(a)}\right\}, \tag{80}$$

ρ_0 being the invariant density, $H_a(s)ds = \int\rho_0(x)\,\mathbb{P}(\tau_x(a)\,\varepsilon\,ds)dx$ being given by Eqs.(52) and (53).

Proof

Observe that the self-adjointness of $H_{\pm}(a)$ gives after an integration by parts

$$\lambda\,\mathbb{E}(e^{-\lambda\tau(a)}) = \lambda\int_0^\infty e^{-\lambda s}\,H_a(s)ds = \lim_{s\uparrow\infty} H_a(s) + \int_0^\infty (1-e^{-\lambda s})(-H_a'(s)ds)\ . \tag{81}$$

Therefore, since the first limit is zero,

$$\lambda\,\mathbb{E}(e^{-\lambda\tau(a)}) = \int_0^\infty (1-e^{-\lambda s})(-H_a'(s)ds)\ , \tag{82}$$

proving the result . //

Poisson Lévy Excursion Measures

Set $d\nu(s) = -H_a'(s)ds/\rho_0(a)$, $t = \sum_j \delta t_j$ and $\int_0^\infty \nu(ds) = \sum_i \Delta_i = \sum_i \nu(ds_i)$, $\Delta_i = \nu(ds_i)$.

Then the above formula gives

$$\mathbb{E}_a\left(e^{-\lambda\gamma^a(t)}\right) = \prod_{\Delta_i,\delta t_j} e^{-\delta t_j\Delta_i}\ e^{\delta t_j\, e^{-\lambda s_i}\,\Delta_i}\ , \tag{83}$$

showing independent contributions to $\mathbb{E}_a\left(e^{-\lambda\gamma^a(t)}\right)$ from disjoint $(\delta t_j \times \Delta_i)$. Therefore, we see that

$$\mathbb{E}_a\left(e^{-\lambda\gamma^a(t)}\right) = \prod_{\Delta_i} e^{-t\Delta_i}\ e^{t\, e^{-\lambda s_i}\,\Delta_i}\ . \tag{84}$$

The contribution to r.h.s. from a typical $t\,\Delta_i = t\,\Delta$ is

$$e^{-t\Delta}\ e^{t\,e^{-\lambda s}\Delta} = e^{-t\Delta}\sum_{n=0}^{\infty} e^{-\lambda sn}\,\frac{(t\Delta)^n}{n!} = \sum_{n=0}^{\infty} e^{-\lambda sn}\left(e^{-t\Delta}\frac{(t\Delta)^n}{n!}\right)\ . \tag{85}$$

Excursions

We identify the excursions of X away from a as the jumps in γ^a. Set $\#(s,t)$ = number of excursions of duration s made before the local time at a equals t.

Corollary

$$\mathbb{P}(\#(s,t) = n) = e^{-td\nu(s)}\,\frac{(td\nu(s))^n}{n!}\ , \qquad \text{for } n = 0, 1, 2, 3, \cdots , \tag{86}$$

where

$$\frac{d\nu(s)}{ds} = -\frac{H_a'(s)}{\rho_0(a)} = \frac{(f_0, H_+^2(a)\, e^{-sH_+(a)} f_0)_{L_2} + (f_0, H_-^2(a)\, e^{-sH_-(a)} f_0)_{L_2}}{\rho_0(a)}, \tag{87}$$

$H_\pm(a)$ being the Dirichlet-Hamiltonians defined by Eq.(44).

Remark

Making the appropriate transcriptions the above results apply to ground-state Nelson diffusions.

The Ups and Downs of Excursions

Since $\mathrm{Leb}\{s < t : X(s) = a\} = 0$, almost surely, for each $t > 0$

$$L^+(t) + L^-(t) = t, \quad \text{almost surely,} \tag{88}$$

where $L^\pm(t) = \mathrm{Leb}\{s < t : X(s) \gtrless a\}$, so that for $\gamma^a(t) = \inf\{u > 0 : L^a(u) > t\}$ we have

$$L^+(\gamma^a(t)) + L^-(\gamma^a(t)) = \gamma^a(t), \quad \text{almost surely.} \tag{89}$$

One can establish subordination properties for $L^\pm(\gamma^a(t))$ in the same way as for $\gamma^a(t)$. Hence, for constants $\Psi^\pm(\lambda)$, for each $\lambda, t > 0$,

$$\mathbb{E}_a\{e^{-\lambda L^\pm(\gamma^a(t))}\} = \exp\{-t\Psi^\pm(\lambda)\}. \tag{90}$$

It follows from the penultimate equation and independence that

$$\Psi^+(\lambda) + \Psi^-(\lambda) = \Psi(\lambda) = \lambda\, \mathbb{E}\{e^{-\lambda\tau(a)}\}/\rho_0(a). \tag{91}$$

We now establish that as a consequence of this result and the Batchelor-Truman formula, Eq. (55),

$$\Psi^\pm(\lambda) = \lambda\, \mathbb{E}_\pm\{e^{-\lambda\tau(a)}\}/\rho_0(a) = \pm\frac{c}{dx}\mathbb{E}\big|_{x=a\pm}\{e^{-\lambda\tau_x(a)}\}. \tag{92}$$

Hence, if we set

$\#^\pm(s,t)$ = number of $\substack{\text{up}\\ \text{down}}$ excursions of duration s upto local time at a equals t,

we obtain

$$\mathbb{P}(\#^\pm(s,t) = n) = e^{-t\,d\upsilon^\pm(s)}\frac{(t\,d\upsilon^\pm(s))^n}{n!}, \quad \text{for } n = 0,1,2,\ldots, \tag{93}$$

where
$$d\upsilon^\pm(s) = -\frac{d_s H_a^\pm(s)}{\rho_0(a)}, \tag{94}$$

with

$$\int_{x \gtrless a} \rho_0(x)\,\mathbb{P}(\tau_x(a) \varepsilon\, ds)\,dx = H_a^{\pm}(s)\, ds. \tag{95}$$

In particular from Eq. (53)

$$\frac{d\upsilon^{\pm}(s)}{ds} = \frac{(f_0, H_{\pm}^2(a)\, e^{-sH_{\pm}(a)} f_0)_{L_2}}{\rho_0(a)} \tag{96}$$

for the Dirichlet Hamiltonians defined in the last section, $f_0 = \rho_0^{1/2}$, ρ_0 being the L^1 invariant density.

Some of the above results are variants of some little known formulae of Lévy (see Ref. (10)), but the Hamiltonian results are new. Our derivation uses a beautiful argument of Chris Rogers involving coloured processes of marks. (See Rogers and Williams Ref. (8)).

<u>Coloured Processes of Marks</u>

We consider a Poisson process of marks made according to the clock $\lambda L^+(t)$. Then

$$\mathbb{P}(\text{no mark occurs before time } \gamma^a(t)) = \mathbb{E}\left\{\exp\left(-\lambda L^+(\gamma^a(t))\right)\right\} = e^{-t\Psi^+(\lambda)}. \tag{97}$$

Hence, if T is the time of the first mark, we have

$$\mathbb{P}(L^a(T) \varepsilon\, dt) = \mathbb{P}(T \varepsilon \gamma^a(dt)) = e^{-t\Psi^+(\lambda)}(1 - e^{-dt\Psi^+(\lambda)}) = \Psi^+(\lambda) e^{-t\Psi^+(\lambda)} dt \tag{98}$$

i.e. $L^a(T)$ is exponential with rate $\Psi^+(\lambda)$.

Suppose now that $\alpha, \lambda_+, \lambda_- > 0$ are all constants. Then, if S is exponential of rate α (independent of X),

$$\int_0^\infty \alpha e^{-\alpha t}\, \mathbb{E}_a\{\exp(-\lambda_+ L^+(t) - \lambda_- L^-(t))\}\, dt = \mathbb{E}_a\{\exp(-\lambda_+ L^+(S) - \lambda_- L^-(S))\}. \tag{99}$$

We now regard S as the time of the first mark in a Poisson process of red marks made according to the clock αt. Also consider a Poisson process of green marks with clock $(\lambda_+ L^+(t) + \lambda_- L^-(t))$. From above

$\mathbb{P}$(first red mark occurs before first green mark)

$$= \int_0^\infty \alpha e^{-\alpha t}\, \mathbb{E}_a\left\{e^{-(\lambda_+ L^+(t) + \lambda_- L^-(t))}\right\} dt$$

$$= \mathbb{E}_a\left\{e^{-(\lambda_+ L^+(S) + \lambda_- L^-(S))}\right\}. \tag{100}$$

The difficulty in obtaining an expression for l.h.s. of above by means of excursion theory is that the first red mark and the first green mark might occur during the same excursion i.e. at the same local time. The remedy for this is in Rogers and Williams (Ref.8).

We consider a Poisson process of plus marks made according to the clock $(\lambda_+ + \alpha)L^+(t)$ and a Poisson process of minus marks made according to the clock $(\lambda_- + \alpha)L^-(t)$. The fact that no excursion contains first plus and minus marks is now obvious. Now colour each plus mark red with probability $\alpha/(\lambda_+ + \alpha)$ and green with probability $\lambda_+/(\lambda_+ + \alpha)$. Likewise colour each minus mark red with probability $\alpha/(\lambda_- + \alpha)$ and green with probability $\lambda_-/(\lambda_- + \alpha)$. The red and green marks have the same structure as before.

Let $T_\pm$ be the time of the first $\genfrac{}{}{0pt}{}{\text{plus}}{\text{minus}}$ mark. Then, arguing as above:

$$L(T_+) \text{ is exponential with rate } \Psi^+(\lambda_+ + \alpha) = q^+,$$

$$L(T_-) \text{ is exponential with rate } \Psi^-(\lambda_- + \alpha) = q^-,$$

and these variables are independent. Because first plus and minus marks never fall in the same excursion

$$\mathbb{P}(T_+ < T_-) = \mathbb{P}(L(T_+) < L(T_-)) = \int_0^\infty q_+ e^{-tq_+} e^{-tq_-} dt = q_+/(q_+ + q_-) \quad . \tag{101}$$

Therefore, we obtain

$\mathbb{P}$(first red mark precedes first green mark)

= $\mathbb{P}$(first plus mark precedes first minus mark and is red)

+ $\mathbb{P}$(first minus mark precedes first plus mark and is red)

$$= \frac{q_+}{(q_+ + q_-)} \frac{\alpha}{(\lambda_+ + \alpha)} + \frac{q_-}{(q_+ + q_-)} \frac{\alpha}{(\lambda_- + \alpha)} \tag{102}$$

$$= \frac{\alpha}{(\Psi^+(\lambda_+ + \alpha) + \Psi^-(\lambda_- + \alpha))} \left\{ \frac{\Psi^+(\lambda_+ + \alpha)}{(\lambda_+ + \alpha)} + \frac{\Psi^-(\lambda_- + \alpha)}{(\lambda_- + \alpha)} \right\}. \tag{103}$$

Setting $\lambda_+ = 0$ and $\lambda_- = \lambda$ and comparing this with the Batchelor-Truman formula gives

$$\frac{\Psi^-(\alpha + \lambda)}{\Psi^+(\alpha)} = \frac{(\alpha + \lambda)\, \mathbb{E}_-(e^{-(\alpha+\lambda)\tau(a)})}{\alpha\, \mathbb{E}_+(e^{-\alpha\tau(a)})} = \frac{\frac{d}{dx}\mathbb{E}(e^{-(\alpha+\lambda)\tau_x(a)})\big|_{x=a_-}}{-\frac{d}{dx}\mathbb{E}(e^{-\lambda\tau_x(a)})\big|_{x=a_+}} \quad . \tag{104}$$

Since $\Psi^+(\lambda) + \Psi^-(\lambda) = \lambda\, \mathbb{E}(e^{-\lambda\bar{\tau}(a)})/\rho_0(a)$, the desired result follows. Therefore, we have proved:-

Proposition

For the one-dimensional time-homogeneous diffusion on $\mathbb{R}$,

$$dX(s) = b(X(s))\,ds + dB(s), \tag{105}$$

for a bounded C^1 drift b, a BM($\mathbb{R}$) process B, with covariance $\mathbb{E}(B(s)B(t)) = \min(s,t)$, with an L^1 invariant measure $\rho_0(x)\,dx$, the number of $\genfrac{}{}{0pt}{}{\text{up}}{\text{down}}$ excursions from the regular point a with $\rho_0(a) \neq 0$, $\#^{\pm}$, satisfies

$$\mathbb{P}(\#^{\pm}(s,t) = n) = e^{-t\,d\upsilon^{\pm}(s)}(t\,d\upsilon^{\pm}(s))^n / n!, \tag{106}$$

n = 0,1,2, ..., where

$$\frac{d\upsilon^{\pm}(s)}{ds} = \rho_0(a)^{-1}(\rho_0^{1/2}, H_{\pm}^2(a)\,e^{-sH_{\pm}(a)}\rho_0^{1/2}))_{L^2}, \quad s > 0, \tag{107}$$

$H_{\pm}(a)$ being the Dirichlet Hamiltonian $H_{\pm}(a) = \lim_{\lambda \uparrow \infty} (-2^{-1}\frac{d^2}{dx^2} + V(x) + \lambda\chi_{\mp}(x))$, $V(x) = 2^{-1}(b'(x) + b^2(x))$, $\chi_{\mp}$ the characteristic function of $\{x : x \lessgtr a\}$.

Remark

For the above ground-state Nelson diffusions, corresponding to the ground-state f_E in the potential V,

$$\frac{d\upsilon^{\pm}(s)}{ds} = \left(f_E, (H_{\pm}(a) - E)^2 e^{-s(H_{\pm}(a)-E)} f_E\right)_{L^2}, \quad s > 0, \tag{108}$$

for $H_{\pm}(a) \overset{\text{def}}{=} \lim_{\lambda \uparrow \infty} (-2^{-1}\frac{d^2}{dx^2} + V(x) + \lambda\chi_{\pm}(x))$, E being the ground-state energy.

Incidentally we have proved:

Corollary

For the one-dimensional time-homogeneous diffusions on $\mathbb{R}$,

$$dX(s) = b(X(s))\,ds + dB(s), \tag{109}$$

for a bounded C^1 drift b, with an L^1 invariant measure $\rho_0(x)\,dx$, for a regular point a with $\rho_0(a) \neq 0$,

$$\int_0^{\infty} e^{-\alpha t}\,\mathbb{E}_a\left\{e^{-\lambda_+ L^+(t) - \lambda_- L^-(t)}\right\} = \frac{\mathrm{Disc}\,|_a\,(\alpha + \lambda)^{-1}\frac{d}{dx}\,\mathbb{E}\{e^{-(\alpha+\lambda)\tau_x(a)}\}}{\mathrm{Disc}\,|_a\,\frac{d}{dx}\,\mathbb{E}\{e^{-(\alpha+\lambda)\tau_x(a)}\}}, \tag{110}$$

where $L^{\pm}(t) = \text{Leb}\,\{s \in [0,t] : X(s) \lesseqgtr a\}$, with $\lambda = \lambda(x) = \lambda_{\pm}$, $x \gtrless a$, respectively, and

$$\text{Disc}\,\big|_a\, f(x) = f(a+) - f(a-).$$

REFERENCES

[1] A. Batchelor and A. Truman, "Hitting, Killing and Capturing in Nelson's Stochastic Mechanics", in Proceedings of the International Conference on Stochastic Processes - Geometry and Physics, Ascona-Locarno, Summer 1988, edited by S. Albeverio and G. Casati et al, World Scientific Press, 1989.

[2] K. Itô and H.P. McKean, Diffusion Processes and Their Sample Paths, Berlin, Springer, 1965

[3] M. Kac, "On some connections between probability theory and differential and integral equations", in Proceedings 2nd Berkeley Symposium on Math. Stat. and Probability, 189-215, California, University of California Press, 1951.

[4] I. Karatzas and S.E. Shreve, Brownian Motion and Stochastic Calculus, New York, Springer, 1988.

[5] E. Nelson, Dynamical Theories of Brownian Motion, Princeton, Princeton University Press, 1967.

[6] E. Nelson, Quantum Fluctuations, Princeton, Princeton University Press, 1984.

[7] M. Reed and B. Simon, Fourier Analysis and Self-adjointness, Vol. 2 of Methods of Modern Mathematical Physics, Academic Press, 1975.

[8] L.C.G. Rogers and D. Williams, Diffusions, Markov Processes and Martingales, Vol. 2 Itô Calculus, Chichester, John Wiley, 1987.

[9] L.C.G. Rogers, "A guided tour through excursions", Bull.L.M.S. # 91, Volume 21, Part 4, July 1989.

[10] B. Simon, Functional Integration and Quantum Physics, New York, 1979.

Aubrey Truman
Department of Mathematics and Computer Science
University College of Swansea
Singleton Park
SWANSEA SA2 8PP
U.K.

David Williams
Department of Pure Mathematics and
Mathematical Statistics
University of Cambridge
16 Mill Lane
CAMBRIDGE CB2 1SB
U.K.

Part III
Diffusion Processes in Stochastic Analysis

Some Regularity Results and Eigenfunction Estimates for the Schrödinger Operator

M. Cranston and Z. Zhao

Abstract. In this paper is given a probabilistic derivation of a gradient estimate for solutions $Lu = -\frac{1}{2}\Delta u + qu = 0$ on a Lipschitz domain D in $\mathbf{R}^d$ when $q \in K_{d+1}^{loc}(D)$ and the first Dirichlet eigenvalue for L on D is positive. An application to the decay of eigenfunctions for L at nodal domain boundaries is discussed.

The purpose of the present note is to derive a regularity result for solutions to Schrödinger's equation. More precisely, let D be a bounded Lipschitz domain in $\mathbf{R}^d$, $d \geq 3$. For $1 \geq \alpha \geq 0$ we say $q \in K_{d+\alpha}^{loc}(D)$ provided

$$\lim_{r \to 0} \sup_{x \in D} \int_{|x-y|<r} \frac{|q|(y)}{|x-y|^{d+\alpha-2}}\, dy = 0 .$$

When $x \in D$ set $\delta_x =$ dist (x, D^c). Our main result is the following.

THEOREM 1. *Suppose the first Dirichlet eigenvalue of $-\frac{1}{2}\Delta + q$ on D is positive and that $q \in K_{d+1}^{loc}(D)$. There is an $r_0 > 0$ such that if $-\frac{1}{2}\Delta u + qu = 0$ on D and Q is a cube centered at x of diameter $\frac{1}{2}\delta_x \wedge r_0$, then*

$$|\nabla u(x)| \leq c \left[r_0 \wedge \frac{\delta_x}{2} \right]^{-1} \sup_Q |u| .$$

Furthermore, when u is positive one has

$$|\nabla u(x)| \leq c \left[r_0 \wedge \frac{\delta_x}{2} \right]^{-1} u(x) .$$

If $q \in K_{d+\alpha}^{loc}(D)$ for $0 < \alpha < 1$ then

$$|u(x) - u(x')| \leq c \left[|x - x'|^\alpha + |x - x'| \left[\frac{\delta_x}{2} \wedge r_0 \right]^{-1} \right] \sup_Q |u| , \; x, x' \in Q .$$

It is known that u need be no better than Hölder for $q \in K_d^{loc}(D)$. Also that $q \in K_d^{loc}(D)$ implies u is Hölder is shown in AS(1982). Our result gives an estimate of the correct Hölder class for u when $q \in K_{d+\alpha}^{loc}(D)$.

The proof relies on the Conditional Gauge Theorem of CFZ(1987) and a coupling due to Lindvall, Rogers (1986). For other gradient estimates obtained via coupling see C(1989).

Denote by (P^x, X), Brownian motion killed on exiting D. By P_z^x we shall mean the measure on Brownian motion paths conditioned to exit D at $z \in \overline{D}$. See Doob (1983) for an exposition on these processes.

Given $q \in K_{d+\alpha}^{loc}(D), 0 \leq \alpha \leq 1$, define the Feynman-Kac functional

$$e_q(t) = \exp\left\{\int_0^t q(X_s)ds\right\} .$$

Finally set $\tau_D = \inf\{t > 0 : X_t \notin D\}$. Then the conditional gauge is $E_z^x[e_q(\tau_D)]$.

The following was proved in CFZ (1987).

CONDITIONAL GAUGE THEOREM. *Suppose D is a bounded Lipschitz domain in $\mathbf{R}^d$ and $q \in K_d^{loc}(\mathbf{R}^d)$. Suppose also that $E^x[e_q(\tau_D)] < \infty$ for some $x \in D$. Then there exist positive constants c_1 and c_2 such that*

$$c_1 \leq E_z^x[e_q(\tau_D)] \leq c_2 , \quad x, z \in \overline{D} .$$

The conclusion also holds if and only if the first Dirichlet eigenvalue for $-\frac{1}{2}\Delta + q$ on D is positive.

The conditional gauge is useful for relating potential theoretic quantities for the perturbed, $-\frac{1}{2}\Delta + q$, operators with like quantities for $-\frac{1}{2}\Delta$. For example, if $G_A^q(x,y)$ and $K_A^q(x,z)$ are the Green function and Poisson kernel relative to some subset A of D for $-\frac{1}{2}\Delta + q$ and $G_A(x,y)$ and $K_A(x,z)$ for $-\frac{1}{2}\Delta$, then

$$K_A^q(x,z) = K_A(x,z) + \int_A G_A(x,y)q(y)K_A^q(y,z)dy . \tag{1}$$

Also,

$$K_A^q(y,z) = E_z^y[e_q(\tau_A)]K_A(y,z) \tag{2}$$

as was shown in CFZ(1987). One last estimate we shall need is due to Widman (1967). Suppose A has a C^2 boundary. Then there is constant c_α so that if $\delta_x = \text{dist}(x, \partial A)$

$$G_A(x,y) \leq c_\alpha \frac{\delta_x^\alpha}{|x-y|^{d-1+\alpha}} . \tag{3}$$

Actually, Widman (1967) showed that for some c

(4) $$G_A(x,y) \le c\frac{\delta(x)}{|x-y|^{d-1}}$$

whereas the bound

(5) $$G_A(x,y) \le c\frac{1}{|x-y|^{d-1}}$$

holds as well. Raising the first inequality to the power α, the second to the power $1-\alpha$ and multiplying the result gives 3).

We shall now specify a special domain which will be moved rigidly about the domain D. Care will be taken with constants in our inequalities to insure they do not change with scaling. Write $x \in \mathbf{R}^d$ as $x = (x_1, \hat{x})$ with $x_1 \in \mathbf{R}$, $\hat{x} \in \mathbf{R}^{d-1}$ and set

$$P_a = \{x = (0, \hat{x}) \in \mathbf{R}^d\ ;\quad |\hat{x}| \le a\}\ .$$

For $0 < a \le 1$, define $B(x,a) = \{y : |x-y| < a\}$ and set

$$U(a) = \bigcup_{x \in P_a} B(x,a)\ .$$

Now it was shown in Zhao (1986) that if $\delta_a(x) = \text{dist}\ (x, U(a)^c)$ and

$$F_a(x,y) = \min\left(\frac{c_d}{|x-y|^{d-2}}\ \cdot\ \frac{c_d\ \delta_a(x)\delta_a(y)}{4|x-y|^{d-2}}\right)$$

then with G_a denoting the Green function for $U(a)$, there is a positive $c = c(d, U(1))$ such that

$$c^{-1}F_1(x,y) \le G_1(x,y) \le cF_1(x,y)\ ;\quad x,y \in U(1)\ .$$

This last implies the so-called $3G$ inequality

$$\frac{G_1(x,y)G_1(y,z)}{G_1(x,z)} \le c\{|x-y|^{2-d} + |y-z|^{2-d}\}\ , x,y,z \in U(1)$$

with a constant c depending on d and $U(1)$ alone. Now if $0 < a \le 1$, then by scaling

$$\delta_a(x) = a^{-1}\delta_1(ax)$$

and

$$G_a(x,y) = a^{2-d}G_1\ \frac{x}{a}, \frac{y}{a}\ .$$

Applying these scalings to (4) and (5) it follows that (3) holds for $U(a) = A$ with a constant independent of a.

Also, with the same $c = c(d, U(1))$ as above,

$$c^{-1}F_a(x,y) \le G_a(x,y) \le cF_a(x,y)\,, \quad x,y \in U(1)$$

and therefore, there is a c, again depending on d and $U(1)$ only, such that

$$\frac{G_a(x,y)G_a(x,z)}{G_a(x,z)} \le c\left(|x-y|^{2-d} + |y-z|^{2-d}\right),\ x,y,\ z \in U\,.$$

Thus, using Khasminski's lemma we have

LEMMA 2. *Given $q \in K^{loc}_{d+\alpha}(D)$ there is an $r_0 > 0$ and $c_1, c > 0$ such that if $U(a)$ is any dilation, rotation and translation of $U(1)$ contained in D with $a < r_0$ then*

$$\int_{U(a)} \frac{|q|(y)}{|x-y|^{d+\alpha-1}}dy < c\,, \quad x \in D \tag{6}$$

and

$$E^x_z\left[e_q(\tau_{U(a)})\right] \le c_1 \quad \text{for } x,z \in \overline{U(a)}\,.$$

Also, if $U^+(a)$ is any translation, rotation and dilation of $U^+(1) = \{x \in U(1) : x_1 > 0\}$ then

$$E^x_z\left[e_q(\tau_{U^+(a)})\right] \le c_1\,, \quad x,z \in U^+(a) \tag{7}$$

with the same bound holding for $U^-(a)$. Finally, given $0 < \alpha \le 1$, there is a constant $c_\alpha > 0$ depending on $U(1)$, d and α alone so that

$$G_a(x,y) \le c_\alpha \frac{\delta_a(x)^\alpha}{|x-y|^{2-d+\alpha}}\,. \tag{8}$$

We next describe a device for coupling Brownian motions which is due to Lindvall, Rogers (1986). With two starting points x and x' let H be the hyperplane perpendicular to the line segment from x to x' which passes through $\frac{x+x'}{2}$. Given a Brownian path X commenced at x, simply reflect it in the hyperplane H to get another Brownian path X' commenced at x'. When discussing probabilities for the pair, the measure $P^{(x,x')}$ will be used, for just X it will be P^x and similarly, for X', $P^{x'}$.

The Feynman-Kac functional along the X' path will be denoted by $e'_q(t)$.

A trivial observation is that the coupling time

$$T = T(X, X') \equiv \inf\{t > 0 : X_t = X'_t\}$$

and the hitting time

$$\sigma_H = \inf\{t > 0 : X_t \in H\}$$

are the same. This observation makes it easy to estimate the probability that T occurs after the two particles leave a subdomain of D. Then taking coupled Brownian motion killed on exiting $U(1)$, $(P^{(x,x')}, (X, X'))$, one has for $U^+ = U^+(1)$, $U^- = U^-(1)$ and

$$\begin{aligned} \tau_{U^+} &= \inf\{t > 0 : X_t \notin U^+\} \\ \tau_{U^-} &= \inf\{t > 0 : X_t \notin U^-\} \\ \text{that} \qquad \tau_{U^+} &= \tau_{U^-}\,. \end{aligned}$$

Also, if we assume U is in its original position, and

$$H = \{x_1 = 0\}$$
$$\text{then } T = T(X, X') = \sigma_H = \inf\{t > 0 : X_t^1 = 0\}$$

and

$$P^{(x,x')}(T > \tau_{U^+} \wedge \tau_{U^-}) = P^x(X_{\tau_{U^+}} \notin H)\,.$$

By an elementary estimate of this harmonic function one gets a positive constant $c > 0$ for which

$$P^{(x,x')}(T > \tau_{U^+} \wedge \tau_{U^-}) \le c x_1. \tag{8}$$

With the same arguments applied to $U(a)$, i.e. any translation, rotation and or dilation of $U(1) = U$,

$$P^{(x,x')}\left(T > \tau_{U^+(a)} \wedge \tau_{U^-(a)}\right) \le c\frac{x_1}{a}\,. \tag{9}$$

Proof (Theorem 1): Take u to be a solution to $(\frac{1}{2}\Delta - q)u = 0$ in D. Given $x \in D$ take the a in the definition of $U^+ = U^+(a)$ to be $\frac{\delta_x}{2} \wedge r_0$ and for another $x' \in D$ with $|x - x'| \ll \delta_x$ align the region U so that the common boundary $\partial\overline{U^+} \cap \partial\overline{U^-} = \Gamma$ passes through $\frac{x+x'}{2}$ lies in the center of $\overline{U^+} \cap \overline{U^-}$

and is orthogonal to the segment from x to x'. Now start a Lindvall-Rogers coupled Brownian pair $((X, X'), P^{(x,x')})$ at x and x', respectively. Note that if $T_\Gamma = \inf\{t > 0 : X_t \in \Gamma\}$ then $\{T_\Gamma < \tau_{U\pm}\} = \{T < \tau_{U\pm}\}$. Thus, dropping a from the notation,

$$\begin{aligned} |u(x) - u(x')| \le E^x \left[\tau_{U^+} < T_\Gamma;\, e_q(\tau_{U^+})|u(X(\tau_{U^+}))|\right] \\ + E^{x'} \left[\tau_{U^-} < T_\Gamma;\, e'_q(\tau_{U^-})|u(X(\tau_{U^-}))|\right] \\ + |E^{(x,x')} \left[T_\Gamma < \tau_{U^+}\, ; |u(X(T_\Gamma))e_q(T_\Gamma) - u(X'(T_\Gamma)e'_q(T_\Gamma)|\right] \\ = I_1 + I_2 + I_3\,. \end{aligned}$$

Now, since $a \le r_0$

$$\begin{aligned} I_1 &= \int_{\partial U^+\setminus\Gamma} K_{U^+}(x,z)|u(z)|E^x_z[e_q(\tau_{U^+})]\sigma(dz) && \text{by 2)} \\ &\le c \sup_{U^+} |u| \int_{\partial U^+\setminus\Gamma} K_{U^+}(x,z)\sigma(dz) && \text{by 7)} \\ &\le \frac{\dot{c}}{a}|x - x'| \sup_{U^+} |u| && \text{by 9)} \end{aligned} \tag{7}$$

since the integral is a harmonic function of $x \in U^+$, vanishes on Γ and is one on $\partial U^+ \setminus \Gamma$. One can use the boundary Harnack principle for this.

Similarly,

$$I_2 \le \frac{c}{a}|x - x'| \sup_{U^-} |u|\,.$$

Finally,

$$\begin{aligned} I_3 &= E^{(x,x')} \left[T_\Gamma < \tau_{U^+};\; u(X(T_\Gamma))\,\big|\left[e_q(T_\Gamma) - e'_q(T_\Gamma)\right]\big]\right| \\ &\le \sup_\Gamma |u| \left[E^x \left[T_\Gamma < \tau_{U^+};\; |e_q(T_\Gamma) - 1|\right] + E^{x'} \left[T_\Gamma < \tau_{U^-};\; |e'_q(T_\Gamma) - 1|\right]\right]\,. \end{aligned}$$

Estimating the first term in the sum on the right gives

$$\begin{aligned} E^x \left[T_\Gamma < \tau_{U^+};\; |e_q(T_\Gamma) - 1|\right] &\le \int_\Gamma |K^q_{U^+}(x,z) - K_{U^-}(x,z)|\sigma(dz) \\ &\le \int_\Gamma \int_{U^+} G_{U^+}(x,y)|q|(y)K^q_{U^+}(y,z)dy\sigma(dz)\,, && \text{by (1)} \\ &\le c \int_{U^+} G_{U^+}(x,y)|q|(y)dy && \text{by (2) and (7)} \\ &\le c|x - x'|^\alpha \int_{U^+} \frac{|q|(y)}{|x - y|^{d+\alpha-1}}dy && \text{by (8) and the} \\ & \qquad \text{fact that } \delta_a(x) = |x - x/2 \\ &\le c|x - x'|^\alpha && \text{by (6)}\,. \end{aligned}$$

The second term in the sum for the upper bound of I_3 satisfies the same bound. Combining all these estimates lead together with $a = r_0 \wedge \frac{\delta_x}{2}$ to

$$|u(x) - u(x')| \leq c \left[|x - x'|^\alpha + \frac{|x - x'|}{(\frac{\delta_x}{2} \wedge r_0)} \right] \sup_U |u| \tag{8}$$

or when $\alpha = 1$

$$|\nabla u(x)| \leq c \left[r_0 \wedge \frac{\delta_x}{2} \right]^{-1} |\sup_U |u| \, . \tag{9}$$

For positive solutions u, note that (8) implies Harnack's inequality and therefore the term $\sup_U |u|$ in (9) may be replaced by $u(x)$. This completes the proof. ∎

As an application we consider the rate of decay of eigenfunctions at the boundary of nodal domains in the case $\alpha = 1$, i.e. $q \in K_{d+1}^{loc}(D)$. Suppose then ϕ_k^q is the kth eigenfunction for $-\frac{1}{2}\Delta + q$ on D with Dirichlet data. Let λ_k be the corresponding eigenvalue and take D_k to be any of the nodal domains for ϕ_k^q, i.e. any connected component of either $\{\phi_k^q > 0\}$ or $\{\phi_k^q < 0\}$.

PROPOSITION 3. *The boundary of D_k is given locally as the graph of a Lipschitz function.*

PROOF: The Lipschitz function is obtained from the implicit function theorem and the observation that $\partial D_k \subset \{\phi_k^q = 0\}$. Thus it suffices to show that ϕ_k^q is Lipschitz. Select a finite cover of ∂D_k by balls $B(x_i, \delta)$, $x_i \in \partial D_k$, $i = 1, \ldots, N$ such that $B(x_i, 2\delta) \subset D$ for $i = 1, \ldots, N$ and δ is so small that with $q^{(k)} = q + \lambda_k^q$

$$E_z^x \left[e_{q^{(k)}}(\tau_{B(x_i,2\delta)}) \right] \leq c, \quad x, z \in \overline{B(x_i, 2\delta)} \, .$$

Then $w_i^{(k)} = 1^{st}$ eigenvalue for $-\frac{1}{2}\Delta + q^{(k)}$ on $B(x_i, 2\delta)$ is positive for each $i = 1, \ldots, N$. Thus by Theorem 1, ϕ_k^q is Lipschitz on each $B(x_i, \delta)$. This completes the proof of the proposition. ∎

The next proposition has been proven by Davies and Simon (1984) when q is bounded and for $q \in K_d^{loc}(D)$ is due to R. Bañuelos (1989).

PROPOSITION 4 (BAÑUELOS). *Suppose $q \in K_d^{loc}(D)$, ∂D is Lipschitz $\lambda_1^q > 0$ and φ_1^q and φ_1 are the first eigenfunctions for $-\frac{1}{2}\Delta + q$ and $-\frac{1}{2}\Delta$ on D. Then there exist positive constants c_1 and c_2 such that*

$$c_1 \varphi_1(x) \leq \varphi_1^q(x) \leq c_2 \varphi_1(x), \quad x \in D \, .$$

Combining Propositions 2 and 3 with the well-known estimate: there exist positive constants $c_1, c_2, \alpha_1, \alpha_2$ such that

$$c_1\delta_x^{\alpha_1} \leq \varphi_1(x) \leq c_2\delta_x^{\alpha_2}$$

we obtain

COROLLARY 4. *If $\lambda_1^q > 0$ with $q \in K_{d+1}^{loc}(D)$ where D has a Lipschitz boundary, then there exist constants $c_{1,k}, c_{2,k}, \alpha_{1,k}, \alpha_{2,k}$ such that the kth eigenfunction φ_k^q for $-\frac{1}{2}\Delta + q$ satisfies*

$$c_{1,k}\ dist(x, \partial D_k)^{\alpha_{1,k}} \leq |\varphi_k(x)| \leq c_{2,k}\ dist(x, \partial D_k)^{\alpha_{2,k}}$$

for x near the boundary of the nodal domain D_k.

Remark. This rate of decay was already known near ∂D.

REFERENCES

[1] M. Aizenman, B. Simon (1982). Brownian motion and Harnack inequality for Schrödinger operators, *Comm. Pure Appl. Math.* **35**, 209–273.

[2] R. Bañuelos (1989). Intrinsic ultracontractivity and eigenfunction estimates for Schrödinger operators, (preprint).

[3] M. Cranston (1989). Gradient estimates using coupling, (preprint).

[4] M. Cranston, E. Fabes, Z. Zhao (1987). Conditional gauge and potential theory for the Schrödinger operator, *TAMS, Vol. 307, No. 1*, 171–194.

[5] E.B. Davies, B. Simon (1984). Ultracontractivity and the heat kernel for Schrödinger operators and Dirichlet Laplacians, *J. Func. Anal.* **59**, 335–395.

[6] J.L. Doob (1983). *Classical Potential Theory and Its Probabilistic Counterpart*, Springer-Verlag, New York.

[7] T. Lindvall, L.C.C. Rogers (1986). Coupling of multidimensional diffusions by reflection, *Annals of Prob., Vol. II,* **3**, 860–872.

[8] K.-O. Widman (1967). Inequalities for the Green function and boundary continuity of the gradient of solutions of elliptic differential equations, *Ann. Inst. Fourier, Grenoble 15,* **1**, 189–258.

[9] Z. Zhao (1986). Green function for the Schrödinger operator and conditional Feynman-Kac gauge. *J. Math. Anal. Appl.* **116**, 309–334 (1986).

Michael C. Cranston
University of Rochester
Mathematics Department
Rochester, NY 14627, U.S.A.

Zhongxin Zhao
Department of Mathematics
University of Missouri
Columbia, MO 65211

On Estimating the Hypercontractive Constant of a Diffusion Process on a Compact Manifold

Jean-Dominique Deuschel

Let M be an N-dimensional, connected, compact, C^∞ Riemannian manifold; we will use $(\ |\ ), \lambda, \nabla$, and Δ to denote the inner-product, normalized measure, gradient and Laplacian. Also $\mathbf{M}_1(M)$ will be the set of probability measures on M and for $\mu \in \mathbf{M}_1(M)$ we write

$$< f >_\mu = \int_M f\, d\mu \qquad f \in L^1(\mu).$$

Set $\Omega = C\big([0,\infty); M\big)$ and define the evaluation map $X_t : \Omega \longrightarrow \Omega$ and the σ-algebra $\mathcal{B}_t$ for $t \in [0,\infty)$ accordingly. For each $x \in M$ there is precisely one $P_x \in \mathbf{M}_1(\Omega)$ which solves the martingale problem for Δ on $C^\infty(M)$ at x. $\{P_x : x \in M\}$ forms a Feller continuous Markov family and we will denote by $\{P_t : t > 0\}$ the associated Markov semigroup. Clearly λ is $\{P_t : t > 0\}$-reversing and if $\mathcal{E}$ is the associated Dirichlet form on $L^2(\lambda)$ then

$$\mathcal{E}(f,g) =< (\nabla f|\nabla g) >_\lambda, \qquad f,g \in C^\infty(M).$$

Because we are working with compact manifolds, there is no question that the diffusion determined by Δ is ergodic. In particular the semigroup $\{P_t, t > 0\}$ is λ-**hypercontractive**: there is $\alpha \in (0,\infty)$ such that

(1) $\quad \|P_t f\|_{L^{q(t,p)}(\mu)} \leq \|f\|_{L^p(\mu)}, \qquad t \in (0,\infty)$, for $f \in C(M)$ and $p \in (1,\infty)$,

where

$$q(t,p) = 1 + (p-1)e^{2\alpha t}.$$

The aim of this paper is to give some estimates of the **hypercontractive constant** α in terms of the geometry of M.

Let us introduce the following rate functions $\mathbf{H}(\cdot|\lambda)$ and $J_\mathcal{E}$ on $\mathbf{M}_1(M)$:

$$\mathbf{H}(\nu|\lambda) = \begin{cases} < f\log(f) >_\lambda & d\nu = f d\lambda \\ \infty & \text{otherwise} \end{cases} \quad \text{and } J_\mathcal{E}(\nu) = \begin{cases} \mathcal{E}(f^{1/2}, f^{1/2}) & d\nu = f d\lambda \\ \infty & \text{otherwise.} \end{cases}$$

$\mathbf{H}(\nu|\lambda)$ is the **relative entropy** of ν with respect to λ and governs the large deviations of the empirical distributions of independent λ-distributed M-valued random variables; whereas $J_\mathcal{E}$ is the rate function of the large deviations of empirical distribution functional of the diffusion process $\mathbf{L}_t : \Omega \longrightarrow \mathbf{M}_1(M)$:

$$\mathbf{L}_t(\omega) \equiv \frac{1}{t}\int_0^t \delta_{X_s(\omega)} ds$$

During the period of this research, the author was partially supported by an NSF grant.

under P_x, cf. Donsker and Varadhan [4]. By L. Gross' well known result [5], (1) is equivalent with a **logarithmic Sobolev inequality** of the form

$$\mathbf{H}(\nu|\lambda) \leq \frac{2}{\alpha} J_{\mathcal{E}}(\nu), \qquad \nu \in \mathbf{M}_1(M). \tag{2}$$

Following D. Bakry and M. Emery [1], let us introduce

$$\Gamma(f,f) \equiv \frac{1}{2}\left[\Delta(\|\nabla f\|^2) - 2(\nabla f|\nabla \Delta f)\right], \qquad f \in C^\infty(M),$$

and set

$$\mathbf{I}(\nu) = \langle f\,\Gamma(\log(f),\log(f))\rangle_\lambda = \langle \Gamma(\log(f),\log(f))\rangle_\nu,$$

for $\nu \in \mathbf{M}_1(M)$ with $\frac{d\nu}{d\lambda} = f \in C^\infty(M;(0,\infty))$.

Next let $d\nu = f d\lambda$ and set $\nu_t(\cdot) = \int_M \nu(dx)P_t(x,\cdot)$. Since λ is $\{P_t : t > 0\}$-reversing it is clear that $d\nu_t = P_t f\, d\lambda$ and by Jensen's inequality $t : [0,\infty) \longrightarrow \mathbf{H}(\nu_t|\lambda)$ is non-increasing. One could also wonder whether the same is true for $t : [0,\infty) \longrightarrow J_{\mathcal{E}}(\nu_t)$. A partial answer to this question in contained in the following result of [1] which will play a key role in our estimates of α.

THEOREM 3. *For $\nu \in \mathbf{M}_1(M)$ with $\nu << \lambda$ set $\nu_t(\cdot) = \int_M \nu(dx)P_t(x,\cdot)$, $t > 0$, then*

$$\frac{d}{dt}\mathbf{H}(\nu_t|\lambda) = -4J_{\mathcal{E}}(\nu_t) \quad \text{and} \quad \frac{d}{dt}J_{\mathcal{E}}(\nu_t) = -\frac{1}{2}\mathbf{I}(\nu_t), \qquad t > 0. \tag{4}$$

Moreover (2) is equivalent with

$$\mathbf{H}(\nu_t|\lambda) \leq e^{-2\alpha t}\mathbf{H}(\nu|\lambda), \qquad \nu \in \mathbf{M}_1(M) \text{ and } t > 0. \tag{5}$$

Finally if $\epsilon > 0$ is such that

$$J_{\mathcal{E}}(\nu_t) \leq \frac{1}{4\epsilon}\mathbf{I}(\nu_t) \qquad t > 0, \tag{6}$$

then (2) holds with $\alpha \geq \epsilon$.

Before we try to use (6) in order to get estimates for α, it is interesting to relate (1) with a **spectral gap** estimate of the form:

$$\|f - < f >_\lambda \|^2_{L^2(\lambda)} \leq \frac{1}{C} < \|\nabla f\|^2 >_\lambda, \qquad f \in C^\infty(M), \tag{7}$$

for some $C > 0$; or equivalently by the Spectral Theorem

$$< \|\nabla f\|^2 >_\lambda \leq \frac{1}{C} < (\Delta f)^2 >_\lambda, \qquad f \in C^\infty(M).$$

Note that the **spectral gap** C always dominates α, cf. B. Simon [6] and in some examples they coincide. Just as 2α is the exponential rate at which the relative entropy decays, C is the exponential rate at which $\{P_t : t > 0\}$ tends in $L^2(\lambda)$ to equilibrium:

$$\|P_t f - \langle f \rangle_\lambda \|_{L^2(\lambda)} \le e^{-Ct} \|f - \langle f \rangle_\lambda \|_{L^2(\lambda)}, \qquad t > 0 \text{ and } f \in C(M).$$

(7) can also be interpreted in terms of variances of central limit theorems for independent λ-distributed M-valued random variables and for the empirical distribution of the process. More precisely introduce the variance operator $\sigma^2 : L^2(\lambda) \longrightarrow [0, \infty)$,

$$\sigma^2(f) = \mathcal{E}(\Delta^{-1}(f - \langle f \rangle_\lambda), \Delta^{-1}(f - \langle f \rangle_\lambda)),$$

and let $\hat{\mathbf{L}}_t(f) : \Omega \longrightarrow \mathbf{R}$

$$\hat{\mathbf{L}}_t(f)(\omega) = \frac{1}{t^{1/2}} \int_0^t (f(X_s(\omega)) - \langle f \rangle_\lambda) ds, \quad t > 0.$$

Using standard martingale techniques, one can easily show that the law of $\hat{\mathbf{L}}_t(f)$ under P_x converges weakly as $t \to \infty$ to the centered Gaussian distribution with variance $\sigma^2(f)$. (7) is then equivalent with

$$\sigma^2(f) \le \frac{1}{C} \|f - \langle f \rangle_\lambda \|^2_{L^2(\lambda)}, \qquad f \in L^2(\lambda).$$

We turn now to the problem of estimating α via (6) using the Bochner-Lichnerowicz-Weitzenböck formula:

$$\Gamma(f, f) = \|\text{Hess} f\|^2 + \text{Ric}(\nabla f, \nabla f), \qquad f \in C^\infty(M),$$

where $\|\text{Hess} f\|$ is the Hilbert-Schmidt norm of the Hessian of f and Ric is the Ricci tensor. Next, let

$$\kappa(x) \equiv \min \left\{ \frac{\text{Ric}(\nabla f, \nabla f)}{\|\nabla f\|^2}(x) : f \in C^\infty(M) \right\}, \quad x \in M,$$

be the lowest eigenvalue of the Ricci curvature tensor at x, and set

$$\tilde{\kappa} = \min_{x \in M} \kappa(x).$$

In view of the above the following estimates were obtained in [3]:

Theorem 8. *Let $C > 0$ be the spectral gap of Δ, then*

$$\alpha \ge \frac{3}{N+2} C + \frac{N}{N+2} \tilde{\kappa}, \tag{9}$$

in particular

$$\alpha \ge \frac{N}{N-1} \tilde{\kappa} \text{ if } \tilde{\kappa} > 0, \qquad \text{and} \qquad \alpha \ge \frac{3\pi^2}{2(N+2)d(M)^2} \text{ if } \tilde{\kappa} \ge 0, \tag{10}$$

where $d(M)$ is the diameter of the manifold M. Moreover

$$\alpha \ge \frac{3}{N+2} \Lambda \left(\frac{N+3}{3} \kappa \right), \tag{11}$$

where

$$\Lambda(V) = \inf \{ \langle \|\nabla \psi\|^2 + V\psi^2 \rangle_\lambda : \psi \in C^\infty(M) \text{ and } \|\psi\|_{L^2(\lambda)} = 1 \}$$

is the lowest eigenvalue of the operator $-\Delta + V$.

Next we will give two different methods of using the estimate (11) . Since

$$\frac{N+3}{N+2}\tilde{\kappa} \le \frac{3}{N+2}\Lambda\left(\frac{N+3}{3}\kappa\right) \le \frac{N+3}{N+2} < \kappa >_\lambda,$$

(11) is useless unless $< \kappa >_\lambda > 0$.

PROPOSITION 12. *Let $C > 0$ be the spectral gap of Δ and set*

$$W(\kappa) = \frac{1}{2}\Big(\|\kappa - < \kappa >_\lambda \|_{L^\infty(\lambda)} + \|\kappa - < \kappa >_\lambda \|_{L^2(\lambda)}\Big),$$

then

$$\alpha \ge \frac{N+3}{N+2}\left\{< \kappa >_\lambda - \frac{N+3}{3C}W(\kappa)^2\right\}. \tag{13}$$

Next for $\tilde{\kappa} \le 0$, set $\beta \in [0,\infty) \longrightarrow \phi(\beta) = \log(< \exp[-\beta\kappa] >_\lambda)$. Assuming that

$$\min_{\beta\ge 0} \phi(\beta) < -\frac{2(N+2)}{3},$$

let $0 < \beta_1 < \beta_2 < \infty$ be such that $\phi(\beta_1) = \phi(\beta_2) = -\frac{2(N+2)}{3}$. If we a priori know that $\alpha \ge \frac{2(N+3)}{3\beta_2}$, then

$$\alpha \ge \frac{2(N+3)}{3\beta_1}. \tag{14}$$

PROOF: (13) will follow from the following estimate

$$\Lambda(V) \ge < V >_\lambda - \frac{W(V)^2}{C}, \qquad V \in C(M). \tag{15}$$

Take $\psi \in C^\infty(M)$ with $\|\psi\|_{L^2(\lambda)} = 1$, then by (7) and Hölder's inequality

$$\begin{aligned}
\langle V\psi^2\rangle_\lambda &= < V >_\lambda + \langle (V - < V >_\lambda)(\psi^2 - < \psi >_\lambda^2)\rangle_\lambda \\
&= < V >_\lambda + \langle (V - < V >_\lambda)(\psi + < \psi >_\lambda)(\psi - < \psi >_\lambda)\rangle_\lambda \\
&\ge < V >_\lambda - \Big(< \psi >_\lambda \|V - < V >_\lambda \|_{L^2(\lambda)} \\
&\quad + \|\psi\|_{L^2(\lambda)}\|V - < V >_\lambda \|_{L^\infty(\lambda)}\Big)\|\psi - < \psi >_\lambda \|_{L^2(\lambda)} \\
&\ge < V >_\lambda - 2W(V)\sqrt{\frac{< \|\nabla\psi\|^2 >_\lambda}{C}}.
\end{aligned}$$

Thus

$$\begin{aligned}
\Lambda(V) \ge \inf\Big\{ &< V >_\lambda - 2W(V)\sqrt{\frac{< \|\nabla\psi\|^2 >_\lambda}{C}} + \langle \|\nabla\psi\|^2\rangle_\lambda \\
&: \psi \in C^\infty(M) \text{ and } \|\psi\|_{L^2(\lambda)} = 1\Big\} \ge < V >_\lambda - \frac{W(V)^2}{C}.
\end{aligned}$$

Next note that ϕ is strictly convex with $\phi(0) = 0, \frac{d}{d\beta}\phi(0) = - < \kappa >_\lambda$ and $\lim_{\beta\to\infty}\phi(\beta) = \infty$ if $\tilde{\kappa} < 0$, respectively 0 if $\tilde{\kappa} = 0$. Also (2) is equivalent with

$$\Lambda(V) \ge -\frac{\alpha}{2}\log(< \exp[-\frac{2}{\alpha}V] >_\lambda), \qquad V \in C(M),$$

cf. §6.1 of [**2**]. Thus (11) implies that

$$\alpha \ge \frac{3}{N+2}\Lambda\left(\frac{N+3}{3}\kappa\right) \ge -\frac{3\alpha}{2(N+2)}\log\left(\left\langle \exp\left[-\frac{2(N+3)}{3\alpha}\kappa\right]\right\rangle_\lambda\right).$$

Therefore $\phi(\frac{2(N+3)}{3\alpha}) \ge -\frac{2(N+2)}{3}$ and (14) holds. ∎

We conclude this note by looking at a gradient perturbation of the Laplacian. More precisely for fixed $U \in C^\infty(M)$ consider the differential operator

$$Lf \equiv \Delta f - (\nabla U | \nabla f), \qquad f \in C^\infty(M),$$

and the measure $\mu \in \mathbf{M}_1(M)$:

$$d\mu = \frac{\exp[-U]}{< \exp[-U] >_\lambda} d\lambda.$$

Note that L is an essentially self-adjoint operator on $L^2(\mu)$ and $\mathcal{E}^U$, the associated Dirichlet form on $L^2(\mu)$, is given by

$$\mathcal{E}^U(f,g) =< (\nabla f | \nabla g) >_\mu, \qquad f,g \in C^\infty(M).$$

Clearly the corresponding semigroup $\{P_t^U, t > 0\}$ is again μ-hypercontractive. Our goal is to derive an estimate of $\alpha(U)$, the hypercontractivity constant, in terms of the spectral gap $C(U)$ using Bakry and Emery's technique as in Theorem 8.

In this context the rôle of Γ is played by

$$\Gamma^U(f,f) \equiv \frac{1}{2}\left[L(\|\nabla f\|^2) - 2(\nabla f | \nabla L f)\right] = \|\mathrm{Hess} f\|^2 + (\mathrm{Ric} + \mathrm{Hess} U)(\nabla f, \nabla f),$$

$f \in C^\infty(M)$. In particular equality (4) in Theorem 3 holds with λ replaced by μ and Γ by Γ^U, cf. [**1**]. For $\delta \in (0,\infty)$ let

$$\kappa_\delta(x) \equiv \min\left\{\frac{(\mathrm{Ric} + \mathrm{Hess} U)(\nabla f, \nabla f) - \delta(\nabla U|\nabla f)^2}{\|\nabla f\|^2}(x) : f \in C^\infty(M)\right\}, \; x \in M,$$

and set

$$\tilde{\kappa}_\delta = \min_{x \in M} \kappa_\delta(x).$$

The following generalizes Theorem 8 for the operator L:

THEOREM 16. *Let $C(U) > 0$ be the spectral gap of L, then for any $\delta \in (0,1)$,*

$$\alpha(U) \geq \frac{3\delta}{N+2\delta} C(U) + \frac{N}{N+2\delta} \tilde{\kappa}_{\frac{\delta}{N(1-\delta)}}. \tag{17}$$

In particular if $\delta = \frac{N}{N+1}$, then

$$\alpha(U) \geq \frac{3}{N+3} C(U) + \frac{N+1}{N+3} \tilde{\kappa}_1.$$

PROOF: From (4) and (6) in Theorem 3, we see that (17) follows from

$$\left\langle f\left(\|\mathrm{Hess}(\log f)\|^2 + (\mathrm{Ric} + \mathrm{Hess} U)(\nabla \log f, \nabla \log f)\right)\right\rangle_\mu \geq 4\epsilon(\delta) \left\langle \|\nabla f^{1/2}\|^2 \right\rangle_\mu,$$

for strictly positive $f \in C^\infty(M)$ with

$$\epsilon(\delta) = \frac{3\delta}{N+2\delta} C(U) + \frac{N}{N+2\delta} \tilde{\kappa}_{\frac{\delta}{N(1-\delta)}}.$$

Using the spectral gap of L and the definition of $\tilde{\kappa}_{\frac{\delta}{N(1-\delta)}}$, the above inequality will be an easy consequence of the following estimate:

$$\begin{aligned}(18)\quad \left\langle f\|\mathrm{Hess}(\log f)\|^2\right\rangle_\mu &\geq \frac{4\delta}{N+2\delta}\Big(2\left\langle\|\mathrm{Hess}f^{1/2}\|^2\right\rangle_\mu + \left\langle(Lf^{1/2})^2\right\rangle_\mu \\ &\qquad - \frac{1}{1-\delta}\left\langle(\nabla U|\nabla f^{1/2})^2\right\rangle_\mu\Big) \\ &= \frac{4\delta}{N+2\delta}\Big(3\left\langle(Lf^{1/2})^2\right\rangle_\mu - 2\left\langle(\mathrm{Ric}+\mathrm{Hess}U)(\nabla f^{1/2},\nabla f^{1/2})\right\rangle_\mu \\ &\qquad - \frac{1}{1-\delta}\left\langle(\nabla U|\nabla f^{1/2})^2\right\rangle_\mu\Big),\end{aligned}$$

cf. Proof of Theorem 1.10 in [**3**].

In order to prove (18), set $h = f^{1/2}$; then integration by parts yields

$$(19)\qquad \left\langle f\|\mathrm{Hess}(\log f)\|^2\right\rangle_\mu = 4\left\langle\|\mathrm{Hess}h\|^2\right\rangle_\mu + 4\left\langle\frac{Lh\|\nabla h\|^2}{h}\right\rangle_\mu,$$

cf. (1.13) of [**3**]. On the other hand

$$\left\langle f(L(\log f))^2\right\rangle_\mu = 4\left[\left\langle(Lh)^2\right\rangle_\mu - 2\left\langle\frac{Lh\|\nabla h\|^2}{h}\right\rangle_\mu + \left\langle\frac{\|\nabla h\|^4}{h^2}\right\rangle_\mu\right].$$

Thus since

$$\begin{aligned}(L(\log f))^2 &\leq \frac{1}{\delta}(\Delta(\log f))^2 + \frac{1}{1-\delta}(\nabla U|\nabla(\log f))^2 \\ &\leq \frac{N}{\delta}\|\mathrm{Hess}(\log f)\|^2 + \frac{4h^{-1/2}}{1-\delta}(\nabla U|\nabla h)^2\end{aligned}$$

$\delta \in (0,1)$, we get

$$4\left\langle\frac{Lh\|\nabla h\|^2}{h}\right\rangle_\mu \geq 2\left\langle(Lh)^2\right\rangle_\mu - \frac{2}{1-\delta}\left\langle(\nabla U|\nabla h)^2\right\rangle_\mu - \frac{N}{2\delta}\left\langle f\|\mathrm{Hess}(\log f)\|^2\right\rangle_\mu.$$

Combining this with (19) yields the desired estimate (18). ∎

REFERENCES

[1] D. Bakry and M. Emery, *Diffusions hypercontractives*, in "Séminaire de Probabilités XIX," Springer Lecture Notes in Mathematics **1123**, 1985, pp. 179–206.

[2] J.D. Deuschel and D.W. Stroock, "Large Deviations," Pure and Applied Mathematics # 137, Academic Press, Boston, 1989.

[3] J.D. Deuschel and D.W. Stroock, *Hypercontractivity and spectral gap of symmetric diffusions with applications to the stochastic Ising model*, J. Funct. Anal. (1990) (to appear).

[4] M.D. Donsker and S.R.S. Varadhan, *Asymptotic evaluation of certain Markov process expectations for large time, III*, Comm. Pure Appl. Math. **29** (1976), 389–461.

[5] L. Gross, *Logarithmic Sobolev inequalities*, Am. J. Math. **97** (1976), 1061–1083.

[6] B. Simon, *A remark on Nelson's best hypercontractive estimate*, Proc. Amer. Math. Soc. **55** (1976), 376–378.

Department of Mathematics, White Hall, Cornell University, Ithaca, NY 14853

Can the Schrödinger Equation be a Boltzmann Equation?

Masao Nagasawa

1. Introduction

A possible answer to the question raised in the title can be given as a solution to the inverse problem of McKean's propagation of chaos for a system of interacting diffusion processes. It yields a sort of Boltzmann equation, which is called McKean-Vlasov equation for the macroscopic spatial distribution of the system. By the inverse problem we mean: Find a microscopic system of interacting diffusion processes for a given (spatial) macroscopic distribution. To solve the problem we postulate that the given distribution agrees with that of an *intermediate* diffusion process. This makes it possible to identify the intermediate diffusion process with the McKean process which is the motion of a *typical* particle in the microscopic system, when the number of particles in it is large enough. This identification can be carried out by equating the drift coefficients of both processes. As is well known the drift coefficient of the (intermediate) diffusion process is given through the duality formula in terms of its distribution density $g(s,x)$. In a stationary case

$$b(x) = \frac{1}{2}\nabla \log g(x),$$

where $b(x)$ denotes the drift coefficient (for general cases see (50) in

---

To the memory of the late Professor Motoo KINOSHITA

Section 6). On the other hand, that of McKean's process is represented in a form of mean field interaction

$$b(x) = \int \boldsymbol{h}(x - y)g(y)dy,$$

where $\boldsymbol{h}(x - y)$ denotes a pair interaction of the system and $g(x)$ is the macroscopic spatial distribution of the system (for general cases see (51) in Section 6). Therefore, the identification of the Schrödinger process and the McKean process yields an important relation between the distribution and the pair interaction

$$\text{(i)} \qquad \tfrac{1}{2}\nabla \log g(x) = \int \boldsymbol{h}(x - y)g(y)dy,$$

which plays a key rôle in our theory (cf. (52) for non-stationary cases). In the above formulation the distribution density g can be arbitrarily chosen; for example,

$$\text{(ii)} \qquad g(x) = |\,\psi(x)\,|^2,$$

where $\psi(x)$ may be a stationary solution of a Schrödinger equation (i.e. an eigenvalue problem). Therefore, if we can find a pair interaction $\boldsymbol{h}(x - y)$ solving the equation (i) and prove the propagation of chaos for a system with this interaction, then it will be allowed to interpret the Schrödinger equation, which provides the distribution $g(x)$ in (ii), as a macroscopic (statistical) equation describing a system of interacting particles. This is a solution to the inverse problem stated above.

If $g(x) > 0$ (of the ground state), there is essentially no problem to follow the above arguments: We can find a pair interaction $\boldsymbol{h}(x - y)$ and apply a classical theorem of McKean (1967). However, if $g(x)$ is not strictly positive (*e.g.* of an excited state), we must handle the problem with great care, because the nodal set of the distribution density $g(x)$ becomes a source of strong singularities of the drift coefficient. Because of the singularities, the equation (i) may have no solution, unless we introduce an additional structure, *segregation of particles*. Furthermore, the classical McKean theorem of the propagation of chaos can not be applied in this case of singular drift.

In Section 2 we define Schrödinger processes which retain a close parallelism between real valued quantities in the theory of diffusion processes and complex valued quantities in quantum mechanics. In Sections 3 and 4 diffusion processes with singular drift and Schrödinger processes will be constructed. They will be characterized in terms of a stochastic variational principle in Section 5. The subject treated in these sections may be of independent interest in the theory of singular diffusion processes. However, those who like have an outline of the main theme can skip Sections 3,4 and 5, assuming the existence of Schrödinger processes, and come back later on. Section 6 is devoted to systems of singular integral equations in connection with Schrödinger processes, and the propagation of chaos for systems of interacting diffusion processes will be discussed in Section 7. Some of applications will be explained in Section 8.

2. Schrödinger processes

Let S be a measurable space. On the product space $[a,b]\times S$, where $-\infty < a < b < +\infty$, we define Schrödinger processes in terms of a pair $(\phi,\widehat{\phi})$ of non-negative measurable functions and a transition density $p(s,x;t,y)$, satisfying

$$\phi(t,x) = \int p(t,x;b,y)\phi(b,y)m(dy), \tag{1}$$

and

$$\widehat{\phi}(t,x) = \int m(dz)\widehat{\phi}(a,z)p(a,z;t,x), \tag{2}$$

where m is a fixed measure (if $S = \mathbf{R}^d$, m is the Lebesgue measure). The transition density is assumed to satisfy the Chapman-Kolmogorov equation ($\phi(s,x)m(dx)$-$a.e.$), but in general not necessarily normalized in order to allow creation and annihilation.

In terms of a given triplet $(\phi,\widehat{\phi},p)$ obeying (1) and (2), a Schrödinger process is characterized by its finite dimensional distributions; namely, in the simplest case of $a < t < b$,

$$m(dz)\widehat{\phi}(a,z)p(a,z;t,x)m(dx)p(t,x;b,y)\phi(b,y)m(dy). \tag{3}$$

General cases can be defined analogously. Therefore, the Schrödinger process is not a Markov process with respect to the transition density $p(s,x;t,y)$, but a Markov field with given boundary values $(\widehat{\phi}(a,.)$, $\phi(b,.))$ on the time axis. However, it turns out to be a Markov process on $D = \{(s,x):\phi(s,x) > 0\}$ with another transition *probability* density $q(s,x;t,y)$ defined on $D\times D$ by

$$q(s,x;t,y) = \frac{1}{\phi(s,x)}p(s,x;t,y)\phi(t,y), \tag{4}$$

and with distribution densities

$$\widehat{\phi}(t,x)\phi(t,x), \quad at \quad t \in [a,b].$$

This is a simple consequence of (3) combined with (1).

In an article published in (1931) Schrödinger considered the following problem: Find a pair $(\phi,\widehat{\phi})$ of functions satisfying (1) and (2) for a given transition density $p(s,x;t,y)$ and fixed initial and final distributions (this will be treated in Section 4). Here we formulate Schrödinger's problem in another way: Find a transition density $p(s,x;t,y)$ satisfying (1) for a given $\phi(t,x)$. Once $p(s,x;t,y)$ is known, $\widehat{\phi}(t,x)$ can be defined by (2) for a non-negative measurable $\widehat{\phi}(a,.)$. In order to make this problem well-posed we must specify, in addition, a parabolic differential operator L, as will be explained in the following.

In connection with a solution ψ of a Schrödinger equation, a pair $(\phi,\widehat{\phi})$ can be defined as follows: Assume that $R = \mathcal{Re}(\log\psi)$ and $S = Im(\log\psi)$ are well-defined, i.e. $\psi = \exp(R + iS)$. Then we set

$$\begin{aligned} \phi(t,x) &= \exp(R(t,x) + S(t,x)), \\ \hat{\phi}(t,x) &= \exp(R(t,x) - S(t,x)). \end{aligned} \tag{5}$$

In short, the Schrödinger process corresponding to a solution ψ of the Schrödinger equation is a diffusion process (X_t,Q) such that

$$\psi\overline{\psi}(t,x)dx = Q[X_t \in dx],$$

where Q is a probability measure on the space of continuous paths defined in terms of finite dimensional distributions, say, by (3). In this case of a given wave function ψ the equations (1) and (2) are real valued counterparts of the well-known unitary transformation of wave functions, which is replaced by a transformation in terms of a semigroup kernel $p(s,x;t,y)$. It is clear that through the first equation (1) we get a *backward* prediction at a moment t, $a < t < b$, assuming $\phi(b,.)$ is prescribed at b and through the second one (2) a *forward* prediction at t assuming $\hat{\phi}(a,.)$ is given at a.

Let us consider a Schrödinger equation in $\mathbf{R}^d$

$$i\frac{\partial\psi}{\partial s} + \tfrac{1}{2}\Delta\psi + ia(s,x)\nabla\psi - V(s,x)\psi = 0, \tag{6}$$

where Δ denotes the symmetric A-Laplacian

$$\Delta = \frac{\partial}{\partial x^j} A^{jk}(s,x) \frac{\partial}{\partial x^k} \tag{7}$$

with a bounded positive definite symmetric diffusion matrix $A^{jk}(s,x) = (\sigma^{jk}(s,x))^2$, $\boldsymbol{a}(s,x)$ a vector potential, and $V(s,x)$ a scaler potential. We assume in addition a gauge condition $\nabla \boldsymbol{a} = 0$, which is harmless for Schrödinger processes, because they are invariant under the gauge transformations (cf. Wakolbinger-Stummer (preprint)).

It is easy to see that the ϕ and $\widehat{\phi}$ defined at (5) satisfy parabolic differential equations in duality

$$\frac{\partial p}{\partial s} + \frac{1}{2}\Delta p + \boldsymbol{a}\nabla p + cp = 0, \quad in\ D, \tag{8}$$

and

$$-\frac{\partial \widehat{p}}{\partial s} + \frac{1}{2}\Delta \widehat{p} - \boldsymbol{a}\nabla \widehat{p} + c\widehat{p} = 0, \quad in\ D, \tag{9}$$

respectively, where $c(t,x)$ is the *reference potential* of ϕ defined by

$$c = -\phi^{-1}L\phi, \tag{10}$$

with a parabolic differential operator

$$L = \frac{\partial}{\partial s} + \frac{1}{2}\Delta + \boldsymbol{a}\nabla. \tag{11}$$

Actually, the relation between c and V is given in terms of R by

$$c(s,x) = V(s,x) - \Delta R(s,x) - (\sigma\nabla R)^2(s,x),$$

or in terms of S

$$= - V(s,x) - 2\frac{\partial S}{\partial s}(s,x) - (\sigma\nabla S)^2(s,x) - 2\boldsymbol{a}\nabla S(s,x).$$

(cf. e.g. Theorem 7.1 in Nagasawa (1989)).

We have chosen the parabolic operator L as in (11) to establish a resemblance between the Schrödinger equation (6) and the diffusion equation (8). Another choice is possible but it will not give a simple relation between c(s,x) and V(s,x) as given above.

Therefore, the wanted transition density $p(s,x;t,y)$ must be the fundamental solution of the diffusion equation (8) with the (singular) creation and annihilation $c(s,x)$ defined at (10). Since there is no clasical existence theorem of nice fundamental solutions for (8) because of the singularity of $c(s,x)$, we will construct a diffusion process with $q(s,x;t,y)$ defined at (4) directly by means of transformations of diffusion processes without relying on the existence of fundamental solutions to (8) in the next section.

If a solution ψ of the Schrödinger equation (6) is already at hand, it is reasonable to construct a Schrödinger process for a given $\phi(t,x)$, namely defined at (5) (then it determines the drift coefficient of the Schrödinger process as will be seen in Section 3). However, if a wave function is not given in advance, it is relevant to proceed the other way round. We bigin with the diffusion equation (8), assuming a potential function $c(t,x)$ is given, and construct ϕ and $\hat{\phi}$. Then, a wave function ψ can be defined in terms of them. This will be discussed in Section 4.

3. Construction of Schrödinger processes for a given $\phi(t,x)$

The contents of this section are essentially adopted from Nagasawa (1989). Assuming sufficient regularity conditions on $\sigma = (\sigma(s,x)^{jk})$ and $\boldsymbol{a} = (a(s,x)^{j})$, which guarantee the existence of a fundamental solution $g(s,x;t,y)$ of

$$Lp = 0, \qquad in\ [a,b] \times \mathbf{R}^d, \tag{12}$$

where L is given at (11) (cf. S. Itô (1952), Friedman (1964)), we define a time inhomogeneous (or space-time) diffusion process $(X_t, P_{(s,x)})$ with $g(s,x;t,y)$ as its transition density with respect to the Lebesgue measure. It is assumed in this section that ϕ is continuous on $[a,b]\times\mathbf{R}^d$, and continuously differntiable once in t and twice in x for $(t,x) \in D = \{(s,x): s \in [a,b],\ \phi(s,x) > 0\}$.

Let us define a multiplicative functional by

$$N_s^t = \exp\left(-\int_s^t \frac{L\phi}{\phi}(v,X_v)dv\right)\frac{\phi(t,X_t)}{\phi(s,X_s)}1_{\{t<T_s\}}, \tag{13}$$

where T_s denotes the first hitting time to the nodal set of ϕ; $N = \{(t,x):\phi(t,x) = 0\}$. Then we have

Theorem 1. (i) *Let* N_s^t *be the multiplicative functional defined at* (13). *Then*

$$P_{(s,x)}[N_s^t] \le 1, \quad for\ (s,x) \in D, \quad s \le t \le b. \tag{14}$$

(ii) *Assume integrability conditions, for* $a \le s \le t \le b$,

$$P_{(s,x)}\left[\exp\left(\int_s^t c(v,X_v)dv\right)\phi(t,X_t);t<T_s\right] < \infty, \tag{15,a}$$

and

$$\int_s^t du P_{(s,x)}\left[\exp\left(\int_s^u c(v,X_v)dv\right)c^+\phi(u,X_u);u<T_s\right] < \infty, \tag{15,b}$$

where c^+ *is the positive part of the reference potential* c *of* ϕ *defined at* (10). *Then, the equality holds in* (14), *i.e.*

$$P_{(s,x)}[N_s^t] = 1, \quad for\ (s,x) \in D, \quad s \le t \le b.$$

If ϕ *is bounded, the conditions* (15,a,b) *can be replaced by a single condition*

$$P_{(s,x)}\left[\exp\left(\int_s^t c(v,X_v)dv\right);t<T_s\right] < \infty, \quad for\ a \le s \le t \le b. \tag{15'}$$

For a proof refer to Theorem 2.1 in Nagasawa (1989), where a proof in the case $A^{jk} = \delta^{jk}$ is given, and the second assertion is shown under an integrability condition

$$\int_s^t du P_{(s,x)}\Big[\exp\Big(\int_s^u c^+(v,X_v)dv\Big)c^+\phi(u,X_u);u<T_s\Big] < \infty.$$

However, we can replace this by a weaker condition (15,b). To do so we show first of all that if p fulfilles the conditions (15,a,b) with $|p|$ in place of ϕ and satisfies

$$Lp + cp = 0, \quad in\ D, \tag{17}$$

then it solves an integral equation

$$p(s,x) = \overline{P}_{(s,x)}[p(t,X_t);t<T_s] + \int_s^t du \overline{P}_{(s,x)}[c^+p(u,X_u);u<T_s], \tag{18}$$

where $\overline{P}_{(s,x)}[F;t<T_s]$ is defined with the negative part c^- of c by

$$\overline{P}_{(s,x)}[F(.);t<T_s] = P_{(s,x)}\Big[\exp\Big(-\int_s^t c^-(v,X_v)dv\Big)F(.);t<T_s\Big], \tag{19}$$

for any bounded $\mathcal{F}_s^t$-measurable function F. This can be done, applying Itô's formula to

$$\exp\Big(-\int_s^t c^-(u,X_u)du\Big)p(t,X_t),$$

(cf. Theorem 4.1, in Nagasawa (1989), and also Appendix of the present paper). Secondly, if the condition (15,a) is satisfied, then $p^o(s,x)$ defined by

$$p^o(s,x) = P_{(s,x)}\Big[\exp\Big(\int_s^t c(v,X_v)dv\Big)\phi(t,X_t);t<T_s\Big] \tag{20}$$

is a solution of (18) with $p(t,x) = \phi(t,x)$. We remark here that from the condition (15,a) with $p(t,x)$ in place of $\phi(t,x)$ follow

$$\overline{P}_{(s,x)}[p(t,X_t);t<T_s] < \infty,$$

and

$$\int_s^t du\overline{P}_{(s,x)}[c^+p(u,X_u);u<T_s] < \infty.$$

Let p_1 and p_2 be solutions of the integral equation (18) under the conditions (15,a,b) with $|p_i(t,.)|$ in place of $\phi(t,.)$, and $p_1(t,.) = p_2(t,.)$. Then $k = |p_1 - p_2|$ satisfies

$$k(s,x) \leq \int_s^t dv\overline{P}_{(s,x)}[c^+k(v,X_v);v<T_s]$$

$$\leq \int_s^t dv\overline{P}_{(s,x)}\Big[\frac{1}{(n-1)!}\Big(\int_s^v c^+(u,X_u)du\Big)^{n-1}c^+k(v,X_v);v<T_s\Big],$$

which vanishes as n tends to infinity under the assumption (15,b) with $|p_i(u,.)|$ in place of $\phi(u,.)$. Therefore, the uniqueness holds for solutions of the integral equations (18). Thus we conclude $\phi(s,x) = p^o(s,x)$, which is nothing but $P_{(s,x)}[N_s^t] = 1$.

Remark. The idea of using the process killed by c^- to prove the uniqueness is due to Sturm.

Remark. Theorem 1 remains valid if $\phi \in H^{1,2}(K)$ for any compact subset K in D, see Aebi (1989).

Applying a theorem of Dynkin (1965) and Kunita-Watanabe (1963), we have a diffusion process $(X_t,Q_{(s,x)},T_s)$ on D such that

$$Q_{(s,x)}[f(t,X_t);t<T_s] = P_{(s,x)}[f(t,X_t)N_s^t]$$

for any bounded measurable function f on D. Under the condition (16) the transformed process $(X_t,Q_{(s,x)},T_s)$ can be identified with the one obtained by the generalized Maruyama-Girsanov transformation of drift (cf. Maruyama (1954), Liptser-Shiryayev (1977)). It is clear that if the equality (16) holds, the transformed process $(X_t,Q_{(s,x)})$ does not hit the nodal set N of ϕ.

Since the transition semigroup of the transformed diffusion process $(X_t,Q_{(s,x)},T_s)$ is of the form $q = p/\phi$, it satisfies a diffusion equation with an additional drift $\sigma^2\nabla\phi/\phi$, which replaces the creation and killing term cp in (8), i.e.,

$$Lq + \sigma^2\frac{\nabla\phi}{\phi}\nabla q = 0,$$

in a weak sense because of (21) in Lemma 1 below.

Lemma 1. *Let p and ϕ be in $C^{1,2}(D)$ and $\phi > 0$ in D. Then $q = p/\phi$ satisfies*

(21) $$Lq + \sigma^2\frac{\nabla\phi}{\phi}\nabla q = \phi^{-1}(Lp + cp) - \phi^{-2}p(L\phi + c\phi), \text{ in } D.$$

If the equality (16) holds, the arguments of the absolute continuity of $Q_{(s,x)}$ with respect to $P_{(s,x)}$ can be applied (cf. e.g. Liptser-Shiryayev (1977)), showing that the process $(X_t,\mathcal{F}_s^t,Q_{(s,x)})$ is a semimartingale satisfying

$$X_t = x + \int_s^t \sigma(v,X_v)dB_v + \int_s^t (\tilde{a} + \sigma^2\frac{\nabla\phi}{\phi})(v,X_v)dv, \quad Q_{(s,x)}\text{-a.s.},$$

where $\tilde{a} = a + \frac{1}{2}\nabla\sigma^2$ and B_t is a Browian motion defined by

$$B_t = \int_s^t \sigma^{-1}(v,X_v)dX_v - \int_s^t (\sigma^{-1}\widetilde{a} + \sigma\frac{\nabla\phi}{\phi})(v,X_v)dv, \quad s \leq t \leq b.$$

Let us define Q by

$$Q[A] = \int \phi(a,x)\widehat{\phi}(a,x)dxQ_{(a,x)}[A],$$

where $\widehat{\phi}(a,x)$ is a measurable function on $\mathbf{R}^d$ satisfying

$$\int \phi(a,x)\widehat{\phi}(a,x)dx = 1.$$

Then, (X_t,Q) is the Schrödinger process for a given ϕ (and a parabolic differential operator L).

Let $p(s,x;t,f\phi)$ be a solution of (18) given by (20) with $f\phi$ in place of ϕ. Then the transition density $p(s,x;t,y)$ can be obtained as a density of $p(s,x;t,f\phi)$ with respect to $\phi(t,y)dy$,

$$p(s,x;t,f\phi) = \int p(s,x;t,y)dy\phi(t,y)f(y),$$

for any bounded continuous function f on $\mathbf{R}^d$.

Remark. For different treatments of diffusion processes with singular drift see also Carlen (1984), Carmona (1985), Zheng-Meyer (1984/85), Blanchard-Golin (1987), Norris (1988), and for a formulation of stochastic mechanics in terms of Schrödinger processes see Zambrini (1986), who assumes c is bounded so that $\phi > 0$.

4. Schrödinger processes for a given potential $c(s,x)$

Let $c(s,x)$ be a measurable function on $[a,b]\times\mathbf{R}^d$ which is continuous on the set $\{(s,x): |c(s,x)| < \infty\}$, and let $(X_t,P_{(s,x)})$, $(s,x)\in[a,b]\times\mathbf{R}^d$, be the diffusion process with drift $\boldsymbol{a}(s,x)$ considered in Section 3. In terms of the first hitting time $\tau_s = \inf\{t > s: | c(t,X_t) | = \infty\}$, our state space D is defined by

$$D = \{(s,x): P_{(s,x)}[\tau_s > 0] = 1\}.$$

We consider the diffusion equation (8) on the space D (resp. (9)). Let (q_a,q_b) be a pair of probability measures on $\mathbf{R}^d$, which satisfy the condition (23) below. We will costruct diffusion processes which have q_a and q_b as the initial and final distributions, respectively. However, we can't use the pair (q_a,q_b) directly for this purpose. Instead, we will find a pair of functions $(\hat{\phi}_a,\phi_b)$ which will be given in Lemma 2 below and use them as entrance and exit laws.

Taking a strictly positive continuous function $k(x)$ with

$$\int k(x)dx = 1,$$

we define a measure p_k on $\mathbf{R}^d \times \mathbf{R}^d$ by

$$p_k(A \times B) = \int k(x)1_A(x)dxP_{(a,x)}\left[\exp\left(\int_a^b c(v,X_v)dv\right)1_B(X_b);b<T_a\right],$$

where T_a denotes the first hitting time to the set $N = D^c$, and set

$$p = \frac{p_k}{p_k(\mathbf{R}^d \times \mathbf{R}^d)},$$

where we require an integrability condition

$$\text{(22)} \qquad p_k(\mathbf{R}^d \times \mathbf{R}^d) = \int k(x)dxP_{(a,x)}\left[\exp\left(\int_a^b c(v,X_v)dv\right);b<T_a\right] < \infty.$$

Let $\mathcal{E}(a,b)$ be the class of all probability measures on $\mathbf{R}^d \times \mathbf{R}^d$ with a given pair (q_a, q_b) of probability measures on $\mathbf{R}^d$ as marginals. We require a condition on the pair (q_a, q_b): There exists a probability measure $q \in \mathcal{E}(a,b)$ such that

$$\text{(23)} \qquad H(q \mid p) < \infty,$$

where $H(q \mid p) = \int dq \log(dq/dp)$ denotes the relative entropy of q with respect to p.

Remark. A sufficient condition to (23) given in Csiszar (1975) is this: If p and $q_a \times q_b$ are equivalent and $\log f \in L^1(q_a \times q_b)$, where f is the density of p with respect to $q_a \times q_b$, then the above condition is fulfilled.

Lemma 2. *Let q_a and q_b be proability measures on $\mathbf{R}^d$ such that there exists some q in $\mathcal{E}(a,b)$ with (23). Then there exists a pair of non-negative measurable functions $(\hat{\phi}_a, \phi_b)$ such that*

$$\text{(24)} \qquad \begin{aligned} q_a(A) &= \int \hat{\phi}_a(x) 1_A(x) dx P_{(a,x)}\Big[\exp\Big(\int_a^b c(v, X_v) dv\Big)\phi_b(X_b); b < T_a\Big], \\ q_b(B) &= \int \hat{\phi}_a(x) dx P_{(a,x)}\Big[\exp\Big(\int_a^b c(v, X_v) dv\Big)\phi_b(X_b) 1_B(X_b); b < T_a\Big], \end{aligned}$$

and $\log \hat{\phi}_a \in L^1(q_a)$ *and* $\log \phi_b \in L^1(q_b)$.

Proof. Let q be in $\mathcal{E}(a,b)$ and $H(q \mid p)$ be the relative entroy of q with respect to p. Then, by Csiszar (1975) (cf. also Föllmer (1988), (1.23) in II) there exists q^o in $\mathcal{E}(a,b)$ such that

$$\min_{q \in \mathcal{E}(a,b)} H(q \mid p) = H(q^o \mid p),$$

and

$$\frac{dq^o}{dp} = \frac{1}{k(x)} \widehat{\phi}_a(x)\phi_b(y)p_k(\mathbf{R}^d \times \mathbf{R}^d),$$

from which (24) follows as its marginals.

Remark. Lemma 2 generalizes a theorem of Fortet (1940), Beurling (1960) and Jamison (1974) for continuous positive kernels. Their theorem is not applicable to our kernels which vanish on a subset N and are not continuous in general.

In terms of $(c,\widehat{\phi}_a,\phi_b)$ we define a probability measure Q by

$$Q[F] = \int \widehat{\phi}_a(x)dxP_{(a,x)}\Big[\exp\Big(\int_a^b c(v,X_v)dv\Big)F(.)\phi_b(X_b);b<T_a\Big]. \tag{25}$$

Then, (X_t,Q) is the Schrödinger process for the given triplet (c,q_a,q_b), as will be shown in the following.

Let $\phi(s,x)$ and $\widehat{\phi}(s,B)$ be defined by

$$\phi(s,x) = P_{(s,x)}\Big[\exp\Big(\int_s^b c(v,X_v)dv\Big)\phi_b(X_b);b<T_s\Big], \tag{26}$$

$$\widehat{\phi}(s,B) = \int \widehat{\phi}_a(x)dxP_{(a,x)}\Big[\exp\Big(\int_a^s c(v,X_v)dv\Big)1_B(X_s);s<T_a\Big].$$

Then, the Markov property of $(X_t,P_{(s,x)})$ applied to (24) implies

$$\int \widehat{\phi}(s,dx)\phi(s,x) = 1,$$

and hence $\phi(s,x) < \infty$, $\widehat{\phi}(s,dx)$-$a.e.$. Moreover, we have by the Markov property of $(X_t,P_{(s,x)})$

(27) $$\phi(s,x) = P_{(s,x)}\Big[\exp\Big(\int_s^t c(v,X_v)dv\Big)\phi(t,X_t);t<T_s\Big],\ for\ a\leq s\leq t\leq b.$$

Let $g^o(s,x;t,y)$, $a \leq s \leq t \leq b$, be defined by

(28) $$g^o(s,x;t,y) = g(s,x;t,y) - P_{(s,x)}[g(T_s,X_{T_s};t,y);T_s<t],$$

in terms of the fundamental solution $g(s,x;t,y)$ of (12) and the diffusion process $(X_t,P_{(s,x)})$ (cf. Nagasawa (1989)). We apply now the backward description. Changing the direction of time, we define the backward (space-time) diffusion process $(X_s^*,P_{(t,y)}^*)$, with the time variable s ($<t$) running backward from t to a, by means of the same fundamental solution $g(s,x;t,y)$. Then, $g^o(s,x;t,y)$ defined at (28) can be given in terms of the backward process as

(28*) $$g^o(s,x;t,y) = g(s,x;t,y) - P_{(t,y)}^*[g(s,x;T_t^*,X_{T_t^*}^*);s<T_t^*].$$

Therefore, it is differentiable once in s and twice in x (resp. in (t,y)).

Remark. Corresponding to (26), we can apply the same arguments to $\widehat{\phi}_a(x)$ in defining $\widehat{\phi}(t,y)$ by means of the backward process.

If we define $p(s,x;t,y)$ and $p^*(s,x;t,y)$ by

$$\int p(s,x;t,y)f\phi(t,y)dy = P_{(s,x)}\Big[\exp\Big(\int_s^t c(v,X_v)dv\Big)\phi f(t,X_t);t<T_s\Big],$$

and

$$\int \widehat{\phi}f(s,x)dxp^*(s,x;t,y) = P_{(t,y)}^*\Big[\exp\Big(\int_s^t c(v,X_v^*)dv\Big)\widehat{\phi}f(s,X_s^*);T_t^*<s\Big],$$

respectively, for any bounded continuous function f on D, then

$$p(s,x;t,y) = p^*(s,x;t,y), \quad \widehat{\phi}(s,x)\phi(t,y)dxdy\text{-}a.e..$$

Thus we can have a regular version of the density $p(s,x;t,y)$, and

$$(29) \qquad \begin{aligned} \phi(s,x) &= \int p(s,x;t,y)dy\phi(t,y), \\ \widehat{\phi}(t,y) &= \int \widehat{\phi}(s,x)dxp(s,x;t,y). \end{aligned}$$

Therefore, (X_t,Q) is a Schrödinger process.

In order to construct a diffusion process $(X_t,Q_{(s,x)})$ in terms of a transformation, we require now the following integrability conditions;

$$(30) \qquad \begin{aligned} &\int g^o(s,x;b,y)\phi_b(y)dy < \infty, \\ &\int_s^b dv\int g^o(s,x;v,y)dy \,|\, c(v,y) \,|\, \phi(v,y) < \infty. \end{aligned}$$

Then $\phi(s,x)$ defined at (26) solves an integral equation

$$(31) \qquad \begin{aligned} p(s,x) = &\int g^o(s,x;b,y)\phi_b(y)dy \\ &+\int_s^b du\int g^o(s,x;u,y)dyc(u,y)p(u,y). \end{aligned}$$

Since $\phi(s,x)$ is a solution of (31), $\phi(t,X_t)$ is right continuous in t, $P_{(s,x)}$-$a.e.$, because the first term of the right-hand side of (31) is space-time harmonic and the second term is the difference of two potentials with c^+p and c^-p, respectively.

A multiplicative functional defined by

$$(32) \qquad N_s^t = \exp\left(\int_s^t c(v,X_v)dv\right)\frac{\phi(t,X_t)}{\phi(s,X_s)} 1_{\{t<T_s\}}$$

is, therefore, right continuous in t, moreover it is a continuous martingale, since (27) implies

$$P_{(s,x)}[N_s^t] = 1.$$

Remark. For the continuity of $\phi(s,x)$ in (s,x) we need additional conditions. Let $\phi_b \in L^1(D_b)$, where $D_b=\{x:(b,x)\in D\}$, $c\phi \in L^1(D)$, and

$$\lim_{\varepsilon\to 0}\sup_{x\in \overline{U}}\int_s^{s+\varepsilon} dv\int_{\overline{V}} g^o(s,x;v,y)dy\,|\,c(v,y)\,|\,\phi(v,y) = 0,$$

for a neighborhood U of any x_0 and any open V containing $\overline{U}$. Then, the $\phi(s,x)$ defined at (26) solves the integral equation (31) and as a solution of the equation (31), it is continuous in (s,x).

In fact, the first term of the right-hand side of the equation (31) and

$$\int_{s+\varepsilon}^{b} dv\int g^o(s,x;v,y)dyc\phi(v,y)$$

are continuous in (s,x) for $\varepsilon > 0$, because $g^o(s,x;v,y)$ is bounded for $v\text{-}s \geq \varepsilon$. Let $\overline{U}\subset V$. Then, we have

$$\sup_{x\in \overline{U}} P_{(s,x)}\,[X_v \in V^c{:}v<T_s] = o(v\text{-}s),$$

by the continuity of paths. Therefore

$$\lim_{\varepsilon\to 0}\sup_{x\in \overline{U}}\int_s^{s+\varepsilon} dv\int_{V^c} g^o(s,x;v,y)dyc\phi(v,y) = 0,$$

and hence, we conclude that $\phi(s,x)$ is continuous in (s,x).

Applying the transformation by means of the multiplicative functional

N_s^t defined at (32), we get a (space-time) diffusion process $(X_t, Q_{(s,x)})$ such that

$$Q_{(s,x)}[f(t,X_t)] = P_{(s,x)}[f(t,X_t)N_s^t],$$

for any bounded continuous function f on D. Finally, we can define a probability measure Q by

$$Q[\mathrm{A}] = \int \widehat{\phi}(a,x)\phi(a,x)dxQ_{(a,x)}[\mathrm{A}],$$

for an arbitrary $\widehat{\phi}(a,x)$ such that

$$\int \widehat{\phi}(a,x)\phi(a,x)dx = 1.$$

Then (X_t, Q) is a Schrödinger process.

Thus we have shown

Theorem 2. (i) *Let $c(t,x)$ be a measurable function defined on $[a,b]\times\mathbf{R}^d$ which is continuous on $\{(s,x): |c(s,x)| < \infty\}$ and satisfies the integrability condition* (22). *Let (q_a, q_b) be a pair of probability measures on $\mathbf{R}^d$ such that there exists a probability measure q on $\mathbf{R}^d\times\mathbf{R}^d$ satisfying* (23) *with the q_a and q_b as its marginals. Then, for the triplet (c,q_a,q_b) there exist a Schrödinger process (X_t, Q) with $\phi(t,x)$ and $\widehat{\phi}(t,x)$, $(t,x) \in [a,b]\times\mathbf{R}^d$, satisfying* (29),

$$q_a(dx) = \widehat{\phi}(a,x)\phi(a,x)dx \quad and \quad q_b(dy) = \widehat{\phi}(b,y)\phi(b,y)dy.$$

(ii) *In addition, assume the condition* (30). *Then, $\phi(t,x)$ solves the integral equation* (31). *Furthermore, by means of the multiplicative functional N_s^t defined at* (32) *a diffusion process $(X_t, Q_{(s,x)})$ is constructed for the given triplet (c,q_a,q_b). If $\phi \in H^{1,2}(K)$ for any compact subset K in D, then the given $c(t,x)$ turns out to be the (weak) reference potential of ϕ, i.e. $c(t,x) = -L\phi/\phi(t,x)$.*

Remark. Assuming that a potential $c(s,x)$ is given we can construct a solution of a Schrödinger equation. Let us define, in terms of the $\phi(t,x)$ and $\widehat{\phi}(t,x)$ in Theorem 2,

$$R = \frac{1}{2}\log(\widehat{\phi}\phi) \text{ and } S = \frac{1}{2}\log(\phi/\widehat{\phi}).$$

Then the complex valued function

$$\psi(t,x) = \exp(R(t,x) + \mathrm{i}S(t,x))$$

is a weak solution of the Schrödinger equation (6) with a potential

$$V = c + \Delta R + (\sigma\nabla R)^2.$$

5. Stochastic variational principle

Schrödinger processes constructed in Sections 3 can be characterized as extremal processes which minimize Yasue's action functional (cf. Theorem 1 in Nagasawa (1990)). The action functional of Yasue (1981) is defined to be symmetric in time reversal. This symmetry is necessary to get dual quantities through the variational principle. However, since Schrödinger processes are time symmetric by nature, a time asymmetric action functional, which will be given below, serves as well.

Let $(X_t, \mathcal{F}^t, Q)$ be a Schrödinger process constructed in Section 3. As admissible processes we consider semimartingales

$$(33) \quad Y_t = X_a + \int_a^t \sigma(v,Y_v)dB_v + \int_a^t \{(\widetilde{\boldsymbol{a}} + \sigma^2\frac{\nabla\phi}{\phi})(v,Y_v) + \sigma\boldsymbol{h}(v,.)\}dv,$$

where $\widetilde{\boldsymbol{a}} = \boldsymbol{a} + \frac{1}{2}\nabla\sigma^2$, $\boldsymbol{h}(t,.)$ is $\mathcal{F}^t$-adapted and

$$Q\left[\int_a^b (\sigma h(v,.))^2 dv\right] < \infty.$$

Since admissible processes Y_t should not hit the nodal set of ϕ and the distributions of Y_b and X_b must coincide, we require

$$\text{(34)} \qquad \lim_{n\to\infty} Q[\log \phi(b\wedge T_n, Y_{b\wedge T_n})] = Q[\log \phi(b, X_b)],$$

where

$$\text{(35)} \qquad T_n = \inf\{s\in [a,b]: \phi(s,Y_s) \leq \tfrac{1}{n}, \text{ or } \int_a^s (\sigma\frac{\nabla\phi}{\phi})^2(u,Y_u)du \geq n\}.$$

Moreover, we consider a pair $(\phi,\hat{\phi})$ such that

$$\int \phi(a,x)\hat{\phi}(a,x)dx = 1,$$

$$\int \phi\hat{\phi}(a,x)dx \,|\log\phi(a,x)| < \infty, \text{ and } \int \phi\hat{\phi}(b,x)dx \,|\log\phi(b,x)| < \infty,$$

and then require

$$Q\left[\int_a^b | c(u,Y_u)| du\right] < \infty,$$

where $c = -L\phi/\phi$ is the reference potential of ϕ.

As an *action functional* we adopt

$$\text{(36)} \qquad \mathrm{I}(Y) = Q\left[\int_a^b \{\tfrac{1}{2}(\sigma\frac{\nabla\phi}{\phi}(v,Y_v) + \sigma h(v,.))^2 - c(v,Y_v)\}dv\right],$$

for $Y \in \mathcal{A}$ = {amissible processes} (or $h \in \mathcal{A}$ = {admissible additional drift coefficients}, where the same notation is used with confusion).

Theorem 3. *Let* I(*Y*) *be the action functional defined at* (36). *Then*

$$\text{I}(Y) = Q\Big[\int_a^b \frac{1}{2}(\sigma \boldsymbol{h})^2 dv\Big] + e_o, \tag{37}$$

where e_o *is a constant defined by*

$$e_o = \int \phi\, \widehat{\phi}(b,x)dx \log\phi(b,x) - \int \phi\, \widehat{\phi}(a,x)dx \log\phi(a,x), \tag{38}$$

and hence the Schrödinger process $(X_t, \mathcal{F}^t, Q)$ *attains the minimum of the action functional*:

$$\text{I}(X) = \min_{Y \in \mathcal{A}} \text{I}(Y) = e_o. \tag{39}$$

Proof. Itô's formula applied to $\log\phi(t,x)$ from a up to $T = T_n$, where T_n is defined at (35), yields

$$\log\phi(T,Y_T) - \log\phi(a,Y_a)$$

$$= \int_a^T \frac{\nabla\phi}{\phi}\sigma dB_s + \int_a^T \Big\{\frac{L\phi}{\phi}(s,Y_s) + \frac{1}{2}\Big(\sigma\frac{\nabla\phi}{\phi}(s,Y_s) + \sigma\boldsymbol{h}\Big)^2 - \frac{1}{2}(\sigma\boldsymbol{h})^2\Big\} ds.$$

Replacing $L\phi/\phi$ by $-c$ according to the definition, taking the expectation of both sides and then applying (34), we have (37).

When we apply theory of large deviations to Schrödinger processes (cf. Föllmer (1988), Dawson-Gorostiza-Wakolbinger (preprint)), we need to formulate the least action principle in terms of the relative entropy. Let us define Q^h by

$$\frac{dQ^h}{dQ} = \exp\Big(\int_a^b \sigma \boldsymbol{h}(s,.)dB_s - \frac{1}{2}\int_a^b (\sigma \boldsymbol{h})^2(s,.)ds\Big), \tag{40}$$

where we consider those $\boldsymbol{h}$'s which satisfy

$$Q^h[\Omega] = 1. \tag{41}$$

By the drift transformation of Maruyama-Girsanov,

$$W_t = B_t - \int_a^t \sigma \boldsymbol{h}(s,.)ds \tag{42}$$

is a Brownian motion with respect to $(\mathcal{F}^t, Q^h)$, and X_t satisfies the stochastic differential equation (33) with W_t in place of B_t:

$$X_t = X_a + \int_a^t \sigma(v, X_v)dW_v + \int_a^t \{(\tilde{a} + \sigma^2 \frac{\nabla\phi}{\phi})(v, X_v) + \sigma \boldsymbol{h}(v,.)\}dv.$$

Therefore $(X_t, \mathcal{F}^t, Q^h)$ is equivalent to the process $(Y_t, \mathcal{F}^t, Q)$ given at (33). Furthermore, we define P_c by

$$\frac{dP_c}{dQ} = \exp\Big(-\int_a^b \sigma\frac{\nabla\phi}{\phi}(v, X_v)dB_v - \int_a^b (\frac{1}{2}(\sigma\frac{\nabla\phi}{\phi})^2 - c)(v, X_v)dv\Big), \tag{43}$$

and set

$$P = \frac{P_c}{P_c(\Omega)}, \tag{44}$$

where we assume

$$P_c(\Omega) < \infty. \tag{45}$$

Remark. The condition (45) is a version of (22) with $\phi\hat{\phi}$ in place of k, because

$$P_c(\Omega) = \int \phi\widehat{\phi}(a,x)\mathrm{d}xP_{(a,x)}[\exp(\int_a^b c(s,X_s)ds)|b<T_a].$$

Combining (40), (42) ,(43) and (44), we have

$$(46) \quad \log \frac{dQ^h}{dP} = \log\frac{dQ^h}{dQ} - \log\frac{dP_c}{dQ} + \log P_c(\Omega)$$

$$= \int_a^b (\sigma\frac{\nabla\phi}{\phi} + \sigma h)dW_v + \frac{1}{2}\int_a^b (\sigma\frac{\nabla\phi}{\phi} + \sigma h)^2 dv - \int_a^b c(v,X_v)dv$$

$$+ \log P_c(\Omega).$$

Therefore, the relative entropy $\mathrm{H}(Q^h|P) = Q^h[\log dQ^h/dP]$ is given by

$$(47) \quad Q^h[\log\frac{dQ^h}{dP}] = Q^h[\frac{1}{2}\int_a^b \{(\sigma\frac{\nabla\phi}{\phi} + \sigma h)^2 - c(v,X_v)\}dv] + \log P_c(\Omega)$$

$$= \mathrm{I}(Y) + \log P_c(\Omega),$$

since $(X_t,\mathcal{F}^t,Q^h)$ and $(Y_t,\mathcal{F}^t,Q)$ are equivalent. Applying Theorem 3 to $\mathrm{I}(Y)$, we have a characterization of Schrödinger processes for a given $\phi(t,x)$ in terms of the minimization problem of the relative entropy.

Theorem 4. *Assume the conditions* (41) *and* (45). *Let* Q^h *and* P *be defined at* (40) *and* (44), *respectively. Then, the relative entropy of the* Q^h *with respect to the* P *is equal to the action functional* $\mathrm{I}(Y)$ *plus a constant* $\log P_c(\Omega)$, *and*

$$Q^h[\log\frac{dQ^h}{dP}] = Q[\int_a^b \frac{1}{2}(\sigma h)^2 dv] + e_o + \log P_c(\Omega),$$

where e_o *is given at* (38), *and hence the Schrödinger process* $(X_t,\mathcal{F}^t,Q)$

(i.e. $h \equiv 0$) *attains the minimum of the relative entropy:*

$$\mathrm{H}(Q \mid P) = \min_{h \in \mathcal{A}} \mathrm{H}(Q^h \mid P)$$

$$= \mathrm{I}(X) + \log P_c(\Omega)$$

$$= e_o + \log P_c(\Omega).$$

Let us consider now the construction of Schrödinger processes in Section 4 for a given triplet (c, q_a, q_b). We can and do assume that processes are defined on the space $(\Omega, \mathcal{F})$ of continuous paths. Assuming the condition (22), we define a probability measure P on $(\Omega, \mathcal{F})$ by

(48) $$P[F] = \int k(x)dxP_{(s,x)}[\exp\Big(\int_a^b c(v, X_v)dv\Big)F(.); b < T_a]/p_k(\mathbf{R}^d \times \mathbf{R}^d).$$

Let $\mathcal{E}$ be the class of probability measures on $(\Omega, \mathcal{F})$ such that the distributions of X_a and X_b coincide with the q_a and q_b, respectively. Then, the first assertion of Theorem 2 can be formulated in terms of the relative entropy in the following form.

Theorem 5. *Assume the integrability condition* (22). *If there is some* $q \in \mathcal{E}(a,b)$ *with* (23), *then there exists* $Q \in \mathcal{E}$ *such that*

$$\inf_{R \in \mathcal{E}} \mathrm{H}(R \mid P) = \mathrm{H}(Q \mid P),$$

and (X_t, Q) *is the Schrödinger process for a given triplet* (c, q_a, q_b).

Remark. As we have seen in Section 4 (cf. Theorem 2), we must show further regularities of the $\phi(t,x)$ defined at (26) in order to specify the drift coefficient of the constructed Schrödinger process for a triplet (c, q_a, q_b).

Remark. The variational principle for Schrödinger processes in terms of the relative entropy is discussed in Föllmer (1988) for $c \equiv 0$, and in Wakolbinger (1989) for bounded Hölder continuous c.

6. A system of singular integral equations (segregation)

The duality formula of a pair of diffusion processes (resp. Schrödinger processes) with respect to a distribution density $g(t,x)$ is given by

$$\frac{b+\widehat{b}}{2} = \frac{1}{2}\nabla\log g \quad \textit{and} \quad \frac{b-\widehat{b}}{2} = \nabla S,$$

where b and $\widehat{b}$ are the drift coefficients of the diffusion processes in duality and S is a free parameter (phase), and hence

$$b = \frac{1}{2}\nabla\log g + \nabla S. \tag{50}$$

On the other hand, McKean's propagation of chaos for a system of interacting diffusion processes yields McKean's process with drift

$$\int \boldsymbol{h}(t,x-y)g(t,y)dy + \boldsymbol{h}_o(t,x), \tag{51}$$

where $\boldsymbol{h}(t,x-y)$ is a pair interaction between particles and $\boldsymbol{h}_o(t,x)$ is an environmental drift field. Therefore, if we identify a Schrödinger process with the McKean process, in other words, equate the drift coefficients of both processes given at (50) and (51), then we get

$$\frac{1}{2}\nabla \log g(t,x) + \nabla S(t,x) = \int \boldsymbol{h}(t,x-y)g(t,y)dy + \boldsymbol{h}_o(t,x). \tag{52}$$

The second step of the inverse problem is to solve the equation (52) and find $\boldsymbol{h}$ and $\boldsymbol{h}_o$. If $\nabla\log g$ and ∇S have no singularity (for example, of the ground state), there is in principle no problem to solve the integral equation (52). However, if the distribution density has zeros, there is, in general, no solution, unless we introduce *segregation of particles.* For example, if g is given in one dimension by

$$g(x) = cx^2\exp(-x^2) \tag{53}$$

which is the distribution of the first excited state of the harmonic oscillator, then $S \equiv 0$ and

$$\begin{aligned} b(x) &= \frac{1}{2}\frac{d}{dx}\log g(x) \\ &= \frac{1}{x} - x. \end{aligned} \tag{54}$$

Because of the singularity at the origin, there is no solution to (52), except a choice of two combinations:

$$h(x) \equiv 0 \quad and \quad h_o(x) = \frac{1}{x} - x\,; \tag{55}$$

or

$$h(x) = -x \quad and \quad h_o(x) = \frac{1}{x}\,.$$

However, the following modified equation

$$\frac{1}{x} = \int_{-\infty}^{0} h_1(x - y)g(y)dy, \quad for \ \ x > 0$$

has a unique solution

$$h_1(x) = \frac{1}{c}\frac{1}{x^4} + 0(\frac{1}{x^2}). \tag{56}$$

This means that if we consider two segregated groups distributed on the left- and right-hand sides, respectively, and $h(x)$ to be a pair interaction between particles belonging to the different groups, then the integral equation (52) turns out to be solvable (Föllmer-Nagasawa, cf, Nagasawa-Tanaka (1985)). The above consideration leads us to the following system of singular integral equations in order to construct a microscopic system of interacting diffusion processes.

Let us call a function $h(x)$ on $(0,\infty)$ *declining*, if it is continuous, non-negative, non-increasing with

$$\lim_{x\to 0} h(x) = \infty,$$

and Lipschitz continuous on $[\varepsilon,\infty)$ for any $\varepsilon > 0$.

Let $h^-(x)$ and $h^+(x)$ be declining functions on $(0,\infty)$. For given w_j and z_k in $C([0,\infty))$, $1 \le j \le m$, $1 \le k \le n$, with

$$\max_{1\le j\le m} w_j(0) \le \min_{1\le k\le n} z_k(0) \tag{57}$$

we consider a system of singular integral equations

$$\begin{cases} x_j(t) = w_j(t) - \displaystyle\int_0^t \frac{1}{n}\sum_{k=1}^{n} h^+(y_k(s) - x_j(s))ds, \ 1\le j\le m, \\ \\ y_k(t) = z_k(t) + \displaystyle\int_0^t \frac{1}{m}\sum_{j=1}^{m} h^-(y_k(s) - x_j(s))ds, \ 1\le k\le n, \end{cases} \tag{58}$$

under a constraint

$$\max_{1\le j\le m} x_j(t) \le \min_{1\le k\le n} y_k(t), \quad \text{for } t \ge 0. \tag{59}$$

If the strict inequality holds in (59), we call the solution *proper*.

Let us call a function $h(x)$ on $(0,\infty)$ *inaccessible,* if

(60) $$x(t) = w(t) + \int_0^t ds h(x(s))$$

has a positive solution on $[0,\infty)$ for $w \in C([0,\infty))$ with $w(0) > 0$.

Lemma 3. *Let $\rho(t)$ be defined by*

(61) $$\rho(t) = \sup\{|w(s)-w(r)|: 0<r<s<T,\ s-r<t\}.$$

Then, any solution $x(t)$ of (60) *is positive for all $t > 0$, if*

(62) $$\rho(t) \leq \int_0^t h(\rho(s))ds, \quad \textit{for sufficiently small } t > 0.$$

(ii) *Assume*

(63) $$h(x) \geq \frac{c}{x^k}, \quad c > 0 \ \textit{and} \ k > 1.$$

Then the equality (62) *holds, if*

(64) $$\rho(t) \leq \{(1+k)ct\}^{\frac{1}{(1+k)}}, \quad \textit{for sufficiently small } t > 0.$$

(iii) *If h satisfies* (63), *then* (64) *holds for almost of all Brownian paths $w(t)$ with $w(0)>0$ and hence* (60) *has a positive solution.*

Theorem 6. *Let $h^-(x)$ and $h^+(x)$ be declining and inaccessible, and w_j and $z_k \in C([0,\infty))$ satisfy the strict inequality in* (57). *Then, there exists a unique proper solution of the system* (58). *There exists a unique (not neccessarily proper) solution of the system* (58), *if* (57) *holds.*

A proof (by Nagasawa and Tanaka) will be published elsewhere.

Corollary. *Assume $h^-(x)$ and $h^+(x)$ are declining and satisfy (63). Let $B_j^-(t)$ and $B_k^+(t)$ be independent one dimensional Brownian motions, and $X_j(0)$ and $Y_k(0)$ be random variables independent of the Brownian motions satisfying*

$$\max_{1\le j\le m} X_j(0) < \min_{1\le k\le n} Y_k(0), \quad a.e..$$

Then the follwoing system of singular SDE's

$$(65) \quad \begin{cases} X_j(t) = X_j(0) + B_j^-(t) - \displaystyle\int_0^t \frac{1}{n}\sum_{k=1}^{n} h^+(Y_k(s) - X_j(s))ds, \ 1\le j\le m, \\ Y_k(t) = Y_k(0) + B_k^+(t) + \displaystyle\int_0^t \frac{1}{m}\sum_{j=1}^{m} h^-(Y_k(s) - X_j(s))ds, \ 1\le k\le n, \end{cases}$$

has a unique proper (pathwise) solution for almost all paths.

Remark. Theorem 5 and its Corollary remain valid, if h^- (resp. h^+) satisfies the stated conditions in an interval $(0,x_0)$, $x_0>0$, and if it is Lipschitz continuous on $[x_0,\infty)$.

7. The propagation of chaos

The components of the solution of the system (65) are not independent of each other, because of the interaction between particles. However, one conjectures that when m and n tend to infinity, $\{X_j(t); j=1,...,m\}$, $t>0$, will become asymptotically independent identically distributed, if the $\{X_j(0); j=1,...,m\}$ were (resp. $\{Y_k(t); k=1,...,n\}$), since the interactions with a large number of particles will average the memory of individual interactions and make the drifts deterministic. This gives a plausible physical explanation of the so called propagation of chaos for the system (65).

To give precise definitions, let $\mathcal{M}(S)$, $S = \mathbf{R}^d$, be the space of all probability measures on S, and $\mathcal{M}(S^n)$ the space of all symmetric probability measures on S^n.

A familiy $\{u_n \in \mathcal{M}(S^n);n=1,2,...\}$ is called *u-chaotic,* if

(66) $$\lim_{n\to\infty} < u_n, f_1\otimes...\otimes f_k\otimes 1\otimes...\otimes 1> = \prod_{j=1}^{k} <u, f_j>,$$

for $f_1,..., f_k \in C_b(S)$, $k = 1,2,...$, where $u \equiv \mathcal{M}(S)$.

The following lemma is well known (cf. e.g. Nagasawa-Tanaka (1987)).

Lemma 4. *A family $\{u_n;n=1,2,...\}$ is u-chaotic, if and only if the weak convergence holds;*

(67) $$\frac{1}{n}\sum_{j=1}^{n} \delta_{X_j^n} \to u, \text{ in probability, as } n \to \infty,$$

where $(X_1^n,...,X_n^n)$ is u_n-distributed and δ_X denotes the point mass at X.

Let $h[x,u]$ be a $\mathbf{R}^d$-valued measurable function on $S\times\mathcal{M}(S)$, and a system of interacting diffusion processes be given by SDE's.

(68) $$X_j^n(t) = X_j^n(0) + B_j(t) + \int_0^t h[X_j^n(s),U^n(s)]ds, \ j=1,...n,$$

where $U^n(t)$ is the empirical distribution of $\{X_j^n(t); j=1,...,n\}$

$$U^n(t) = \frac{1}{n}\sum_{j=1}^{n} \delta_{X_j^n(t)},$$

$B_j(t)$'s are independent d-dimensional Brownian motions, and $X_j^n(0)$'s are random variables independent of the Brownian motions $B_j(t)$.

The *McKean process* corresponding to the system (68) is defined by

$$X(t) = X(0) + B(t) + \int_0^t \boldsymbol{h}[X(s),u(s)]ds, \tag{69}$$

where B(t) is a d-dimensional Brownian motion, $u(t)$ is the probability distribution of $X(t)$, and $X(0)$ is a random variable which is u-distributed and independent of the Brownian motion $B(t)$.

We say that the *propagation of chaos* holds for the system (68), if the family $\{u_n(t);n=1,2,...\}$ of the distributions $u_n(t)$ of $(X_1^n(t),...,X_n^n(t))$ in (68) is $u(t)$-chaotic, where $u(t)$ is the distribution of the McKean process $X(t)$ at (69), whenever the family $\{u_n(0);n=1,2,...\}$ of the initial distributions $u_n(0)$ of $(X_1^n(0),...,X_n^n(0))$ is u-chaotic.

Suppose the propagation of chaos holds and let n tend to infinity in (68), formally. Then, we infer that the *SED* (69) holds, because of (67). That is, the system (68) of interacting diffusion processes converges to an infinite collection of independent copies of the McKean processes given at (69). In order to make the above formal arguments rigorous, we need some kind of continuity of $h[x,u]$ in (x,u) (cf. Nagasawa-Tanaka (1987,b), Oelschläger (1984)).

To show that the system (68) includes the case of segregation of particles, let us consider $h[x,u]$, in one dimension, of the form

$$h[x,u] = \int h(x,y,u)u(dy), \tag{70}$$

where $h(x,y,u)$ is given as follwos: For $u \in \mathcal{M}(R)$ we first define the segregating front $\gamma(u)$ by

$$\gamma(u) = \min\{x;\ u((-\infty,x)) \le \theta \le u((-\infty,x])\},$$

with a fixed $0 < \theta < 1$, and then set, denoting $A = \{y;\ -\infty < y \le \gamma(u)\}$ and $B = \{y;\ \gamma(u) < y < \infty\}$,

$$(71)\qquad h(x,y,u) = \begin{cases} \frac{1}{\theta} b_{11}(x,y) 1_A(y) + \frac{1}{1-\theta} b_{12}(x,y) 1_B(y), & x \leq \gamma(u), \\ \\ \frac{1}{\theta} b_{21}(x,y) 1_A(y) + \frac{1}{1-\theta} b_{22}(x,y) 1_B(y), & x > \gamma(u). \end{cases}$$

Let $(X_1(t),X_2(t),\ldots,X_{n+m}(t))$ be a solution of the system (68) with the $h[x,u]$ of (70) and (71) with $\theta = m/(m+n)$, and apply the order statistics (cf. Nagasawa-Tanaka (1987,a)) to the solution. Then , we get a system

$$(X_1(t),X_2(t),\ldots X_m(t),Y_1(t),Y_2(t),\ldots,Y_n(t))$$

segregated in the sense that

$$\max_{1\leq j\leq m} X_j(t) \leq \min_{1\leq k\leq n} Y_k(t),$$

and they constitute a system of interacting diffusion processes with a moving reflecting boundary set between the two groups $\{X_j(t); j=1,2,\ldots,m\}$ and $\{Y_k(t);k=1,2,\ldots,n\}$. For a proof see Nagasawa-Tanaka (1987,a,b), in which the propagation of chaos for the system (68) with $h[x,u]$ at (70) is shown under the assumption that $b_{jk}(x,y)$, $j,k=1,2$, in (71) are bounded continuous. The propagation of chaos for a system of interacting diffusion processes with unbounded $b_{jk}(x,y)$, $j,k=1,2$, with singularities is an open problem.

Putting $h_{11}\equiv 0$, $h_{12}(x,y)=-h(y-x)$, $h_{21}(x,y)=h(x-y)$, and $h_{22}\equiv 0$ in (71), we get the system (65) (since $X_j(t)$ and $Y_k(t)$ do not meet, because of the singularity of $h(x)$, we can delete the additional moving reflecting boundary condition which appears through the order statistcs). Since the propagation of chaos for the system (65) is still not well settled, we consider another model putting a fixed reflecting boundary at the origin: For $j,k=1,2,\ldots,n$,

$$(72)\qquad \begin{cases} X_j(t) = X_j(0) + B_j^-(t) - \displaystyle\int_0^t \frac{1}{n}\sum_{k=1}^{n} h(Y_k(s) - X_j(s))ds - a_j^-(t), \\ \\ Y_k(t) = Y_k(0) + B_k^+(t) + \displaystyle\int_0^t \frac{1}{n}\sum_{j=1}^{n} h(Y_k(s) - X_j(s))ds + a_k^+(t), \end{cases}$$

where $a_j^-(t)$ and $a_k^+(t)$ are non-negative, non-decreasing continuous processes which make the origin a reflecting boundary of the processes $X_j(t)$ and $Y_k(t)$, respectively. If $h(x)$ is a non-increasing continuous function on $(0,\infty)$ which may diverge at the origin, the propagation of chaos holds for the system (72) (cf. Nagasawa-Tanaka (1986)).

If the singularity of $h(x)$ is strong enough at the origin, so that the origin becomes inaccessible for the McKean process, then it satisfies

$$(73)\qquad \begin{cases} X(t) = X(0) + B^-(t) - \displaystyle\int_0^t ds \int h(y - X(s))u_{y(s)}(dy), \\ \\ Y(t) = Y(0) + B^+(t) + \displaystyle\int_0^t ds \int h(Y(s) - x)u_{x(s)}(dx), \end{cases}$$

where $u_{x(t)}$ and $u_{y(t)}$ denote the distributions of $X(t)$ and $Y(t)$, respectively.

As we have considered in Section 6, if $X(0)$ and $Y(0)$ are distributed on $(-\infty,0)$ and $(0,\infty)$, respectively, by means of $g(x)$ given at (53) and if we choose $h(x) = h_1(x) - x$, where $h_1(x)$ is the one given at (56) and $h_o \equiv 0$ (resp. one of the combinations in (55)), then the McKean process coincides with

$$(74)\qquad X(t) = B(t) + \int_0^t b(X(s))ds,$$

where $b(x)$ is the drift coefficient in (54), that is, the McKean process agrees with the Schrödinger process determined by the distribution of the first excited state of the harmonic oscillator.

8. Some applications

(i) *A Statistical interpretation of Schrödinger equations*

A Schrödinger process corresponding to a Schrödinger equation can not across over the nodal set of its distribution density and is trapped in one of the regions surrounded by the nodal set. Let us consider, for example, a hydrogen atom. If the Schrödinger process describes the motion of a *single* electron in the atom, there must be various kinds of different states of the hydorogen atom in an excited state depending on the location of the electron. This is not plausible but the validity of this implication should be checked experimentally.

If we apply our statistical model to the Schrödinger equation, we should interprete it as a "Boltzmann equation" for a macroscopic distribution of a system of interacting diffusion processes. For those interacting diffusion processes we can consider paths of virtual photons, obtained through a diffusion approximation, confined in the atom. Because of the creation and annihilation of virtual photons and virtual electrons we can not specify the location of an electron in the hydrogen atom. However, the probability that we find an electron somewhere in the atom must be proportional to the spatial distrbution of virtual photons, that is, it agrees with the distribution given in terms of a solution of the Schrödinger equation of a hydrogen atom. This is, in a sense, a revival of the chemists' favorite picture (originally due to Schrödinger), namely, *cloud of an electron* (or *charge*) in a hydrogen atom, but now what we have is *cloud of* (virtual) *photons*, instead of an electron (or charge).

(ii) *An eigenvalue problem*

Consider a Schrödinger equation

(75) $$\frac{1}{2}\sigma^2\frac{d^2\psi}{dx^2} + (\lambda - k\,|\,x\,|)\psi = 0, \quad k > 0.$$

Setting $x = \alpha y$, $\mu = \lambda(k\alpha)^{-1}$ and $\alpha = (\sigma^2/2k)^{1/3}$, we get

(76) $$\frac{d^2\psi}{dy^2} + (\mu - |\,y\,|)\psi = 0.$$

The eigenvalues of (76) are the zeros of

$$J_{1/3}(\tfrac{2}{3}\mu^{3/2}) + J_{-1/3}(\tfrac{2}{3}\mu^{3/2}) \quad and \quad J_{2/3}(\tfrac{2}{3}\mu^{3/2}) - J_{-2/3}(\tfrac{2}{3}\mu^{3/2})$$

(cf. Titchmarsh (1962), pp.90-92). A table of the eigenvalues is given in Nagasawa (1981), Nagasawa-Yasue (1982). The potential

$$V(x) = k\,|\,x\,|, \quad k > 0,$$

in (75) induces a constant attractive force toward the origin independent of the distance from the origin. Therefore, the equation (75) is for a *string* model. A Statistical model based on (75) has been applied to the mass spectrum of mesons and to septation of *Escherichia coli.*

The mass spectrum of mesons

By means of our statistical theory a meson can be interpreted as a system of glueons (=string) distributed linearly according to a distribution defined in terms of an eigenfunction of (75) and, in addition, two quarks. Through the identification of the ground state with π-meson which is the lightest of all mesons, the constants in (75) are determined to be

$$\{(\sigma k)^2/2\}^{1/3} = \text{mass of } \pi\text{-meson}/\mu_1 = 136.99336 \text{ Mev.}$$

Then, the mass spectrum of mesons is computed numerically based on this statistical string model in Nagasawa-Yasue (1982). Some of them are

		Mass (Mev)	*computed*
Case 1.	$\pi^{\pm}$	*139.5669*	*139.5669* (postulated)
Case 2.	$K^{\pm}$	*493.67*	*495*
Case 3.	η	*549*	*545*
Cses 4.	ρ	*769*	*756*
Case 5.	ω	*783*	*771*
Case 6.	K^{*}	*892*	*894*
Case 7.	η'	*958*	*944*
....	....	......	......
Case 39.	D	*1869*	*1863*
....	....	......	......
Case 42.	$F^{\pm}$	*2021*	(*1986*)
....	....	......	......
Case 58.	J/ψ	*3097*	*3097*
Case 59.	$\chi(0)$	*3415*	*3390*
Case 60.	$\chi(1)$	*3510*	*3517*
Case 61.	$\chi(2)$	*3556*	*3571*

A comparison with given experimental data shows good agreement except the case 42 of $F^{\pm}$ (now called $D_s^{\pm}$). For this case, "according to the "Data Booklet 1982", the observed mass was *2021* ±*15* Mev. On the other hand the mass computed by the rule (25) of Nagasawa-Yasue (1982) is (c,ϕ_{12},s) = *1986* Mev, which is too light to be identified with *2021* Mev. Therefore, it was not possible to identify $F^{\pm}$ with our composite model of mesons. However, in the "Data Booklet 1984" the mass of $F^{\pm}$ is corrected to be *1971* ±*16* Mev, which agrees with our predicted value *1986* Mev" (cf. Nagasawa (1985)). According to an up to date "Stable Particle Summary Table (1988)" the mass of $D_s^{\pm}$ is *1969.3* ±*1.1* Mev.

Moreover, we made, based on our statistical string model, a speculative prediction on the mass of $B^{\pm}$ (resp. B^0) which was not known in 1982. The prediction says that it must appear between *5000* and *5500* Mev (cf. Nagasawa-Yasue (1982)). In the "Stable Particle Summary Table (1988)" the mass of $B^{\pm}$ is given to be *5277.6* ±*1.4* Mev, as predicted.

Therefore, we conclude that the substantial part of our statistical composite model of mesons is justified, that is, *a considerable portion of the mass of a meson is contributed by its glueon distribution* (*not only by the mass of quarks*), in contrast to conventional meson models in which it is assumed that the masses of mesons are from quarks, neglecting glueon distributions.

Septation of Escherichia coli

When the string model is applied to an *E.coli*, the minimum cell length of it is identified with the distribution of the ground state of (75), where the *E.coli* is considered as a system of molecules tangling together in the *E.coli*. Let $\mu_1 < \mu_2 <\ldots$ be eigenvalues of (75). Since

$$2\mu_1 < \mu_2,$$

the *E.coli* splits into two cells before reaching the first exicted state corresponding to μ_2. We can interprete this as the normal septation of an *E.coli*. There is a mutant *E.coli* such that "At $41°C$ the septation mechanism is switched off in the mutant *E.coli* which can grow in this higher temperature" and "at high temperature ($41°C$) it can survive, and grow expontentially without septation, becoming 20-30 (sometimes several hundred) times longer and will eventually die out. A notable peculiarity is that if the temperature is lowered to $30°C$, in 2-3 h (the cell length becomes four to eight times longer), the *E.coli* can and will start septation again after ca. 20-30 *min*. If it is, for example, four times longer, the most frequent septation site is 1:3." This fact can be well explained by comparing eigenvalues.

In fact, we have

$$\mu_4 < \mu_1 + \mu_3 < 2\mu_1 + \mu_2 < 2\mu_2 < \mu_5,$$

which shows that the most frequent septation site is 1:3, the next frequent 1:1:2, and the least frequent 2:2 (Experimental data given by Yamada and

Hirota, cf. Table 1 in Nagasawa (1981)).

(iii) *Segregation of a population*

The applicability of a statistical theory to segregation of a population in an environment is discussed in Nagasawa (1980). Consider, for example, "there is a feeding place in a mountain where we find a population of monkeys (we assume that the movement of a monkey can be approximated as a sample path of a diffusion process, since he is hopping and jumping from one place to another). The feeding place attracts monkeys and most of them move toward the place and will eventually be distributed with the peak density at the centre" and " in a state of higher excitation the population cannot keep itself as a single party but splits into two groups at least. Once it is devided into two parties, it can take one of three possible equilibrium distributions." We call this *segregation of a population.* Let us mathematize it introducing two kind of quantities K and Q which measure the strength of *excition*, and a (macroscopic) *environmental potential* $V(x)$. Here K and Q are given by

$$K = \frac{1}{2}(\frac{1}{2}\nabla g/g)^2 \ and \ Q = \frac{1}{2}(-\frac{1}{2}\Delta g)/g,$$

respectively, by means of a population density g. Macroscopic equilibrium states of a population are defined by a relation

$$K(x) + Q(x) + V(x) = E, \ constant.$$

Then, if we assume $g = |\psi|^2$, the above equation of equilibrium is reduced to an eigenvalue problem

$$\frac{1}{2}\Delta\psi(x) + (E - V(x))\psi(x) = 0.$$

(cf. Nagasawa (1980)). Therefore, the population must be segregated into several groups, if it is heighly excited.

(iv) *Titus-Bode Law*

Our statistical model is applied to the formation of the orbits of planets, and a statistical interpretation of Titus-Bode law is given by Albeverio, S., Blanchard, Ph., Høgh-Krohn, R., (1984).

Appendix

We have treated two integral equations (18) and (31). However, they are in fact equivalent. Therefore, the uniqueness of non-negative solutions of (31) holds under the conditions (15,a) and (15,b). For a proof we follow an idea of Uchiyama. Let us denote for $s \leq u \leq t$

(A1) $$\mathbf{g}_s^u f(s,x) = P_{(s,x)}[f(u,X_u):u < T_s],$$

and

(A2) $$\overline{\mathbf{g}}_s^u f(s,u) = P_{(s,x)}\left[e^{-c^-[s,u]} f(u,X_u):u < T_s\right],$$

where

$$c^-[s,u] = \int_s^u c^-(r,X_r)dr,$$

with $c = c^+ - c^-$, and consider non-negative solutions of

(A3) $$p(s) = \mathbf{g}_s^t f + \int_s^t du\, \mathbf{g}_s^u(cp),$$

and

(A4) $$p(s) = \overline{\mathbf{g}}_s^t f + \int_s^t du\, \overline{\mathbf{g}}_s^u(c^+p).$$

In the following first we assume c^- in (A2) is bounded, and then apply it later on to $c_n^- = c^- \wedge n$ of an unbounded c^-.

Let p be a nonnegative solution of (A3) and set

(A5) $$y(u) = \overline{\mathbf{g}}_s^{t-u} p$$

$$= \mathrm{I} + \mathrm{II},$$

where, because of the Markov property of X_t,

$$\mathrm{I} = \overline{\mathbf{g}}_s^{t-u}(\mathbf{g}_{t-u}^t f) = P_{(s,x)}[\mathrm{e}^{-c^-[s,t-u]} f(t,X_t) : t < T_s],$$

and

$$\mathrm{II} = \int_{t-u}^t dv P_{(s,x)}[\mathrm{e}^{-c^-[s,t-u]} c(v,X_v) p(v,X_v) : v < T_s].$$

Therefore, we have

$$\frac{dI}{du} = P_{(s,x)}[c^-(t-u,X_{t-u})\mathrm{e}^{-c^-[s,t-u]} f(t,X_t) : t < T_s],$$

and

$$\frac{dII}{du} = P_{(s,x)}[\mathrm{e}^{-c^-[s,t-u]} c(t-u,X_{t-u}) p(t-u,X_{t-u}) : t-u < T_s]$$

$$+ \int_{t-u}^t dv\, P_{(s,x)}[c^-(t-u,X_{t-u})\, \mathrm{e}^{-c^-[s,t-u]} c(v,X_v) p(v,X_v) : v < T_s]$$

$$= \left(\frac{dII}{du}\right)_1 + \left(\frac{dII}{du}\right)_2,$$

and hence

$$\frac{dI}{du} + \left(\frac{dII}{du}\right)_2 = P_{(s,x)}\left[c^-(t-u,X_{t-u})\mathrm{e}^{-c^-[s,t-u]}\left\{\mathrm{P}_{(t-u,X_{t-u})}[f(t,X_t) : t < T_{t-u}]\right.\right.$$

$$+\int_{t-u}^{t} dv\, P_{(t-u,X_{t-u})}[c(v,X_v)p(v,X_v):v<T_{t-u}]\}:t-u<T_s]$$

$$= P_{(s,x)}[c^-(t-u,X_{t-u})e^{-c^-[s,t-u]}p(t-u,X_{t-u}):t-u<T_s],$$

which implies

$$\frac{dy}{du}=\left(\frac{dII}{du}\right)_1+\{\frac{dI}{du}+\left(\frac{dII}{du}\right)_2\}$$

$$= P_{(s,x)}[c^+(t-u,X_{t-u})e^{-c^-[s,t-u]}p(t-u,X_{t-u}):t-u<T_s].$$

Accordingly, because $p(t)=f(t)$,

$$p(s)-\overline{\mathbf{g}}_s^t p=\int_0^{t-s} du\,\frac{dy}{du}$$

$$=\int_s^t du\ P_{(s,x)}[e^{-c^-[s,u]}c^+(u,X_u)p(u,X_u):u<T_s]$$

$$=\int_s^t du\,\overline{\mathbf{g}}_s^u(c^+p),$$

which proves that p satisfies (A4).

Now cancelling the boundedness assumption on c^-, we define a sequence of operators $\overline{\mathbf{g}}(n)_s^u$ by (A2) with $c_n^-=c^-\wedge n$ in place of c^-. Let p be a non-negative solution of (A3). Then, applying the above arguments, we have

$$\text{(A6)}\qquad p(s)=\overline{\mathbf{g}}(n)_s^t p+\int_s^t du\,\overline{\mathbf{g}}(n)_s^u(c^+p)-\int_s^t du\overline{\mathbf{g}}(n)_s^u\{(c^--c_n^-)p\}.$$

Since $c^- - c_n^- \downarrow 0$ as n $\to\infty$ and also $\overline{\mathbf{g}}(n)_s^u f\downarrow \overline{\mathbf{g}}_s^u f$ for non-negative f, the third term of the right-hand side of (A6) vanishes as $n\to\infty$ and the first and second terms converge to the ones without n. This proves that a non-negative solution of (A3) satisfies (A4) with an unbounded c.

Acknowledgement. I am indebted to Th. Sturm for valuable comments and discussions.

References

Aebi, R. (1989): MN-transformed a-diffusion with singular drift. Doctoral Dissertation at the University of Zürich.

Aebi, R. (preprint): A Sobolev space to establish Schrödinger processes.

Albeverio, S., Blanchard, Ph., Høgh-Krohn, R., (1984): A stochastic model for the orbits of planets and satellites: An interpretation of Titius-Boode law. Asterisuqe.

Beurling, A. (1960): An automorphism of product measures. Ann. Math. 72, 189-200.

Blanchard, P., Golin, S. (1987): Difffusion propcesses with singular drift fields. Commun. Math. Phys. 109, 421-435.

Carlen, E.A. (1984): Conservative diffusions. Commun. Math. Phys. 94, 293-315.

Carmona, R. (1985): Probabilistic construction of Nelson processes. Taniguchi Symp. PMMP Katata, 55-81.

Csiszar, I.(1975): I-divergence geometry of probability distribution and minimization problems. The Ann. Probability. 3,146-158.

Dawson, D., Gorostiza, L., Wakolbinger, A. (preprint): Schrödinger processes and large deviations.

Dynkin, E.B. (1965): Markov processes. Springer.

Föllmer, H. (1988): Random fields and diffusion processes. Ecole d'été de Saint Flour XV-XVII (1985-87), Lect. Notes Math. 1362, Springe

Fortet, R. (1940): Résolution d'un système d'equation de M. Schrödinger J. Math. Pures Appl. IX, 83-95.

Friedman, A. (1964): Partial differential equations of parabolic type. Prentice-Hall, Inc.

Ikeda. N., Watanabe, S., (1989): Stochastic differential equations and diffusion processes, Second edition. Kodansha Ltd, North-Holland Publ. Co.

Itô, S. (1957): Fundamental solutions of parabolic differential equations and boundary value problems. Japan.J.Math.27, 55-102.

Jamison, B., (1974): Reciprocal processes. Z.Wahrsch. Verw. Geb. 30,65-86.

Jamison, B., (1975): The Markov processes of Schrödinger. Z. Wahrscheinl. Verw. Geb. 32, 323-331.

Kunita, H., Watanabe, T. (1963): Note on transformations of Markov processes connected with multiplicative functionals. Mem.Fac.Sci. Kyushu Univ. Ser. A. 17, 181-191

Kusuoka, S., Tamura, Y. (1984): Gibs measures for mean field potentials. J. Fac. Sci. Univ. Tokyo Sect. IA Math. 31, 223-245.

Lipster, R.S., Shiyayev, A.N. (1977): Statistics of random processes I. General theory. Springer.

Maruyama, G. (1954): On the transition probability functions of the Markov processes. Nat. Sci. Rep. Ochanomizu Univ. 5, 10-20.

McKean, H.P. (1966): A class of Markov processes associated with non-linear parabolic equations. Proc. Natl. Acad. Sci. 56, 1907-1911

McKean, H.P. (1967): Propagation of Chaos for a class of nonlinear parabolic equations. Lecture Series in Differential Equations, Catholic Univ., 41-57.

Nagasawa, M. (1969): Markov processes with creation and annihilation Z. Wahrscheinl. Verw. Geb. 14, 49-60.

Nagasawa, M. (1980): Segrgation of a population in an environment. J. Math. Biol. 9, 213-235.

Nagasawa, M. (1981): An application of segregation model for septation of Escherichia Coli. J. Theoret. Biol. 90, 445-455.

Nagasawa, M. (1985): Macroscopic, intermediate, microscopic and mesons. Lect. Notes Phys. 262, 427-437. Springer.

Nagasawa, M. (1989): Transformations of diffusion and Schrödinger processes. Probab. Th. Rel. Fields, 82, 109-136.

Nagasawa, M. (1990): Stochastic variational principle of Schrödinger processes. Seminar on stochastic processes. ed. Cinlar, Chung, Getoor. Birkhäuser.

Nagasawa, M. (1990): A statistical model of segregation of a population. Stochastic model-ling in Biology. ed. P. Tautu.

Nagasawa, M., Yasue, K. (1982): A statistical model of mesons. Publ. de l'Inst. rech. Math. Ava. (CNRS) 33,1-48, Univ. Strasbourg.

Nagasawa, M., Tanaka, H. (1985): A diffusion process in a singular meandrift-field. Z. Wahrschl. Verw. geb. 68, 247-269.

Nagasawa, M.,Tanaka, H. (1986): Propagation of chaos for diffusing particles of two types with singular mean field interaction. Probab.Th.Rel.Fields,71,69-83.

Nagasawa, M., Tanaka, H. (1987,a): Diffusion with interactions and collisions between coloured particles and the propagation of chaos. Probab. Th. Rel. Fields, 74, 161-198.

Nagasawa, M., Tanaka, H. (1987,b): A proof of the propagation of chaos for diffusion processes with drift coefficients not of average form. Tokyo J.Math. 10, 403-418.

Norris, J.R. (1988): Constructions of diffusions with a given density. In:Stochastic Calculs in application, (ed. Norris, J.R). 69-77. Longman.

Oelschläger, K. (1984): A martingale approach to the law of large numbers for weakly interacting stochastic processes. Ann. Probab. 12, 458-479.

Oelschläger, K. (1985): A law of large numbers for modelately interacting diffusion processes. Z. Wahrscheilichkeitstheor. Verw.Geb. 69, 279-322.

Oelschläger, K. (1989): Many-particle systems and the continuum description of their dynamics. Univ. Heidelberg.

Schrödinger, E. (1931): Ueber die Umkehrung der Naturgesetze. Sitzungsberichte der preussischen Akademie der Wissenschaften Physicalisch Mathematische Klasse, 144-153.

Schrödinger, E.(1932): Théorie relativiste de l'électron et l'interprétation de la méchanique quantique. Ann. Inst. H. Poincaré 2, 269-310.

Sznitman, A.S. (1984): Non-linear reflecting diffusion processes and propagation of chaos and flutuations associated. J.Funct. Anal. 56,311-336.

Sznitman, A.S., (1989): Topics in propagation of chaos. Ecole d'été de probabilités de Saint Flour.

Tanaka, H., (1984): Limit theorems for certain diffusion processes with interaction. In:Stcoachstic Analysis (ed. Itô,K), 469-488, Kinokuniya, Nort-Holland.

Titchmarsh, E.C., (1962): Eigenfunction expansion associated with second order differential equations, Part 1, 2nd edn. Oxford: Clareddom Press.

Wakolbinger, A., Stummer, W. (preprint): Schrödinger processes and stochastic Newton equations.

Wakolbinger, A. (preprint): A simplified variational characterization of Schrödinger processes.

Yasue, K. (1981): Stochastic calculs of variations. J. Funct. Anal. 41,327-340.

Zambrini, J.C. (1986): Stochastic mechanics according to Schrödinger. Phys. Rev. A. 33, 1532-1548.

Zhao, Z. (1986): Scnrödinger conditional Brownian motion and stochastic culculus of variations. Stochastics 18, 1-15.

Zheng, W.A., Meyer, P.A. (1984/85): Sur la construction de certain diffusions. Sem. de Probab. XX. Lect. Notes math. 1294, 334-337. Springer.

Zheng, W.A., (1985): Tightness results for laws of diffusion processes. Application to stochastic mechanics Ann. Inst. H. Poincaré B21. 103-124.

Masao NAGASAWA
Institut für Angewandte Mathematik
der Universität Zürich
Rämistrasse 74, CH-8001 Zürich
Switzerland

A Stochastic Approach to Moving Boundary Problems

Bernt Øksendal*

Summary

Moving boundary problems arise for example in the study of fluid flow in porous media. Using optimal stopping of an associated diffusion, a (stochastic) weak concept of a solution of a moving boundary problem is introduced. This allows the use of methods from stochastic analysis to investigate weak/variational and classical solutions.

In particular, we prove the existence of a variational solution $\{p(x,t), W_t\}_{t\geq 0}$ of the moving boundary problem

$$\begin{cases} div(k(x)\nabla_x p(x,t)) = -f(x,t) & \text{for } x \in W_t,\ t \geq 0 \\ p(\cdot,t) \in H_0^1(W_t) & \text{for } t \geq 0 \\ \theta_0|\partial W_t \cdot \frac{d}{dt}(\partial W_t) = -k\nabla_x p|\partial W_t & \text{for } t \geq 0 \end{cases}$$

where W_0 is a given bounded open set in $\mathbf{R}^n$, $\theta_0(x)$ and $f(x,t)$ are bounded measurable functions and we assume that

(i) $k(x) \geq$ is a Muckenhoupt A_2 weight. (In particular, this allows k to have zeroes)

and

(ii) $W_0 \subset W_t$ for all $t \geq 0$. (This holds, for example, if $f(x,t) \geq 0$ for all x, t)

*) This work is supported by a grant from the VISTA program "Fluid flow in stochastic reservoirs"

§1. Introduction

To describe the flow of an incompressible fluid in a porous medium the following basic equation is used:

(1.1) Darcy's law:

$$\vec{q} = -\frac{k}{\mu} \nabla p$$

where $\vec{q} = \vec{q}(x,t)$ denotes the seepage - velocity of the fluid at the point $x \in \mathbf{R}^3$ and at time t, $k = k(x) \geq 0$ is the permeability of the medium, μ is the viscosity of the fluid and $p = p(x,t)$ is the pressure of the fluid. (Here and throughout ∇ denotes the gradient with respect to x).

(1.2) We also use the continuity equation

$$\frac{\partial \theta}{\partial t} = -\mathrm{div}(\rho \vec{q}) + f$$

where $\theta = \theta(x,t)$ is the degree of saturation (i.e. the fluid weight per unit volume of the medium) ρ is the density of the fluid and $f = f(x,t)$ gives the fluid source/sink (depending on the sign of f) rate at the point x and at time t. Combining (1.1) and (1.2) we get

$$\frac{\partial \theta}{\partial t} = \mathrm{div}(k \nabla p) + f, \tag{1.3}$$

where for simplicity we have set the value of the two constants ρ, μ equal to 1.

In addition we need a relation between θ and p. We will assume that at every instant t $\theta(x,t)$ assumes one of only 2 possible values, 0 or $\theta_0(x) > 0$, corresponding to zero saturation ("dry" region) or complete saturation ("wet" region). Thus we put

$$W_t = \{x; \theta(x,t) = \theta_0(x)\} \text{ (the wet region)} \tag{1.4}$$

As we will explain below the interpretation of (1.3) then becomes:

$$Lp := \mathrm{div}(k \nabla p) = -f \quad \text{for } x \in W_t \tag{1.5}$$

$$p = 0 \quad \text{for } x \in \partial W_t \quad \text{(the boundary of } W_t) \tag{1.6}$$

$$\theta_0|_{\partial W_t} \frac{d}{dt}(\partial W_t) = -k \nabla p|_{\partial W_t} \quad (t \geq 0) \tag{1.7}$$

Thus *the moving boundary problem* is the following:

Given measurable functions $k : \mathbf{R}^n \to [0,\infty), \theta_0 : \mathbf{R}^n \to (0,\infty)$ and $f : \mathbf{R}^{n+1} \to \mathbf{R}$ and given a bounded initial domain W_0, find a function $p(x,t)$ and a family of domains $\{W_t\}_{t\geq 0}$ such that (1.5)-(1.7) hold. Following Gustafsson [17] we call a solution (p, W_t) of (1.5)-(1.7) a *classical* solution.

Remark. The exact meaning of the "expansion velocity" $\frac{d}{dt}(\partial W_t)$ in (1.7) is the following: Suppose that locally at some time t_0 and some point $y \in \partial W_{t_0}$ the domains W_t can be described by

$$W_t = \{x; \phi(x) < t\},$$

where ϕ is some smooth level function with $\nabla\phi \neq 0$. Then we define

$$\frac{d}{dt}(\partial W_t) = \frac{\nabla\phi}{|\nabla\phi|^2} \text{ at } y. \tag{1.8}$$

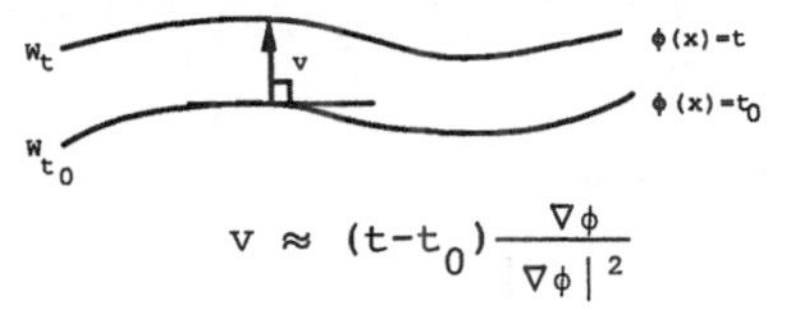

In this paper we are primarily interested in the case where the permeability $k(x)$ is allowed to vary rapidly from point to point, so we want to impose as few restrictions on $k(x)$ as possible. In particular, we do not want to assume in the set-up that $k(x)$ is smooth, only that $k(x)$ is (Borel) measurable (and that $k(x) \geq 0$).
This means that (1.3) and (1.5)-(1.7) should be interpreted in the distribution sense. More precisely, let $H_0 = H_0(W_t)$ denote the closure of $C_0^\infty(W_t)$ in the norm

$$\|\psi\|_{H_0}^2 = \int_{W_t} |\psi|^2 k dx + \int_{W_t} |\nabla\psi|^2 k dx$$

The natural variational (or distributional) interpretation of (1.3) is that $p(\cdot,t)\epsilon H_0(W_t)$ (where W_t is open) for each t and

$$\begin{aligned} &-\int\int \theta(x,t)\psi(x)\phi'(t)dxdt \\ &= -\int\int \nabla p\cdot\nabla\psi\cdot k\phi dxdt + \int\int f\psi\phi dxdt, \end{aligned} \tag{1.9}$$

for all $\phi(t) \in C_0^\infty(\mathbf{R}), \psi(x) \in C_0^\infty(\mathbf{R}^3)$

(Here - and in the rest of this paper - $dx, dt \cdots$ means Lebesgue measure on $\mathbf{R}^n$ for the appropriate n. In general we let $C_0^\infty(U)$ denote the family of infinitely differentiable functions with compact support in the set $U \subset \mathbf{R}^n$, while C_0^∞ means $C_0^\infty(\mathbf{R}^n)$ for the appropriate n).

Using (1.4) we rewrite the left hand side of (1.9) as

$$-\int \Big(\int_{W_t} \theta_0(x)\psi(x)dx \Big) \phi'(t)dt = \int \frac{d}{dt} \Big(\int_{W_t} \theta_0(x)\psi(x)dx \Big) \phi(t)dt \tag{1.10}$$

Since this holds for all $\phi \in C_0^\infty$ we conclude that

$$\frac{d}{dt} \Big(\int_{W_t} \theta_0(x)\psi(x)dx \Big) = -\int \nabla p \cdot \nabla \psi \cdot k dx + \int_{W_t} f\psi dx \text{ for all } \psi \in C_0^\infty \tag{1.11}$$

In fact, this argument gives that supp $f(\cdot, t) \subset \overline{W}_t$ and, if f is continuous,

$$\begin{aligned} \int_{W_t} \theta_0(x)\psi(x)dx - \int_{W_0} \theta_0(x)\psi(x)dx &= -\lim_{\psi_j \to \psi} \int_0^t \Big(\int \nabla p(x,s) \cdot \nabla \psi_j(x) dx \Big) ds \\ &+ \int_0^t \Big(\int_{W_t} f(x,s)\psi(x)dx \Big) ds \end{aligned} \tag{1.12}$$

for all $\psi \in C_0^\infty$ and all choices of $\psi_j \in C_0^\infty(W_t)$ such that $\psi_j \to \psi | W_t$ in $L^2(W_t)$ as $j \to \infty$. This is the interpretation of (1.3) we will use. In particular, suppose $t \to W_t$ is *left continuous*, in the sense that for all compact sets $K \subset W_t$ there exists $\epsilon > 0$ such that $K \subset W_s$ for all $s \in (t - \epsilon, t]$. Then by (1.12) we get

$$\int \nabla p \cdot \nabla \psi k dx = \int f \psi dx \quad \text{for all } \psi \in C_0(W_t), \tag{1.13}$$

which is the variational interpretation of (1.5). To see that (1.12) also contains (1.7), we assume that ∂W_s is smooth and write (for $\psi \in C_0^\infty$)

$$\int_{W_s} \nabla p \cdot \nabla \psi k dx = \int_{\partial W_s} \psi \cdot \frac{\partial p}{\partial n} k d\sigma - \int_{W_s} \psi \cdot L p dx, \tag{1.14}$$

where $d\sigma$ denotes surface measure on ∂W_s. Since $Lp = -f$ in W_s, the substitution of (1.14) in (1.12) gives

$$\frac{d}{dt} \Big(\int_{W_t} \theta_0(x)\psi(x)dx \Big) = -\int_{\partial W_t} \psi \cdot \frac{\partial p}{\partial n} \cdot k d\sigma, \quad \forall \psi \in C_0^\infty \tag{1.15}$$

which is the variational formulation of (1.7).

Thus we may regard both (1.5) and (1.7) as consequences of the one condition (1.3), i.e. of (1.12). However, it is convenient to split (1.12) into the corresponding "inner" and "boundary" part, thereby also allowing a localization of the boundary part. This leads to the following:

DEFINITION 1.1. Let $k : \mathbf{R}^n \to (0,\infty), \theta_0 : \mathbf{R}^n \to (0,\infty)$ and $f : \mathbf{R}^n \times \mathbf{R} \to \mathbf{R}$ be measurable functions, $k \in L^1(dx)$ and θ_o and f bounded, f continuous, and let W_0 be a bounded domain in $\mathbf{R}^n$. We say that $\{p(x,t), W_t\}_{t\geq 0}$ is a variational solution of the moving boundary problem if the following, (1.16)-(1.18), hold:

(1.16) $$\int_{W_t} \nabla p \cdot \nabla \psi k dx = \int_{W_t} f \psi dx \quad \text{for all} \quad \psi \in C_0^\infty(W_t), t \geq 0$$

(1.17) $W_t \subset \mathbf{R}^n$ is open and $p(\cdot,t) \in H_0(W_t)$ for all $t \geq 0$

(1.18) For $\psi \in C_0^\infty$ we have

$$\frac{d}{dt}\Big(\int_{W_t} \theta_0(x)\psi(x)dx\Big) = -\int \nabla p(x,t) \cdot \nabla\psi(x)k(x)dx + \frac{d}{dt}\Big(\int_0^t \int_{W_t} f(x,s)\psi(x)dxds\Big)$$

(in the sense of (1.12)).

If (1.18) is interpreted in distribution sense with respect to t, we call the solution *weak* variational. If (1.18) is interpreted in the (strong) sense that the t-derivative exists for each t, we call the solution *strong* variational.

The purpose of this paper is twofold. First, we introduce the concept of a *stochastic solution*, defined quite generally. Second, we prove that under mild conditions the stochastic solution actually constitutes a variational solution outside W_0 and we give conditions which guarantee that the stochastic solution is a classical solution.

Regarding the first part, it is now well known through the works of Baiocchi [1], Duvaut [6], Elliott [9], Gustafsson [16], [17] and subsequently Begehr & Gilbert [2], [3] that if $p(x,t) \geq 0$ is a classical solution and we define

(1.19) $$u(x,t) = \int_0^t p(x,s)ds$$

then for each t the function $u(\cdot,t)$ is a solution of a certain variational inequality denoted by $\mathcal{U}_t$. In general, if for each t a solution $u(\cdot,t)$ of $\mathcal{U}_t$ exists, then $u(x,t)$ is called a *weak* solution of the moving boundary problem. This concept was introduced by Gustaffson in [16], [17]. We refer to these works for more information about such weak solutions. We remark that it is well known that under certain conditions optimal stopping problems and variational inequalities are equivalent [4], so in such cases our concept of a stochastic weak solution coincides with the weak one. However, as mentioned before we are interested in studying moving boundary problems with as few regularity/ellipticity conditions on the permeability $k(x)$ as possible and in such a general set-up we no longer necessarily have this equivalence. Moreover, by introducing a stochastic approach we can benefit from an efficient machinery from stochastic analysis and Dirichlet forms.

In particular, we prove the existence of a variational solution of the moving boundary problem under the assumptions that (see Theorem 3.4 below)

(1.20) $k(x)$ is Muckenhoupt A_2 weight (see (2.3) below)

(in particular, this allows $k(x)$ to have zeroes)

$$W_0 \subset W_t \qquad \text{for alle} \quad t \tag{1.21}$$

(A sufficient, but not necessary, condition for (1.21) is that $f(x,t) \geq 0$ for all x, t).

Remark. This result extends directly to the anistropic case where $k(x) \geq 0$ is replaced by a matrix $[k_{ij}(x)]_{1 \leq i,j \leq n}$, provided that (1.20) is replaced by

(1.20)' There exists a Muckenhoupt A_2 weight $w(x) \geq 0$ and a constant $C > 0$ such that

$$\frac{1}{C} w(x)|\xi|^2 \leq \sum_{i,j} \xi_i \xi_j k_{ij}(x) \leq C w(x)|\xi|^2$$

for all $\xi = (\xi_1, ..., \xi_n) \in \mathbb{R}^n$ and all x.

Theorem 3.4 represents a substantial extension of the (isotropic case of the) main result of Begehr & Gilbert in [3], where the corresponding assumptions are

$$k(x) \in C^{1+\alpha}, \tag{1.22}$$

$$(\exists \lambda > 0) \qquad \lambda \leq k(x) \leq \lambda^{-1} \qquad \text{for all} \quad x \tag{1.23}$$

and

$$f(x,t) = f(x) \geq 0 \qquad \text{for all} \quad x \tag{1.24}$$

§2. Construction of the stochastic solution

Let

$$L\phi = \operatorname{div}(k \bigtriangledown \phi) \tag{2.1}$$

be the operator in (1.5). First we construct a diffusion $(X_s, P^x, \zeta); s \geq 0, x \in \mathbb{R}^n$ whose generator is L. Here P^x denotes the law of $\{X_s\}_{s\geq 0}$ starting at x and $\zeta \leq \infty$ is the life time of X_s.

A sufficient condition for the existence of such a stochastic process is the Hamza condition

$$\frac{1}{k(x)} \in L^1_{loc}(dx)\ a.e.(dx) \text{ outside } Z, \tag{2.2}$$

where

$$Z = \{x; k(x) = 0\}, dx = \text{ Lebesgue measure on } \mathbb{R}^n.$$

If (2.2) holds then the symmetric bilinear form

$$\mathcal{E}(u,v) = \int_{\mathbb{R}^n} \bigtriangledown u^T \bigtriangledown v k dx; \quad u, v \in C_0^\infty$$

(regarded as a densely defined form on $K = L^2(\mathbb{R}^n, dx)$) is *closable* and hence constitutes a regular Dirichlet form with generator L, i.e,

$$\mathcal{E}(u,v) = -(Lu, v)_K; \quad u \in \mathcal{D}(L), v \in C_0^\infty$$

(Here $(\cdot,\cdot)_K$ denotes the usual inner product in K, i.e.

$$(\phi,\psi)_K = \int_{\mathbb{R}^n} \phi(x)\psi(x)dx; \quad \phi,\psi \in C_0^\infty)$$

Therefore there exists a diffusion $\{X_s\}$ whose generator coincides with L. See [14] for details.

Note that condition (2.2) is very weak. For example, it suffices that $k(x)$ is lower semicontinuous. In such a generality we cannot get enough information about X_t for our purposes, so we will from now on assume that $k(x)$ satisfies the following stronger condition

$$\begin{gathered} k(x) \text{ is a } \textit{Muckenhoupt } A_2 \textit{ weight}, \text{i.e.} \\ \sup_B\Big(\frac{1}{|B|}\int_B k(x)dx\Big)\Big(\frac{1}{|B|}\int_B \frac{1}{k(x)}dx\Big) < \infty, \end{gathered} \tag{2.3}$$

the sup being taken over all balls $B \subset \mathbb{R}^n$ and $|B| = \int_B dx$ is the volume of B.

The A_p weights were originally introduced by Muckenhoupt in connection with weighted maximal function inequalities. The concept has turned out to be important in many other connections also, for example in the potential theory for degenerate elliptic equations. For example, in [10], [11] and [12] several fundamental results are established regarding the potential theory for

$$L\phi = \text{div}(k \bigtriangledown \phi)$$

when k is an A_2-weight as in (2.3). We will apply some of these results in this paper.

Fix a bounded open set $U \subset \mathbf{R}^n$ such that $U \supset W_0$ and let ζ denote the first exit time for U for X_t. From now on we assume that there exists $T > 0$ such that

$$W_t \cup \text{ supp } f(\cdot, t) \subset U \text{ for all } t < T \tag{2.4}$$

and we consider only the time interval $0 \leq t < T$. The *Green operator* (of X_t in U), is defined by

$$Gv(x) = G_U v(x) = E^x[\int_0^\zeta v(X_t)dt] \text{ for } v \text{ lower bounded,} \tag{2.5}$$

where E^x denotes expectation with respect to P^x.

This definition is the stochastic equivalent to the variational definition of the Green operator in [11, Theorem 1]. There $G = G_U : H_0(U)^* \to H_0(U)$ is defined by the relation

$$\mathcal{E}_U(G(F), v) = F(v) \quad \text{for } v \in H_0(U), F \epsilon H_0(U)^*, \tag{2.6}$$

where $\mathcal{E}_U(\phi, \psi) = \int_U \bigtriangledown \phi \cdot \bigtriangledown \psi \cdot k dx; \phi, \psi \in C_0^\infty(v)$ is the Dirichlet form corresponding to the process X killed at the first exit time from U and $H_0(U)^*$ denotes the dual of $H_0(U)$. Now $H_0(U)^*$ can be identified with the space of distributions

$$F = f_0 - \text{div} \vec{f}$$

where $\vec{f} = (f_1, \cdots, f_n)$ and $\frac{f_i}{k} \in L^2(U, kdx), i = 0, 1, \cdots, n$, with the action of F on $v \in H_0(U)$ given by

$$F(v) = \int_U v f_0 dx + \sum_{i=1}^n \int_U f_i \frac{\partial v}{\partial x_i} dx$$

In particular, choosing $T = f(\cdot, t)$ we get from (2.6) the useful relation

$$\mathcal{E}_U(G_U f(\cdot, t), v) = \int_U f(\cdot, t) v(x) dx, \quad v \in H_0(U). \tag{2.7}$$

For a more detailed explanation of the equivalence of the two definitions (2.5) and (2.6) see for example [20,§3].

We can now apply an important result about the Green operator from [10, Lemma 3.6]:

(2.8) If v is bounded, then Gv is continuous.

By a version of the Hille-Yosida theorem (see e.g. [5, p. 252]) it follows that

(2.9) X_t is *Feller-continuous*,

i.e. $x \to E^x[u(X_t)]$ is continuous for each $t \geq 0$ and for each bounded, continuous function u.

We are now ready to start the construction of the (weak) stochastic solution of the moving boundary problem:

Define

$$\eta = \begin{cases} 0 & \text{on} \quad W_0 \\ \theta_0 & \text{outside} \quad W_0 \end{cases} \tag{2.10}$$

Then by (2.8) the function $G\eta$ is continuous. Choose a continuous function g_0 such that

$$\begin{cases} g_0 < G\eta & \text{on} \quad W_0 \\ g_0 = G\eta & \text{outside} \quad W_0 \end{cases} \tag{2.11}$$

Before we proceed we recall some basic notions and results from the theory of optimal stopping:

If g is a lower bounded, continuous function on $\mathbf{R}^n$ (*the reward function*) we define

$$g^*(x) = \sup_\tau E^x[g(X_\tau)],$$

the sup being taken over all $\{\mathcal{M}_t\}$-stopping times τ, where $\mathcal{M}_t$ is the σ-algebra generated by $\{X_s(\cdot); s \leq t\}$.

A function, denoted by $\hat{g}$, is called *the least X_t-superharmonic majorant* of g if $\hat{g}$ is X_t-superharmonic (i.e. $\hat{g}$ is lower semi continuous and $\hat{g}(x) \geq E^x[g(X_\tau)]$ for all M_t-stopping times τ), $\hat{g} \geq g$ and if h is any X_t-superharmonic function such that $h \geq g$ then $h \geq \hat{g}$.

If $V \subset \mathbf{R}^n$ is a Borel set we let

$$\tau_V = \inf\{t > 0; X_t \notin V\}$$

denote the first exit time from V for X_t.

The fundamental theorem of optimal stopping states the following:

THEOREM A. Assume that (2.4) holds. Then we have:

(i) $\hat{g} = g^*$

(ii) Define $D = \{x; g(x) < \hat{g}(x)\}$ (the continuation region). Assume that $\tau_0 < \infty$ a.s. P^x and that g is bounded. Then τ_D is an optimal stopping time in the sense that

$$g^*(x) = E^x[g(X_{\tau_D})]$$

A proof of Theorem A (which only requires that the process is Feller continuous) can be found in [19, Th. 10.9]. Note that the existence of $\hat{g}$ (which is not obvious) is a part of the statement (i) and that since $\hat{g}$ is lower semicontinuous the set D must be open.

We now apply this to our function g_0 above:

LEMMA 2.1.

$$\hat{g}_0 = G\eta.$$

Proof. Since $G\eta$ is X_t-superharmonic and $G\eta \geq g_0$ we clearly have

$$G\eta \geq \hat{g}_0 \tag{2.12}$$

On the other hand, by the strong Markov property we have that $G\eta$ is X_t-harmonic in W_0, so

$$G\eta(x) = E^x[G\eta(X_\sigma)]$$

where $\sigma = \tau_{W_0}$. Therefore

$$G\eta(x) \leq \sup_\tau E^x[g_0(X_\tau)] = g_0^*(x) \tag{2.13}$$

By Theorem A (i) we have $g_0^* = \hat{g}$ so Lemma 2.1 follows from (2.12) and (2.13).

We conclude that we can identity the starting region W_0 as the continuation region for the optimal stopping problem for g_0:

COROLLARY 2.2.

$$W_0 = \{x; g_0(x) < \hat{g}_0(x)\}$$

The idea is to extend this identification to work for all $t \geq 0$:

For $t \geq 0$ define

$$g_t(x) = g_0(x) - \int_0^t Gf(x,s)ds \tag{2.14}$$

and let

$$g_t^*(x) = \sup_\tau E^x[g_t(X_\tau)] \tag{2.15}$$

We now have all the ingredients for the weak stochastic solution concept:

DEFINITION 2.3. Let

$$w(x,t) = g_t^*(x) - g_t(x)$$

and put

$$D_t = \{x; g_t(x) < g_t^*(x)\};\ t \geq 0.$$

Then $(w(x,t), D_t)_{t\geq 0}$ is called *the stochastic solution* of the moving boundary value problem.

§3. When does a stochastic solution give a variational solution?

To justify the terminology of Definition 2.3 we now proceed to show that under reasonable conditions we have that

$$p(x,t) := \frac{\partial w}{\partial t}(x,t) \text{ and } W_t := D_t$$

actually is a variational solution of the moving boundary problem (1.16)-(1.18).

First we establish some useful (basically well known) auxiliary results.

LEMMA 3.1. Let $h > 0$. Then

$$g_t^* - \int_t^{t+h} G|f|(x,s)ds \leq g_{t+h}^* \leq g_t^* + \int_t^{t+h} G|f|(x,s)ds$$

Proof. Note that if $h > 0$ then

$$\hat{g}_t = (g_{t+h} + \int_t^{t+h} Gf(x,s)ds)^\wedge \leq \hat{g}_{t+h} + \int_t^{t+h} G|f|(x,s)ds$$

and

$$\hat{g}_{t+h} = (g_t - \int_t^{t+h} Gf(x,s)ds)^\wedge \leq \hat{g}_t + \int_t^{t+h} G|f|(x,s)ds$$

COROLLARY 3.2. If $f(x,t) \geq 0$ for all x, t then $w(x,t)$ and D_t are increasing in t

Proof. Choose $h > 0$. Then by Lemma 3.1

$$w(x,t+h) = g_{t+h}^*(x) - g_{t+h}(x) \geq g_t^*(x) - \int_t^{t+h} Gf(x,s)ds - g_0(x) + \int_0^{t+h} Gf(x,s)ds$$

$$= g_t^*(x) - g_t(x) = w(x,t).$$

It follows that $D_t = \{x; w(x,t) > 0\} \subset \{x; w(x,t+h) > 0\} = D_{t+h}$.

COROLLARY 3.3. $w(x,\cdot)$ is Lipschitz-continuous for each x.

Proof. $|w(x,t+h) - w(x,t)| \le 2 \int_t^{t+h} G|f|(x,s)ds \le hM$ for all $h > 0$, for some constant M depending only on f and the domain U, using (2.8).

In particular, $\frac{\partial w}{\partial t}(x,t)$ exists for a.a.t. We proceed to show that $\frac{\partial w}{\partial t}$, if it exists, is related to p in (1.13) or (1.5):

LEMMA 3.4. Assume that

$$\frac{\partial w(x,t)}{\partial t} \text{ exists for } t = t_0 \text{ and some } x \in D_{t_0}.$$

Then

$$(\frac{\partial w}{\partial t})_{t=t_0} = Gf(x,t_0) - (Gf)^{(t_0)}(x,t_0), \tag{3.1}$$

where in general $h^{(t)}(x) = E^x[h(X_{\tau_t})]$ denotes the X-harmonic extension of a given function h on ∂D_t to the interior D_t and we have put

$$\tau_t = \inf\{s > 0; X_s \notin D_t\}(= \tau_{D_t}).$$

Proof. For all $t \ge 0$ define the X-harmonic measure $\lambda_x^{(t)}$ on ∂D_t by

$$\lambda_x^{(t)}(F) = P^x[X_{\tau_t} \in F]; F \subset \partial D_t$$

Then if $\frac{\partial w}{\partial t}(x,t_0)$ exists, we have

$$\begin{aligned}
(\frac{\partial w}{\partial t})_{t=t_0} &= \frac{\partial}{\partial t}(g_t^*)_{t=t_0} - \frac{\partial}{\partial t}(g_t)_{t=t_0} \\
&= \frac{\partial}{\partial t}(\int_{\partial D_t} g_t(y)d\lambda_x^{(t)}(y))_{t=t_0} + Gf(x,t_0) \\
&= \int_{\partial D_{t_0}} \frac{d}{dt}(g_t(y))_{t=t_0} d\lambda_x^{(t_0)}(y) + \frac{d}{dt}(\int_{\partial D_t} g_{t_0}(y)d\lambda_x^{(t)}(y))_{t=t_0} + Gf(x,t_0) \\
&= -\int_{\partial D_{t_0}} Gf(y)d\lambda_x^{(t_0)}(y) + Gf(x,t_0) = Gf(x,t_0) - (Gf)^{(t_0)}(x,t_0).
\end{aligned}$$

In this argument we have used that

$$t \to \int_{\partial D_t} g_{t_0}(y)d\lambda_x^{(t)}(y)$$

is maximal for $t = t_0$ (since D_{t_0} is the continuation region for g_{t_0}) and therefore the derivative is zero if it exists.

We now define

$$p(x,t) = Gf(x,t) - (Gf)^{(t)}(x,t) \quad \text{for all } x \text{ and } t \tag{3.2}$$

To see that $p(\cdot,t)$ actually solves the boundary value problem (1.13) with $W_t := D_t$, we use the strong Markov property to rewrite p as follows: (Put $\tau_t = \tau$ for simplicity)

$$\begin{aligned} p(x,t) &= E^x[\int_0^\zeta f(X_s)ds] - E^x[E^{X_\tau}[\int_0^\zeta f(X_r)dr]] \\ &= E^x[\int_0^\zeta f(X_s)ds] - E^x[E^x[\int_\tau^\zeta f(X_s)ds]] \\ &= E^x[\int_0^\tau f(X_s)ds] = G_{D_t}f(x,t), \end{aligned} \tag{3.3}$$

using the notation of (2.5).

Thus $p(\cdot,t)$ coincides with the Green operator for the domain D_t applied to f and this function is by construction in $H_0(D_t)$ (see (2.6)). Moreover by (2.7) and (3.3) we get

$$\int_{W_t} \nabla p \cdot \nabla \psi k dx = \mathcal{E}_t(p,\psi) = \mathcal{E}_t(G_{D_t}f,\psi) = \int_{D_t} f\psi dx$$

for all $\psi \in C_0^\infty(D_t)$, where $\mathcal{E}_t = \mathcal{E}_{D_t}$. Hence (1.15) holds.

Note that from Corollary 3.3 and Lemma 3.4 we get that for all x $\frac{\partial w}{\partial t}(x,t)$ exists for a.a.t and

$$w(x,t) = \hat{g}_0 - g_0 + \int_0^t G_{D_s}f(x,s)ds \quad \text{for all } t \tag{3.4}$$

We are now ready to prove one of the main results of this paper:

THEOREM 3.4. Let θ_0, f be as in Definition 1.1, U as in (2.4) and suppose that $k \in A_2$, i.e. k satisfies (2.3).

Suppose that

$$W_0 \subset D_t \quad \text{for all} \quad t \geq 0. \tag{3.5}$$

Then

$$p(x,t) := G_{D_t} f(x,t) \text{ and } W_t := D_t = \{x; w(x,t) > 0\}$$

constitute a (weak) variational solution of the moving boundary problem (1.16)- (1.18).

Proof. We have already established that (1.16) and (1.17) hold and that $D_0 = W_0$ (Corollary 2.2). It remains to prove that (1.18) holds.

Fix $t \geq 0$, $x_0 \in \partial D_t$, let B be a ball centered at x_0 and choose $\psi \in C_0^\infty(B)$.

As above put $\mathcal{E}_t(\cdot,\cdot) = \mathcal{E}_{D_t}(\cdot,\cdot)$ and let $(\cdot,\cdot)_t$ denote the inner product in $K_t = L^2(D_t; dx)$. Then since $\eta = \theta_0$ outside W_0, $\eta = 0$ on W_0 and $W_0 \subset D_t$ we have

$$\int_{D_t} \theta_0 \psi dx - \int_{W_0} \theta_0 \psi dx = \int_{D_t} \eta \psi dx = (\eta, \psi)_t. \tag{3.6}$$

To compute $(\eta,\psi)_t$ we approximate $\psi|D_t$ in $L^2(D_t)$ by $\psi_j \in C_0^\infty(D_t \cap B)$. For such ψ_j we have

$$\begin{aligned}(\eta, \psi_j)_t = \mathcal{E}_t(G\eta, \psi_j) &= \mathcal{E}_t(g_0, \psi_j) + \mathcal{E}_t(G\eta - g_0, \psi_j) \\ &= \mathcal{E}_t(g_t, \psi_j) + \mathcal{E}_t(\int_0^t G f ds, \psi_j)\end{aligned}$$

Now if $\psi_j \in \mathcal{D}_L$ (the domain of the generator L) then

$$\begin{aligned}\mathcal{E}_t(\int_0^t G f ds, \psi_j) &= -(L\psi_j, \int_0^t G f ds)_t = -\int_0^t (L\psi_j, G f)_t ds \\ &= \int_0^t \mathcal{E}_t(G f, \psi_j) ds = \int_0^t (f, \psi_j)_t ds,\end{aligned} \tag{3.7}$$

and since $\mathcal{D}_L$ is dense in C_0^∞ (in the $H_0(U)$-norm) (3.7) extends to all $\psi_j \in C_0^\infty$.

By Theorem A we have $g_t^*(x) = E^x[g_t(X_{\tau_{D_t}})]$, which is X-harmonic in D_t. Therefore

$$g_t^*(x) = E^x[g_t(X_\sigma)] = E^x[\overline{g}_t(X_\sigma)] \text{ for } x \in D_t \setminus \overline{D}_0,$$

where

$$\sigma = \inf\{s > 0; X_s \notin D_t \setminus \overline{D}_0\} \text{ and } \overline{g}_t = G\eta - \int_0^t G f ds.$$

Hence $g_t^* = G_{D_t \setminus \overline{D}_0}(L\overline{g}_t) + \overline{g}_t$ in $D_t \setminus \overline{D}_0$, so

$$\begin{aligned}\mathcal{E}_t(g_t^*, \psi_j) &= \mathcal{E}_t(G_{D_t \setminus \overline{D}_0}(L\overline{g}_t), \psi_j) + \mathcal{E}_t(\overline{g}_t, \psi_j) \\ &= (L\overline{g}_t, \psi_j) + \mathcal{E}_t(\overline{g}_t, \psi_j) \\ &= -\mathcal{E}_t(\overline{g}_t, \psi_j) + \mathcal{E}_t(\overline{g}_t, \psi_j) = 0\end{aligned} \tag{3.8}$$

Therefore, using (3.4) we may write

$$\begin{aligned}\mathcal{E}_t(g_t,\psi_j) &= \mathcal{E}_t(g_t,\psi_j) - \mathcal{E}_t(g_t^*,\psi_j) = -\mathcal{E}_t(w_t,\psi_j)\\ &= -\mathcal{E}_t(\hat{g}_0 - g_0,\psi_j) - \mathcal{E}_t(\int_0^t \frac{\partial w}{\partial s}(x,s)ds,\psi_j)\\ &= -\int_0^t \mathcal{E}_t(\frac{\partial w}{\partial s},\psi_j)ds,\end{aligned} \tag{3.9}$$

by adapting the same argument as in (3.7).

Substituting (3.9) and (3.7) in (3.6) we get

$$\int_{D_t} \theta_0\psi dx - \int_{W_0} \theta_0\psi dx = -\lim_{\psi_j\to\psi} \int_0^t \mathcal{E}_t(\frac{\partial w}{\partial s},\psi_j)ds + \int_0^t (f,\psi)_t ds, \tag{3.10}$$

and (1.18) (i.e. (1.12)) follows.

■

The proof above also leads to a sufficient condition that the stochastic solution is a *strong* variational solution. From (3.10) we see that the left hand side is differentiable for all t if the functions $f(x,s)$ and $G_{D_s}f(x,s)$ are both s-continuous for each x. If we assume that $f(x,\cdot)$ is continuous then a sufficient condition for the continuity of $s \to G_{D_s}f(x,s)$ is the following:

(3.11) $t \to \tau_t(\omega)$ is continous in probability (measure) with respect to P^x, for all x.

Condition (3.11) states that in some weak (stochastic) sense the domains D_t vary continuously with t. A situation where (3.11) does not hold- and where one cannot expect to find a classical solution of the moving boundary problem - is indicated on the figure below.

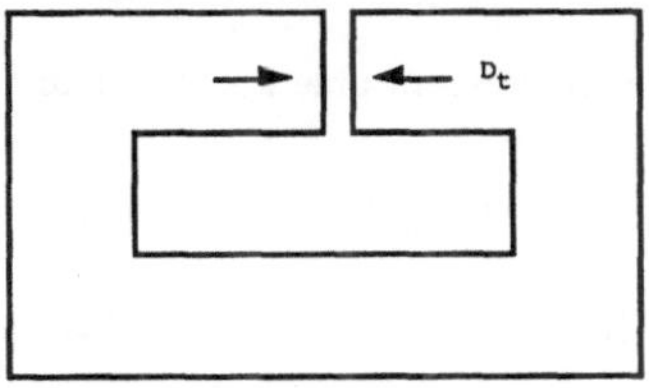

This gives the following result

THEOREM 3.5. Let θ_0, f, U, k, D_t be as in Theorem 3.4 and assume in addition that

(3.11) $s \to \tau_s(w)$ is continuous in P^x-measure, for all x

and

(3.12) $$\text{supp } f(\cdot, s) \subset \overline{D}_s \text{ for all } s$$

Then $s \to w(x, s)$ is continuously differentiable for all x and

$$p(x,t) = \frac{\partial w}{\partial t}(x,t) = G_{D_t} f(x,t), W_t = D_t = \{x; w(x,t) > 0\}$$

constitute a strong variational solution of (1.16)-(1.18).

Finally we return to the classical problem (1.5)-(1.7) and ask for sufficient conditions that our general stochastic solution gives rize to a classical solution. Applying Theorem 3.4 we obtain the following result:

(In the following $C^{k,\alpha}$ will denote the functions whose derivatives up to $k'th$ order are Hölder continuous with exponent $\alpha > 0$)

THEOREM 3.6 Suppose the following holds:

(3.13) $$k(x) \in C^{1,\alpha} \text{ for some } \alpha > 0$$

(3.14) $$\inf_{x \in K} k(x) > 0 \text{ for all compacts } K \subset \mathbf{R}^n$$

(i.e. $L = \text{div}(k\text{grad})$ is (locally) uniformly elliptic)

(3.15) $$\text{For all } t \geq 0 \text{ there exists } \alpha > 0 \text{ such that } f(\cdot, t) \in C^{0,\alpha}$$

In addition, assume that (3.5) and (3.12) hold and that

(3.16) $$D_t \text{ is a } C^{2,\alpha} \text{ domain for each } t$$

Then $t \to w(x, t)$ is continuously differentiable for all x and the function

$$p(x,t) = \frac{\partial w}{\partial t}(x,t) = G_{D_t} f(x,t)$$

together with the sets $D_t = \{x; w(x,t) > 0\}$ solve the classical moving boundary problem (1.15)-(1.17).

Proof of Theorem 3.6: Properties (1.5)-(1.7) follow from known regularity results of solutions of elliptic boundary value problems, (see e.g. [15] or [18]) combined with Theorem 3.5.

Acknowledgements

I am grateful to B. Gustafsson for valuable communication.

This research has been supported by VISTA, a research cooperation between the Norwegian Academy of Science and Letters and Den norske stats oljeselskap a.s. (Statoil).

REFERENCES

[1] C. Baiocchi: Free boundary problems in the theory of fluid flow through porous media. In proc. ICM 1974, Vol. 2, Canadian Mathematical Congress 1975, 237-243.

[2] H. Begehr & R.P.Gilbert: Hele- Shaw type flows in $\mathbb{R}^n$. Nonlinear Analysis 10 (1986), 65-85.

[3] H. Begehr & R.P. Gilbert: Non-Newtonian Hele-Shaw flows in $n \geq 2$ dimensions. Nonlinear Analysis 11 (1987), 17-47.

[4] A. Bensoussan & J.L. Lions: Application og Variational Inequalities in Stochastic Control. North-Holland 1982.

[5] C. Constantinescu & C. Cornea: Potential Theory on Harmonic Spaces. Springer-Verlag 1972.

[6] G. Duvaut: The solution of a two-phase Stefan problem by a variational inequality. In Ockendon & Hodgkins (editors): Moving Boundary Problems in Heat Flow and Diffusion. Clarendon Press 1975, pp.173-181.

[7] E.B. Dynkin: Markov Processes, Vol. I. Springer-Verlag 1965.

[8] E.B. Dynkin & R.J. Vanderbei: Stochastic waves. Trans. AMS 275 (1983), 771-779.

[9] C.M. Elliott: On a variational inequality formulation of an electrotechnical machinery moving boundary problem and its approximation by the finite element method. J. Inst. Math. Appl. 25 (1980), 121-131.

[10] E.B. Fabes, D. Jerison & C.E. Kenig: The Wiener test for degenerate elliptic equations. Ann. l'instiut Fourier 32 (1982), 151-182.

[11] E.B. Fabes, D. Jerison & C.E. Kenig: Boundary behaviour of solutions to degenerate elliptic equations. In W. Beckner, A.P. Calderon, R. Fefferman and P.W. Jones (editors): Conference on harmonic analysis in honor of Antoni Zygmund, Vol. II . Wadsworth Math. Ser. 1983, pp. 577-589.

[12] E.B. Fabes, C.E. Kenig & R. Serapioni: The local regularity of solutions of degenerate elliptic equations. Comm. PDE 7 (1982), 77-116.

[13] A. Friedman: Stochastic Differential Equations and Applications, Vol. II. Academic Press 1976.

[14] M. Fukushima: Dirichlet Forms and Markov Processes. North-Holland 1980.

[15] D. Gilbarg & N.S. Trudinger: Elliptic Partial Differential Equations of Second Order. Springer-Verlag 1983.

[16] B. Gustafsson: Variational inequality formulation of a moving boundary problem in rock mechanics. TRITA-MAT-1984-8, Royal Institute of Technology, S-100 44

Stockholm.

[17] B. Gustafsson: Applications of variational inequalities to a moving boundary problem for Hele Show flows. Siam J. Math. Anal. 16 (1985), 279-300.

[18] C. Miranda: Partial Differential Equations of Elliptic Type. Springer-Verlag 1970.

[19] B. Øksendal: Stochastic Differential Equations. Universitext, Springer-Verlag. Second edition 1989.

[20] B. Øksendal: Weighted Sobolev inequalities and harmonic measure associated with quasiregular functions. To appear in Comm. PDE.

[21] B. Øksendal: The high contact principle in optimal stopping and stochastic waves. In Cinlar, Chung & Getoor (editors): Seminar on Stochastic Processes 1989. Birkhäuser 1990, pp. 177-192.

Bernt Øksendal
Department of Mathematics
University of Oslo
Box 1053 Blindern
N-0316 Oslo 3, Norway

Monotonicity Methods for White Noise Driven Quasi-Linear SPDEs

R. Buckdahn and E. Pardoux

Abstract

We establish existence and uniqueness for nonlinear elliptic and parabolic SPDEs driven by white noise, with nonlinearities of monotone type.

1. Introduction

The aim of this paper is to establish existence and uniqueness results both for nonlinear elliptic stochastic partial differential equations of the type :

$$(1.1)\qquad \begin{cases} -\Delta u(x) + f(u(x)) = \dot{W}(x),\ x \in D \\ u|_{\partial D} = 0 \end{cases}$$

where D is an open bounded subset of $\mathbb{R}^k$ ($k = 1, 2, 3$), $\dot{W}$ denotes white noise, and f is an increasing function; and for nonlinear parabolic stochastic partial differential equations of the type :

(1.2)

$$\begin{cases} \dfrac{\partial u}{\partial t}(t,x) - \dfrac{\partial^2 u}{\partial x^2}(t,x) + f(u)(t,x) = \dfrac{\partial^2 W}{\partial t \partial x}(t,x) + g(t,x);\ t \geq 0,\ 0 \leq x \leq 1 \\ u(0,x) = u_0(x),\ 0 \leq x \leq 1; u(t,0) = u(t,1) = 0 \end{cases}$$

where $\frac{\partial^2 W}{\partial t \partial x}$ denotes the second order mixed derivative of the brownian sheet $\{W_{tx}\}$, i.e. space–time white noise, and

$$f = f_1 + f_2$$

where f_1 is increasing and f_2 is Lipschitz.

We shall establish existence and uniqueness of a solution for these two classes of equations.

Note that a lot is known about equation (1.1) in case of a linear function f, see e.g. Benfatto, Gallavotti, Nicolò [2], Rozanov [7]. It is known that the solution is a process in the ordinary sense iff k – the dimension of the space variable x – is less than or equal to three. This is the reason for our restriction to $k \leq 3$. There seems to be no literature on nonlinear elliptic SPDEs with the white noise on the right side. For nonlinear elliptic PDEs with measures as right side or boundary condition, we refer to Boccardo, Gallouët [1] and Röckner, Zegarlinski [6]. Our approach will consist in generalizing the use of monotonicity by Lions [4] to solve certain classes of deterministic nonlinear PDEs. Let us note that our results could be extended to more general PDE operators, and other types of boundary conditions.

Moreover, in the case $k = 1$, the nonlinear function f can depend also on $\frac{\partial u}{\partial x}$, see Nualart, Pardoux [5] where it is shown that the solution $(u, \frac{\partial u}{\partial x})$ is a Markov field iff f is linear. The positive part of that result is still true for $k = 2, 3$ (see e.g. Rozanov [8]), and we suspect that the negative part also generalizes, but we have not been able to prove it yet.

Concerning the parabolic equation (1.2), let us note that existence and uniqueness is known in case f is Lipschitz, even with a non constant diffusion coefficient, see e.g. Walsh [9]. However, our result with f_1 monotone increasing and not necessarily Lipschitz seems to be new.

2. Elliptic equations

The aim of this section is to study the equation :

$$(2.1) \qquad \begin{cases} -\Delta u(x) + f(u)(x) = g(x) + \dot{W}(x),\ x \in D \\ u|_{\partial D} = 0 \end{cases}$$

where D is a *bounded* domain of $\mathbf{R}^k$, $k = 1, 2$ or 3, whose boundary ∂D is supposed to be regular in the sense of potential theory, $g \in L^2(D)$, $\dot{W}$ denotes "white noise" and

$$f(u)(x) = f(x, u(x))$$

where$f(x,r)$ is a measurable function of $(x,r) \in D \times \mathbf{R}$ which satisfies properties to be stated below. We first note that we can and shall w.l.o.g. assume that $f(x, 0) = 0$. We moreover assume that :

(2.2) $\qquad r \to f(x, r)$ is continuous and non decreasing, for any $x \in D$

and moreover

(2.3) $\qquad f$ is locally bounded

Note that (2.2) together with $f(x,0) = 0$ imply that $rf(x,r) \geq 0$, and moreover from (2.3) $f(x,u(x))$ is bounded whenever $u(x)$ is bounded.

Let us now give a more rigorous formulation of equation (2.1), i.e. a "weak formulation" , which is as follows. An a.s. bounded function u from D into $\mathbf{R}$ is said to satisfy the weak form of (2.1) if for any $\phi \in C_b^2(D) \cap C(\bar{D})$, which vanishes on ∂D,

(2.4)

$$-\int_D u(x)\Delta\phi(x)\,dx + \int_D f(u)(x)\phi(x)\,dx = \int_D g(x)\phi(x)\,dx + \int_D \phi(x)\,dW_x$$

where $\{W_x; x \in \mathbf{R}^k\}$ is a standard Wiener process with k–dimensional parameter, i.e. it is an a.s. continuous Gaussian random field with zero mean and covariance operator defined by

$$E[W_x W_y] = x \wedge y$$

$(x \wedge y = (x_1 \wedge y_1) \cdots (x_k \wedge y_k))$.

We shall work with another equivalent formulation, which we call an "integral formulation". Before introducing that formulation, we need to introduce the kernel associated to the linear version of the above equation. Consider the elliptic PDE :

$$\begin{cases} -\Delta v(x) = \varphi(x),\ x \in D \\ \quad v|_{\partial D} = 0 \end{cases}$$

That equation defines a linear continuous mapping from $L^2(D)$ into itself which can be written as :

$$v(x) = \int_D K(x,y)\varphi(y)\,dy$$

We shall now give an expression for $K(x,y)$ in the three cases $k = 1,2,3$.

In case $n = 1$, $D = (0,t)$,

$$K(x,y) = x \wedge y - \frac{xy}{t}$$

In case $n = 2$, we have

$$K(x,y) = -\frac{1}{2\pi}\log|x-y| + \frac{1}{2\pi}E_x\left[\log|B_\tau - y|\right]$$

where $\{B_t;\ t \geq 0\}$ denote the standard two–dimensional Brownian motion starting from x under P_x, and τ is the exit time from D.

In case $n = 3$, we use the same notations as for $n = 2$ (but now $\{B_t; t \geq 0\}$ denotes the three–dimensional standard Brownian motion). We have :

$$K(x,y) = (4\pi|x-y|)^{-1} - E_x\left[(4\pi|B_\tau - y|)^{-1}\right]$$

Lemma 2.1. *In the three cases $k = 1,2,3$, the random field*

$$\left\{v(x) = \int_D K(x,y)\, dW_y,\ x \in \bar{D}\right\}$$

possesses an a.s. continuous modification.

Proof: We consider successively the three cases. If $D = (0,t)$,

$$v(x) = \frac{x}{t}\int_0^t W_s\, ds - \int_0^x W_s\, ds$$

and the results is obvious in this case.

Consider now the case $k = 2$. Let

$$\bar{v}(x) = \int_D \log|x-y|\, dW_y$$

We have for $x, z \in D$, $\varepsilon > 0$:

$$E\left[|\bar{v}(x) - \bar{v}(z)|^2\right] = \int_D \left|\log|x-y| - \log|z-y|\right|^2 dy$$

$$\leq |x-z|^{2-\varepsilon}\int_D \left|\log|x-y| - \log|z-y|\right|^\varepsilon \left(\int_0^1 \frac{d\theta}{\theta|x-y| + (1-\theta)|z-y|}\right)^{2-\varepsilon} dy$$

$$\leq |x-z|^{2-\varepsilon}\int_D \left|\log|x-y| - \log|z-y|\right|^\varepsilon \left(\frac{1}{|x-y|} + \frac{1}{|z-y|}\right)^{2-\varepsilon} dy$$

$$\leq c_\varepsilon |x-z|^{2-\varepsilon}\left(\int_D \left|\log|x-y| - \log|z-y|\right|^{q\varepsilon} dy\right)^{1/q}$$

$$\times\left(\int_D \frac{dy}{|x-y|^{(2-\varepsilon)p}} + \int_D \frac{dy}{|z-y|^{(2-\varepsilon)p}}\right)^{1/p}$$

where $\frac{1}{p} + \frac{1}{q} = 1$, $1 < p < (2-\varepsilon)^{-1}2$. It is not hard to deduce from these inequalities :

$$E\left[|\bar{v}(x) - \bar{v}(z)|^2\right] \leq \bar{c}'_\varepsilon |x-z|^{1-\varepsilon},\ x, z \in D$$

But $\bar{v}(x) - \bar{v}(z)$ is a zero mean Gaussian random variable, and consequently for any integer r and $\varepsilon > 0$,

$$E\left[|\bar{v}(x) - \bar{v}(z)|^r\right] \leq c'_\varepsilon |x - z|^{(1-\varepsilon)r},\ x, z \in D$$

Choosing $r > 2$, we conclude from Kolmogorov's Lemma that $\bar{v}$ possesses an a.s. continuous modification. Having chosen that modification, we deduce that $E_x[\bar{v}(B_\tau)]$ is continuous on $\bar{D}$, and also :

$$\begin{aligned} v(x) &= \int_D K(x,y)\, dW_y \\ &= (2\pi)^{-1/2}\left(E_x[\bar{v}(B_\tau)] - \bar{v}(x)\right) \end{aligned}$$

where E_x means here integrating with respect to the law of B_τ, $\{W_y\}$ being fixed.

We consider finally the case $n = 3$, which is analogous to the last case. It suffices to check that $\bar{v}(x) = \int_D |x-y|^{-1}\, dW_y$ possesses an a.s. continuous modification. For $x, z \in D$, $0 < \varepsilon < 1/4$,

$$\begin{aligned} E\left[|\bar{v}(x) - \bar{v}(z)|^2\right] &= \int_D \left|\frac{1}{|x-y|} - \frac{1}{|z-y|}\right|^2 dy \\ &\leq \int_D \left|\frac{1}{|x-y|} - \frac{1}{|z-y|}\right|^{\frac{5}{4}+\varepsilon} \left|\frac{|z-y| - |x-y|}{|x-y||z-y|}\right|^{\frac{3}{4}-\varepsilon} dy \\ &\leq |x-z|^{\frac{3}{4}-\varepsilon}\left(\int_D \left|\frac{1}{|x-y|} - \frac{1}{|z-y|}\right|^{\frac{5}{2}+2\varepsilon} dy\right)^{1/2}\left(\int_D \frac{dy}{(|x-y||z-y|)^{\frac{3}{2}-2\varepsilon}}\right)^{1/2} \\ &\leq c_\varepsilon |x-z|^{\frac{3}{4}-\varepsilon}\left(\int_D \frac{dy}{|x-y|^{\frac{5}{2}+2\varepsilon}} + \int_D \frac{dy}{|z-y|^{\frac{5}{2}+2\varepsilon}}\right)^{1/2} \\ &\quad \times \left(\int_D \frac{dy}{|x-y|^{3-4\varepsilon}}\right)^{1/4}\left(\int_D \frac{dy}{|z-y|^{3-4\varepsilon}}\right)^{1/4} \end{aligned}$$

It then follows :

$$E\left[|\bar{v}(x) - \bar{v}(z)|^2\right] \leq c'_\varepsilon |x - z|^{\frac{3}{4}-\varepsilon},\ x, z \in D$$

Again from well–known properties of Gaussian random variables, we deduce that for any integer p,

$$E\left[|\bar{v}(x) - \bar{v}(z)|^p\right] \leq c'_\varepsilon |x - z|^{(\frac{3}{8}-\frac{\varepsilon}{2})p},\ x, z \in D$$

We conclude using Kolmogorov's Lemma. □

Remark 2.2. *It follows from the proof of Lemma 2.1 that in case $k = 1$, v has Lipchitz paths, in case $k = 2$, the paths of v are Hölder continuous of exponent $1 - \varepsilon$, $\forall \varepsilon > 0$, and in case $k = 3$ they are Hölder continuous of exponent $\frac{3}{8} - \varepsilon$, $\forall \varepsilon > 0$.* □

We shall say that an a.s. bounded function u from D into $\mathbf{R}$ solves the "integral form" of equation (2.1) if :

(2.5)

$$u(x) + \int_D K(x,y) f(u)(y)\, dy = \int_D K(x,y) g(y)\, dy + \int_D K(x,y)\, dW_y, \quad x \in D$$

Note that if u satisfies (2.5), u is a.s. continuous on $\bar{D}$, and vanishes on ∂D.

Lemma 2.3. (2.4) *and* (2.5) *are equivalent.*

Proof: Suppose that u satisfies (2.5). Then u is a.s. continuous on $\bar{D}$. Let $\phi \in C^\infty(\mathbf{R}^k)$, with compact support in D. Multiply (2.5) by $\Delta\phi(x)$ and integrate over D. Using the identity

$$-\int_D \Delta\phi(x) K(x,y)\, dx = \phi(y)$$

we deduce (2.4) for smooth ϕ; the genaral case follows by density.

Suppose now that u satisfies (2.4). Choose

$$\phi(y) = \int_D \psi(x) K(x,y)\, dx$$

with $\psi \in C^\infty(D)$. Noting that

$$-\Delta\phi(y) = \psi(y),$$

we conclude that

$$\int_D u(x)\psi(x)\, dx + \int_D \int_D \psi(x) K(x,y) f(u)(y)\, dx dy$$
$$= \int_D \int_D \psi(x) K(x,y) g(y)\, dx dy + \int_D \int_D \psi(x) K(x,y)\, dW_y dx$$

from which (2.5) follows. □

We shall rewrite equation (2.5) as :

$$u + Kf(u) = Kg + K\dot{W}$$

Let us denote by $\|\cdot\|$ and $(\cdot,\cdot)$ the usual norm and scalar product in $L^2(D)$. The existence and uniqueness proof will rely on the following :

Lemma 2.4. *There exists a constant $a > 0$ such that for any $\varphi \in L^2(D)$,*

$$(K\varphi, \varphi) \geq a\|K\varphi\|^2$$

Proof: For $\varphi \in L^2(D)$, $K\varphi$ is the unique element of the Sobolev space $H_0^1(D)$ which solves the equation :

$$\begin{cases} -\Delta v(x) = \varphi(x), \; x \in D \\ \quad v|_{\partial D} = 0 \end{cases}$$

Multiplying the above equation by v and integrating by parts, we obtain :

$$\sum_{i=1}^{k} \|\frac{\partial v}{\partial x^i}\|^2 = (\varphi, v)$$

However, from Poincaré's inequality, see Gilbarg, Trudinger [3, p. 157], there exists a constant $a > 0$ such that for any $v \in H_0^1(D)$:

$$\sum_{i=1}^{k} \|\frac{\partial v}{\partial x^i}\|^2 \geq a\|v\|^2$$

The result follows. □

We are now in a position to establish the main result of this section :

Theorem 2.5. *Let D be a bounded domain of $\mathbf{R}^k$, $1 \leq k \leq 3$, with a regular boundary, let f satisfy (2.2) and (2.3), and $g \in L^2(D)$. Then equation (2.5) possesses a unique solution which is a.s. continuous on $\bar{D}$.*

Proof: *Uniqueness.* Let u, v be two solutions. Then

$$u - v + K[f(u) - f(v)] = 0 \tag{2.6}$$

Multiplying that equation by $f(u) - f(v)$, we obtain :

$$(u - v, f(u) - f(v)) + (K[f(u) - f(v)], f(u) - f(v)) = 0$$

It follows from Lemma 2.4, (2.2) and (2.6) that :

$$a\|u - v\|^2 \leq 0$$

from which uniqueness follows. Moreover, the argument clearly implies the following stronger uniqueness statement : if u and v satisfy (2.5) on a measurable subset $\tilde{\Omega}$ of Ω, then

$$u(x) = v(x) \quad \text{a.e. on } D \times \tilde{\Omega}.$$

Existence. Step 1. We suppose in this first step that f satisfies (2.2) and is *bounded*. Let $\{\dot{W}^n,\ n \in \mathbb{N}\}$ be a sequence of processes with trajectories in $L^2(D)$, which is such that :

$$K\dot{W}^n \to K\dot{W} \text{ in } L^2(\Omega \times D), \text{ as } n \to \infty$$

For each $n \in \mathbb{N}$, we consider the elliptic PDE:

$$\begin{cases} -\Delta u^n + f(u^n) = g + \dot{W}^n \\ u^n|_{\partial D} = 0 \end{cases} \tag{2.7}$$

The existence of a unique solution $u^n \in H_0^1(D)$ for equation (2.7) follows from Lions [4, Theorem 2.1 p. 171]. Clearly,

$$u^n + Kf(u^n) = Kg + K\dot{W}^n \tag{2.8}$$

and

$$u^n - u^m + K[f(u^n) - f(u^m)] = K[\dot{W}^n - \dot{W}^m]$$

Multiplying by $f(u^n) - f(u^m)$, we get :

$$\begin{aligned}(u^n - u^m, f(u^n) - f(u^m)) + (K[f(u^n) - f(u^m)], f(u^n) - f(u^m)) \\ = (K[\dot{W}^n - \dot{W}^m], f(u^n) - f(u^m))\end{aligned}$$

By (2.2) and Lemma 2.4, we obtain :

$$a\|u^n - u^m\|^2 \le (K[\dot{W}^n - \dot{W}^m], f(u^n) - f(u^m) + 2a(u^n - u^m))$$

Since $E[\|K(\dot{W}^n - \dot{W}^m)\|^2]$ tends to 0 as $n, m \to \infty$ and f is bounded, $\{u^n\}$ is a Cauchy sequence in $L^2(\Omega \times D)$. Define $u = \lim_n u^n$. Since f is bounded, $f(u^n) \to f(u)$ in $L^2(\Omega \times D)$ as $n \to \infty$. Existence follows by taking the limit in (2.8).

Existence. Step 2. We now suppose that f satisfies (2.2) and (2.3) and is *bounded from below*. Let

$$f_n(x, r) = f(x, r) \wedge n.$$

For each $n \in \mathbb{N}$, let u^n denote the unique solution (constructed in the above step) of the equation

$$u^n + Kf_n(u^n) = Kg + K\dot{W}.$$

It follows from Lemma 2.6 below that the sequence $\{u_n(x),\ n \in \mathbb{N}\}$ is decreasing for any $x \in D$, hence converges in $\mathbb{R} \cup \{-\infty\}$. Let

$$\Omega_n = \{\sup_x f(x, u^0(x)) \le n\}$$

On Ω_n, $f(u^m) \leq f(u^0) \leq n$, hence $f_m(u^m) = f(u^m)$ on Ω_n, for any $m \geq n$. Then for any $m \geq n$, u^m is the unique solution on Ω_n to our equation

$$u + Kf(u) = Kg + K\dot{W}.$$

Consequently, $u^m = u^n$ on Ω_n, for $m \geq n$, and clearly $u^n \to u$ where u solves our equation.

Existence. Step 3. We now assume that f satisfies (2.2) and (2.3). Let

$$f_n(x,r) = f(x,r) \vee (-n).$$

We can proceed as in the proof of the second step, constructing this time an increasing sequence $\{u^n\}$, and defining $\Omega_n = \{\inf_x f(x, u^0(x)) \geq -n\}$. □

Lemma 2.6. *Let f and h both satisfy (2.2) and (2.3), and moreover :*

$$h(x,r) \leq f(x,r),\ x \in D, r \in \mathbf{R}.$$

Let u and v be a.s. continuous random fields on $\bar{D}$, solutions of respectively :

$$u + Kf(u) = Kg + K\dot{W}$$
$$v + Kh(v) = Kg + K\dot{W}$$

Then

$$u(x) \leq v(x),\ x \in D.$$

Proof: From our assumptions,

$$u - v + K(f(u) - h(v)) = 0.$$

Consequently, $u - v \in H_0^1(D)$ and

$$-\Delta(u - v) + f(u) - h(v) = 0.$$

Multiplying the above identity by $(u - v)^-$ and integrating by parts, we obtain :

$$\|\nabla(u - v)^+\|^2 + (f(u) - h(v), (u - v)^+) = 0$$

But on the set $(u - v)^+ > 0$, $f(u) \geq f(v) \geq h(v)$. Hence $(u - v)^+ = 0$. □

Remark 2.7. *Note that the Theorem is still true if f, instead of being nondecreasing, satisfies :*

$$(f(x,r) - f(x,z))(r - z) \geq -\alpha|r - z|^2,\ \forall x \in D,\ r, z \in \mathbf{R}$$

provided $\alpha < a$, a being the constant appearing in Lemma 2.4, i.e. f could be the sum of an increasing function satisfying (2.3) and a Lipschitz function with a Lipschitz constant strictly smaller than a. □

3. Parabolic equations

We now want to study the equation:

(3.1)

$$\begin{cases} \frac{\partial u}{\partial t}(t,x) - \frac{\partial^2 u}{\partial x^2}(t,x) + f(u)(t,x) = \frac{\partial^2 W}{\partial t \partial x}(t,x) + g(t,x);\ t \geq 0,\ 0 \leq x \leq 1 \\ u(0,x) = u_0(x),\ 0 \leq x \leq 1; u(t,0) = u(t,1) = 0 \end{cases}$$

where $\frac{\partial^2 W}{\partial t \partial x}$ denotes the "space–time white noise", $g \in L^2_{\text{loc}}(\mathbf{R}_+ \times (0,1))$, $u_0 \in C([0,1])$ vanishes at the two endpoints, and

$$f(u)(t,x) = f_1(t,x;u(t,x)) + f_2(t,x;u(t,x))$$

satisfies properties which we now state. First note that because of the forcing term g, we can and shall assume w.l.o.g. that

$$f_1(t,x;0) = f_2(t,x,0) = 0.$$

$f_1(t,x;r)$, $f_2(t,x;r)$ are measurable functions of $(t,x,r) \in \mathbf{R}_+ \times [0,1] \times \mathbf{R}$ satisfying :

(3.2)

$r \longrightarrow f_1(t,x;r)$ is continuous and non decreasing, for any $(t,x) \in \mathbf{R}_+ \times [0,1]$,

(3.3) f_1 is locally bounded,

and there exists a constant c s.t. for any $(t,x,r,z) \in \mathbf{R}_+ \times [0,1] \times \mathbf{R} \times \mathbf{R}$,

(3.4) $$|f_2(t,x;r) - f_2(t,x;z)| \leq c|r - z|.$$

We shall write $f_1(u)$, $f_2(u)$ for $f_1(\cdot, u(\cdot))$, $f_2(\cdot, u(\cdot))$.

It is shown in Walsh [9] that the two following formulations of equation (3.1) are equivalent (at least under the assumption that u – and then also $f(u)$ – is locally bounded) :

(3.5)

$$\begin{cases} \int_0^1 u(t,x)\phi(x)\,dx - \int_0^t \int_0^1 u(s,x)\frac{\partial^2 \phi}{\partial x^2}(x)\,dxds + \int_0^t \int_0^1 f(u)(s,x)\phi(x)\,dxds \\ \quad = \int_0^1 u_0(x)\phi(x)\,dx + \int_0^t \int_0^1 \phi(x)\,dW_{sx} + \int_0^t \int_0^1 g(s,x)\phi(x)\,dxds, \\ \qquad \forall t \geq 0,\ \phi \in C^2([0,1]) \cap C_0([0,1]) \end{cases}$$

($C_0([0,1])$ denotes the set of continuous functions from $[0,1]$ into $\mathbf{R}$ which vanish at 0 and 1.)

(3.6)

$$\begin{cases} u(t,x) + \int_0^t \int_0^1 G_{t-s}(x,y) f(u)(s,y)\,dyds = \int_0^1 G_t(x,y)u_0(y)\,dy \\ + \int_0^t \int_0^1 G_{t-s}(x,y)\,dW_{sy} + \int_0^t \int_0^1 G_{t-s}(x,y)g(s,y)\,dtds,\ t \geq 0,\ 0 \leq x \leq 1 \end{cases}$$

where $(W_{tx}; t \geq 0, 0 \leq x \leq 1)$ denotes the standard Brownian sheet, defined on a probability space $(\Omega, \mathcal{F}, P)$, and $G_t(x,y)$ is the fundamental solution of the heat equation with Dirichlet boundary conditions, i.e. for any $\varphi \in C_0([0,1])$,

$$v(t,x) = \int_0^1 G_t(x,y)\varphi(y)\,dy$$

is the unique solution of

$$\begin{cases} \dfrac{\partial v}{\partial t}(t,x) - \dfrac{\partial^2 v}{\partial x^2}(t,x) = 0,\ t > 0,\ 0 < x < 1 \\ v(0,x) = \varphi(x),\ 0 \leq x \leq 1;\ v(t,0) = v(t,1) = 0,\ t \geq 0 \end{cases}$$

It is shown in Walsh [8] that the random field

$$\left\{ \int_0^t \int_0^1 G_{t-s}(x,y)\,dW_{sy};\ t \geq 0;\ 0 \leq x \leq 1 \right\}$$

possesses an a.s. continuous version.

We shall rewrite (3.6) as follows:

$$u + Gf(u) = Hu_0 + G\dot{W} + Gg \tag{3.7}$$

Note that for $\varphi \in L^2_{\text{loc}}(\mathbf{R}_+ \times (0,1))$, $v = G\varphi$ is the unique solution of :

$$\begin{cases} \dfrac{\partial v}{\partial t}(t,x) - \dfrac{\partial^2 v}{\partial x^2}(t,x) = \varphi(t,x),\ t > 0,\ 0 < x < 1 \\ v(0,x) = 0,\ v(t,0) = v(t,1) = 0 \end{cases} \tag{3.8}$$

Let $(\cdot,\cdot)$ and $|\cdot|$ denote the usual scalar product and norm on $L^2(0,1)$, $((\cdot,\cdot))_t$ and $\|\cdot\|_t$ the usual scalar product and norm on $L^2((0,t)\times(0,1))$, and $|||\cdot|||_t$ denote the norm on $C([0,t];L^2(0,1))$ defined by :

$$|||v|||_t^2 = \frac{1}{2}|v(t,\cdot)|^2 + \|v\|_t^2.$$

We shall make repeated use of the following:

Lemma 3.1. *For any* $\varphi \in L^2((0,t)\times(0,1))$,

$$((G\varphi,\varphi))_t \geq |||G\varphi|||_t^2$$

Proof : $G\varphi$ is the unique $v \in L^2(0,t;H_0^1(0,1))$ which solves (3.8). Multiplying (3.8) by $v(t,x)$ and integrating by parts, we obtain:

$$\frac{1}{2}|v(t,\cdot)|^2 + \|\frac{\partial v}{\partial x}\|_t^2 = ((\varphi, v))_t$$

But elementary inequalities yield:

$$u \in H_0^1(0,1) \Rightarrow |u| \leq \left|\frac{du}{dx}\right|$$

The result follows. □

We can now establish the

Theorem 3.2. *Under conditions (3.2), (3.3), and (3.4), equation (3.6) has a unique solution*

$$u \in C(\mathbf{R}_+ \times [0,1]) \text{ a.s.}$$

Proof : *Uniqueness* Let u and v be two solutions. Then the difference satisfies :

$$u - v + G[f(u) - f(v)] = 0$$

Multiplying by $f(u) - f(v)$, and using the monotonicity of f_1 and Lemma 3.1, we obtain :

$$((u - v, f_2(u) - f_2(v)))_t + |||u - v|||_t^2 \leq 0$$

from which we deduce, with the help of (3.3),

$$|u(t) - v(t)|^2 \leq k \int_0^t |u(s) - v(s)|^2 \, ds$$

The result follows from Gronwall's Lemma. As in the elliptic case, we note that the same proof shows that whenever u and v are two solutions of the equation on $[0,t] \times [0,1] \times \bar{\Omega}$, then $u = v$ a.e. on $[0,t] \times [0,1] \times \bar{\Omega}$.

Existence. Step 1. We suppose, in addition to the above assumptions, that f_1 is *bounded.* Let $\{W_{sy}^n,\ n \in \mathbb{N}\}$ denote a sequence of smooth random fields which are such that

$$\int_0^t \int_0^1 G_{t-s}(x,y)\, dW_{sy}^n \to \int_0^t \int_0^1 G_{t-s}(x,y)\, dW_{sy}$$

in $L^2(\Omega \times (0,1))$ for each $t > 0$ and in $L^2(\Omega \times (0,T) \times (0,1))$ for any $T > 0$, as $n \to \infty$. If we replace W by W^n, equation (3.1) has a.s. a unique solution u^n in $L^2_{\text{loc}}(\mathbf{R}_+; H^1(0,1))$, see Lions [4, Theorem 1.2, page 162], which satisfies in particular :

$$(3.7_n) \qquad u^n + Gf(u^n) = Hu_0 + G\dot{W}^n + Gg.$$

Let us show that $\{u^n\}$ is a Cauchy sequence in $L^2(\Omega \times (0,t) \times (0,1))$. We clearly have :

$$((u^n - u^m, f(u^n) - f(u^m)))_t + ((G[f(u^n) - f(u^m)], f(u^n) - f(u^m)))_t$$
$$= ((G[\dot{W}^n - \dot{W}^m], f(u^n) - f(u^m)))_t$$

We deduce from Lemma 3.1, the monotonicity of f_1 and (3.4) :

$$\begin{aligned}|||u^n - u^m|||_t^2 \leq &(u^n(t) - u^m(t), G[\dot{W}^n - \dot{W}^m](t)) \\ &+ ((G[\dot{W}^n - \dot{W}^m], f(u^n) - f(u^m) + 2(u^n - u^m)))_t \\ &+ c||u^n - u^m||_t^2\end{aligned}$$

From the assumption of the sequence $\{W^n\}$ and the boundedness of f_1,

$$\begin{aligned}\varepsilon(n,m) =& \frac{1}{2}E(|G[\dot{W}^n - \dot{W}^m](t)|^2) + E\left(((f_1(u^n) - f_1(u^m), G[\dot{W}^n - \dot{W}^m]))_t\right) \\ &+ E(|||G[\dot{W}^n - \dot{W}^m]|||_t^2)\end{aligned}$$

tends to zero as $n, m \to \infty$. But from the above estimate and (3.3), there exists a constant c such that :

$$E\left(|u^n(t) - u^m(t)|^2\right) \leq c\left(\varepsilon(n,m) + \int_0^t E\left(|u^n(s) - u^m(s)|^2\right) ds\right)$$

Then from Gronwall's lemma,

$$E\left(|u^n(t) - u^m(t)|^2\right) \leq c\varepsilon(n,m)e^{ct}$$

We know have that there exists $u \in L^2(\Omega \times (0,t) \times (0,1))$ for all $t > 0$, such that $u^n \to u$ and $f(u^n) \to f(u)$ in $L^2(\Omega \times (0,t) \times (0,1))$. It remains to take the limit in (3.7_n). Note that it is shown in Walsh [9] that $u \in C(\mathbf{R}_+ \times [0,1])$ a.s.

Existence. Step 2 and 3. Those steps are completely analogous to the corresponding steps in the proof of Theorem 2.5, given the following Lemma. □

Lemma 3.3. *Let φ and ψ be measurable mappings : $\mathbf{R}_+ \times [0,1] \times \mathbf{R} \to \mathbf{R}$ which satisfy (3.2) and (3.3) and such that*

$$\varphi(t,x,r) \geq \psi(t,x,r) \quad a.e. \text{ on } \mathbf{R}_+ \times [0,1] \times \mathbf{R}.$$

Let u and v belong to $C(\mathbf{R}_+ \times [0,1])$ a.s. and satisfy respectively :

$$u + G\varphi(u) + Gf_2(u) = Hu_0 + G\dot{W} + Gg$$

$$v + G\psi(v) + Gf_2(v) = Hu_0 + G\dot{W} + Gg$$

Then $\quad u(t,x) \leq v(t,x) \quad$ *a.e. on* $\mathbf{R}_+ \times [0,1] \times \Omega$.

Proof : It follows from the assumptions that

$$u - v + G[\varphi(u) - \psi(v)] + G[f_2(u) - f_2(v)] = 0.$$

Then $u-v \in L^2(O,T;H_0^1(0,1)) \cap C([0,T];L^2(0,1))$ for any $T > 0$ and solves :

$$\begin{cases} \dfrac{\partial}{\partial t}(u-v) - \dfrac{\partial^2}{\partial x^2}(u-v) + \varphi(u) - \psi(v) + f_2(u) - f_2(v) = 0 \\ (u-v)(0,x) = 0, \; (u-v)(t,0) = (u-v)(t,1) = 0 \end{cases}$$

Multiplying this equation by $(u-v)^+$, we deduce that :

$$\frac{1}{2}\frac{d}{dt}\|(u-v)^+(t)\|^2 + \|\frac{\partial}{\partial x}(u-v)^+(t)\|^2 + (\varphi(u(t)) - \psi(v(t)), (u-v)^+(t)) \\ + (f_2(u(t)) - f_2(v(t)), (u-v)^+(t)) = 0.$$

However, if $(u-v)^+ \neq 0$, $\varphi(u) \geq \varphi(v) \geq \psi(v)$. Hence the second and third terms in the above identity are non negative, and from the Lipschitz property of f_2,

$$\frac{d}{dt}\|(u-v)^+(t)\|^2 \leq 2c\|(u-v)^+(t)\|^2, \qquad (u-v)^+(0) = 0.$$

Hence $(u-v)^+(t) = 0$, $\forall t \geq 0$.

Acknowledgements. *The second author wants to thank A. Bensoussan for a valuable discussion which yielded an improvement over an earlier version of this paper.*

Note added in proof. *In a recent preprint, Dembo and Zeitouni study an equation similar to our elliptic SPDE, under the assumption that the nonlinear perturbation f is Lipschitz with a Lipschitz constant smaller than the constant α of Lemma 2.4 (see Remark 2.7 above). In a forthcoming paper by I. Gyöngy and E. Pardoux, the above results will be generalized in the parabolic case.*

Bibliography

[1] L. Boccardo, T. Gallouët : Nonlinear elliptic and parabolic equations involving measure data, *J. of Funct. Anal.* **87**, 149–169, 1989.

[2] G. Benfatto, G. Gallavotti, F. Nicolò : Elliptic equations and Gaussian processes, *J. of Funct. Anal.* **36**, 343–400, 1980.

[3] D. Gilbarg, N.S. Trudinger : *Elliptic partial differential equations of second order*, Springer 1977.

[4] J.L. Lions : *Quelques méthodes de résolution des problèmes aux limites non linéaires*, Dunod 1969.

[5] D. Nualart, E. Pardoux : Second order stochastic differential equations with boundary conditions, *Preprint.*

[6] M. Röckner, B. Zegarlinski : The Dirichlet problem for quasi-linear partial differential operators with boundary data given by a distribution, *Bibos preprint* 188/86.

[7] Y. Rozanov : *Markov Random Fields*, Springer 1982.

[8] Y. Rozanov : Markov random fields and stochastic partial differential equations, *Mat. Sb.* **103**, 590–613, 1977.

[9] J. Walsh : An introduction to stochastic partial differential equations, in *Ecole d'été de Probabilités de St Flour XIV*, Lecture Notes in Math. **1180**, Springer 1986.

R. Buckdahn
Sektion Mathematik
Humboldt Universität
DDR 1087 Berlin

E. Pardoux
Mathématiques, URA 225
Université de Provence
F 13331 Marseille cedex 3

A Probabilistic Approach to the Heat Equation for the $\bar{\partial}$-Neumann Problem

Naomasa Ueki

§1. Introduction

In this note, we consider the heat equation for the $\bar{\partial}$-Neumann problem. This is an initial boundary value problem whose boundary consition includes an imaginary directional differentiation. The main purpose of this note is to give a probabilistic expression (2.12) below of the heat kernel for the $\bar{\partial}$-Neumann problem on a strongly pseudoconvex Siegel domain D, which is one of the most fundamental complex manifolds with boundary. In this case, by using the methods of the partial differential equations, Stanton gave an explicit formula of the kernel on the space of $(0,q)$-forms [**8**]. We give here an explicit representation of the heat kernel on the space of (p,q)-forms in terms of the Malliavin calculus.

In [**6**], Malliavin has considered a similar problem in a domain of the complex plane. He constructed the solution of the heat equation for the $\bar{\partial}$-Neumann problem on a half plane by using a method of singular perturbations and pointed out that a method related to the Fourier transform can be applied to this problem. As we will state in §3 below, by combining a method in the Malliavin calculus with the Fourier transform of the Dirac δ-function, we can construct the heat kernel of this problem. To describe the main idea of our method, we first consider the heat equation associated with the $\bar{\partial}_b$-Laplacian on the Heisenberg group H_n:

$$\left(\frac{\partial}{\partial t} + \Box_\alpha\right) F = 0 \tag{1.1}$$

where

This work was supported by Grant-in-Aid for Scientific Research, The Ministry of Education, Science and Culture.

$$\square_\alpha = -\frac{1}{4}\sum_{j=1}^{2n} X_j^2 - i\alpha\frac{\partial}{\partial u}, \qquad -n < \alpha < n,\ \alpha \in \mathbb{Z}.$$

For the notations and notions, see §2 below and Folland-Stein [**2**]. Let $k_t^\alpha(Y, Y')$, $Y, Y' \in H_n$, be the heat kernel of (1.1). In case of $\alpha = 0$, the kernel $k_t^0(Y, Y')$ can be represented in terms of the generalized expectation in the sense of Watanabe [**10**]:

$$k_t^0(0,(x,u)) = E\left[\delta_{(x,u)}\left(\frac{x(t)}{\sqrt{2}}, S_0(t)\right)\right] \tag{1.2}$$

where $x(t)$ is the $(2n)$-dimensional Brownian motion and $S_0(t)$ is a sum of the stochastic areas of Lévy given by (2.8) below. Since

$$\delta_u(v) = \frac{1}{2\pi}\int_{-\infty}^{\infty} d\lambda\ e^{i\lambda(u-v)},$$

as Gaveau [**3**] has showed, the right hand side of (1.2) can be rewritten in the form

$$k_t^0(0,(x,u)) = \int_{-\infty}^{\infty} d\lambda\ f_t^{0,0}(x,u,\lambda) \tag{1.3}$$

where

$$f_t^{0,0}(x,u,\lambda) = \frac{1}{2\pi}e^{i\lambda u}E\left[e^{-i\lambda S_0(t)}\delta_x\left(\frac{x(t)}{\sqrt{2}}\right)\right]. \tag{1.4}$$

In general, we may formally expect that

$$k_t^\alpha(0,(x,u)) = E\left[\delta_{(x,u)}\left(\frac{x(t)}{\sqrt{2}}, S_0(t) - i\alpha t\right)\right]. \tag{1.5}$$

However, since $\delta_{(x,u)}$ is the Dirac δ function on $\mathbb{R}^{2n+1}$ and $(x(t)/\sqrt{2}, S_0(t) - i\alpha t)$ does not lie on $\mathbb{R}^{2n+1}$, we do not know how to give a correct mathematical sense to the right hand side of (1.5). But, as in the case of Gaveau [**3**], the function defined by (1.6) below is also a candidate of the kernel $k_t^\alpha(0,(x,u))$:

$$\int_{-\infty}^{\infty} d\lambda\ f_t^{0,\alpha}(x,u,\lambda) \tag{1.6}$$

where

$$f_t^{0,\alpha}(x,u,\lambda) = \frac{1}{2\pi} e^{i\lambda u - t\lambda\alpha} E\left[e^{-i\lambda S_0(t)}\delta_x\left(\frac{x(t)}{\sqrt{2}}\right)\right] \tag{1.7}$$
$$= \frac{1}{2\pi^{n+1}}\left(\frac{\lambda}{\sinh(t\lambda)}\right)^n \exp(i\lambda u - t\lambda\alpha - |x|^2\lambda\coth(t\lambda)).$$

In fact, it is easily seen by using the last line of the above expression that the integral in (1.6) is well defined and gives the desired heat kernel of (1.1) (cf. [8]). Keeping the above observations in mind, we can show that similar methods can be applied to the construction of the heat kernel for the $\bar{\partial}$-Neumann problem on the domain D and so we obtain the representation (2.12) below of the kernel. However, in our case, to show the convergence of integrals appeared in the representation of the heat kernel, we need many steps of discussions including the proof of (2.14) below, which plays a crucial role. Hence, for simplicity, we here give only the proof of the representation of the heat kernel for the $\bar{\partial}$-Neumann problem on a half plane, which Malliavin also considered. The details of the proof of (2.12) will appear in [9].

§2. Results in the strongly pseudoconvex Siegel domain

In this section, we state our results in the strongly pseudoconvex Siegel domain. Let $D = \{(z,w) \in \mathbb{C}^n \times \mathbb{C}; \text{Im } w > |z|^2\}$. This D is called a strongly pseudoconvex Siegel domain. We use the global coordinate $(x^1, x^2, \cdots, x^{2n}, u, r)$ defined by

$$\begin{aligned} &z^j = x^j + ix^{n+j}, \quad j = 1, 2, \cdots, n, \\ &\text{Re } w = u, \\ &\text{Im } w - |z|^2 = r \ (>0). \end{aligned} \tag{2.1}$$

Then the domain D can be identified with the product $H_n \times \mathbb{R}_+$, and $(x^1, x^2, \cdots, x^{2n}, u)$ and r are coordinates of H_n and $\mathbb{R}_+$, respectively. This H_n is the Heisenberg group whose group law is given by

$$(x', u')(x, u) = (x' + x, u' + u + 2\sum_{j=1}^{n}(x'^{n+j}x^j - x'^j x^{n+j})). \tag{2.2}$$

We define the vector fields $X_1, X_2, \cdots, X_{2n}$ on D by

$$(2.3)\qquad X_j = \frac{\partial}{\partial x^j} + 2x^{n+j}\frac{\partial}{\partial u}, \qquad X_{n+j} = \frac{\partial}{\partial x^{n+j}} - 2x^j\frac{\partial}{\partial u}, \quad j = 1, 2, \cdots, n.$$

We give D the Hermitian metric so that the system of vector fields $\{X_1/\sqrt{2}, X_2/\sqrt{2}, \cdots, X_{2n}/\sqrt{2}, \partial/\partial u, \partial/\partial r\}$ constitutes an orthonormal frame of the tangent bundle TD. This domain D is one of the most fundamental Hermitian manifolds with boundary.

We define the complex differential 1-forms $\{\omega^1, \omega^2, \cdots, \omega^{n+1}\}$ by

$$(2.4)\qquad \omega^j = dx^j + idx^{n+j}, \quad j = 1, 2, \cdots, n, \qquad \omega^{n+1} = \partial r.$$

Then $\{\omega^1, \omega^2, \cdots, \omega^{n+1}\}$ constitutes a unitary frame of the holomorphic cotangent bundle $\Lambda^{1,0}(T^*D)$. We define the projectors P and Q on $\Lambda^{p,q}(T^*D)$ by

$$(2.5)\qquad P = \mathrm{ext}(\bar{\omega}^{n+1})\mathrm{int}(\omega^{n+1}), \qquad Q = \mathrm{int}(\omega^{n+1})\mathrm{ext}(\bar{\omega}^{n+1}).$$

Then the heat equation for the $\bar{\partial}$-Neumann problem is the following equation on the space of (p, q)-forms:

$$(2.6)\qquad \begin{cases} \left(\dfrac{\partial}{\partial t} + \Box\right)F(t, X) = 0, & t > 0,\ X \in D, \\ \lim\limits_{t \searrow 0} F(t, X) = f(X), & \\ PF(t, X) = Q\left(\dfrac{\partial}{\partial r} - i\dfrac{\partial}{\partial u}\right)F(t, X) = 0 & \text{on } bD, \end{cases}$$

where bD is the boundary of the domain D and $\Box$ is the complex Laplacian acts as follows:

$$(2.7)\qquad \begin{aligned} \Box = & -\frac{1}{4}\sum_{j=1}^{2n} X_j^2 - \frac{1}{2}\left(\frac{\partial^2}{\partial u^2} + \frac{\partial^2}{\partial r^2}\right) + i(n - 2C)\frac{\partial}{\partial u} \\ & - \sum_{j=1}^{2n} \frac{A_j}{\sqrt{2}} X_j - B. \end{aligned}$$

C, A_j and B are endomorphisms on the bundle $\Lambda^{p,q}(T^*D)$ whose coefficients of the frame $\{\omega^{i_1}\wedge\cdots\wedge\omega^{i_p}\wedge\bar\omega^{j_1}\wedge\cdots\wedge\bar\omega^{j_q}\}$ are independent of the point D. Especially, C is expressed as

$$C = qQ + (q-1)P.$$

In (2.7), we assume that the differential operators act only on the coefficients of the frame $\{\omega^{i_1}\wedge\cdots\wedge\omega^{i_p}\wedge\bar\omega^{j_1}\wedge\cdots\wedge\bar\omega^{j_q}\}$. The equation (2.6) has the similar form to that of the heat equation for the $\bar{\partial}$-Neumann problem ([**1**],[**5**],[**7**]) except for *the imaginary directional differentiation* in the boundary condition. This imaginary directional differentiation is the particular point of the equation (2.6). Our purpose is to construct the fundamental solution (heat kernel) $h_t(X, X')$ of the equation (2.6).

Now, let (W_0^{2n}, P) be the (2n)-dimensional classical Wiener space and $x(t) = (x^1(t), x^2(t), \cdots, x^{2n}(t))$ be the element of W_0^{2n}. We define a process $S_0(t)$ by

$$S_0(t) = \sum_{j=1}^{n}\int_0^t (x^{n+j}(s)dx^j(s) - x^j(s)dx^{n+j}(s)) \tag{2.8}$$

and an $\mathrm{End}(\Lambda^{p,q}(T^*D))$-valued process $M(t)$ by the solution of the following stochastic integral equation:

$$M(t) = I + \int_0^t M(s)\left\{\sum_{j=1}^{2n} A_j dx^j(s) + B ds\right\}. \tag{2.9}$$

Keeping the obserbation mentioned in §1 in mind, we set, for any $\alpha \in \mathbb{Z}$ and $(t, x, u, \lambda) \in [0,\infty)\times H_n \times \mathbb{R}$,

$$\mathbf{f}_t^\alpha(x,u,\lambda) = \frac{1}{2\pi} e^{i\lambda u - t\lambda\alpha} E\left[e^{-i\lambda S_0(t)} M(t)\delta_x\left(\frac{x(t)}{\sqrt{2}}\right)\right]. \tag{2.10}$$

where $E[\,\cdot\,]$ in the right hand side means the generalized Wiener functional expectation (cf. [**10**]). Next, we set

$$\begin{aligned} \mathbf{p}_t^\alpha(x,u) &= \int_{-\infty}^{\infty} d\lambda\ \mathbf{f}_t^\alpha(x,u,\lambda)\exp\left(-\frac{t}{2}\lambda^2\right), \\ \mathbf{q}_t^\alpha(x,u,r) &= -\frac{2}{\sqrt{\pi}}\int_{-\infty}^{\infty} d\lambda\ \lambda \mathbf{f}_t^\alpha(x,u,\lambda)e^{r\lambda}\int_{\sqrt{\frac{t}{2}}(\frac{r}{t}+\lambda)}^{\infty} d\mu\ e^{-\mu^2} \end{aligned} \tag{2.11}$$

and

$$e_t^{\pm}(r,r') = \frac{1}{\sqrt{2\pi t}}\left\{\exp\left(-\frac{(r-r')^2}{2t}\right) \pm \exp\left(-\frac{(r+r')^2}{2t}\right)\right\}.$$

Then our results are stated as follows:

Main Theorem. *The heat kernel $h_t(X,X')$ for the $\bar\partial$-Neumann problem on the (p,q)-forms is expressed as follows:*

$$\begin{aligned} &h_t((x,u,r),(x',u',r')) \\ &= \mathbf{p}_t^{n-2q+2}((x,u)^{-1}(x',u'))e_t^{-}(r,r')P \\ &\quad + \mathbf{p}_t^{n-2q}((x,u)^{-1}(x',u'))e_t^{+}(r,r')Q \\ &\quad + \mathbf{q}_t^{n-2q}((x,u)^{-1}(x',u'),r+r')Q. \end{aligned} \tag{2.12}$$

This $h_t((x,u,r),(x',u',r'))$ is the restriction to $\bar D$ of a rapidly decreasing function of (x,u,r) (resp. (x',u',r')) for each fixed $(t,x',u',r') \in (0,\infty)\times D$ (resp. $(t,x,u,r) \in (0,\infty)\times D$) uniformly with respect to $t \in [t_0,t_1]$ for any $0 < t_0 \leqq t_1$.

This theorem has been proved in the $(0,q)$-form case by Stanton [**8**]. In the $(0,q)$-form case, $M(t)$ becomes identity. So by using the formula of Lévy, as we stated in §1, we can represent $\mathbf{f}_t^{\alpha}$ in (2.10) more explicitly:

$$\begin{aligned} \mathbf{f}_t^{\alpha}(x,u,\lambda) &= f_t^{0,\alpha}(x,u,\lambda) \\ &= \frac{1}{2\pi^{n+1}}\left(\frac{\lambda}{\sinh(t\lambda)}\right)^n \exp(i\lambda u - t\lambda\alpha - |x|^2\lambda\coth(t\lambda)). \end{aligned} \tag{2.13}$$

In the proof of the main theorem, the following lemma plays a crucial role:

Key Lemma. *For any $\alpha,\beta \in \mathbf{Z}_+^{2n}$, $k \in \mathbf{Z}_+$ and $0 < t_0 < t_1$, the following estimate holds;*

$$\begin{aligned} &\varlimsup_{\lambda\to\pm\infty} \sup_{t_0\leqq t\leqq t_1} \frac{1}{t|\lambda|}\log \sup_{x\in\mathbf{R}^{2n}} \Big\| x^{\beta}\partial_x^{\alpha} E\Big[S_0(t)^k e^{-i\lambda S_0(t)} \\ &\qquad \times M(t)\delta_x\left(\frac{x(t)}{\sqrt{2}}\right)\Big]\Big\| \\ &\leqq -n, \end{aligned} \tag{2.14}$$

where $\|\cdot\|$ *is the Hilbert-Schmidt norm in* $\mathrm{End}(\Lambda^{p,q}(T^*D))$.

By using this lemma, we can prove that the expression (2.12) is well-defined. Next, by using the Feynman-Kac formula, we see the following:

$$\frac{\partial}{\partial t}\mathrm{f}_t^\alpha((x,u)^{-1}(x',u'),\lambda) = -\Box_\lambda^\alpha \mathrm{f}_\lambda^\alpha((x,u)^{-1}(x',u'),\lambda) \tag{2.15}$$

where

$$\begin{aligned}\Box_\lambda^\alpha = &-\frac{1}{4}\sum_{j=1}^{n}\left\{\left(\frac{\partial}{\partial x^j} - 2i\lambda x^{n+j}\right)^2 + \left(\frac{\partial}{\partial x^{n+j}} + 2i\lambda x^j\right)^2\right\} \\ &-\sum_{j=1}^{n}\left\{\frac{A_j}{\sqrt{2}}\left(\frac{\partial}{\partial x^j} - 2i\lambda x^{n+j}\right) + \frac{A_{n+j}}{\sqrt{2}}\left(\frac{\partial}{\partial x^{n+j}} + 2i\lambda x^j\right)\right\} \\ &+ B + \lambda\alpha.\end{aligned}$$

Furthermere, by refering the key lemma, we obtain

$$\left(\frac{\partial}{\partial t} + \Box\right) h_t((x,u,r),(x',u',r')) = 0. \tag{2.16}$$

By repeating straightforward calculations, we can show that the expression (2.12) satisfies the boundary condition. For the initial condition, we need further devices. The detail of the matters in this section will appear in [9].

§3. The case of a half plane

In this section, we apply our methods to the $\bar{\partial}$-Neumann problem on a half plane, which Malliavin also considered in [6]. Let $M = \{z = (x,y) \in \mathbf{C} : x \geqq 0\}$ with the canonical metric. On the space of (0,0)-forms, the heat equation for the $\bar{\partial}$-Neumann problem is as follows:

$$\begin{cases} \dfrac{\partial F(t,z)}{\partial t} = \dfrac{1}{2}\left(\dfrac{\partial^2}{\partial x^2} + \dfrac{\partial^2}{\partial y^2}\right) F(t,z), & t > 0,\ z \in M, \\ \lim\limits_{t \searrow 0} F(t,z) = f(z), & \\ \left(\dfrac{\partial}{\partial x} + i\dfrac{\partial}{\partial y}\right) F(t,z) = 0 & \text{on } bM. \end{cases} \tag{3.1}$$

This is exactly the heat equation with the oblique boundary condition except for *the imaginary directional differentiation.*

Let $X(t,x)$ be a reflecting Brownian motion starting at a point $x \geqq 0$ and $\phi(t,x)$ be the local time at the point 0. Then as same as in the case of the oblique boundary condition, if $Y(t,y)$ is a Brownian motion starting at a point y independent of $X(t,x)$, the heat kernel $h_t((x,y),(x',y'))$ is expected to be

$$h_t((x,y),(x',y')) = E\left[\delta_{(x',y')}(X(t,x), Y(t,y) + i\phi(t,x))\right]. \tag{3.2}$$

However, since $\delta_{(x',y')}$ is the Dirac δ function on $\mathbb{R}^2$ and $(X(t,x), Y(t,y) + i\phi(t,x))$ does not lie on $\mathbb{R}^2$, we do not know how to give a correct mathematical sense to the right hand side of (3.2).

If we take the Fourier transformation of the both hand sides of (3.2) and exchange the order of integrations formally, then we have

$$\begin{aligned}
&\frac{1}{\sqrt{2\pi}}\int_{-\infty}^{\infty} dy' e^{-i\lambda y'} h_t((x,y),(x',y)) \\
&= \frac{1}{\sqrt{2\pi}}\int_{-\infty}^{\infty} dy' e^{-i\lambda y'} E\left[\delta_{(x',y')}(X(t,x), Y(t,y) + i\phi(t,x))\right] \\
&= \frac{1}{\sqrt{2\pi}} E\left[e^{-i\lambda Y(t,y)} e^{\lambda\phi(t,x)} \middle| X(t,x) = x'\right] P(X(t,x) = x').
\end{aligned} \tag{3.3}$$

The right hand side of (3.3) is well-defined. By using the Fourier inversion theorem, we have

$$\begin{aligned}
&h_t((x,y),(x',y')) \\
&= \frac{1}{2\pi}\int_{-\infty}^{\infty} d\lambda e^{-i\lambda y'} E\left[e^{-i\lambda Y(t,y)}\right] E\left[e^{\lambda\phi(t,x)} \middle| X(t,x) = x'\right] P(X(t,x) = x').
\end{aligned} \tag{3.4}$$

Hence the problem has reduced to show that the integral of the right hand side of (3.3) is well defined. The term $E[e^{\lambda\phi(t)}|X(t) = x']$ diverges exponentially as λ tends to $\pm\infty$. But the term $E[e^{-i\lambda Y(t)}] = \exp(-i\lambda y - \frac{\lambda^2}{2}t)$ decays very well as λ tends to $\pm\infty$. After all, if $x + x' > 0$, the right hand side of (3.4) converges and gives the desired heat kernel. In fact, the right hand side of (3.4) can be rewritten in the form

$$\begin{aligned}
&\frac{1}{2\pi t}\left\{\exp\left(-\frac{(x-x')^2}{2t}\right) + \exp\left(-\frac{(x+x)^2}{2t}\right)\right\}\exp\left(-\frac{(y-y')^2}{2t}\right) \\
&+ \frac{i}{\pi}\frac{\partial}{\partial y}\left\{\frac{1}{x+x'+i(y-y')}\exp\left(-\frac{(x+x')^2+(y-y')^2}{2t}\right)\right\}.
\end{aligned} \tag{3.5}$$

We can easily check that this satisfies the heat equation (3.1).

In our discussion, the important point is that we can dissolve the problem of the imaginary value by only using the simple Fourier transformation in one valuable. This is due to the fact that the metric, differential and complex structures are *constant in the direction that we must consider the imaginary value.* So this fact is a key point in our discussion. The strongly pseudoconvex Siegel domain in the last section is one of the most fundamental complex manifolds with boundary satisfying this condition.

On the other hand, we can also consider the right hand side of (3.2) as follows: in Gelfand-Silov [**4**], the following series expansion in a space of complex generalized functions is discussed:

$$\delta(z+ib)=\sum_{j=0}^{\infty}\frac{(ib)^j}{j!}\delta^{(j)}(z),\quad z\in\mathbb{C},\ b\in\mathbb{R}. \tag{3.6}$$

As was pointed out by Ikeda, if we formally apply this expansion to the right hand side of (3.2), we obtain the following expression:

$$\begin{aligned}&E\left[\delta_{(x',y')}(X(t,x),Y(t,y)+i\phi(t,x))\right]\\ &=\int_0^{\infty}db\sum_{j=0}^{\infty}E\left[\frac{(ib)^j}{j!}\frac{\partial^j}{\partial y^j}\delta_{y'}(Y(t,y))\right]P(X(t,x)=x',\ \phi(t,x)=b).\end{aligned} \tag{3.7}$$

The each term of the right hand side of (3.7) can be given a correct mathematical sense (see [**7**], [**10**]). In the above simple case, we easily see that the value of (3.4) and (3.7) coincide.

Acknowledgement

The author thanks Professor P. Malliavin and Professor G. Ben Arous for helpful advices at the conference.

References

[1] H. Airault, *Perturbations singulières et solutions stochastique de problèmes de D. Neumann-Spencer*, J. Math. Pures Appl. **55** (1976), 233–268.

[2] G. B. Folland and E. M. Stein, *Estimates for the $\bar{\partial}_b$ complex and analysis on the Heisenberg group*, Comm. Pure. Appl. Math. **27** (1974), 429–522.

[3] B. Gaveau, *Principe de moindre action, propagation de la chaleur, estimées sous elliptiques sur certains groupes nilpotents*, Acta Math. **139** (1977), 1–63.

[4] I. M. Gelfand and G. E. Shilov, "Generalized functions I," Academic Press, New York/London, 1964.

[5] N. Ikeda and S. Watanabe, "Stochastic differential equations and diffusion processes, second edition," Kodansha/North-Holland, Tokyo/Amsterdam, 1989.

[6] P. Malliavin, *Sur une résolution stochastique de certains problèmes de dérivée oblique dans le semi-espace*, C. R. Acad. Sc. Paris **283** (1976), 515–518.

[7] I. Shigekawa, N. Ueki and S. Watanabe, *A probabilistic proof of the Gauss -Bonnet-Chern theorem for manifolds with boundary*, Osaka J. Math. **26** (1989), 897–930.

[8] N. K. Stanton, *The heat equation for the $\bar{\partial}$-Neumann problem in a strictly pseudoconvex Siegel domain*, J. Analyse Math. **38** (1980), 67–112.

[9] N. Ueki, *A probabilistic construction of the heat kernel for the $\bar{\partial}$-Neumann problem on a strongly pseudoconvex Siegel domain*, (preprint).

[10] S. Watanabe, *Analysis of Wiener functionals (Malliavin calculus) and its applications to heat kernels*, Ann. Probab. **15** (1987), 1–39.

Department of Mathematics
Faculty of Science
Osaka University
Toyonaka, Osaka,
560 Japan

An Equivalence Theorem for Schrödinger Operators and its Applications

Z. Zhao

ABSTRACT We prove an equivalence theorem for a Schrödinger operator $H = -\frac{\Delta}{2} + V$ in a bounded domain D in $R^d (d \geq 1,\ V \in K_d^{\text{loc}})$, which equates various conditions on (D, V) including (a) Finiteness or boundedness of the stopped Feynman-Kac gauge (b) Positivity of all eigenvalues of H in D (c) Exponential decay of the norm $||T_t||$ of the local Schrödinger semigroup with t (d) Existence of a solution of $Hu = 0$ with a positive lower bound in D; and provides three applications of the main theorem: (i) An alternate proof of the Allegretto-Piepenbrink theory (ii) Strict monotonity of the principal eigenvalue with domains (iii) Uniform convergence of the solutions $\{u_n\}$ of $\left(-\frac{\Delta}{2} + V_n\right) u_n = 0$ with $u_{n|_{\partial D}} = f\ (n \geq 1)$ as the potentials of $\{V_n\}$ converge with a Kato type norm.

Let D be a bounded domain in $R^d (d \geq 1)$, $V \in K_d^{\text{loc}}$ (see [S] for definition) and let $\{X_t\}$ be the d-dimensional Brownian motion.

Define the stopped Feynman-Kac gauge u_0 and the local Schrödinger semigroup $\{T_t\}$:

$$u_0(x) = E^x \left[e_V(\tau_D)\right], \; x \in D \tag{1}$$

and

$$T_t f(x) = E^x \left[t < \tau_D, \; e_V(t) f(X_t)\right], \; t > 0, \; x \in D, \tag{2}$$

where $\tau_D = \inf\{t > 0 : \; X_t \notin D\}$ and

$$e_V(t) = \exp\left(-\int_0^t V(X_s)ds\right), \; 0 \leq t \leq \infty.$$

We consider the Schrödinger operator $H = -\frac{\triangle}{2} + V$ in D as an infinitesimal generator of $\{T_t\}$ in $L^2(D)$ and as a differential operator in the homogeneous Schrödinger equation

$$Hu \equiv -\frac{\triangle}{2}u + Vu = 0 \text{ in } D. \tag{3}$$

Any solution of (3) is in the distributional sense. Due to Theorem 1.5 in [A-S], we assume any solution of $Hu = 0$ is continuous in D.

Let $G_D(.,.)$ be the Green function of $-\frac{\triangle}{2}$ in D and G_D be the integral operator with the kernal $G_D(.,.)$, i.e. for any $f \in \mathcal{B}_b(D) \equiv \{$ all bounded Borel measurable functions in $D\}$, define

$$G_D f(x) = E^x \left[\int_0^{\tau_D} f(X_t)dt\right]. \tag{4}$$

Put

$$|||V||| \equiv \sup_{x \in D} \int_D G_D(x,y)|V(y)|dy \tag{5}$$

For any $V \in K_d^{\text{loc}}$, it is easy to see by definition that $|||V||| < \infty$. A lower bound of the gauge u_0 follows from Jensen's inequality:

$$u_0(x) \geq \exp(-|||V|||). \tag{6}$$

Consideration of the upper bound of u_0 leads to the following definition:

Definition 1. (D, V) is said gaugeable iff u_0 is bounded on $\overline{D}$.

Before proving the equivalence theorem we provide a complete treatment for the existence and uniqueness of the Dirichlet problem for the general case, which is given by Chung and Rao in [C-R] for a bounded and Hölder continuous V. The following existence theorem can be proved by the same method as Theorem 2.1 in [C-R]. A related existence theorem is given in [A-S] Theorem A.4.1.

Theorem 1 If (D, V) is gaugeable, then for any $f \in \mathcal{B}_b(\partial D)$,

$$u(x) = E^x\left[e_V(\tau_D)f(X(\tau_D))\right] \tag{7}$$

is a solution of $Hu = 0$ in D, and if, in addition, D is regular and $f \in C(\partial D)$, then we have for each $z \in \partial D$.

$$\lim_{x \to z} u(x) = f(z). \tag{8}$$

The uniqueness of the Dirichlet problem for a general $V \in K_d^{\text{loc}}$ is subtle. We need a preliminary proposition, which is also a key to the equivalence theorem.

Proposition 1 Let D_1 and D_2 be two bounded domains with $D_1 \cap D_2 \neq \emptyset$. Suppose for each $i = 1, 2$, (D_i, V) is gaugeable and for any solution $\psi \in C(\overline{D}_i)$ of $Hu = 0$ in D_i,

$$\psi(x) = E^x \left[e_V(\tau_{D_i}) \psi(X(\tau_{D_i})) \right], \; x \in D_i. \tag{9}$$

Then for $D = D_1 \cup D_2$, we have

(a) If (D, V) is gaugeable, then for any solution $\psi \in C(\overline{D})$ of $Hu = 0$ in D,

$$\psi(x) = E^x \left[e_V(\tau_D) \psi(X(\tau_D)) \right], \; x \in D. \tag{10}$$

(b) If there exists a solution $\psi > 0$ in $C(\overline{D})$ of $Hu = 0$ in D, then (D, V) is gaugeable.

Proof. Put $T_0 = 0$, and for $k \geq 1$.

$$T_{2k-1} = T_{2k-2} + \tau_{D_1} \circ \theta_{T_{2k-2}}$$

$$T_{2k} = T_{2k-1} + \tau_{D_2} \circ \theta_{T_{2k-1}}.$$

We first prove that for any solution ψ in $C(\overline{D})$ of $Hu = 0$ in D,

$$\psi(x) = E^x \left[e_V(T_n) \psi(X(T_n)) \right], \; n \geq 0. \tag{11}$$

Obviously (11) is true for $n = 0$. Suppose (11) is true for $n = 2k (k \geq 0)$. Then noticing that $X(T_{2k}) \in D_1$ on $(T_{2k} < T_{2k+1})$ we have by (9) and the strong Markov property,

$$E^x \left[e_V(T_{2k+1}) \psi(X(T_{2k+1})) \right] = E^x \left[T_{2k} = T_{2k+1}, \; e_V(T_{2k+1}) \psi(X(T_{2k+1})) \right]$$
$$+ E^x \left[T_{2k} < T_{2k+1}, e_V(T_{2k}) E^{X(T_{2k})} [e_V(\tau_{D_1}) \psi(X(\tau_{D_1}))] \right]$$
$$= E^x \left[T_{2k} = T_{2k+1}, e_V(T_{2k}) \psi(X(T_{2k})) \right] + E^x \left[T_{2k} < T_{2k+1}, e_V(T_{2k}) \psi(X(T_{2k})) \right]$$
$$= E^x \left[e_V(T_{2k}) \psi(X_{2k}) \right] = \psi(x).$$

Similarly we can prove that if (11) holds for $n = 2k+1$, then it holds for $n = 2k+2$. Hence it holds for all $n \geq 0$ by induction.

Since $T_n \uparrow$ and $T_n \leq \tau_D$, we have $\lim_{n\to\infty} T_n = T \leq \tau_D$. If $T < \tau_D$, then since $X(T_{2k-1}) \in \partial D_1$ and $X(T_{2k}) \in \partial D_2$ for all $k \geq 1$, we have by continuity

$$X(T) \in (\partial D_1) \cap (\partial D_2) \subseteq (D_1 \cup D_2)^c = D^c$$

thus $T \geq \tau_D$, which is a contradiction. Therefore, almost surely

$$\lim_{n\to\infty} T_n = \tau_D. \tag{12}$$

Now suppose (D, V) is gaugeable. Then for each $x \in D$, $E^x [e_V(\tau_D)] < \infty$. By (6),

$$\begin{aligned} E^x \left[e_V(\tau_D) \middle| F_{T_n}\right] &= e_V(T_n) E^{X(T_n)} [e_V(\tau_D)] \\ &\geq e^{-|||V|||} e_V(T_n), \ n \geq 1. \end{aligned}$$

Hence for any solution $\psi \in C(\overline{D})$, the family $\{e_V(T_n)|\psi(X(T_n))| : n \geq 1\}$ is uniformly integrable under P^x. This allows us to take the limit under E^x in (11) as $n \to \infty$. Thus (10) follows from (12) and continuity of ψ and the Brownian paths. This proves (a).

Now suppose there exists a solution $\psi > 0$ in $C(\overline{D})$. Then

$$c = \min_{x\in\overline{D}} \psi(x) > 0.$$

Thus by (11) for all $n \geq 0$.

$$\psi(x) \geq c E^x [e_V(T_n)]. \tag{13}$$

It follows from (12), (13) and Fatou's lemma that

$$E^x [e_V(\tau_D)] \leq c^{-1} \psi(x) < \infty. \tag{14}$$

Thus by the gauge theorem (see [Z] Theorem 7), (D, V) is gaugeable. (b) is proved.

□

The following proposition has been given in [C-L-W] Proposition 3. Both the result and proof there still hold for $V \in K_d^{\text{loc}}$.

Proposition 2 Suppose

$$E^x\left[\exp\left(\int_0^{\tau_D} |V(X_s)|ds\right)\right] < \infty \text{ in } D.$$

Then the only function in $C(\overline{D})$ satisfying $Hu = 0$ and $u\big|_{\partial D} = 0$ is the zero function.

We now prove the uniqueness theorem of the Dirichlet problem for a gaugeable (D, V).

Theorem 2. Let D be a bounded domain such that (D, V) is gaugeable. Then for any solution $\psi \in C(\overline{D})$ of $H\psi = 0$ in D, we have

$$\psi(x) = E^x\left[e_V(\tau_D)\psi(X(\tau_D))\right], \; x \in D. \tag{15}$$

Proof Since $V \in K_d^{\text{loc}}$, there is a number $r_0 > 0$ such that for any ball $B \subseteq D$ with a radius $\leq r_0$,

$$\sup_{x\in B} E^x\left[\int_0^{\tau_B} |V(X_s)|ds\right] < \frac{1}{2}.$$

Then by Khasminskii's lemma (see Lemma B.1.2 [S]),

$$\sup_{x\in B} E^x\left[\exp\int_0^{\tau_B} |V(X_s)|ds\right] \leq 2.$$

Thus by Theorem 1 and Proposition 2, (B, V) is gaugeable and for any solution ψ in $C(\overline{B})$ of $Hu = 0$ in B, (10) holds for B. It is easy to construct a sequence of $\{D_n\}$ such that $\overline{D}_n \subseteq D$, $D_n \uparrow D$ and for each n, $D_n = \bigcup_{k=1}^{k_n} B_k$, where B_k are balls with radius $\leq r_0$. Thus by Proposition 1 (a) and mathematical induction,

$$\psi(x) = E^x \left[e_V(\tau_{D_n}) \psi(X(\tau_{D_n})) \right]. \tag{16}$$

Since (D, V) is gaugeable, the same argument as Proposition 1 shows that the family $\{e_V(\tau_{D_n})|\psi(X(\tau_{D_n}))| : n \geq 1\}$ is uniformly integrable under E^x. Thus letting $n \to \infty$ in (16) yields (15) by continuity.

□

The next theorem is a part of the equivalence theorem, but for a general domain D.

Theorem 3 For any domain D in R^d, if there exists a solution ψ of $Hu = 0$ with $c \equiv \inf_{x \in D} \psi(x) > 0$, then for each $x \in D$,

$$E^x \left[e_V(\tau_D) \right] \leq c^{-1} \psi(x) < \infty. \tag{17}$$

Proof As shown in the proof of Theorem 2, for any domain D in R^d, we can construct a sequence $\{D_n\}$ such that $D_n \uparrow D$ and for each n, $\overline{D}_n \subseteq D$, $D_n = \bigcup_{k=1}^{k_n} B_k$, where each B_k is a ball satisfying

$$\sup_{x \in B_k} E^x \left[\exp \int_0^{\tau_{B_k}} |V(X_s)| ds \right] < \infty.$$

Then by Proposition 1 (a), (b) and Proposition 2, for each $n \geq 1$, we have

$$\begin{aligned} \psi(x) &= E^x \left[e_V(\tau_{D_n}) \psi(X(\tau_{D_n})) \right] \\ &\geq c E^x \left[e_V(\tau_{D_n}) \right]. \end{aligned}$$

Since $\tau_{D_n} \uparrow \tau_D$, it follows from Fatou's lemma that

$$E^x\left[e_V(\tau_D)\right] \leq c^{-1}\psi(x) < \infty.$$

□

For any domain D in R^d, it can be shown by the same argument as in [A-S] that for each $t > 0$, T_t given in (2) is a linear bounded operator in $L^p(D)$ $(1 \leq p \leq \infty)$ with

$$||T_t||_p \leq ||T_t||_\infty \tag{18}$$

and

$$||T_t||_\infty = ||T_t 1||_\infty = \sup_{x \in D} E^x\left[t < \tau_D;\ e_V(t)\right],$$

here and below, $||\cdot||_p (1 \leq p \leq \infty)$ denotes the norm in $L^p(D)$.

Now let D be a bounded domain. Then $\{T_t\}$ as a strongly continuous operator semigroup in $L^2(D)$, has an infinitesimal generator $A = \frac{\triangle}{2} - V$ in its definition domain with eigenvalues $\lambda_1 < \lambda_2 < \ldots \lambda_n < \ldots$. Let $\lambda(D, V)$ denote the principal eigenvalue λ_1. We also have

$$\lambda(D, V) = \inf_{\substack{||\varphi||_2 = 1 \\ \varphi \in C_c^\infty(D)}} \left[\frac{1}{2}\int_D \nabla\varphi \cdot \nabla\varphi dx + \int_D V\varphi^2 dx\right], \tag{19}$$

where $C_c^\infty(D) = \{$ all infinitely differentiable functions in D with compact support $\}$.

Theorem 4 Let D be a bounded domain in $R^d (d \geq 1)$ and $V \in K_d^{\text{loc}}$. Then the following conditions on (D, V) are equivalent:

(i) (D, V) is gaugeable, i.e. $\sup_{x \in \overline{D}} E^x\left[e_V(\tau_D)\right] < \infty$.

(ii) $E^x\left[e_V(\tau_D)\right] \not\equiv \infty$ for $x \in D$.

(iii) $\lambda(D, V) > 0$.

(iv) There exists a $\alpha > 0$ such that $||T_\alpha||_\infty < 1$.

(v) There exist $\beta > 0$ and $\lambda > 0$ such that for all $t > 0$,

$$||T_t||_\infty \leq \beta e^{-\lambda t}.$$

(vi) There exists a solution ψ of $Hu = 0$ in D with

$$0 < \inf_{x \in D} \psi(x) \leq \sup_{x \in D} \psi(x) < \infty.$$

(vii) There exists a solution ψ of $Hu = 0$ in D with

$$\inf_{x \in D} \psi(x) > 0.$$

Remark In the case that V is bounded and Hölder continuous, the equivalence of (i), (iii) and (vi) has been proven in [C-L-W]. For $V \in K_d^{\text{loc}}$, the implication (iii) $\Rightarrow$ (i) has been proven in [A-S]. The implication "(ii) $\Rightarrow$ (i)" is called the gauge theorem, which has been proven in [C-R] for bounded V and in [Z] for $V \in K_d^{\text{loc}}$.

Proof We shall prove the theorem by completing the two implication cycles:

$$\text{(i)} \Rightarrow \text{(vi)} \Rightarrow \text{(vii)} \Rightarrow \text{(ii)} \Rightarrow \text{(i)}$$

and

$$\text{(i)} \Rightarrow \text{(iv)} \Rightarrow \text{(iii)} \Rightarrow \text{(v)} \Rightarrow \text{(ii)} \Rightarrow \text{(i)}.$$

(i) $\Rightarrow$ (vi): By Theorem 1 with $f \equiv 1$, we have that $u_0(x)$ is a solution of $Hu = 0$ in D. The upper bound is given by the assumption (i) and the lower bound is given by (6).

(vi) ⇒ (vii): Obvious.

(vii) ⇒ (ii): This is due to Theorem 3 (17).

(ii) ⇒ (i): This is given by Theorem 7 in [Z].

(i) ⇒ (iv): By (6) and the Markov property, for each $x \in D$ and $t > 0$,

$$\begin{aligned}(T_t 1)(x) &= E^x\left[t < \tau_D,\ e_V(t)\right] \\ &\le e^{|||V|||} E^x\left[t < \tau_D,\ e_V(t) E^{X(t)}[e_V(\tau_D)]\right] \\ &= e^{|||V|||} E^x\left[t < \tau_D,\ e_V(\tau_D)\right].\end{aligned} \tag{20}$$

Since $\tau_D < \infty$, $1_{(t<\tau_D)} \downarrow 0$ as $t \uparrow \infty$ P^x-a.s.. It follows from (i), (20) and the dominated convergence theorem that

$$\lim_{t\uparrow\infty} E^x\left[t < \tau_D,\ e_V(t)\right] = 0. \tag{21}$$

Also by condition (i) and (20), the function $x \to (T_t 1)(x)$ is bounded for all $t > 0$ and $x \in D$. Then by (21) and the bounded convergence theorem we have

$$\lim_{t\uparrow\infty} ||T_t 1||_1 = 0. \tag{22}$$

It can be proved by Hölder inequality and duality argument that for each fixed $t > 0$, $1 \leqq p,\ q \leqq \infty$, T_t is a linear bounded operator from $L^p(D)$ to $L^q(D)$, i.e., its norm

$$||T_t||_{p,q} < \infty \tag{23}$$

(see [A-S] (1.16)).

For $t > 1$, we have

$$||T_t||_\infty = ||T_t 1||_\infty = ||T_1(T_{t-1}1)||_\infty \le ||T_1||_{1,\infty} ||T_{t-1}1||_1.$$

Therefore (22) and (23) imply

$$\lim_{t\uparrow\infty} ||T_t||_\infty = 0,$$

from which (iv) follows.

(iv) $\Rightarrow$ (iii): For each $t > 0$, since T_t is a bounded symmetric operator in $L^2(D)$, we have

$$||T_t||_2 = \exp\left[-\lambda(D,V)t\right]. \tag{24}$$

Now if (iv) holds, then by (18) and (24), for an $\alpha > 0$,

$$\exp\left[-\lambda(D,V)\alpha\right] < 1.$$

Thus $\lambda(D,V) > 0$.

(iii) $\Rightarrow$ (v): By Theorem 4.5 in [A-S], for $V \in K_d^{\mathrm{loc}}$, there exists $b > 0$ such that for each $0 < t \le b$,

$$\sup_{x\in D}\left(\int_0^t |V(X_s)|ds\right) < \frac{1}{2}.$$

Then by Khasminskii's lemma, we have for $0 < t \le b$

$$||T_t||_\infty \le 2. \tag{25}$$

For $t > b$, by (23) and (24),

$$\begin{aligned} ||T_t||_\infty &\le ||T_{\frac{b}{2}}||_{\infty,2}||T_{t-b}||_2||T_{\frac{b}{2}}||_{2,\infty} \\ &= ||T_{\frac{b}{2}}||_{\infty,2}||T_{\frac{b}{2}}||_{2,\infty}\exp\left[-\lambda(D,V)(t-b)\right]. \end{aligned} \tag{26}$$

Since we assume $\lambda(D,V) > 0$, (v) follows from (25) and (26).

(v) $\Rightarrow$ (ii): Since $|V| \in L^1(D)$, we have by (18), condition (v) and Fubini's theorem,

$$\begin{aligned} &\int_D \left[\int_0^\infty E^x[t < \tau_D;\ e_V(t)|V(X_t)|]dt\right] dx \\ &= \int_0^\infty ||T_t(|V|)||_1 dt \le \int_0^\infty ||T_t||_1 dt ||(|V|)||_1 < \infty. \end{aligned}$$

Therefore for almost every $x \in D$,

$$\int_0^\infty E^x\left[t < \tau_D,\ e_V(t)|V(X_t)|\right] dt < \infty. \tag{27}$$

This allows us to use Fubini's theorem in the following equalities:

$$\begin{aligned} E^x\left[e_V(\tau_D)\right] &= E^x\left[1 + \int_0^{\tau_D} \frac{d}{dt}[e_V(t)]dt\right] \\ &= 1 - E^x\left[\int_0^{\tau_D} e_V(t)V(X_t)dt\right] \\ &\le 1 + \int_0^\infty E^x\left[t < \tau_D,\ e_V(t)|V(X_t)|\right] dt < \infty. \end{aligned}$$

for a.s. $x \in D$. That implies (ii).

□

We now turn to applications of the main theorem. The first one is an alternate proof of the Allegretto-Piepenbrink theory for a general domain D and a $V \in K_d^{\mathrm{loc}}$. (see Theorem C.8.1 and the related references in [S]).

Definition 2 For any domain D in R^d, $H = -\frac{\Delta}{2} + V \ge 0$ in D iff for any $\varphi \in C_c^\infty(D)$,

$$\frac{1}{2}\int_D \nabla\varphi \cdot \nabla\varphi dx + \int_D V\varphi^2 dx \ge 0.$$

Theorem 5. Let D be a domain in R^d and $V \in K_d^{\mathrm{loc}}$. Then $H = -\frac{\Delta}{2} + V \ge 0$ in D if and only if there exists a solution $\psi > 0$ of $Hu = 0$ in D.

Proof. Take a sequence of bounded domains $\{D_n\}$ such that $D_n \uparrow D$ and $\overline{D}_n \subseteq D$ for each $n \ge 1$.

We first assume there exists a solution $\psi > 0$ of $Hu = 0$ in D. By the remark after (3), ψ is continuous in D. Then for each $n \geq 1$, $\inf_{x \in \overline{D}_n} \psi(x) > 0$. Hence by Theorem 4, $\lambda(D_n, V) > 0$.

For any $\varphi \in C_c^\infty(D)$ we can choose an $m \geq 1$ such that $\operatorname{Supp}(\varphi) \subseteq D_m$. Thus by (19),

$$\begin{aligned}&\frac{1}{2}\int_D \nabla\varphi \cdot \nabla\varphi dx + \int_D V\varphi^2 dx \\ &= \frac{1}{2}\int_{D_m} \nabla\varphi \cdot \nabla\varphi dx + \int_{D_m} V\varphi^2 dx \\ &\geq \|\varphi\|_2^2 \lambda(D_m, V) \geq 0.\end{aligned}$$

Hence $H \geq 0$ in D by Definition 2.

We now assume $H \geq 0$ in D. Then for each $n \geq 1$, by Definition 2 and (19), we have $\lambda(D_n, V + \frac{1}{n}) \geq \frac{1}{n} > 0$. By Theorem 4, there exists a solution $\psi_n > 0$ in D_n satisfying

$$-\frac{\Delta}{2}\psi_n + V\psi_n + \frac{1}{n}\psi_n = 0 \text{ in } D_n. \tag{28}$$

Pick a point x_0 in D_1, put $u_n(x) = \psi_n(x_0)^{-1}\psi_n(x)$, $x \in D_n$. Then u_n also satisfies (28) with $u_n(x_0) = 1$.

As in the proof of Theorem C.8.1 in [S], by Harnack's inequality, we can choose a subsequence $\{u_{n_k}\}$ such that for any $\varphi \in C_c^\infty(D)$, $(u_{n_k}, \varphi) \to (u, \varphi)$ as $k \to \infty$ and u is a positive locally bounded function in D and satisfies $Hu = 0$ by passing to the limit in the distributional form of (28).

□

The following property about strict monotonity of the principal eigenvalue with domains becomes more nontrivial when we allow the potential V to take both signs.

Theorem 6 Let $V \in K_d^{\text{loc}}$ and D_1, D_2 be two bounded domains with $\overline{D}_1 \subseteq D_2$. Then

$$\lambda(D_1, V) > \lambda(D_2, V).$$

Proof Put $c = \lambda(D_2, V)$. Then we have $-\frac{\triangle}{2} + V - c \geq 0$ in D_2. So by Theorem 5, there exists a solution $\psi > 0$ in D_2 that satisfies

$$-\frac{\triangle}{2}\psi + V\psi - c\psi = 0 \text{ in } D_2.$$

Since $\overline{D}_1 \subseteq D_2$ and ψ is continuous, $\inf_{x \in D_1} \psi(x) > 0$. Thus by Theorem 4, we have

$$\lambda(D_1, V - c) > 0,$$

$$\text{i.e.} \qquad \lambda(D_1, V) > c = \lambda(D_2, V).$$

□

Our last application is about continuity of solutions of the Dirichlet problem with potentials when the boundary function is fixed. The norm $|||V|||$ given in (5) proves to be suitable for this continuous dependence. More precisely, we have

Theorem 7 Let D be a bounded domain, $V \in K_d^{\text{loc}}$, and $f \in C(\partial D)$. We assume (D, V) is gaugeable and put

$$u(V;\ x) = E^x \left[e_V(\tau_D) f(X(\tau_D)) \right]. \tag{29}$$

Let $\{V_n\} \subseteq K_d^{\text{loc}}$ with $|||V_n - V||| \to 0$ as $n \to \infty$. Then for sufficiently large n, (D, V_n) is gaugeable and

$$||u(V_n, \cdot) - u(V, \cdot)||_\infty \to 0 \text{ as } n \to \infty \tag{30}$$

Proof Since (D,V) is gaugeable, by Theorem 4, $\lambda(D,V) > 0$. By Lemma A.4.3. in [A-S], there is an integer $m > 1$ such that

$$A = \sup_{x \in D} E^x \left[e_{\frac{m}{m-1}V}(\tau_D) \right] < \infty$$

Choose $N \geq 1$ such that for each $n \geq N$

$$|||V_n - V||| < \frac{1}{2m}.$$

Then by Hölder inequality and Khasminskii's lemma,

$$\begin{aligned} &E^x \left[e_{V_n}(\tau_D) \right] = E^x \left[e_V(\tau_D) e_{V_n - V}(\tau_D) \right] \\ &\leq \left\{ E^x \left[e_{\frac{m}{m-1}V}(\tau_D) \right] \right\}^{\frac{m-1}{m}} \left\{ E^x \left[e_{m(V_n - V)}(\tau_D) \right] \right\}^{\frac{1}{m}} \\ &\leq A^{\frac{m-1}{m}} \left(1 - m|||V_n - V||| \right)^{-\frac{1}{m}} < \infty. \end{aligned}$$

Hence for such n, (D, V_n) are gaugeable.

$$\begin{aligned} |u(V_n, x) - u(V, x)| &\leq ||f||_\infty E^x \left[|e_{V_n}(\tau_D) - e_V(\tau_D)| \right] \\ &= ||f||_\infty E^x \left[e_V(\tau_D) \, |e_{V_n - V}(\tau_D) - 1| \right] \\ &\leq ||f||_\infty A^{\frac{m-1}{m}} \left\{ E^x \left[|e_{V_n - V}(\tau_D) - 1|^m \right] \right\}^{\frac{1}{m}}. \end{aligned} \tag{31}$$

Noticing $\sum_{k=0}^{m} (-1)^{m-k} \binom{m}{k} = 0$, we have by using Khasminskii's lemma again,

$$\begin{aligned} &E^x \left[|e_{V_n - V}(\tau_D) - 1|^m \right] \\ &\leq E^x \left\{ \left[e_{-|V_n - V|}(\tau_D) - 1 \right]^m \right\} \\ &= \sum_{k=0}^{m} (-1)^{m-k} \binom{m}{k} E^x \left[e_{-k|V_n - V|}(\tau_D) \right] \\ &= \sum_{k=1}^{m} (-1)^{m-k} \binom{m}{k} \left\{ E^x \left[e_{-k|V_n - V|}(\tau_D) \right] - 1 \right\} \\ &\leq \sum_{k=1}^{m} \binom{m}{k} \left[\frac{1}{1 - k|||V_n - V|||} - 1 \right] \longrightarrow 0 \text{ as } n \to \infty. \end{aligned} \tag{32}$$

Then (30) follows from (31) and (32).

□

Acknowledgement. The main idea and argument of the part (a) of Proposition 1 and Theorem 2 belong to K.L. Chung, with whom the author should share the credit for this paper.

References

[A-S] M. Aizenman and B. Simon, *Brownian motion and Harnack inequality for Schrödinger operators*, Comm. Pure Appl. Math. **35** (1982), 209–273.

[C-L-W] K.L. Chung, P. Li and R.J. Williams, *Comparison of probability and classical methods for the Schrödinger equation*, Exposition. Math **4** (1986), 271–278.

[C-R] K.L. Chung and M. Rao, *Feynman-Kac functional and Schrödinger equation*, Seminar on Stochastic Processes, Birkhäuser, Boston, 1981.

[S] B. Simon, *Schrödinger semigroups*, Bull. Amer. Math. Soc. **7** (1982), 447–526.

[Z] Z. Zhao, *Conditional gauge and unbounded potential*, Z. Wahrsch. Verw. Gebiete. **65** (1983), 13–18.

Z. Zhao
Department of Mathematics
University of Missouri
Columbia, MO 65211

Part IV
Diffusion Processes in Geometry

Stochastic Control Problems in Symmetric Cones and Spherical Functions

T.E. Duncan and H. Upmeier

0. Introduction

While there has been a sizable amount of research in stochastic optimal control, there are only a relatively few examples of controlled diffusions where the optimal control is expressed explicitly as a function of the state of the system. The relative scarcity of explicitly solvable stochastic control problems is not surprising from the fact that completely integrable Hamiltonian systems are not generic.

In this paper we announce the construction of a class of explicitly solvable stochastic control problems in certain curved state spaces, more precisely, Riemannian symmetric cones M of negative curvature. The controlled diffusion process comes from a non-euclidean Brownian motion in M whose infinitesimal generator is the invariant Laplace-Beltrami operator Δ_M. Writing

$$M = G/K$$

for a suitable semi-simple non-compact Lie group G and a maximal compact subgroup K, we consider K-invariant eigenfunctions of Δ_M, the so-called spherical functions. The formulation of the corresponding control problem

Research partially supported by NSF Grants ECS-8718026 and DMS 8702371.

and its explicit solution are then based on a detailed study of these spherical functions.

For symmetric spaces of rank 1, the so-called "hyperbolic spaces," such a study has been done in [2]. To motivate our study of spherical functions on spaces of higher rank, we first briefly review the results of [2], giving the first example of an explicitly solvable stochastic control problem in a non-compact manifold with non-zero sectional curvature.

1. A Stochastic Control Problem in Hyperbolic Three Space.

Consider the open unit ball $B_1(0) \subset \mathbb{R}^3$ that is,

$$B_1(0) = \{y \in \mathbb{R}^3 : |y| < 1\}$$

where $|\cdot|$ is the usual Riemannian metric in $\mathbb{R}^3$. If $B_1(0)$ is given the Riemannian structure

$$ds^2 = 4(1 - |y|^2)^{-2}(dy_1^2 + dy_2^2 + dy_3^2) \tag{1.1}$$

then this space has constant negative sectional curvature -1. This space is a geometric model for real hyperbolic three space

$$\mathbb{H}^3(\mathbb{R}) \cong SL(2,\mathbb{C})/SU(2) \simeq SO_0(3,1)/SO(3).$$

It is well known that any space of constant negative curvature is globally trivial, that is, it is globally diffeomorphic to Euclidean space. Geodesic polar coordinates at the origin, denoted by 0, provide such a global trivialization, that is,

$$\exp_0 Y \mapsto (r, \theta_1, \theta_2)$$

where $Y \in T_0\mathbb{H}^3(\mathbb{R})$, $r = |Y|_0$ with $|\cdot|_0$ the Riemannian metric at 0, and (θ_1, θ_2) are coordinates of the unit vector $Y/|Y|_0$. In these coordinates the Riemannian metric (1.1) is

$$ds^2 = dr^2 + (\sinh r)^2 \, d\sigma^2$$

where $d\sigma^2$ is the Riemannian metric on the unit sphere in $T_0\mathbb{H}^3(\mathbb{R})$. Furthermore in these coordinates the Laplace-Beltrami operator $\Delta_{\mathbb{H}^3(\mathbb{R})}$ is

$$\Delta_{\mathbb{H}^3(\mathbb{R})} = \frac{\partial^2}{\partial r^2} + 2 \coth r \frac{\partial}{\partial r} + (\sinh r)^2 \Delta_{S^2}$$

where Δ_{S^2} is the Laplace Beltrami operator on the unit sphere in $T_0\mathbb{H}^3(\mathbb{R})$.

The radial part of $\Delta_{\mathbb{H}^3(\mathbb{R})}$, denoted $\mathfrak{L}(\Delta_{\mathbb{H}^3(\mathbb{R})})$, in the geodesic polar coordinates at 0 is

$$\mathfrak{L}(\Delta_{\mathbb{H}^3(\mathbb{R})}) = \frac{\partial^2}{\partial r^2} + 2 \coth r \frac{\partial}{\partial r}$$

The stochastic control problem in $\mathbb{H}^3(\mathbb{R})$ is to control a Brownian motion by a control in the drift of the Brownian motion so that this controlled diffusion stays near the origin. Specifically the infinitesimal generator of the controlled diffusion is

$$\frac{1}{2} \Delta_{\mathbb{H}^3(\mathbb{R})} + u(t, r) \frac{\partial}{\partial r}. \tag{1.2}$$

Let $(Y(t), t \geq 0)$ be the diffusion process whose infinitesimal generator is (1.2). The dependence of this diffusion on the control has been suppressed for notational convenience. The cost functional for the stochastic control problem is

$$J(U) = E_\alpha \int_0^T \cosh|Y(t)|_0 + \frac{\sinh^2 |Y(t)|_0}{\cosh |Y(t)|_0} U^2(t)dt \tag{1.3}$$

where $(Y(t), t \in [0,T])$ is described in the geodesic polar coordinates at 0, $|Y(0)|_0 = \alpha > 0$ and $T > 0$ has the property that

$$\sup_{0 \leq t \leq T} g(t) \leq \frac{1}{2}$$

where g is the unique, positive solution of the Riccati differential equation

$$\frac{dg}{dt} + g - \frac{1}{4}g^2 + 1 = 0 \tag{1.4}$$

$$g(T) = 0.$$

An admissible control at time t is a Borel measurable function of Y(t) that is a smooth function on $\mathbb{H}^3(\mathbb{R}) - \{0\}$ such that the solution of

$$dX(t) = \coth X(t)dt + U(t, X(t))dt + dW(t) \tag{1.5}$$

$$X(0) = \alpha$$

exists and is unique in a sample path sense where $\alpha > 0$ and $(W(t), t \in [0,T])$ is a real-valued standard Brownian motion. It easily follows from the radial part of $\Delta_{\mathbb{H}^3(\mathbb{R})}$ that $X(t) = |Y(t)|_0$ a.s. Since the cost functional for U fixed only depends on $|Y(t)|_0$ the control only depends on $|Y(t)|_0 = X(t)$.

The Hamilton-Jacobi or dynamic programming equation for the stochastic control problem (1.2 - 1.3) is

$$0 = \frac{\partial W}{\partial s} + \min_{v \in \mathbb{R}} \left[\frac{1}{2}\frac{\partial^2 W}{\partial r^2} + \coth r \frac{\partial W}{\partial r} + v \frac{\partial W}{\partial r} + \cosh r + \frac{\sinh^2 r}{\cosh r} v^2 \right] \tag{1.6}$$

with the boundary condition $W(s,r) = 0$ for $(s, r) \in \{T\} \times \mathbb{H}^3(\mathbb{R})$. Performing the minimization in (1.6) the candidate for an optimal control is

$$U^*(s,r) = \frac{-1}{2f(r)} \frac{\partial W(s,r)}{\partial r} \tag{1.7}$$

where

$$f(r) = \frac{\sinh^2 r}{\cosh r}.$$

A verification theorem (p. 159 [5]) states that if (1.6) has a smooth solution satisfying the boundary conditions then it gives a solution to the stochastic optimal control problem.

It is elementary to verify that

$$W(s,r) = g(s) \cosh r$$

where g satisfies (1.4) is a solution to (1.6) satisfying the boundary condition. It can be shown that the solution of the stochastic differential equation (1.5) using the optimal control (1.7) has one and only one solution in a sample path sense. Basically it is required to show that this solution does not hit the origin which is accomplished by comparison of the solution with the two-dimensional Bessel process.

Recall that the so-called two-dimensional Bessel process is a real-valued process with the infinitesimal generator

$$\frac{1}{2}\left(\frac{d^2}{dr^2} + \frac{1}{r}\frac{d}{dr}\right).$$

It is a process that does not hit r = 0 at a positive time and it can be defined as the strong solution of the scalar stochastic differential equation

$$dZ(t) = \frac{1}{2Z(t)}\,dt + dB(t) \tag{1.8}$$

The preceding discussion has provided an outline of the proof of the following proposition whose complete proof is given in [2].

1.1 PROPOSITION. *The stochastic control problem described in (1.2 - 1.3) has an optimal control* U^* *that is*

$$U^*(s, r) = -\frac{1}{2}\, g(s) \coth r$$

where $s \in [0, T]$, r *is given in geodesic polar coordinates at* 0, *and* g *is the unique positive solution of the Riccati differential equation (1.4).*

To obtain other examples of explicitly solvable stochastic control problems it is natural to search for other functions that have the properties of cosh r that are used to solve the control problem (1.2 - 1.3). Specifically it is important to find other functions that are monotone increasing functions of r or more generally increasing functions along rays from the origin and that are eigenfunctions of the radial part of the Laplace-Beltrami operator. This latter property is more usefully described as being a spherical function.

2. Symmetric Cones and Their Spherical Functions

The Riemannian symmetric spaces have been classified by É. Cartan [6]. Every irreducible symmetric space is either of compact type, euclidean type or non-compact non-euclidean type. Further, there is a canonical "duality" between the compact and the non-compact (non-euclidean) type. Among the non-compact spaces, two classes are of special importance, e.g., in the theory of automorphic forms and group representations: the symmetric cones in $\mathbb{R}^n$ and the bounded symmetric domains in $\mathbb{C}^n$. We will consider curved state spaces of these types for our stochastic control problems. Since every bounded symmetric domain can be realized as a tube domain (or, more generally, a Siegel domain [9]) over a symmetric cone, we will consider in this paper only the symmetric cones which are slightly easier to describe than bounded symmetric domains.

The basic examples of symmetric cones are the spaces

$$\Lambda_{\mathbb{K}} = GL(r,\mathbb{K})/U(r,\mathbb{K}), \tag{2.1}$$

where r is a positive integer (called the rank of $\Lambda_{\mathbb{K}}$) and $\mathbb{K}$ is one of the real division algebras $\mathbb{R}$, $\mathbb{C}$ or $\mathbb{H}$ (quaternions). Another example is the forward light cone

$$\Lambda_n = \{x \in \mathbb{R}^n: x_1 > \sqrt{x_2^2 + \cdots + x_n^2}\} \tag{2.2}$$

of dimension $n \geq 3$ ($n = 4$ is the standard forward light cone). Writing

$$\Lambda_n = \mathbb{R}_+^* \times SO_0(n-1, 1)/SO(n-1)$$

we see that Λ_n has rank r = 2. It can be shown that up to an exceptional cone of rank 3 and dimension 27 (related to the octonions), (2.1) and (2.2) exhaust the list of all irreducible symmetric cones.

Apart from the Lie theoretic description every symmetric cone Λ has also a more geometric *Jordan algebraic* description [1]: Λ can be a realized as an open convex sharp cone in a real vector space X endowed with a Jordan algebra product

$$(x,y) \mapsto x \circ y$$

and, in fact, the closure of Λ coincides with the set of squares in X:

$$\overline{\Lambda} = \{x^2: x \in X\}.$$

The basic example is the open half-line $\Lambda = \{x \in \mathbb{R}: x > 0\}$ in $X := \mathbb{R}$, endowed with the usual product. In this picture the cones (2.1) come from the Jordan algebra

$$X_{\mathbb{K}} = \{x \in \mathbb{K}^{r \times r}: x^* = x\}$$

of all self-adjoint r×r-matrices with entries in $\mathbb{K}$, with Jordan product

$$x \circ y := \frac{1}{2}(xy + yx)$$

given by the anti-commutator. Similarly, the forward light cone (2.2) comes from a Jordan algebra structure on $X = \mathbb{R}^n$ called a "spin factor" [1]. Using the realization of Λ as the positive cone of a Jordan algebra X, the Lie theoretic description

$$\Lambda = G/K$$

of Λ as a symmetric space becomes more concrete:

$$G = \mathrm{Aut}(\Lambda) \subset GL(X)$$

is the group of all linear transformations of X leaving Λ invariant, and

$$K = \mathrm{Aut}(X) = \{g \in \mathrm{Aut}(\Lambda): g(e) = e\}$$

is the isotropy subgroup at the unit element $e \in \Lambda$ of X, which agrees with the Jordan algebra automorphism group of X. For example, we have

$$\mathrm{Aut}(\Lambda_{\mathbb{K}}) = \{x \mapsto g\,x\,g^*: g \in GL(r,\mathbb{K})\}$$

and

$$\mathrm{Aut}(X_{\mathbb{K}}) = \{x \mapsto g\,x\,g^*: g \in U(r,\mathbb{K})\}.$$

Fix a scalar product $(x|y)$ on X which is invariant under Aut(X) and is normalized by putting $(e|e) = r$. Every (formally real) Jordan algebra X of rank r has a basic polynomial function $N: X \to \mathbb{R}$ of degree r which is called the *determinant* of X (or norm function) since it generalizes Cramer's rule

$$x^{-1} = \frac{\mathrm{grad}_x N}{N(x)}$$

to the non-associative setting. More generally, fixing a system $e_1,\ldots,e_r$ of mutually orthogonal minimal projections in X, one has a Peirce decomposition

$$X = \sum_{1 \le i \le j \le r} X_{ij} \tag{2.3}$$

into mutually orthogonal subspaces and one can define the ℓ-th minor ($1 \le \ell \le r$) as

$$N_\ell(x) := N_{X_\ell}(P_\ell x). \tag{2.4}$$

Here N_{X_ℓ} is the determinant of the Jordan algebra

$$X_\ell := \sum_{r-\ell+1 \le i \le j \le r} X_{ij} \tag{2.5}$$

of rank ℓ, and $P_\ell: X \to X_\ell$ is the orthogonal "Peirce projection." (For more details, cf. [9]). We have $N_r(x) = N(x)$ and $N_1(x) = (x|e_r)$ for all $x \in X$.

2.1 PROPOSITION. *For every partition* $\underline{m} = (m_1,\dots,m_r)$ *of integers* $m_1 \geq m_2 \geq \dots \geq m_r \geq 0$, *the function*

$$\varphi_{\underline{m}}(x) := \int_K N_{\underline{m}}(kx)dk, \tag{2.6}$$

where

$$N_{\underline{m}}(x) := N_1^{m_1-m_2}(x) \cdot N_2^{m_2-m_3}(x) \cdots N_r^{m_r}(x)$$

is a spherical function on Λ.

Proof. Let G = NAK be the Iwasawa decomposition of G = Aut(Λ) [6]. Using an explicit root decomposition of the Lie algebra $\mathfrak{g}$ of G in terms of the Peirce subspaces X_{ij} (cf. (2.3)) it is shown in [14] that the polynomials $N_{\underline{m}}(x)$ are invariant under the action of N,

$$N_{\underline{m}}(gx) = N_{\underline{m}}(x) \qquad (g \in N)$$

and transform by a character $\chi_{\underline{m}}$ under the action of A,

$$N_{\underline{m}}(gx) = \chi_{\underline{m}}(g)N_{\underline{m}}(x) \qquad (g \in A).$$

Now the assertion follows from [8; p. 55, Theorem 4]. Q.E.D.

Since the stochastic control problem considered in Section 4 is invariant under the action of K, it is sufficient to consider the intersection $\Lambda \cap X_0$ of Λ with the "diagonal" subspace

$$X_0 := \sum_{i=1}^{r} X_{ii} = \mathbb{R}\langle e_1,\dots,e_r\rangle \tag{2.7}$$

spanned by the projections $e_1,\dots,e_r$. By K-invariance, there is a unique expansion

$$\varphi_{\underline{m}}(x) = \sum_{\underline{\mu} \leq \underline{m}} c_{\underline{\mu}}\, m_{\underline{\mu}}(x) \tag{2.8}$$

valid for $x = x_1e_1 + \cdots + x_re_r \in X_0$, where for any partition $\underline{\mu} \leq \underline{m}$ (lexicographic ordering, cf. [11]), we define the polynomial (monomial symmetric function)

$$m_{\underline{\mu}}(x) = \sum_{\alpha} x_1^{\alpha_1} \ldots x_r^{\alpha_r}$$

summed over all distinct permutations $\alpha = (\alpha_1, \ldots, \alpha_r)$ of $\mu = (\mu_1, \ldots, \mu_r)$ [11; p. 11].

2.2 LEMMA. *Let* X *be a formally-real Jordan algebra of rank* r, *with norm function* N. *Let* $a_1, \ldots, a_r \in X$ *be elements of rank* ≤ 1. *Then*

$$N(x_1a_1 + \cdots + x_ra_r) = x_1 \cdots x_r\, N(a_1 + \cdots + a_r)$$

for all $x_1, \ldots, x_r \in \mathbb{R}$.

Proof. Write

$$N(x_1a_1 + \cdots + x_ra_r) = \sum_{\mu} c_\mu x^\mu$$

where $\mu = (\mu_1, \ldots, \mu_r) \in \mathbb{N}^r$ is a multi-index with $|\mu| = \mu_1 + \cdots + \mu_r = r$, $c_\mu \in \mathbb{R}$ and

$$x^\mu := x_1^{\mu_1} \ldots x_r^{\mu_r}.$$

For any proper subset $I \subset \{1, \ldots, r\}$, the sum $\sum_{i \in I} x_ia_i \in X$ has rank $< r$, which implies

$$N(\sum_{i \in I} x_ia_i) = 0.$$

Taking $x_j := 0$ whenever $j \notin I$, we obtain $c_\mu = 0$ for all μ having support I (i.e., $\mu_j = 0$ for $j \notin I$). Since $|\mu| = r$, this gives $c_{1, \ldots, 1}$ as the only non-zero coefficient. Q.E.D.

The main properties of spherical functions needed for the solution of our control problem (analogous to the properties of cosh(r) in section 1) follow from the following

2.3 PROPOSITION. *The expansion (2.8) of* $\varphi_{\underline{m}}(x)$ *has positive coefficients* $c_{\underline{\mu}} \geq 0$.

Proof. For each fixed $\ell \leq r$ and $k \in K$, it follows from (2.4) that

$$N_\ell(kx) = N_\ell(x_1(ke_1) + \cdots + x_r(ke_r)) = N_{X_\ell}(x_1 P_\ell(ke_1) + \cdots + x_r P_\ell(ke_r)).$$

Since N_ℓ is ℓ-homogeneous in x, the monomials occurring in its expansion cannot have more than ℓ variables. Let $I \subset \{1, \ldots, r\}$ have order $\leq \ell$. Putting $x_j := 0$ if $j \notin I$, we get from Lemma 2.2 applied to the Jordan algebra X_ℓ of rank ℓ:

$$N_\ell(kx) = N_{X_\ell}\Big(\sum_{i\in I} x_i P_\ell(ke_i)\Big) = \prod_{i\in I} x_i \; N_{X_\ell}\Big(\sum_{i\in I} P_\ell(ke_i)\Big)$$

since the elements $P_\ell(ke_i)$, $i \in I$, have rank ≤ 1 in X_ℓ (The group K preserves the rank but the Peirce projections P_ℓ may decrease it). Since $e_i \in \bar{\Lambda}$ for all i, we get $P_\ell(ke_i) \in \bar{\Lambda}$ and thus

$$N_{X_\ell}\Big(\sum_{i\in I} P_\ell(ke_i)\Big) \geq 0.$$

It follows that

$$N_\ell(kx) = \sum_{|I|=\ell} c_I \prod_{i\in I} x_i \tag{2.9}$$

with $c_I \geq 0$. Taking products in (2.9) and integrating over K (which makes the polynomials symmetric in $x_1, \ldots, x_r$), the assertion follows.

Q.E.D.

3. The Laplacian and Its Radial Part

On any Riemannian manifold M, the invariant Brownian motion is the diffusion process whose infinitesimal generator is the Laplace-Beltrami operator Δ_M. If $M = G/K$ is a symmetric space of rank r, Δ_M belongs to the algebra of all G-invariant differential operators on M, denoted by $\mathrm{Diff}_G(M)$, which is a polynomial algebra in r generators $\Delta_1, \ldots, \Delta_r$ [6]. By definition, the spherical functions are K-invariant joint eigenfunctions of these operators. The *radial part* $\mathfrak{L}(D)$ of an operator $D \in \mathrm{Diff}_G(M)$ is a differential operator on the "diagonal subspace" $A{\cdot}e \subset M = G{\cdot}e$ ($G = NAK$ the Iwasawa decomposition) which is determined by the property

$$\mathfrak{L}(D)\tilde{f} := \widetilde{Df}$$

for all K-invariant smooth functions f on M. Here $\tilde{f}$ denotes the restriction of f to $A{\cdot}e$. In the special case of symmetric cones $\Lambda \subset X$ of rank r considered in Section 4, the radial parts of the generators $\Delta_1^{\Lambda}, \ldots, \Delta_r^{\Lambda}$ of $\mathrm{Diff}_G(\Lambda)$ (which are of order $1, 2, \ldots, r$ respectively) have a particularly simple form [12]: Using a system $e_1, \ldots, e_r$ of mutually orthogonal minimal projections in X (cf. Section 2) and the base point $e := e_1 + \cdots + e_r \in \Lambda$ (the unit element of X), we have

$$A{\cdot}e = \Lambda \cap X_0 = \{e^{t_1}e_1 + \cdots + e^{t_r}e_r \colon t_1, \ldots, t_r \in \mathbb{R}\}$$

(cf. (2.7)). Now put

$$x_i := e^{t_i} \tag{3.0}$$

and

$$D_i := x_i \frac{\partial}{\partial x_i}. \tag{3.1}$$

Define

$$\omega := \prod_{i<j} (x_i - x_j) = \sum_{\pi} (-1)^{\pi} x^{\pi\delta}, \tag{3.2}$$

where $\delta := (r-1, r-2, \ldots, 1, 0)$ is the "Vandermonde partition" and, for every permutation π of $\{1, \ldots, r\}$, we put

$$x^{\pi\delta} := x_1^{(\pi\delta)_1} \cdots x_r^{(\pi\delta)_r}. \tag{3.3}$$

Using this notation, one can show (cf. [12])

3.1 PROPOSITION. *The radial parts* $\mathfrak{L}(\Delta_k^\Lambda)$, $1 \le k \le r$, *are given by the polynomial identity*

$$\sum_{k=0}^{r} t^k \mathfrak{L}(\Delta_k^\Lambda) = \frac{1}{\omega} \sum_{\pi} (-1)^\pi x^{\pi\delta} \prod_{j=1}^{r} \left[1 + t\left(\frac{2}{s} D_j + (\pi\delta)_j\right)\right] \tag{3.4}$$

Here t *is an indeterminate and* $s := \dim X_{ij}$ *for* $i < j$. *(Note that* $\alpha := \frac{2}{s}$ *in [12].) Further,* π *runs over all permutations of* $\{1, \ldots, r\}$.

As a special case of (3.4), we have

$$\mathfrak{L}(\Delta_1^\Lambda) = \frac{2}{s} E + \frac{(r-1)r}{2},$$

where $E = \sum_{i=1}^{r} D_i$ is the Euler operator. For $\Lambda = \Lambda_{\mathbb{K}}$ (cf. (2.1)), we have $s = \dim_{\mathbb{R}} \mathbb{K}$, whereas the forward light cone $\Lambda = \Lambda_n$ (cf. (2.2)) has $s = n - 2$. In case $\mathbb{K} = \mathbb{C}$, i.e. $s = 2$, formula (3.4) simplifies to

$$\begin{aligned}\sum_{k=0}^{r} t^k \mathfrak{L}(\Delta_k^\Lambda) &= \frac{1}{\omega} \sum_{\pi} (-1)^\pi x^{\pi\delta} \prod_{j=1}^{r} (1 + t(D_j + (\pi\delta)_j)) \\ &= \frac{1}{\omega} \sum_{\pi} (-1)^\pi \prod_{j=1}^{r} (1 + tD_j) x^{\pi\delta} \\ &= \frac{1}{\omega} \prod_{j=1}^{r} (1 + tD_j)\omega,\end{aligned}$$

since $x^{-\pi\delta} D_j x^{\pi\delta} = D_j + (\pi\delta)_j$. Thus we have

$$\mathfrak{L}(\Delta_k^\Lambda) = \omega^{-1} \sigma_k(D_1, \ldots, D_r)\, \omega$$

in case $s = 2$, where σ_k is the k-th elementary symmetric polynomial. The Laplace-Beltrami operator Δ_Λ corresponds essentially to $k = 2$. Its radial part can also be computed directly via Harish-Chandra's formula (cf. [7; p.335])

$$\mathfrak{L}(\Delta_\Lambda) = \delta^{-1} \sum_{1 \le i,j \le r} g^{ij} \frac{\partial}{\partial t_i} \delta \frac{\partial}{\partial t_j},$$

where $t_1, \ldots, t_r$ are coordinates on the abelian subspace $\mathfrak{a} \approx \mathbb{R}^r$ of the Iwasawa decomposition $\mathfrak{g} = \mathfrak{k} \oplus \mathfrak{a} \oplus \mathfrak{n}$ of the Lie algebra $\mathfrak{g}$ of $G = \mathrm{Aut}(\Lambda)$, with respect to a basis $\{H_1, \ldots, H_r\}$ of $\mathfrak{a}$, (g^{ij}) is the inverse of the matrix of inner products

$$(g_{ij}) := ((H_i|H_j))_{1 \le i,j \le r}$$

for the modified Killing form of $\mathfrak{g}$, and

$$\delta := \prod_{\lambda} \sinh(\lambda)^{m_\lambda}$$

is the function on $\mathfrak{a}$, coming form the positive roots λ of (G,K) and their multiplicity m_λ. For the cone Λ, realized in the Jordan algebra X, the subspace $\mathfrak{a}$ is spanned by the vector fields

$$H_i = e_i \circ x \frac{\partial}{\partial x}, \qquad 1 \le i \le r.$$

Here $x \circ y$ is the Jordan product and $e_1, \ldots, e_r$ are chosen as in Section 2. One can show that

$$(H_i|H_j) = \frac{sr}{2} \delta_{ij} \qquad \text{(Kronecker symbol)}$$

and

$$\delta = (x_1 \cdots x_r)^{-\frac{s}{2}(r-1)} \cdot \omega^s$$

in the coordinates $x_i = e^{t_i}$. A straightforward calculation gives

3.2 PROPOSITION. *For every irreducible symmetric cone of rank* r, *the Laplace-Beltrami operator* Δ_Λ *has the radial part*

$$(3.5)\qquad \frac{sr}{2}\cdot \mathfrak{L}(\Delta_\Lambda) = \sum_{k=1}^{r}\left(x_k^2\frac{\partial^2}{\partial x_k^2} + \left(1-\frac{s}{2}(r-1)\right)x_k\frac{\partial}{\partial x_k}\right) + s\sum_{i\neq j}\frac{x_i^2}{x_i-x_j}\frac{\partial}{\partial x_i}$$

$$= \sum_{k=1}^{r}\left(\frac{\partial^2}{\partial t_k^2} - \frac{s}{2}(r-1)\frac{\partial}{\partial t_k}\right) + s\sum_{i\neq j}\frac{1}{1-e^{t_j-t_i}}\frac{\partial}{\partial t_i}.$$

Comparing with (3.4), we have

3.3 PROPOSITION. *We have*

$$\frac{sr}{2}\,\mathfrak{L}(\Delta_\Lambda) = -\frac{s^2}{2}\,\mathfrak{L}(\Delta^{\wedge}_2) + \alpha E + E^2 + \beta,$$

where α *and* β *are constants.*

Proof. Applying (3.4) to $k=2$, we obtain

$$\mathfrak{L}(\Delta^{\wedge}_2) = \frac{1}{\omega}\sum_{\pi}(-1)^{\pi}x^{\pi\delta}\sum_{i<j}\left(\frac{2}{s}D_i + (\pi\delta)_i\right)\left(\frac{2}{s}D_j + (\pi\delta)_j\right)$$

$$= \frac{4}{s^2}\sum_{i<j}D_iD_j + \frac{2}{s\omega}\sum_{\pi}(-1)^{\pi}x^{\pi\delta}\sum_{i<j}\left(D_i(\pi\delta)_j + (\pi\delta)_iD_j\right) + \sum_{i<j}(\pi\delta)_i(\pi\delta)_j.$$

Now

$$2\sum_{i<j}D_iD_j = E^2 - \sum_{i=1}^{r}D_i^2 = E^2 - \sum_{i=1}^{r}x_i^2\frac{\partial^2}{\partial x_i^2} - E$$

and

$$2\sum_{i<j}(\pi\delta)_i(\pi\delta)_j = \left(\sum_i(\pi\delta)_i\right)^2 - \sum_i(\pi\delta)_i^2$$

$$= \left(\frac{(r-1)r}{2}\right)^2 - \frac{(r-1)r(2r-1)}{6}.$$

For the middle term, we have

$$\sum_{i<j}\left(D_i(\pi\delta)_j + (\pi\delta)_iD_j\right) = E\,\frac{(r-1)r}{2} - \sum_i D_i(\pi\delta)_i.$$

For every $i \in \{1, \ldots, r\}$ one has

$$\sum_{j\neq i} \frac{1}{x_i - x_j} = \frac{1}{\omega}\frac{\partial\omega}{\partial x_i} = \frac{1}{\omega}\sum_{\pi} (-1)^{\pi} \frac{\partial}{\partial x_i} x^{\pi\delta}.$$

Multiplying by $x_i^2 \frac{\partial}{\partial x_i}$ and summing over i, we obtain

$$\begin{aligned}\sum_{j\neq i} \frac{x_i^2}{x_i - x_j}\frac{\partial}{\partial x_i} &= \frac{1}{\omega} \sum_{\pi,i} (-1)^{\pi} (\pi\delta)_i\, x^{\pi\delta} D_i \\ &= \frac{1}{\omega} \sum_{\pi} (-1)^{\pi} x^{\pi\delta} \sum_i D_i(\pi\delta)_i.\end{aligned}$$

Collecting terms, we get

$$\begin{aligned}-\frac{s^2}{2}\mathfrak{L}(\Delta_2^{\Lambda}) = \sum_{i=1}^{r} x_i^2 \frac{\partial^2}{\partial x_i^2} + s\sum_{j\neq i}\frac{x_i^2}{x_i - x_j}\frac{\partial}{\partial x_i} + (1 - \frac{s}{2}(r-1)r)E - E^2 \\ - \frac{s^2}{4}\left[\left(\frac{(r-1)r}{2}\right)^2 - \frac{(r-1)r(2r-1)}{6}\right].\end{aligned}$$

Q.E.D.

4. Stochastic Control Problems in a Symmetric Cone

In this section some solvable stochastic control problems are described using the spherical polynomials (cf. (2.6) and (2.8)) in the symmetric cones. The proofs of these results will only be sketched here. Let $\Lambda = G/K$ be a fixed symmetric cone in the generalized half space realization. The controlled diffusion has infinitesimal generator

$$\frac{\kappa}{2}\Delta_{\Lambda} + \sum_{i=1}^{r} u_i(s,\underline{t}) \frac{\partial}{\partial t_i} \tag{4.1}$$

where $\underline{t} = (t_1, ..., t_r)$, $\kappa = \frac{rd}{2}$, $d = \dim X_{ij}$ and X_{ij} is given in (2.3).

Fix a partition $\underline{m}$ and let $\varphi_{\underline{m}}(x)$ be the spherical polynomial (2.8) associated with $\underline{m}$ that is a homogenous polynomial of degree $|\underline{m}|$

$$\varphi_{\underline{m}}(x) = \sum_{\underline{\mu}\leq\underline{m}} c_{\underline{m}}^{\underline{\mu}}\, q_{\underline{\mu}}(x)$$

where $q_{\underline{\mu}}$ is the monomial symmetric function induced from $\underline{\mu}$. Using the definition of the monomial symmetric function we have

$$\varphi_{\underline{m}}(x) = \sum a_{\underline{m}}(\underline{i})\, x_1^{i_1} \cdots x_r^{i_r}$$

where $\underline{i} = (i_1, \ldots, i_r)$ and $a_{\underline{m}}(\underline{i}) = a_{\underline{m}}(\underline{j})$ from the definition of the monomial symmetric functions if $\underline{j}$ is a permutation of $\underline{i}$. For $j \in \{1, \ldots, r\}$ define $a_{\underline{m}}^j(\underline{i})$ as

$$a_{\underline{m}}^j(\underline{i}) = \frac{i_j}{|m|}\, a_{\underline{m}}(\underline{i})$$

and

$$\varphi_{\underline{m}j}(x) = \sum a_{\underline{m}}^j(\underline{i})\, x_1^{i_1} \cdots x_r^{i_r}$$

where the sum is over distinct r-tuples such that $a_{\underline{m}}^j(\underline{i}) \neq 0$. Clearly

$$\varphi_{\underline{m}}(x) = \sum_{j=1}^{r} \varphi_{\underline{m}j}(x)$$

The cost functional for the stochastic control problem is

$$J_{\underline{m}}(U) = E_{X(0)} \int_0^T \hat{\varphi}_{\underline{m}}(X(t)) + \sum_{j=1}^{r} f_j(X(t))\, U_j^2(t)\, dt \tag{4.2}$$

where

$$\hat{\varphi}_{\underline{m}}(\underline{t}) = \varphi_{\underline{m}}(e^{t_1}, \ldots, e^{t_r});$$

$$\hat{\varphi}_{\underline{m}j}(\underline{t}) = \varphi_{\underline{m}j}(e^{t_1}, \ldots, e^{t_r});$$

$$f_j(\underline{t}) = \frac{[\partial_j \hat{\varphi}_{\underline{m}}(\underline{t})]^2}{\hat{\varphi}_{\underline{m}j}(\underline{t})};$$

$$\partial_j = \frac{\partial}{\partial t_j};$$

and $(X(t),\ t \in [0,T])$ is the controlled diffusion projected to the positive Weyl chamber $\{(t_1, \ldots, t_r) : t_1 > t_2 > \cdots > t_r\}$ by the Weyl group invariance of $\hat{\varphi}_{\underline{m}}$.

The infinitesimal generator of $(X(t),\ t \in [0,T])$ is

$$\frac{\kappa}{2}\,\mathfrak{L}(\Delta_\Lambda) + \sum_{i=1}^{r} u_i(s,\underline{t})\,\frac{\partial}{\partial t_i}$$

$$= \frac{1}{2}\sum_{k=1}^{r}\left[\frac{\partial^2}{\partial t_k^2} - \frac{d}{2}(r-1)\frac{\partial}{\partial t_k} + d\sum_{j\neq k}\frac{1}{1-\exp(t_j - t_k)}\frac{\partial}{\partial t_k}\right]$$

$$+ \sum_{k=1}^{r} u_k(s,\underline{t})\,\frac{\partial}{\partial t_k}$$

The associated family of stochastic differential equations is

$$dX_k(s) = \left[\frac{-d}{4}(r-1) + \frac{d}{2}\sum_{j\neq k}\frac{1}{1-\exp(X_j(s)-X_k(s))} + U_k(s, X(s))\right]ds \tag{4.3}$$

$$+\ dB_k(s)$$

for $k = 1, \ldots, r$, $X(0) = (X_1(0), \ldots, X_r(0))$, $X_1(0) > X_2(0) > \cdots > X_r(0)$, $X(s) \in \mathfrak{a}^+$ and $(B_1(s), \ldots, B_r(s);\ s \geq 0)$ is a standard r-dimensional Brownian motion.

Let $(Z(t) \in \Lambda,\ t \in [0, T])$ be the controlled diffusion with the infinitesimal generator (4.1). An admissible control at time t is a Borel measurable function of $Z(t)$, such that the stochastic differential equation whose generator is (4.1) exists and is unique in a sample path sense. By the Weyl group invariance of the cost functional it suffices to consider controls at time t that are measurable functions of $X(t)$ that is the solution of (4.3) so that this stochastic differential equation has one and only one solution in a sample path sense.

Initially it is necessary to show that the stochastic control problem (4.1 - 4.2) is well posed, specifically that there is at least one control that gives a finite value to $J_{\underline{m}}(U)$ in (4.2).

4.1 LEMMA. *Let* $(\tilde{X}(t),\ t \geq 0)$ *be the diffusion process in the positive Weyl chamber* $\mathfrak{a}^+ = \{\underline{t} : t_1 > t_2 > \cdots > t_r\}$ *with the infinitesimal generator* $\frac{1}{2}\mathfrak{L}(\Delta_\Lambda)$*, that is, the unique strong solution of (4.2) with* $U(t) \equiv 0$. *Then*

$$E_{X(0)} \int_0^T \hat{\varphi}_{\underline{m}}(\tilde{X}(t))\, dt < \infty. \tag{4.4}$$

To verify the inequality (4.4) it suffices to compare the distance of the diffusion to the origin with a Brownian motion in a hyperbolic space of the same dimension as Λ with a suitable sectional curvature and then use a result in [3].

The solution of the stochastic control problem (4.1 - 4.2) is given now.

4.2 THEOREM. *The stochastic control problem described by (4.1 - 4.2) has an optimal control that is*

$$U_i^*(s, \underline{t}) = \frac{-\hat{\varphi}_{\underline{m}j}(\underline{t})}{2\partial_i \hat{\varphi}_{\underline{m}}(\underline{t})}\, g(s) \tag{4.5}$$

where $s \in [0, T]$, $\underline{t} \in \mathfrak{a}^+$, $\partial_j = \frac{\partial}{\partial t_j}$ *and* g *is the unique positive solution of*

$$\frac{dg}{ds} + \frac{1}{2}\chi_{\underline{m}} g - \frac{1}{4} g^2 + 1 = 0 \qquad g(T) = 0 \tag{4.6}$$

where $\chi_{\underline{m}}$ *is the eigenvalue of* $\varphi_{\underline{m}}$ *for* $\kappa\mathfrak{L}(\Delta_\Lambda)$.

A sketch of the proof of this result is given here while the precise details will appear elsewhere [4]. The basic structure of the proof follows by analogy with the verification of Proposition 1.1 for the stochastic control problem in $\mathbb{H}^3(\mathbb{R})$. The Hamilton-Jacobi equation for the stochastic control problem (4.1 - 4.2) is

$$0 = \frac{\partial W}{\partial s} + \min_{v \in \mathbb{R}^r}\left[\frac{\kappa}{2}\mathfrak{L}(\Delta_\Lambda)W + \sum_{i=1}^r v_i \frac{\partial W}{\partial x_i} + \hat{\varphi}_{\underline{m}}(\underline{t}) + \sum_{i=1}^r f_i(\underline{t}) v_i^2\right] \tag{4.7}$$

with the boundary condition $W(s, \underline{t}) = 0$ for $(s, \underline{t}) \in \{T\} \times \Lambda$. A smooth solution of (4.7) satisfying the boundary condition is

$$W(s, \underline{t}) = \hat{\varphi}_{\underline{m}}(\underline{t})\ g(s)$$

where g is the solution of (4.4). It is necessary to show that (4.3) with the control obtained from W has one and only one solution in the sample path sense. Since the coefficients of the family of stochastic differential equations (4.3) with the controls obtained from W are locally smooth in $\mathfrak{a}^+$, a unique strong solution can be constructed at least locally. To globalize this construction it is basically only necessary to verify that this controlled diffusion process does not hit the walls of the positive Weyl chamber. This verification is accomplished by a comparison of the scalar process that describes the distance from a wall of the Weyl chamber and the so-called two dimensional Bessel process.

REFERENCES

1. H. Braun, M. Koecher: Jordan-Algebren, Springer, 1966.

2. T. E. Duncan: A solvable stochastic control problem in hyperbolic three space, Systems and Control Lett. 8(1987), 435-439.

3. T. E. Duncan: Some solvable stochastic control problems in noncompact symmetric spaces of rank one, to appear in Stochastics.

4. T. E. Duncan, H. Upmeier: Explicitly solvable stochastic control problems in symmetric spaces of higher rank, preprint.

5. W. H. Fleming, R. W. Rishel: Deterministic and Stochastic Optimal Control, Springer-Verlag, Berlin-New York, 1975.

6. S. Helgason: Differential Geometry and Symmetric Spaces, Acad. Press, 1962.

7. S. Helgason: Groups and Geometric Analysis, Acad. Press, 1984.

8. S. Lang: $SL_2(\mathbb{R})$, Springer, 1985.

9. O. Loos: Jordan pairs, Lect. Notes in Math 460, Springer, 1975.

10. O. Loos: Bounded Symmetric Domains and Jordan Pairs, Univ. of California, Irvine, 1977.

11. I. G. Macdonald: Symmetric Functions and Hall Polynomials, Oxford Univ. Press, 1979.

12. I. G. Macdonald: Commuting differential operators and zonal spherical functions, Springer Lect. Notes in Math 1271 (1987), 189-199.

13. H. Upmeier: Symmetric Banach Manifolds and Jordan C*-Algebras, North Holland, 1985.

14. H. Upmeier: Jordan algebras and harmonic analysis on symmetric spaces, Amer. J. Math. 108(1986), 1-25.

Department of Mathematics
University of Kansas
Lawrence, KS 66045

Applications of Semigroup Domination

Steven Rosenberg

§1. Introduction. The classical Bochner technique in Riemannian geometry shows the nonexistence of solutions for elliptic equations of geometric interest given certain positive curvature conditions. The strong form of semigroup domination leads to extensions of these vanishing theorems to cases where the curvature is mostly positive. In this paper, we use semigroup domination together with geometric estimates for Schrödinger operators from [ER II] to prove several such results. In particular, we extend Lichnerowicz's vanishing theorem from the case of Riemannian spin manifolds with positive scalar curvature to metrics with the scalar curvature positive off a set of small volume. We also improve results of Berger and Ebin on minimal hypersurfaces and deformations of Einstein metrics and a result of Bochner on Killing fields. Finally, using results of Gilkey, we discuss the general case of differential operators with principal symbol given by the metric tensor.

§2. Semigroup domination and the $\hat{A}$-genus. Let M be a Riemannian manfold, let E be a hermitian vector bundle over M, and let $D : \Gamma(E) \to \Gamma(E)$ be a self adjoint second order differential operator on sections of E. D has a *Weitzenböck formula* if there exists a unitary connection ∇ on E such that $D = \nabla^*\nabla + R$, where R is a zeroth order operator; i.e. $R \in Hom(E, E)$. The most familiar example for D is the Laplacian on p-forms.

Since R is a symmetric endomorphism in each fiber of E, we may define a function $R' : M \to \mathbf{R}$ by setting $R'(x)$ to be the lowest eigenvalue of R acting on the fiber of E at x. Although all the data above is assumed smooth, R' need only be continuous. The strong form of semigroup domination is

$$\left|e^{-tD}(\psi)(x)\right| \leq (dim\ E) \cdot e^{-t(\Delta+R')}|\psi|(x).$$

Here ψ is an L^2 section of E, Δ is the Laplacian on functions on M and the norms are measured with respect to the metrics on M and E. Various forms of semigroup domination are used in [DL], [HSU], [M]; a probabilistic proof of the strong form is in [R].

Now let M be an oriented compact manifold. Given the formal factorization $p(M) = \prod(1 + x_i^2)$ of the total Pontrjagin class of M, the $\hat{A}$-polynomial of M is defined to be the characteristic class $\prod \frac{\frac{x_i}{2}}{sinh(\frac{x_i}{2})}$ expressed as a rational combination of the Pontrjagin classes. The $\hat{A}$-genus $\hat{A}(M)$ of M is defined by evaluating the $\hat{A}$-polynomial on the fundamental class of M; alternatively, the $\hat{A}$-genus is the unique

Research partially supported by the NSF.

rational cobordism invariant which is multiplicative on product manifolds and which is 0 for $\mathbf{CP}^n$, n odd, and the rational number given by evaluating a generalized Bernoulli polynomial $\frac{2^{2n}}{(2n)!}D_{2n}^{(n)}(x_1, ..., x_{2n})$ on the fundamental class of $\mathbf{CP}^n$, n even [H, §1.9]. If M is a spin manifold, $\hat{A}(M)$ is an integer. This is explained by (and was one of the main motivations behind) the Atiyah-Singer index theorem. Choose a Riemannian metric g on M. Let $\not{\partial} : S \to S$ be the Dirac operator acting on smooth sections of the bundle of spinors; $\not{\partial}$ is defined to be the composition of the Levi-Civita connection ∇ (promoted to S) with Clifford multiplication. Under the splitting $S = S^+ \oplus S^-$ of S into positive and negative spinors, $\not{\partial}$ splits into the sum of two operators $\not{\partial}^\pm : S^\pm \to S^\mp$. Then the index theorem applied to the Dirac operator gives

$$\textit{index } \not{\partial}^+ = \textit{dim ker } \not{\partial}^+ - \textit{dim ker } \not{\partial}^- = \hat{A}(M)$$

since $(\not{\partial}^+)^* = \not{\partial}^-$.

By a calculation due to Lichnerowicz, the spinor Laplacian $L = \not{\partial}^*\not{\partial}$ satisfies $L = \nabla^*\nabla + \frac{s}{4}$, where s is the scalar curvature of the metric. Since $\nabla^*\nabla$ is always a nonnegative operator, $s > 0$ implies that L and hence $\not{\partial}^\pm$ have no kernel. This gives Lichnerowicz's theorem:

Theorem 2.1: *Let M be a compact spin manifold which admits a metric g of positive scalar curvature. Then $\hat{A}(M) = 0$. In fact, for the metric g we have $\ker \not{\partial}^\pm = 0$.*

We can strengthen this result to the case when the scalar curvature is positive off a set of small volume. Let $\mathcal{M} = \mathcal{M}(n, K, D, V)$ be the class of Riemannian n-manifolds with Ricci curvature bounded below by K, diameter at most D and volume at least V.

Theorem 2.2: *Let $(M, g) \in \mathcal{M}(n, K, D, V)$ and pick $S_0 > 0$. There exists $\epsilon = \epsilon(S_0, \mathcal{M}) > 0$ such that if the scalar curvature is bounded below by S_0 except on a set of volume less than ϵ, then $\hat{A}(M) = 0$. In fact, $\ker \not{\partial}^\pm = 0$ on every finite cover of M with the pullback metric, and the L^2 kernel of $\not{\partial}^\pm$ vanishes on every infinite cover of M with the pullback metric.*

Proof: Applying semigroup domination to a spinor ψ yields

$$\left|e^{-tL}(\psi)(x)\right| \leq (\textit{dim } S) \cdot e^{-t(\Delta + \frac{s}{4})}|\psi|(x). \tag{2.3}$$

If ψ is a harmonic spinor, the left hand side is constant in t, while the right hand side decays as $t \to \infty$ if $\Delta + \frac{s}{4} > 0$. Thus $\Delta + \frac{s}{4} > 0$ implies that $\ker \not{\partial}^\pm = 0$. The following result [ER II, Prop. 1.2] shows that the hypotheses in the theorem force $\Delta + \frac{s}{4} > 0$.

Proposition 2.4: *Let $W : M \to \mathbf{R}$ be continuous and choose $B > 0$ and $b < 0$. Then there exists $a \in (0, vol(M))$ such that*

(i) $W \geq B$ except on a set W_-

(ii) $W \geq b$ on W_-

(iii) $vol(W_-) \leq a$

implies $\Delta + W > 0$. *Moreover* $a = a(B, b, n, K, D, V)$, *with* n, k, D, V *as above.*

The proof of 2.4 in [ER] shows that a can be explicitly estimated in terms of its parameters. In our case, the lower bound on the Ricci curvature gives a lower bound on the scalar curvature.

For the statement about finite covers, we note that the lowest eigenvalue of $\Delta + \frac{s}{4} > 0$ equals minus the quantity $\nu(M) = \overline{\lim_{t\to\infty}}\, ln\, \mathbf{E}[exp(-\int_0^t \frac{s}{4}(x_r)dr)]$, with $\mathbf{E}$ expectation with respect to the Wiener measure for Brownian motion starting at $x_0 \in M$ [A]. Since Brownian motion on a finite cover M' of M with respect to the pullback metric covers Brownian motion on M, $\nu(M) = \nu(M')$. Thus $\Delta + \frac{s}{4} > 0$ on M if and only if $\Delta + \frac{s}{4} > 0$ on M'.

If M' is an infinite cover, we still have $\nu(M) = \nu(M')$, and $-\nu(M')$ is a lower bound for the infimum of the spectrum of $\Delta + \frac{s}{4}$ on $L^2(M')$ (cf. [ER I, Prop. 4B]). Applying (2.3) to an L^2 harmonic spinor ψ, we see as before that the L^2 kernel of $\not{\partial}^{\pm}$ must vanish.

As an application, let $f : M \to \mathbf{R}$ be a smooth function on M which changes sign. According to a theorem of Kazdan and Warner [KW], within each conformal class of metrics on M there is a metric $g = g_f$ with scalar curvature s_g equal to f (for $dim\ M \geq 3$). Very little is known about the qualitative geometry of this metric. However, if M is spin and the $\hat{A}$-genus of M is nonzero, then by 2.2 $vol_g(\{s_g < \alpha\}) > \epsilon(\alpha, \mathcal{M})$, where $(M, g) \in \mathcal{M}$ and $\alpha \in (0, f_{max})$. In other words, even if the set $\{f < \alpha\}$ is small as measured by an original metric, this set cannot be too small as measured by g.

§3. Minimal hypersurfaces and Einstein metrics. In this section we generalize two results of Berger and Ebin [BE] by weakening a positivity condition on the curvature term in a Weitzenböck formula to almost positivity.

Let (M^n, g) be a minimal hypersurface in S^{n+1} with the standard metric. Define an endomorphism $K : S^2T^*M \to S^2T^*M$ of the second symmetric power of the cotangent bundle by $(Kh)_{ab} = R_{a\ell}h^{\ell}_{b} + R_{b\ell}h^{\ell}_{a} - 2R_{a\ell bm}h^{\ell m}$. Here $h = \sum h_{ab}dx^a \otimes dx^b$ in local coordinates, and $R_{a\ell}$, $R_{a\ell bm}$ are the components of the Ricci and Riemann curvature tensor respectively. K is the symmetrization of the curvature term in the Weitzenböck formula $\Delta^2 = \nabla^*\nabla + R^2$ for the Laplacian on two forms. Using work of Simons [S], Berger and Ebin show that the second fundamental form h of (M, g) satisfies $(\nabla^*\nabla + \frac{K}{2})h = 0$, where ∇ is the Levi-Civita connection promoted to S^2T^*M. Thus M is an equator of S^{n+1} if $K > 0$. Moreover, $K > 0$ at $x \in M$ if the sectional curvatures are positive at x.

Let $\mathcal{N}(n, C, D, V)$ be the class of Riemannian n-manifolds with sectional curvature of absolute value less than C, diameter at most D, and volume at least V.

Theorem 3.1: *Let* $(M, g) \in \mathcal{N}(n, C, D, V)$ *and pick* $C_0 > 0$. *There exists* $\epsilon = \epsilon(C_0, \mathcal{N}) > 0$ *such that if* (M, g) *is a minimal hypersurface in* S^{n+1} *with sectional curvature greater than* C_0 *except on a set of volume less than* ϵ, *then* (M, g) *is an equator of* S^{n+1}.

Of course, we say that the sectional curvature is greater than C_0 at a point x if the sectional curvatures of all two planes in T_xM are greater than C_0.

Proof: Fix a point $x \in M$ and a lower bound Y for the sectional curvatures at x. Pick an orthonormal basis $\{e_a\}$ of T_xM in which h diagonalizes, and let $\sigma(e_a, e_b)$ be the sectional curvature of the plane spanned by e_a, e_b. [BE] show that $\langle Kh, h\rangle = \sum_{a,b} \sigma(e_a, e_b)(h_{aa} - h_{bb})^2$. Since $\sum h_{aa}^2 = \langle h, h\rangle$ and $tr\ h = \sum h_{aa} = 0$, we easily get $\langle Kh, h\rangle > 2nY\langle h, h\rangle$. Thus the lower bound $-C$ on the sectional curvatures of M gives a lower bound for K on all of M and the lower bound C_0 gives an improved lower bound except on a set of volume less than ϵ. By 2.4, $\Delta + \frac{K'}{2} > 0$ for ϵ small enough, so by semigroup domination h must vanish and (M, g) must be an equator.

Next we produce a vanishing results for deformations of Einstein metrics. Recall that a metric g is Einstein if the Ricci tensor R_{ij} satisfies $R_{ij} = k \cdot g_{ij}$ for some constant k. Assume that there is a family $g(t)$ of Einstein metrics and define the symmetric two tensor $h = \left.\frac{d}{dt}\right|_{t=0} g(t)$. If $k(t) > 0$ for all t, we can renormalize the family so that $k(t) \equiv 1$. Berger and Ebin show that in this case h is in the kernel of the operator $\nabla^*\nabla + L$, where $(Lh)_{ab} = -\sum_{\ell,m} R_{a\ell bm}h^{\ell m}$.

We assume that the metric has been scaled so that all sectional curvatures are bounded above by one. We now show that a lower bound on the sectional curvature gives a lower bound on L. The argument extends [BE, Lemma 7.4].

Lemma 3.2: *Pick $F \leq 1$. There is a dimension constant $A = A(n)$ such that if the sectional curvatures of (M,g) are greater than F at a point $x \in M$, then the endomorphism L at x satisfies*

$$\frac{1}{2}\langle Lh, h\rangle \geq A^2\left(\frac{4}{n} + F - 1\right)\langle h, h\rangle$$

if n is even, and

$$\frac{1}{2}\langle Lh, h\rangle \geq A^2\left(\frac{4n}{n^2-1} + F - 1\right)\langle h, h\rangle$$

if n is odd.

More precisely, the constant A depends on n and the sign of $\frac{4}{n} + F - 1$ $\left(\frac{4n}{n^2-1} + F - 1\right)$ if n is even (odd).

Proof: As before, we diagonalize h with respect to an orthonormal basis $\{e_a\}$ and set $h_{aa} = x_a$, $\sigma(e_a, e_b) = \alpha_{ab}$. Since $tr\ h = 0$ [BE, Lemma 7.1],

$$\begin{aligned}\frac{1}{2}\langle Lh, h\rangle &= -\sum_{a\neq b} \alpha_{ab}x_a x_b = \left(\sum_a x_a\right)^2 - \sum_{a\neq b} \alpha_{ab}x_a x_b \\ &= \sum_a x_a^2 + \sum_{a\neq b}(1 - \alpha_{ab})x_a x_b.\end{aligned}$$

Relabel the x_a by letting y_i (z_j) denote the x's which are nonnegative (negative) and set $A = \sum_i y_i = -\sum_j z_j$. Then

$$\begin{aligned}\frac{1}{2}\langle Lh,h\rangle &\geq \sum_i y_i^2 + \sum z_j^2 + \sum_{i,j}(1-F)y_i z_j \\ &= \sum_i y_i^2 + \sum z_j^2 + (1-F)(\sum_i y_i)(\sum_j z_j) \\ &\geq \frac{A^2}{p} + \frac{A^2}{n-p} - (1-F)A^2,\end{aligned}$$

where p is the number of y_i's. If n is even, the last line is at least $A^2(\frac{4}{n} + F - 1)$. We may assume that $\langle h,h\rangle = 1$, in which case the functions $A = A(p)$ take on nonzero maxima and minima on the set $\{\sum x_a^2 = 1\} \cap \{\sum x_a = 0\}$. Setting $A(n)$ to be the maximum (minimum) of the maxima (minima) of $A(p)$, $p = 1,\ldots,n-1$, if $\frac{4}{n} + F - 1$ is positive (negative) finishes the proof for n even. The case n odd is similar.

Remark: The inequality $1-\alpha_{ab} \geq \epsilon \geq 0$ in the proof in [BE] should be $0 \leq 1-\alpha_{ab} \leq \epsilon$.

Theorem 3.3: *Let $(M,g) \in \mathcal{N}(n,C,D,V)$ be an Einstein metric with sectional curvatures at most one. Pick $F \in (1-\frac{4}{n},1]$ if n is even and $F \in (1-\frac{4n}{n^2-1},1]$ if n is odd. There exists $\epsilon = \epsilon(F,\mathcal{N}) > 0$ such that if the sectional curvatures of (M,g) are greater than F except on a set of volume less than ϵ, then (M,g) is isolated in the space of Einstein metrics on M. Any finite cover of M with the pullback metric is also isolated in the space of Einstein metrics.*

Remark: Berger and Ebin prove this result when the sectional curvatures of M are all greater than F.

The proof follows from semigroup domination and Proposition 2.4, since the sectional curvatures greater than F at some $x \in M$ and (M,g) Einstein imply that $k = k(t) = 1$, and since the lower bound on sectional curvature gives a lower bound on the endomorphism L.

§4. **Killing Fields.** Let (M,g) admit a one-parameter family of isometries. The vector field X associated to this family is a Killing vector field, which means $X_{j;k} + X_{k;j} = 0$ and $X^i{}_{;i} = 0$. Now $\langle(\nabla^*\nabla - R)X, Y\rangle + \langle\delta X, \delta Y\rangle = 0$, where X and Y are arbitrary vector fields and R is the Ricci curvature as an endomorphism of TM; cf. [YB, p. 52] for $X = Y$, which gives the equation above by polarization. If X is a Killing field, $\delta X = 0$ and we must have $(\nabla^*\nabla - R)X = 0$. This shows that a compact manifold with negative Ricci curvature admits no nontrivial family of isometries. More generally, we have:

Theorem 4.1: *Let $(M,g) \in \mathcal{M}(n,K,D,V)$ and pick $F < 0$. There exists $\epsilon = \epsilon(F,\mathcal{M}) > 0$ such that if the Ricci curvature of M is less than F except on a set* *of volume less than ϵ, then the identity is isolated in the space of isometries of (M,g).*

As an example, consider a finite volume cusp of a hyperbolic surface truncated at two heights. Note that the metric has an S^1 symmetry. Identify the boundary circles to form a torus M and smooth the hyperbolic metric to a metric g at the glued boundaries in an S^1 invariant manner. Since the smooth metric still has a one paramter family of isometries, we see that the set of positive curvature has volume too big (relative to its class $\mathcal{M}$) for the theorem to apply. Of course the volume of the set of positive curvature can be made arbitrarily small while keeping the volume and diameter bounded by shrinking the region of smoothing. This shows that the curvature bounds are essential in the theorem above and more fundamentally in Proposition 2.4.

§5. General Differential Operators. Let $D : \Gamma(E) \to \Gamma(E)$ be a second order differential operator acting on sections of the hermitian bundle E. Gilkey [G] proved the following:

Theorem 5.1: *If the leading order symbol of D is given by the metric tensor, then there is a unique connection ∇ on E such that $R = \nabla^*\nabla - D$ is a zeroth order operator.*

In the previous sections the connection ∇ was always the Levi-Civita connection, but this need not be the case in general. The arguments in the previous sections give a general result:

Theorem 5.2: *Let $(M,g) \in \mathcal{M}(n,K,D,V)$ and let D, R be as above. For $A > 0$, $a < 0$ there exists $\epsilon = \epsilon(A,a,\mathcal{M}) > 0$ such that if $R' > a$ on M and the volume of the set $\{R' < A\}$ is less than ϵ, then D has vanishing kernel. The pullback of D has vanishing kernel on all finite covers of M, and the L^2 kernel of the pullback vanishes on all infinite covers of M.*

REFERENCES

[A] L. Arnold: "A formula connecting sample and moment stability of linear stochastic systems," *SIAM J. Appl. Math.* **44** (1984), 793-802.

[BE] M. Berger & D. Ebin: "Some decompositions of the space of symmetric tensors on a Riemannian manifold," *J. Differential Geometry* **3** (1969), 379-392.

[DL] H. Donnelly & P. Li: "Lower bounds for the eigenvalues of Riemannian manifolds," *Michigan Math. J.* **29** (1982), 149-161.

[ER I] K. D. Elworthy & S. Rosenberg: "Generalized Bochner theorems and the spectrum of complete manifolds," *Acta Appl. Math.* **12** (1988), 1-33.

[ER II] __________: "Manifolds with wells of negative curvature," to appear in *Inventiones Math.*

[G] P. B. Gilkey: "The spectral geometry of real and complex manifolds," Proc. Sympos. Pure Math. **27**, Amer. Math. Soc., Providence, R.I., 1973, 269-280.

[HSU] H. Hess, R. Schrader & D. Uhlenbrock "Domination of semigroup and a generalization of Kato's inequality," *Duke Math. J.*, **44** (1977), 893-904.

[H] F. Hirzebruch: *Topological Methods in Algebraic Geometry. 3rd edition.* Springer-Verlag, Berlin, 1978.

[KW] J. Kazdan & F. Warner: "Existence and conformal deformations of metrics with prescribed Gaussian and scalar curvatures," *Ann. of Math.* **101** (1975), 317-331.

[M] P. Malliavin: "Formules de la moyenne, calcul de perturbations et theorèmés d'annulation pour les formes harmoniques," *J. Functional Analysis* **17** (1974), 274-292.

[R] S. Rosenberg: "Semigroup domination and vanishing theorems," in *Geometry of Random Motion* (R. Durrett & M. Pinsky, eds.), Contemporary Mathematics **73**, American Mathematical Society, Providence, R.I., 1988, 287-302.

[S] J. Simons: "Minimal varieties in Riemannian manifolds, *Ann. of Math.* **88** (1968), 62-105.

[YB] K. Yano & S. Bochner: *Curvature and Betti Numbers.* Princeton University Press, Princeton, N.J., 1953.

Department of Mathematics, Boston University, Boston, MA 02215

A Degenerating Sequence of Riémannian Metrics on a Manifold and their Brownian Motions

Nobuyuki Ikeda[1] and Yukio Ogura[2]

1. Introduction

In recent development of geometry, Lipschitz and Hausdorff distances are often introduced in the relevant spaces and the structure of closure with respect to these topology are extensively studied. M. Gromov [8] is one of the most stimulating literature in this fields, and various works of theoretical and applicative interest followed. In the study of eigenvalue problem for the Laplace-Beltrami operators, K.Fukaya[4] gave the notion of measured Hausdorff topology in the space of Riemannian manifolds whose dimensions and sectional curvatures are bounded by a universal constant, and proved that the corresponding eigenvalues are continuous in its closure space with respect to this topology. It is remarkable that this theory deduces some convergence theorems of eigenvalues for a collapsing sequence of Riemannian manifolds.

On the other hand, the study of convergence of stochastic processes has been one of the main subjects in probability therory. Its most interesting aspect has been that nice processes such as Brownian motions or stable processes are obtained as the limits of the convergent stochastic processes. However, in the recent study such as in [12] or [13], we encounter the cases where the limit processes are not so standard. For example, in the drive-driven process theory in [13], the dimension of the state space of the limit process is smaller than those of the converging ones by the reason that the sample paths for the convergence processes move more and more rapidly in some directions. The basic routine to start with the processes on the whole space and to obtain the limit process for the driven process on its lower dimensional subspace is very similar to that for collapsing sequence of manifolds. Indeed, there is a strong relation between the convergence of Riemannian metrics and that of the corresponding Brownian motions.

The object of this paper is to clarify the relation of the two subjects mentioned above. Actually, the setting in the collapsing theory is more general than that in probability theory, in the sense that the manifolds are not necessarily fixed. However, once the basic manifold is fixed, the sequences of Riemannian metrics dealt in the

[1]Work supported in part by the Grant-in-Aid for Scientific Research 01540180
[2]Work supported in part by the Grant-in-Aid for Scientific Research 01540189

probability theory are more general than those dealt in the collapsing theory. In this paper, we first give a new topology in the space of Riemannian metrics on a manifold, and then show the continuity of the corresponding heat semigroups with respect to this topology. For this, we will make a systematic use of the theory of Dirichlet forms, especially of the one developed in [1]. An advantage of our method is that the convergence of eigenvalues and eigenfunctions is derived almost directly from our theorems.

The arrangement of this paper is the following. In Section 1, we will state the definitions and the results. Our main theorem is Theorem 2. Theorem 3 gives the limit theorem for the corresponding processes. Theorem 4 is an application of Theorem 2 to the eigenvalue problem. In Section 3, we give three examples where the underlying manifold is a product manifold. In Example 2 where it is a 2-dimensional torus, we will illustrate the relation between our results and those by geometricians. Sections 4 is devoted to the proofs of the convergence theorems. In the proof of a key Lemma 4.1 below, we will show that, due to the boundedness of the Dirichlet norms of the semigroups for the convergent metrics, their gradients in the directions for the diverging term go to zero as the metrics degenerate. This implies that the domain of the Dirichlet form for the limit semigroup is restricted to those functions which are constant on the integral manifolds for those directions. In Section 5, we will prove Theorem 4.

Acknowledgment : The authors are grateful to Professor J.Hano for his valuable discussions.

2. Definitions and Results

Let $(\widetilde{M},\tilde{g})$ be a d-dimensional compact orientable Riemannian manifold with the volume element $\tilde{\mu}$. Let also

$$\mathcal{R} = \{g : g \ \textit{ is a symmetric nonnegative continuous } (0,2)\textit{-tensor on } \widetilde{M}\},$$

$$\mathcal{R}_d = \{g \in \mathcal{R} : g \ \textit{ is positive definite on } T(\widetilde{M})_q \ \textit{ for all } \ q \in \widetilde{M}\}.$$

We introduce a distance function d on $\mathcal{R}$ by

$$d(g_1,g_2)_q = \sup\{|g_1(X,Y)_q - g_2(X,Y)_q| : X,Y \in T_q(\widetilde{M}),\ |\tilde{g}(X,X)_q| \le 1,\ |\tilde{g}(Y,Y)_q| \le 1\}, \qquad \textit{for } g_1,\ g_2 \in \mathcal{R}$$

and $d(g_1,g_2) = \sup\{d(g_1,g_2)_q : q \in \widetilde{M}\}$. Let

$$\mathcal{M}(\widetilde{M}) = \{\mu : \mu \ \textit{ is a nonnegative measure on } (\widetilde{M},\mathcal{B}(\widetilde{M})) \textit{ and is absolutely continuous w.r.t. } \tilde{\mu} \textit{ with continuous and positive Radon-Nikodym derivative } d\mu/d\tilde{\mu}\}$$

and

$$d_{\mathcal{M}(\widetilde{M})}(\mu_1,\mu_2) = \sup\{|d\mu_1/d\tilde{\mu}(q) - d\mu_2/d\tilde{\mu}(q)| : q \in \widetilde{M}\}.$$

For each $g \in \mathcal{R}_d$ and $\mu \in \mathcal{M}(\widetilde{M})$, we define a symmetric form

$$\mathcal{E}^{g,\mu}(u,v) = \int_{\widetilde{M}} g^*(du,dv)_q \mu(dq), \qquad u,v \in C^\infty(\widetilde{M}).$$

This is closable, and the smallest closed extension $(\mathcal{E}^{g,\mu}, \mathcal{D}[\mathcal{E}^{g,\mu}])$ is a regular Dirichlet space on $L_2(\mu)$ (see [5]). We denote the associated semigroup and resolvent by $\{T_t^{g,\mu} : t > 0\}$ and $\{G_\alpha^{g,\mu} : \alpha > 0\}$ respectively. Notice that $\{T_t^{g,\mu} : t > 0\}$ has a Feller version, and there corresponds a unique Hunt process $X^{g,\mu} = (q_{g,\mu}(t), P_q)$ on $\widetilde{M}$. We also set

$$\mathcal{E}_\alpha^{g,\mu}(u,v) = \mathcal{E}^{g,\mu}(u,v) + \alpha(u,v)_\mu, u,v \in \mathcal{D}[\mathcal{E}^{g,\mu}], \alpha > 0,$$

where $(u,v)_\mu = \int_{\widetilde{M}} u(q)v(q)\mu(dq)$ and $\|u\|_\mu = (u,u)_\mu^{1/2}$. Now our first theorem is

Theorem 1 *Let* $\{(g_n,\mu_n)\}_{n=1}^\infty \subset \mathcal{R}_d \times \mathcal{M}(\widetilde{M})$, $(g,\mu) \in \mathcal{R}_d \times \mathcal{M}(\widetilde{M})$ *and*

$$\lim_{n\to\infty} \{d(g_n,g) + d_{\mathcal{M}(\widetilde{M})}(\mu_n,\mu)\} = 0.$$

Then

$$\lim_{n\to\infty} \|T_t^{g_n,\mu_n}u - T_t^{g,\mu}u\|_{\tilde{\mu}} = 0, \quad u \in L_2(\tilde{\mu}),\ t > 0. \tag{2.1}$$

When $\mu = \mu_g := \Omega_g/\mathrm{vol}(\widetilde{M},g)$, where Ω_g is the volume element of $(\widetilde{M},g)$, then we simply denote $\mathcal{E}^{g,\mu_g}(u,v)$ by $\mathcal{E}^g(u,v)$. Also we denote $T_t^g = T_t^{g,\mu_g}, G_\alpha^g = G_\alpha^{g,\mu_g}$ and $X^g = X^{g,\mu_g}$. Then we have the next corollary immediately.

Corollary 1 *Let* $\{g_n\}_{n=1}^\infty \subset \mathcal{R}_d, g \in \mathcal{R}_d$, *and* $\lim_{n\to\infty} d(g_n,g) = 0$. *Then*

$$\lim_{n\to\infty} \|T_t^{g_n}u - T_t^g u\|_{\tilde{\mu}} = 0, \quad u \in L_2(\tilde{\mu}),\ t > 0.$$

Remark 2.1 Let ρ_g be the distance function induced by a $g \in \mathcal{M}(\widetilde{M})$. Then, under the assumptions of Theorem 1, the metric spaces $(\widetilde{M},\rho_{g_n})$ converges to $(\widetilde{M},\rho_g)$ both in Lipschitz and Hausdorff distance. However, in the following case, our topology is different from those distances. We will discuss on this subject in more detail in Example 2 of the next section.

In the case where the limit metric g degenerates, the situation is more complicated. In order to deal with this case, we need some restriction. For a $g \in \mathcal{R}$, we set $\mathcal{D}(g)_q = \{X \in T_q(\widetilde{M}) : g(X,Y)_q = 0 \text{ for all } Y \in T_q(\widetilde{M})\}$, and assume that

(A.1) *the distribution* $\mathcal{D}(g) : q \to \mathcal{D}(g)_q$ *is a* (d-d')*-dimensional completely integrable differentiable distribution for some* $1 \le d' \le d$.

Recall that a necessary and sufficient condition for a differentiable distribution to be completely integrable is that the Lie brackets $[X,Y]$ of all vector fields $X, Y \in \mathcal{D}(g)$

belong to $\mathcal{D}(g)$. Further, for each $p \in \widetilde{M}$, there exists a maximal connected integral manifold F_p of $D(g)$ which includes the point p. Thus we can define an equivalent relation $\sim$ by $q \sim q'$ if q and q' belong to the same F_p. Denote $M_g = \widetilde{M}/\sim$ and the natural map by ψ, i.e., $\psi : \widetilde{M} \to M_g$. Our next assumption is

(A.2) *M_g is a d'-dimensional differentiable compact manifold with the quotient topology, ψ is differentiable and $\psi_*(T_q(\widetilde{M})) = T_{\psi(q)}(M_g)$, $q \in \widetilde{M}$.*

We then set

$$\mathcal{R}_r = \{g \in \mathcal{R} : \ g \quad satisfies\ the\ conditions\ (A.1)\ and\ (A.2)\}.$$

Let $g \in \mathcal{R}_r, q \in \widetilde{M}$ and $\mathcal{D}(g)_q^{\perp} = \{\omega \in T_q^*(\widetilde{M}) : \langle \omega, X\rangle = 0 \text{ for all } X \in \mathcal{D}(g)_q\}$. We then have

$$\psi^*(T^*_{\psi(q)}(M_g)) \ \subseteq \ \mathcal{D}(g)_q^{\perp} \ = \ \phi_g(T_q(\widetilde{M})), \tag{2.2}$$

where ϕ_g is a linear operator from $T_q(\widetilde{M})$ to $T_q^*(\widetilde{M})$ given by

$$\langle \phi_g(X), Y\rangle_q \ = \ g(X,Y)_q, \qquad Y \in T_q(\widetilde{M}). \tag{2.3}$$

Thereby, we can define a bilinear form g^* on $\mathcal{D}(g)_q^{\perp}$ by

$$g^*(\phi_g(X), \phi_g(Y))_q \ = \ g(X,Y)_q, \qquad X, Y \ \in \ T_q(\widetilde{M}). \tag{2.4}$$

Now let $\mu \in \mathcal{M}(\widetilde{M})$. Then the symmetric form on $L_2(\psi_*\mu)$ which we want is

$$\mathcal{E}^{g,\mu}(u,v) \ = \ \int_{\widetilde{M}} g^*(\psi^* du, \psi^* dv)_q \mu(dq), \quad u, v \ \in \ C^\infty(M_g). \tag{2.5}$$

In order to see that the symmetric form (2.5) is closable and its minimal closed extension gives a Dirichlet form, we shall define a contraction operator $\pi_{g,\mu}$ from $L_2(\mu)$ to $L_2(\psi_*\mu)$. For each $u \in L_2(\mu)$, the σ-additive set functions $u^{\pm}\mu(dq) := u^{\pm}(q)\mu(dq)$ are absolutely continuous with respect to the measure $\mu(dq)$ where u^+ and u^- are the positive and the negative part of u respectively. Hence $\psi_* \circ u^{\pm}\mu \prec \psi_*\mu$, and we obtain the Radon-Nikodym densities $\pi_{g,\mu}u(p) := (d\psi_* \circ u^+\mu / d\psi_*\mu)(p) - (d\psi_* \circ u^-\mu/d\psi_*\mu)(p)$. Notice that

$$\int_{\widetilde{M}} f \circ \psi(q) \pi_{g,\mu} u \circ \psi(q) \mu(dq) \ = \ \int_{\widetilde{M}} f \circ \psi(q) u(q) \mu(dq), \quad f \ \in \ B(M_g),$$

so that $\|\pi_{g,\mu}u\|_{L_2(\psi_*\mu)} \le \|u\|_{L_2(\mu)}$, u $\in L_2(\mu)$. Further, as will be shown at the end of this section, we have

$$\pi_{g,\mu}u(p) \ = \ \int_{F_p} u(r)\mu_p(dr), \qquad p \in \ M_g, \tag{2.6}$$

for some probability measure μ_p on F_p.

Now we transfer the bilinear form g^* on $\mathcal{D}(g)_q^{\perp}$ to that on $T^*_{\psi(q)}(M_g)$ by the operator ψ^*, and then average it by μ_p :

$$h^*(\theta, \eta)_p \ := \ \int_{F_p} g^*(\psi^*\theta, \psi^*\eta)_q \mu_p(dq), \quad \theta, \eta \ \in \ T_p^*(M_g).$$

It then follows that

$$\mathcal{E}^{g,\mu}(u,v) = \int_{M_g} h^*(du,dv)_p \psi_*\mu(dp), \qquad u,v \in C^\infty(M_g).$$

Since h^* is a positive symmetric continuous (2,0)-tensor field on M_g, it then follows that the symmetric form $\mathcal{E}^{g,\mu}$ on $L_2(\psi_*\mu)$ is closable and its minimal closed extension gives a Dirichlet form. Thus we can define the Dirichlet space $(\mathcal{E}^{g,\mu}, \mathcal{D}[\mathcal{E}^{g,\mu}])$ on $L_2(\psi_*\mu)$, the associated semigroup $\{T_t^{g,\mu} : t > 0\}$ and the resolvent $\{G_\alpha^{g,\mu} : \alpha > 0\}$ by the usual way. Notice that $\{T_t^{g,\mu} : t > 0\}$ has a Feller version and there corresponds a unique Hunt process $X^{g,\mu} = (p_{g,\mu}(t), P_p)$ on M_g. Further, the (2,0)-tensor h^* can be converted to a (0,2)-tensor h on M_g.

Let now $\{g_n\}_{n=1}^\infty \subset \mathcal{R}_d, g \in \mathcal{R}_r$ and $\lim_{n\to\infty} d(g_n, g) = 0$. Then (2.4) implies

$$\lim_{n\to\infty} g_n^*(\phi_{g_n}(X), \phi_{g_n}(Y)) = g^*(\phi_g(X), \phi_g(Y)).$$

However, this does not in general ensure

$$\lim_{n\to\infty} g_n^*(\phi_g(X), \phi_g(Y)) = g^*(\phi_g(X), \phi_g(Y)).$$

Thus we shall assume

(B.1) $\lim_{n\to\infty} g_n^*(\psi^*\theta, \psi^*\eta)_q = g^*(\psi^*\theta, \psi^*\eta)_q, \qquad \theta, \eta \in T^*_{\psi(q)}(M_g).$

On the other hand, due to (A.1) with Frobenius Theorem, we can choose a cubic coordinate system $(U; y^1, y^2, ..., y^d)$ such that, for each $p = (x^1, x^2, ..., x^d) \in U$, the subspace

$$\{q = (y^1, y^2, .., y^{d'}, .., y^d) \in U : y^1 = x^1, y^2 = x^2, .., y^{d'} = x^{d'}\}$$

is an integral manifold of $\mathcal{D}$ (see [15]). Further, we can choose a finite covering $\{U^{(k)}\}_{k=1}^L$ of $\widetilde{M}$ with such cubic coordinate neighborhood satisfying $U^{(k)} \subset \{-1 < y^i < 1 \text{ for } 1 \le i \le d\}$. The next assumption is rather technical. We will give a reasonable sufficient condition for it in a forthcoming paper.

(B.2) *With the cubic coordinate systems in the above, we can find a sequence $K_n \to \infty$ such that, for each $(\xi_1, \xi_2, ..., \xi_d) \in \mathbf{R}^d$,*

$$\sum_{i,j=1}^d g_n^*(dx^i, dx^j)_q \xi_i \xi_j \ge K_n \sum_{i=d'+1}^d |\xi_i|^2, \quad q \in U^{(k)}, \ 1 \le k \le L, \tag{2.7}$$

and

$$\lim_{n\to\infty} \frac{1}{K_n^{1/2}} \max\{|g_n^*(dx^i, dx^j)_q| : d' + 1 \le i \le d, \\ 1 \le j \le d', q \in U^{(k)}, 1 \le k \le L\} = 0. \tag{2.8}$$

We then obtain the next

Theorem 2 *Let* $\{g_n, \mu_n)\}_{n=1}^{\infty} \subset \mathcal{R}_d \times \mathcal{M}(\widetilde{M})$, $(g, \mu) \in \mathcal{R}_r \times \mathcal{M}(\widetilde{M})$,

$$\lim_{n\to\infty}\{d(g_n, g) + d_{\mathcal{M}(\widetilde{M})}(\mu_n, \mu)\} = 0$$

and assume that Conditions (B.1), (B.2) *are satisfied. Then*

$$\lim_{n\to\infty} \|T_t^{g_n,\mu_n}u - \psi^* \circ T_t^{g,\mu} \circ \pi_{g,\mu}u\|_{\tilde{\mu}} \quad = 0, \qquad u \in L_2(\tilde{\mu}),\ t > 0. \tag{2.9}$$

Remark 2.2 Under the assumptions of Theorem 2, the laws of $(\psi(q_{g_n,\mu_n}(\cdot)), P_q)$ converge weakly to that of $(p_{g,\mu}(\cdot), P_{\psi(q)})$ in the space of $C([0,\infty) \to M_g)$.

The proof of Remark 2.2 is just a repeat of the arguments in [13] and will be omitted. It gives enough information about the behavior of $\{(\psi(q_{g_n,\mu_n}(\cdot)), P_q\}$ but less about the behavior of $\{(q_{g_n,\mu_n}(\cdot), P_q)\}$ themselves. In order to study the behavior of $\{q_{g_n,\mu_n}(t)\}$, we give a realization of sample paths for the process corresponding to the semigroup $\{\psi^* \circ T_t^{g,\mu} \circ \pi_{g,\mu}\}$. Let $\{p_{g,\mu}(t;p)\}$ be a realization of the Hunt process $X^{g,\mu}$ starting at $p \in M_g$ on a probability space $(\Omega, \mathcal{F}, P)$ and $\{\xi(t;p) : t \geq 0, p \in M_g\}$ be an independent system of random variables taking the values in M_g which is also independent of $\{p_{g,\mu}(t;p)\}$ and has the stationary law

$$E[v(\xi(t;p))] \;=\; \int_{F_p} v(q)\mu_p(dq), \qquad v \;\in\; C(F_p), \quad t \geq 0,\ p \in M_g \tag{2.10}$$

(see [12]). More precisely, $\xi(t_1, p_1), \cdots, \xi(t_N, p_N)$ are mutually independent for each $t_1 < t_2 < \cdots < t_N$ and distinct $p_1, \cdots, p_N \in M_g$ and they are independent of $\{p_{g,\mu}(t;p)\}$. We then consider the pair $(p_{g,\mu}(t;\psi(q)), \xi(t; p_{g,\mu}(t;\psi(q))))$ as a point in $\widetilde{M}$ in the natural way, and denote it by $q(t;q)$. As is shown in Lemma 4.2 below, the law of $(q(t;q), P)$ coincides with that induced by $\psi^* \circ T_t^{g,\mu} \circ \pi_{g,\mu}$. Since the process $\{(q(t;q), P)\}$ is not continuous, one can not expect the convergence of $\{(q_{g_n,\mu_n}(t), P_q)\}$ to $\{(q(t;q), P)\}$ weakly in $C([0,T] \to M)$ nor in $D([0,T] \to M)$. However we obtain the convergence of finite dimensional distributions.

Theorem 3 *Under the assumptions of Theorem 2, the finite dimensional distributions of* $\{(q_{g_n,\mu_n}(t), P_q)\}$ *converge to those of* $\{(q(t;q), P)\}$, *i.e., for each* $0 < t_1 < t_2 < \cdots < t_N$ *and* $u_1, u_2, \cdots, u_N \in C(\widetilde{M})$,

$$\begin{aligned} &\lim_{n\to\infty} E_q[u_1(q_{g_n,\mu_n}(t_1))u_2(q_{g_n,\mu_n}(t_2)) \cdots u_N(q_{g_n,\mu_n}(t_N))] \\ &= E[u_1(q(t_1;q)u_2(q(t_2;q)) \cdots u_N(q(t_N;q))], \quad q \in \widetilde{M}. \end{aligned} \tag{2.11}$$

We will next give an application of Theorem 2 to the eigenvalue problem.

Definition Let $(\mathcal{E},\mathcal{D}[\mathcal{E}])$ be a Dirichlet form on a Hilbert space H. Then λ is called an *eigenvalue* of $(\mathcal{E},\mathcal{D}[\mathcal{E}],H)$ and $\varphi \in H$ its *normalized eigenfunctions,* if $\varphi \in \mathcal{D}[\mathcal{E}], \mathcal{E}(\varphi, \psi) = \lambda(\varphi, \psi)_H$ for all $\psi \in \mathcal{D}[\mathcal{E}]$ and $\|\varphi\|_H = 1$.

Let now $0 = \lambda_0^{(n)} < \lambda_1^{(n)} \leq \cdots$ be the eigenvalues of the Dirichlet system $(\mathcal{E}^{g_n,\mu_n}, \mathcal{D}[\mathcal{E}^{g_n,\mu_n}], L_2(\mu_n))$, and $\varphi_0^{(n)}, \varphi_1^{(n)}, \cdots$ be their corresponding normalized independent eigenfunctions. Without loss of generality, we may assume that they are orthonormal for each $n \in \mathbf{N}$. Let also $0 = \lambda_0 < \lambda_1 \leq \cdots$ be the eigenvalues of the Dirichlet system $(\mathcal{E}^{g,\mu}, \mathcal{D}[\mathcal{E}^{g,\mu}], L_2(\psi_*\mu))$, and

$$E(\lambda) := \{\varphi \in \mathcal{D}[\mathcal{E}^{g,\mu}] : \mathcal{E}^{g,\mu}(\varphi, \psi) = \lambda(\varphi, \psi)_\mu \text{ for all } \psi \in \mathcal{D}[\mathcal{E}^{g,\mu}]\}.$$

We then have

Theorem 4 *Under the assumptions of Theorem 2, it holds that*

$$\lim_{n\to\infty} \lambda_k^{(n)} = \lambda_k, \quad k = 0, 1, 2, \cdots, \tag{2.12}$$

$$\lim_{n\to\infty} d_{\tilde{\mu}}(\varphi_k^{(n)}, \psi^*(E(\lambda_k))) = 0, \quad k = 0, 1, 2, \cdots, \tag{2.13}$$

where $d_{\tilde{\mu}}$ is the metric induced by the norm $\| \ \|_{\tilde{\mu}}$.

Before closing this section, we shall prove (2.6).

Proof of (2.6) Without loss of generality, we may assume that $\tilde{g}$ is bundle like in the sense of Reinhart (see [14]). Then, with the cubic coordinate system in the above, we have

$$\tilde{g} = \sum_{i,j=1}^{d'} \tilde{g}_{ij}(\psi(p))dy^i dy^j \oplus \sum_{\alpha,\beta=1}^{d-d'} \tilde{g}_{\alpha\beta}(q)\theta^\alpha\theta^\beta,$$

where $dy^1, ..., dy^{d'}, \theta^1, ..., \theta^{d-d'}$ is a basis of the cotangent space $T_q^*(\widetilde{M})$ such that $\langle\theta^\alpha, X\rangle = 0$ for all X perpendicular $\mathcal{D}(g)_q$. It is then clear that the measure μ_p is given by

$$\tilde{\mu}_p(dr) = \sqrt{\widetilde{G}_{22}(\iota_p(r))}|\iota_p^*(\theta^1 \wedge ... \wedge \theta^{d-d'})| / \int_{\widetilde{F}_p} \sqrt{\widetilde{G}_{22}(\iota_p(r))}|\iota_p^*(\theta^1 \wedge ... \wedge \theta^{d-d'})|,$$

where $\widetilde{G}_{22}(q) = |\det(\tilde{g}_{\alpha,\beta}(q))|$, ι_p is the inclusion operator of F_p into $\widetilde{M}$ and $|\iota_p^*(\theta^1 \wedge ... \wedge \theta^{d-d'})|$ stands for the measure induced by $\iota_p^*(\theta^1 \wedge ... \wedge \theta^{d-d'})$.

For a general $\mu \in \mathcal{M}(\widetilde{M})$, we have only to set

$$\mu_p(dr) = \frac{d\mu}{d\tilde{\mu}}(\iota_p(r))\frac{d\psi_*\tilde{\mu}}{d\psi_*\mu}(p)\tilde{\mu}_p(dr).$$

The existence of the Radon-Nikodym derivatives in the above is easily checked. ∎

3. Examples of product manifold

In this section, we will consider the case where $\widetilde{M}$ is a product manifold, which is a typical example of Theorem 2. This is also a simple case in the drive-driven process theory.

Example 1 Let M and N be orientable compact smooth manifolds with dimension d'

and $d-d'$ respectively. On the product space $\widetilde{M} = M \times N$, we consider a Riemannian metric

$$\tilde{g} = g_{ij}^M(x,p')dx^i dx^j \oplus g_{\alpha\beta}^N(p,y)dy^\alpha dy^\beta, \qquad q = (p,p') = (x,y) \in U \times V,$$

where (U,x) and (V,y) are coordinate systems of M and N respectively, and $g_{ij}^M(x,p')$ and $g_{\alpha\beta}^N(p,y)$ are symmetric positive definite functions which are continuous in $q \in \widetilde{M}$ and satisfy the coordinate consistency condition

$$\bar{g}_{ij}^M(\bar{x},p') = g_{k\ell}^M(x,p')\frac{\partial x^k}{\partial \bar{x}^i}\frac{\partial x^\ell}{\partial \bar{x}^j}, \qquad (x,p') = (\bar{x},p') \in (U \cap \bar{U}) \times N,$$

$$\bar{g}_{\alpha\beta}^N(p,\bar{y}) = g_{\gamma\delta}^N(p,y)\frac{\partial y^\gamma}{\partial \bar{y}^\alpha}\frac{\partial y^\delta}{\partial \bar{y}^\beta}, \qquad (p,y) = (p,\bar{y}) \in M \times (V \cap \bar{V})$$

(here and hereafter, we use the Einstein convention $a_{ij}b^{ij} = \sum_{i,j} a_{ij}b^{ij}$). In other words, we are given continuous positive (0,2)-tensors $g_{p'}^M$ [g_p^N] on M [N] for each $p' \in N$ [resp. $p \in M$], and $\tilde{g}$ is given by

$$\tilde{g}(X,Y)_{(p,p')} = g_{p'}^M(\psi_*^M X, \psi_*^M Y) + g_p^N(\psi_*^N X, \psi_*^N Y), \qquad X,\ Y \in T_{(p,p')}(\widetilde{M}),$$

where ψ^M and ψ^N are the projection operators $M \times N \to M$ and $M \times N \to N$ respectively. We denote by $\{U^{(k)}\}$ and $\{V^{(\ell)}\}$ finite coverings of coordinate neighborhoods of M and N respectively. The partitions of unity subordinate to them are denoted by $\{\phi_k^M\}$ and $\{\phi_\ell^N\}$.

We consider a sequence of symmetric nonnegative continuous (0,2)-tensors

$$g_n = g_{ij}^M(x,p')dx^i dx^j \oplus n^{-2} g_{\alpha\beta}^N(p,y)dy^\alpha dy^\beta, \qquad q = (p,p') = (x,y) \in U \times V.$$

It is then clear that $g_n \in \mathcal{R}_d$ and $\lim_{n\to\infty} d(g_n, g) = 0$, where $g = g_{ij}^M(x,p')dx^i dx^j$. We will check that $g \in \mathcal{R}_r$. Let $\iota_{p'}^M$ and ι_p^N be the inclusion operators

$$\iota_{p'}^M(p) = (p,p'),\ p \in M \quad \textit{and} \quad \iota_p^N(p') = (p,p'),\ p' \in N.$$

Then $\mathcal{D}_{(p,p')}(g) = \iota_{p*}^N(T_{p'}(N))$ so that $\mathcal{D}(g) : q \to \mathcal{D}(g)_q$ is a $(d-d')$-dimensional completely integrable distribution. For each $q = (p,p') \in \widetilde{M}$, the maximal integral manifold of $\mathcal{D}$(g) including q is $F_p = \iota_p^N(N)$ and $M_g = \widetilde{M}/\!\sim$ is identified with the manifold M. Thus the natural map $\psi : (p,p') \to p$ is differentiable and $\psi_*(\xi^i X_i + \lambda^\alpha Y_\alpha) = \xi^i X_i$, where $X_i = \partial/\partial x^i$ and $Y_\alpha = \partial/\partial y^\alpha$ considered as elements of $T_q(\widetilde{M})$ in the left hand side of the equality. Hence $\psi_*(T_q(\widetilde{M})) = T_{\psi(q)}(M)$ and we have $g \in \mathcal{R}_r$.

It is clear that

$$g_n^*(\theta^i,\theta^j)_q = g^{M,ij}(x,p'), \qquad g_n^*(\eta^\alpha,\eta^\beta)_q = n^2 g^{N,\alpha\beta}(p,y),$$

$$g_n^*(\theta^i,\eta^\alpha)_q = 0, \qquad q = (p,p') = (x,y) \in U \times V,$$

where $\theta^i = dx^i, \eta^\alpha = dy^\alpha$ and $[g^{M,ij}]$, $[g^{N,\alpha\beta}]$ are the inverse matrices of $[g^M_{ij}]$, $[g^N_{\alpha\beta}]$ respectively. Hence Conditions (B.1) and (B.2) are clearly satisfied. Further, $\phi = \phi_g$ is given by $\phi(X_i)_{(x,p')} = g^M_{ij}(x,p')\theta^j$ and $\phi(Y_\alpha) = 0$ so that $g^*(\theta^i, \theta^j)_q = g^{M,ij}(x,p')$ for $1 \le i,j \le d'$.

Let next $\mu_n = \mu_{g_n} = \Omega_{g_n}/\mathrm{vol}(\widetilde{M}, g_n)$. Then

$$\mu := \mu_n = \sqrt{\widetilde{G}(x,y)}dxdy/\mathrm{vol}(\widetilde{M},\tilde{g}), \qquad q=(x,y) \in U \times V,$$

where $\widetilde{G}(x,y) = (\det[g^M_{ij}(x,p')])(\det[g^N_{\alpha\beta}(p,y)])$ and $dx = dx^1 dx^2 \cdots dx^{d'}$, $dy = dy^1 dy^2 \cdots dy^{d-d'}$. Note that the symmetric form associate to (g_n, μ_n) is given by

$$\mathcal{E}^{g_n,\mu_n}(u,v) = \int_{\widetilde{M}} \{g^{M,ij}(x)(X_i u)_x (X_j v)_x$$

$$+ n^2 g^{N,\alpha\beta}(y)(Y_\alpha u)_y (Y_\beta v)_y\} \sqrt{\widetilde{G}(x,y)}dxdy/\mathrm{vol}(\widetilde{M},\tilde{g}),, \quad u,\ v \in C^\infty(\widetilde{M}),$$

Now, it holds that $d_{\mathcal{M}(\widetilde{M})}(\mu_n, \mu) \to 0$ and

$$\mathcal{E}^{g,\mu}(u,v) = \sum_{k,\ell} \int_{\widetilde{M}} \phi^M_k(x)\phi^N_\ell(y) g^{M,ij}(x,y)(X_i u)_x (X_j v)_x \sqrt{\widetilde{G}(x,y)}dxdy/\mathrm{vol}(\widetilde{M},\tilde{g}),$$

$$u,v \in C^\infty(\widetilde{M}).$$

Further,

$$\psi_*\mu(dx) = \frac{1}{\mathrm{vol}(\widetilde{M},\tilde{g})}(\sum_\ell \int_N \phi^N_\ell(y)\sqrt{\widetilde{G}(x,y)}dy)dx,$$

so that $\pi_{g,\mu}u(p) = \int_N u(q)\mu_p$, where

$$\mu_x(dy) = \sqrt{\widetilde{G}(x,y)}dy/(\sum_\ell \int_N \phi^N_\ell(y)\sqrt{\widetilde{G}(x,y)}dy), \quad (x,y) \in U \times V.$$

Hence, we have

$$h^*(\theta^i,\theta^j)_x = \sum_\ell \int_N \phi^N_\ell(y) g^{M,ij}(x,y)\sqrt{\widetilde{G}(x,y)}dy/(\sum_\ell \int_N \phi^N_\ell(y)\sqrt{\widetilde{G}(x,y)}dy),$$

and

$$\mathcal{E}^{g,\mu}(u,v) = \sum_k \int_M \phi_k(x) h^*(\theta^i,\theta^j)_x (X_i u)_x (X_j v)_x \psi_*\mu(dx) \quad u,v \in C^1(M).$$

We now obtain (2.9) for the semigroups corresponding the items in the above by Theorem 2.

The next example is a special case of the previous one. But everything is simpler and we can give a discussion on the relation with the results in geometry.

Example 2 (Two-dimensional torus) Let $\widetilde{M} = T^2 = S^1 \times S^1 = \{(e(x), e(y)) : x,y \in \mathbf{R}\}$ with $x = e(x) \equiv e^{ix}$ and

$$\tilde{g} = a(x,e(y))^2 dx^2 \oplus b(e(x),y)^2 dy^2, \quad q=(x,y)=(e(x),e(y)) \in M,$$

where a and b are positive and sufficiently smooth functions in $\widetilde{M}$. Then we have

$$g_n = a(x, e(y))^2 dx^2 \oplus b(e(x), y)^2 dy^2/n^2, \quad q = (x, y) = (e(x), e(y)) \in \widetilde{M},$$

and $g = a(x, e(y))^2 dx^2$. Thus, letting $\mu_n = \mu_{g_n} = \Omega_{g_n}/\mathrm{vol}(\widetilde{M}, g_n)$, we have

$$\mu = \mu_n = ab(x, y)dxdy / \int_{\widetilde{M}} ab(x, y)dxdy,$$

$$\mathcal{E}^{g_n, \mu_n}(u, v) = \frac{1}{\mathrm{vol}(\widetilde{M}, \tilde{g})} \int_{\widetilde{M}} \{\frac{\partial u}{\partial x}\frac{\partial v}{\partial x}\frac{b}{a} + n^2 \frac{\partial u}{\partial y}\frac{\partial v}{\partial y}\frac{a}{b}\} dxdy$$

and

$$\mathcal{E}^{g, \mu}(u, v) = \frac{1}{\mathrm{vol}(\widetilde{M}, \tilde{g})} \int_{\widetilde{M}} u'(x)v'(x)\frac{b}{a}(x, y)dxdy.$$

Further,

$$\psi_*\mu(dx) = (\int_{S^1} ab(x, y)dy)dx/\mathrm{vol}(\widetilde{M}, \tilde{g})$$

and

$$\pi_{g,\mu}u(x) = \int_{S^1} uab(x, y)dy / \int_{S^1} ab(x, y)dy.$$

We thus have

$$h^*(dx, dx)_x = \int_{S^1} \frac{b}{a}(x, y)dy / \int_{S^1} ab(x, y)dy,$$

and obtain (2.9) for the corresponding semigroups. The (2,0)-tensor h^* on S^1 can be converted to a (0,2)-tensor h through $h(\partial/\partial x, \partial/\partial x)_x = 1/h^*(dx, dx)_x$.

Let $(M_n, \rho_n, \mu_n), n \in \mathbf{N}$ be the spaces $(\widetilde{M}, g_n)$ with the measures μ_n considered as measured metric spaces. Also let metric space (S^1, ρ_o, μ_o) be the measured metric space where ρ_o is the distance function corresponding to the (0,2)-tensor

$$\underline{h}(\frac{\partial}{\partial x}, \frac{\partial}{\partial x})_x = \min\{a(x, y) : y \in S^1\}$$

and $\mu_o = \psi_*\mu$. We then have the following assertion.

Lemma 3.1 *The measured metric spaces (M_n, ρ_n, μ_n)converge to (S^1, ρ_o, μ_o) in the measured Hausdorff topology, i.e., there exist a sequence $\delta(n) \downarrow 0$ and measurable maps $\psi_n : M_n \to S^1$ such that the $\delta(n)$-neighborhood of $\psi_n(M_n)$ coincides with S^1,*

$$|\rho_n(p, q) \ - \ \rho_o(\psi_n(p), \psi_n(q))| \ < \ \delta(n), \quad p, q \ \in \ M_n \tag{3.1}$$

and the image measures $\psi_{n*}\mu_n$ weakly converge to μ_o.

Proof Let $\psi_n = \psi, n \in \mathbf{N}$, where ψ is the map given in the above. Then we have $\psi_n(M_n) = \psi(\widetilde{M}) = S^1$ and $\psi_{n*}\mu_n = \psi_*\mu = \mu_o$ for all $n \in \mathbf{N}$. Hence we have only to show (3.1).

Since the inequality $\rho_o(\psi_n(p), \psi_n(q)) \leq \rho_n(p, q), p, q \in M_n$ is clear, we will show that

$$\rho_n(p, q) \ \leq \ \rho_o(\psi_n(p), \psi_n(q)) \ + \ Kn^{-1}, \quad p, q \ \in \ M_n, \tag{3.2}$$

for some positive constant K. Let $p = (x, y) = (e(x), e(y))$, $q = (x', y') = (e(x'), e(y'))$ $\in M_n$, with $0 \leq x' - x \leq \pi$. Let also $L = [(x' - x)n] + 1$ and $x = t_0 < t_1 < \cdots < t_{2L+2} = x'$ be the partition of the interval $[x, x']$ satisfying $t_{j+1} - t_j = 1/(2L+2), j = 0, 1, \cdots, 2L+1$. Let $x_k = e(t_{2k})$ and choose y_k so that

$$a(x_k, y_k) = \min\{ a(x_k, y) : y \in S^1\}, \; k = 0, 1, 2, \cdots, L,$$

and $y_{L+1} = y'$. Define $c(t) = (c_1(t), c_2(t))$, $x \leq t \leq x'$ by

$$c_1(t) = \begin{cases} x_k, & t_{2k} \leq t < t_{2k+1}, \\ e(2(t - t_{2k+1}) + t_{2k}), & t_{2k+1} \leq t < t_{2k+2}, \end{cases}$$

$$c_2(t) = \begin{cases} \gamma_2(t; y_k, y_{k+1}), & t_{2k} \leq t < t_{2k+1}, \\ y_{k+1}, & t_{2k+1} \leq t < t_{2k+2}, \end{cases}$$

for $k = 0, 1, 2, \ldots, L$, where $\gamma_2(t; y_k, y_{k+1})$ is a curve in S^1 combining y_k and y_{k+1}. We then have

$$\begin{aligned} \rho_n(p, q) &\leq \int_{c[x,x']} ds_n \\ &\leq \sum_{k=0}^{L-1} \int_{t_{2k+1}}^{t_{2k+2}} a(e(2(t - t_{2k+1}) + t_{2k}), y_k) 2dt + K_1 n^{-1}, \end{aligned} \tag{3.3}$$

for some constant $K_1 > 0$, where ds_n is the line element for g_n. Further,

$$a(x, y_k) = a(x_k, y_k) + \frac{\partial}{\partial x} a(\xi_1, y_k) i\xi_1 (t - t_{2k+1}),$$

for $x = e(t)$ with $t_{2k+1} \leq t < t_{2k+2}$, where ξ_1 is an element in $\{e(\tau) : t_{2k+1} \leq \tau \leq t\}$, and similarly,

$$\underline{a}(x) = a(x, \underline{y}) = a(x_k, \underline{y}) + \frac{\partial}{\partial x} a(\xi_2, \underline{y}) i\xi_2 (t - t_{2k}),$$

for some $\underline{y} = \underline{y}(x)$. Since $a(x_k, y_k) \leq a(x_k, \underline{y})$, this implies

$$a(x, y_k) \leq \underline{a}(x) + 2K_2(t - t_{2k}),$$

with $K_2 = \max\{|\partial a(x, y)/\partial x| : (x, y) \in M\}$. Substituting this into (3.3), we obtain

$$\rho_n(p, q) \leq \sum_{k=0}^{L-1} \int_{t_{2k+1}}^{t_{2k+2}} \underline{a}(e(2(t - t_{2k+1}) + t_{2k})) 2dt + K_3 n^{-1},$$

which proves (3.2). ∎

Remark 3.1 If $a(x, y)$ is independent of y, we have $h(\partial/\partial x, \partial/\partial x) = \underline{h}(dx, dx) = a(x)^2$, and vice versa. This condition is equivalent to that the Gauss curvatures K_n of (M_n, g_n) are uniformly bounded in n, and the example is reduced to a special case dealt with in the general theory in [4].

Remark 3.2 We had to restrict ourselves to the case of Example 2 to obtain the uniform estimate (3.1). In the general case of Example 1, we could only prove $\lim_{n\to\infty} |\rho_n(p,q) - \rho_o(\psi_n(p), \psi_n(q))| = 0, p, q \in M_n$.

Example 3 (Skew product diffusion) Skew products of diffusion processes were utilized by Itô and McKean [11] for constructing Brownian motions on $\mathbf{R}^n$ or those on S^n, and later were extensively used by Ikeda and Watanabe [9] for the study of the diffusion processes of their special features. Recently, Fukushima and Oshima [6] gave a systematic approach to the skew products from the Dirichlet space theory. We note that the same notion is referred to as warped products in geometry. Indeed, the skew product of diffusion processes on a manifold is just the Brownian motion associated to the warped product metric(see [2]). Here we will give an asymptotic behavior for a simple process which is a little more general than a skew product, but can be dealt with in the same way as it.

Let M, N and $\widetilde{M}$ be those in Example 1. In this example, we assume that $\tilde{g}$ is the product Riemannian metric, that is g^M_{ij} are independent of p' whereas $g^N_{\alpha\beta}$ of p. Then the volume element $\tilde{\mu}$ of $\tilde{g}$ is given by $\tilde{\mu} = \mu^M \times \mu^N$, where μ^M and μ^N are the volume elements of g^M and g^N respectively. We then have Brownian motions $X^M = (p^M(t), P^M_p)$ and $X^N = (p^N(t), P^N_{p'})$ on M and N corresponding to g^M and g^N respectively. The Brownian motion $X = (q(t), P_q)$ on $\widetilde{M}$ corresponding to $\tilde{g}$ is realized by the product of X^M and X^N ; $q(t) = (p^M(t), p^N(t)), P_{(p,p')} = P^M_p \times P^N_{p'}$. Let next $m(q)$ be a positive continuous function on $\widetilde{M}$ and set $\nu = m\tilde{\mu}$. Then it is a smooth measure and there corresponds a unique positive continuous additive functional $A^\nu_t = \int_0^t m(q(s))ds$. The Dirichlet form for the process $X^\nu = (q^\nu(t), P_q)$ with $q^\nu(t) = (p^M(A^\nu_t), p^N(t))$ is the extension of the symmetric form

$$\mathcal{E}^\nu(u,v) = \int_{\widetilde{M}} \{g^{M,ij}(x)(X_i u)_x (X_j v)_x m(x,y)$$

$$+ g^{N,\alpha\beta}(y)(Y_\alpha u)_y (Y_\beta v)_y\} \sqrt{\widetilde{G}(x,y)}dxdy, \quad u,\ v \in C^\infty(\widetilde{M}),$$

on $L_2(\tilde{\mu})$. In other words,

$$\mathcal{E}^\nu(u,v) \ = \ \int_{\widetilde{M}} \{g^{M*}(\iota^{M*}_{p'} du, \iota^{M*}_{p'} dv)_p m(p,p') \ + \ g^{N*}(\iota^{N*}_p du, \iota^{N*}_p dv)_{p'}\}\tilde{\mu}(dpdp')$$

(see Example 1 for the definition of $\iota^M_{p'}$ and ι^N_p).

Now let $X^{(n)} = (q^{(n)}(t), P_q)$ with $q^{(n)}(t) = (p^M(A^{\nu/n}_{nt}), p^N(nt))$. Then the corresponding Dirichlet form on $L_2(\tilde{\mu})$ is given by

$$\mathcal{E}^{(n)}(u,v) = \int_{\widetilde{M}} \{g^{M*}(\iota^{M*}_{p'} du, \iota^{M*}_{p'} dv)_p m(p,p') \ + \ n g^{N*}(\iota^{N*}_p du, \iota^{N*}_p dv)_{p'}\}\tilde{\mu}(dpdp').$$

In our notation, this is equal to $\mathcal{E}^{g_n,\tilde{\mu}}$ with g_n given by

$$g_n(X,Y)_{(p,p')} = m(p,p')^{-1} g^M(\psi^M_* X, \psi^M_* Y) \oplus n^{-1} g^N(\psi^N_* X, \psi^N_* Y),$$

$$X,\ Y \in T_{(p,p')}(M).$$

We thus obtain that the semigroups $T_t^{(n)}$ for $X^{(n)}$ converge to the semigroup $\psi^* \circ S_t \circ \pi$, where S_t corresponds to the symmetric form

$$\mathcal{E}(u,v) = \int_M g^{M*}(du,dv)_p \pi m(p) \mu^M(dp), \qquad u,\ v \in C^\infty(M),$$

on $L_2(\mu^M/\mu^N(N))$ with $\pi w(p) = \int_N w(p,p')\mu^N(dp')/\mu^N(N)$. The process for the semigroup $\{S_t\}$ is nothing but $(p^M(A_t^{\psi_* \nu'}), P_q^M)$, where $\nu' = \nu/(\mu^N(N))^2$ and $A_t^{\psi_* \nu'} = \int_0^t \pi m(p^M(s))ds$. When $m(p,p')$ is independent of p, the function πm is constant and $A_t^{\psi_* \nu'} = t\pi m$. Finally, we note that the above procedure should work for more general smooth measure ν, or even for more general Dirichlet form $\mathcal{E}^\nu$ such as the one dealt with in [6].

4. Proof of Theorems 1, 2 and 3

We first introduce the Sobolev space $H_\beta(\widetilde{M})$ on $\widetilde{M}$ following [7 : p.6, pp.27-30]. For a real β and $u, v \in C^\infty(\widetilde{M})$, let

$$(u,v)_\beta = \sum_k \int_{\mathbf{R}^d} (1+|\xi|^2)^\beta (\widehat{u\phi_k})^*(\xi) \widehat{v\phi_k}(\xi) d\xi,$$

$\|u\|_\beta^2 = (u,u)_\beta$, where

$$\widehat{u\phi_k}(\xi) \;=\; \int_{U(k)} e^{-i\xi\cdot x}(u\varphi_k)(x)dx/(2\pi)^d$$

and $(v\phi_k)^*(\xi)$ is the complex conjugate of $\widehat{v\phi_k}(\xi)$, $\{\phi_k\}$ being a partition of unity subordinate to $\{U^{(k)}\}$. Note that, although this norm is not invariant under a change of coordinate functions, it determines a unique topology(see ibid.). Thus we denote by $H_\beta(\widetilde{M})$ the completion of $C^\infty(\widetilde{M})$ with respect to the norm $\| \ \|_\beta$. Notice that, if $(g,\mu) \in \mathcal{R}_d \times \mathcal{M}(\widetilde{M})$, then $H_1(\widetilde{M}) = \mathcal{D}[\mathcal{E}^{g,\mu}]$ and the norm $\|u\|_1$ is equivalent to $\mathcal{E}_\alpha^{g,\mu}(u,u)^{1/2}$ for each $\alpha > 0$. Further, $\|u\|_0 = \|u\|_{\tilde{\mu}}$.

Proposition 4.1 *Let $r > 1$ and $\mathcal{R}_d(r) = \{g \in \mathcal{R}_d : d(0,g) \le r\}, \mathcal{M}(\widetilde{M},r) = \{\mu \in \mathcal{M}(\widetilde{M}) : 1/r \le d\mu/d\tilde{\mu}(p) \le r, p \in \widetilde{M}\}$. Then $\{T_t^{g,\mu} : (g,\mu) \in \mathcal{R}_d(r) \times \mathcal{M}(\widetilde{M},r)\}$ is relatively compact in the strong topology in $L_2(\tilde{\mu})$, i.e., for each infinite subset $\Lambda \subset \mathcal{R}_d(r) \times \mathcal{M}(\widetilde{M},r)$, we can find a sequence $\{(g_n,\mu_n)\} \subset \Lambda$ and a nonnegative contraction semigroup T_t on $L_2(\tilde{\mu})$ such that*

$$\lim_{n\to\infty} \|T_t^{g_n,\mu_n}u - T_t u\|_{\tilde{\mu}} = 0, \qquad u \in L_2(\tilde{\mu}),\ t > 0. \tag{4.1}$$

Proof Note first that we can choose a sequence $\{(g_m,\mu_m)\} \subset \Lambda$ and a $\mu^* \in \mathcal{M}$ such that $\mu_m \to \mu^*$ weakly. Further, in view of the inequality

$$\frac{1}{r}\int_{\widetilde{M}} u(p)^2 d\tilde{\mu}(p) \;\le\; \int_{\widetilde{M}} u(p)^2 d\mu(p) \le r\int_{\widetilde{M}} u(p)^2 d\tilde{\mu}(p), \ \mu \in \mathcal{M}(\widetilde{M},r),$$

we have $L_2(\tilde{\mu}) = L_2(\mu_m) = L_2(\mu^*)$, $m \in \mathbf{N}$. Recall next the spectral representation

$$(T_t^{g,\mu}u,v)_\mu = \int_0^\infty e^{-\lambda t} d(E_\lambda^{g,\mu}u,v)_\mu, \qquad u,v \in L_2(\mu),$$

with

$$(total\ variation\ of\ \ \{(E_\lambda^{g,\mu}u,v)_\mu : \lambda \geq 0\}) \leq \|u\|_\mu \|v\|_\mu \leq r\|u\|_{\tilde{\mu}}\|v\|_{\tilde{\mu}}.$$

Thereby, due to the separability of $L_2(\tilde{\mu})$, we can find a subsequence $\{(g_n,\mu_n)\} \subset \{(g_m,\mu_m)\}$ independent of $u, v \in L_2(\tilde{\mu})$ and a spectral resolution $\{E_\lambda : \lambda \geq 0\}$ such that

$$\lim_{n\to\infty}(E_\lambda^{g_n,\mu_n}u,v)_{\mu_n} = (E_\lambda u,v)_{\tilde{\mu}}, \qquad u,v \in L_2(\tilde{\mu}),$$

for each continuity point λ of the right hand side. Hence it follows that

$$\lim_{n\to\infty}(T_t^{g_n,\mu_n}u,v)_{\mu_n} = (T_t u,v)_{\tilde{\mu}}, \quad u,v \in L_2(\tilde{\mu}),\ t > 0, \tag{4.2}$$

where $T_t u \in L_2(\tilde{\mu})$ is defined by $(T_t u,v)_{\tilde{\mu}} = \int_0^\infty e^{-\lambda t} d(E_\lambda u,v), u,v \in L_2(\tilde{\mu})$.

We next show that

$$g^*(du,du) \geq \tilde{g}^*(du,du)/r, \quad u \in C^\infty(\widetilde{M}), \quad g \in \mathcal{R}_d(r). \tag{4.3}$$

For this let

$$0 < \lambda_1(g,q) \leq \lambda_2(g,q) \leq \cdots \leq \lambda_d(g,q)$$

be the eigenvalues of the linear operator A_g on $T_q(\widetilde{M})$ defined by

$$g(X,Y)_q = \tilde{g}(A_g X,Y)_q, \quad Y \in T_q(\widetilde{M}).$$

It then follows that $\lambda_d(g,q) \leq r, q \in \widetilde{M}, g \in \mathcal{R}_d(r)$, which verifies (4.3). Notice that (4.3) implies

$$\mathcal{E}_\alpha^{g,\mu}(u,u) \geq \mathcal{E}_\alpha^{\tilde{g},\tilde{\mu}}(u,u)/r^2 \ \geq\ \delta\|u\|_1^2, \quad u \in H_1(\widetilde{M}), \quad (g,\mu) \in \Lambda,$$

for some $\delta > 0$.

Now fix $t > 0$ and $u \in L_2(\tilde{\mu})$. In view of the inequality

$$\mathcal{E}_\alpha^{g,\mu}(T_t^{g,\mu}u, T_t^{g,\mu}u) \leq \frac{1}{2t}\{(1+\alpha)(u,u)_\mu - (T_t^{g,\mu}u, T_t^{g,\mu}u)_\mu\}, \tag{4.4}$$

we then have that the sequence $\{T_t^{g_n,\mu_n}u\}_{n=1}^\infty$ is bounded in $H_1(\widetilde{M})$. Hence, for any subsequence $\{(g_k,\mu_k)\} \subset \{(g_n,\mu_n)\}$, we can find a subsequence $\{(g_\ell,\mu_\ell)\} \subset \{(g_k,\mu_k)\}$ and a $u^* \in H_0(\widetilde{M})$ such that

$$\lim_{\ell\to\infty} \|T_t^{g_\ell,\mu_\ell}u - u^*\|_0 = 0,$$

which is equivalent to

$$\lim_{\ell\to\infty} \ \|T_t^{g_\ell,\mu_\ell}u - u^*\|_{\tilde{\mu}} = 0.$$

This with (4.2) verifies $T_t u = u^*$ and (4.1). ∎

Proof of Theorem 1 Note first that $\{(g_n,\mu_n)\} \cup \{(g,\mu)\} \subset \mathcal{R}_d(r) \times \mathcal{M}(\widetilde{M},r)$ for some $r>1$. Thereby the only thing we have to do is to identify T_t in Proposition 4.1 with $T_t^{g,\mu}$. Let $G_\alpha u = \int_0^\infty e^{-\alpha t} T_t u dt$. Then (4.1) ensures

$$\lim_{n\to\infty} \|G_\alpha^{g_n,\mu_n} u - G_\alpha u\|_0 = 0, \qquad u \in L_2(\tilde{\mu}),\ \alpha > 0. \tag{4.5}$$

Further, we have

$$\mathcal{E}^{g_n,\mu_n}(G_\alpha^{g_n,\mu_n} u, v) + \alpha (G_\alpha^{g_n,\mu_n} u, v)_{\mu_n} = (u,v)_{\mu_n},\ u \in L_2(\mu_n),\ v \in \mathcal{D}[\mathcal{E}^{g_n,\mu_n}], \tag{4.6}$$

and

$$\|G_\alpha^{g_n,\mu_n} u\|_{\tilde{\mu}}/r \le \|G_\alpha^{g_n,\mu_n} u\|_{\mu_n} \le \|u\|_{\mu_n}/\alpha \le r\|u\|_{\tilde{\mu}}/\alpha.$$

Put $v = G_\alpha^{g_n,\mu_n} u$ in (4.6), we see that the sequence $\{G_\alpha^{g_n,\mu_n} u\}_{n=1}^\infty$ is bounded in $H_1(\widetilde{M})$. Hence, we can find a subsequence $\{(g_n,\mu_n)\}$, for which we use the same symbol, and a $u^* \in H_1(\widetilde{M})$ such that $(G_\alpha^{g_n,\mu_n} u, v)_1 \to (u^*, v)_1, v \in H_1(\widetilde{M})$. However, due to the inequality

$$|(u,v)_1|^2 \le \sum_k \Big(\int_{\mathbf{R}^d} |\widehat{u\phi_k}(\xi)|^2 d\xi\Big)\Big(\int_{\mathbf{R}^d} (1+|\xi|^2)^2 |\widehat{v\phi_k}(\xi)|^2 d\xi\Big) \le \|u\|_0^2 \|v\|_2^2$$

for $u \in H_0(\widetilde{M})$ and $v \in H_2(\widetilde{M})$, (4.5) ensures $(G_\alpha^{g_n,\mu_n} u, v)_1 \to (G_\alpha u, v)_1, v \in H_2(\widetilde{M})$. Hence we have $G_\alpha u = u^*$ and $(G_\alpha^{g_n,\mu_n} u, v)_1 \to (G_\alpha u, v)_1, v \in H_1(\widetilde{M})$.

On the other hand, from our assumption that $g \in \mathcal{R}_d$, we have $\inf\{\lambda_1(g_n,q) : n \in \mathbf{N}, q \in \widetilde{M}\}$ is positive, so that

$$\sup\{g_n^*(du,du)_q/\tilde{g}^*(du,du)_q : n \in \mathbf{N}, q \in \widetilde{M}\}$$

is finite for each $u \in C^1(\widetilde{M})$ with $du \ne 0$. In view of $\{\mu_n\} \subset \mathcal{M}(\widetilde{M},r)$, we thus obtain

$$\mathcal{E}_\alpha^{g_n,\mu_n}(G_\alpha^{g_n,\mu_n} u, v) \to \mathcal{E}_\alpha^{g,\mu}(G_\alpha u, v), \quad u \in L_2(\widetilde{M}),\ v \in H_1(\widetilde{M}).$$

This with (4.6) implies

$$\mathcal{E}^{g,\mu}(G_\alpha u, v) + \alpha(G_\alpha u, v)_\mu = (u,v)_\mu, \qquad u \in L_2(\mu),\ v \in H_1(\widetilde{M}),$$

and $G_\alpha u = G_\alpha^{g,\mu} u$. Hence $T_t u = T_t^{g,\mu} u$, which completes the proof. ∎

We next proceed to the proof of Theorem 2. As in the proof of Theorem 1, we can find a semigroup $\{T_t\}$, a resolvent $\{G_\alpha\}$ on $L_2(\tilde{\mu})$ and a subsequence of $\{(g_n,\mu_n)\}_{n=1}^\infty$, which we use the same symbol, such that

$$\lim_{n\to\infty} \|G_\alpha^{g_n,\mu_n} u - G_\alpha u\|_0 = 0, \qquad u \in L_2(\tilde{\mu}),\ \alpha > 0, \tag{4.7}$$

and

$$\lim_{n\to\infty} (G_\alpha^{g_n,\mu_n} u, v)_1 = (G_\alpha u, v)_1, \quad v \in H_1(M). \tag{4.8}$$

Lemma 4.1 *It holds that*

$$G_\alpha u = \psi^* \circ \pi_{g,\mu} \circ G_\alpha u = G_\alpha \circ \psi^* \circ \pi_{g,\mu} u, \qquad u \in L_2(\tilde{\mu}). \tag{4.9}$$

Proof Denoting $u_n = T_t^{g_n,\mu_n} u$, we have from (4.4) that

$$\mathcal{E}_\alpha^{g_n,\mu_n}(u_n, u_n) \leq \frac{1}{2t}(1+\alpha)\|u\|_{\mu_n}^2.$$

Further (2.7) implies

$$g_n^*(du_n, du_n) = g_n^*(\omega^i, \omega^j) X_i u_n X_j u_n \geq K_n \sum_{i=d'+1}^{d} |X_i u_n|^2$$

with $\omega^i = dx^i$ and $X_i = \partial/\partial x^i$. Thus we have

$$\lim_{n\to\infty} \sum_k \int_{\widetilde{M}} \sum_{i=d'+1}^{d} |X_i u_n|^2(q)\phi_k(q)\mu(dq) = 0. \tag{4.10}$$

On the other hand, for $y = (x, y^{d'+1}, .., y^d), z = (x, z^{d'+1}, .., z^d) \in F_x \cap U^(k)$, we have

$$u_n(y) - u_n(z) = \sum_{i=d'+1}^{d} \int_{z^i}^{y^i} X_i u_n(x, z^{d'+1}, .., z^{i-1}, \xi^i, y^{i+1}, .., y^d) d\xi^i.$$

Since the measure μ_p has a density function with respect to the measure $dy^{d'+1}\ldots dy^{d-d'}$, this ensures

$$\int_{M_g} \int_{F_p \cap U^{(k)}} \int_{F_p \cap U^{(k)}} |u_n(q) - u_n(r)|^2 \mu_p(dq)\mu_p(dr)\psi_*\mu(dp)$$

$$\leq K \sum_k \int_{M_g} \int_{F_p} \sum_{i=d'+1}^{d} |X_i u_n|^2(q)\phi_k(q)\mu_p(dq)\psi_*\mu(dp)$$

for some positive K independent of u_n. Hence it follows from (4.10) that, for $\psi_*\mu$-almost all $p, T_t u(q) = T_t u(r)$ for $\mu_p \times \mu_p$-almost all $q, r \in F_p$. Due to the inequality

$$\int_{\widetilde{M}} |v(q) - \pi_{g,\mu} v \circ \psi(q)|^2 \mu(dq)$$

$$\leq \int_M \int_{F_p} \int_{F_p} |v(q) - v(r)|^2 \mu_p(dq)\mu_p(dr)\psi_*\mu(dp),$$

this implies $T_t u = \psi^* \circ \pi_{g,\mu} T_t u$ in $L_2(\mu)$. Hence we have $G_\alpha u = \psi^* \circ \pi_{g,\mu} \circ G_\alpha u$ in $L_2(\tilde{\mu})$.

Furthermore, noting that $G_\alpha^{g_n,\mu_n}$ is self-adjoint on $L_2(\mu_n)$, we also see that G_α is self-adjoint on $L_2(\mu)$, so that

$$(u, \psi^* \circ \pi_{g,\mu} \circ G_\alpha v)_\mu = (\psi^* \circ \pi_{g,\mu} u, G_\alpha v)_\mu = (G_\alpha \circ \psi^* \circ \pi_{g,\mu} u, v)_\mu.$$

This proves $G_\alpha u = G_\alpha \circ \psi^* \circ \pi_{g,\mu} u$. ∎

Proof of Theorem 2 Note first

$$\mathcal{E}^{g_n,\mu_n}(G_\alpha^{g_n,\mu_n}u, \psi^* \circ \pi_{g,\mu}v) = \sum_k \sum_{i=1}^{d} \sum_{j=1}^{d'} \int_{\widetilde{M}} g_n^{ij}(X_i u_n)(X_j \pi_{g,\mu}v)\phi_k(q)\mu(dq),$$

where $u_n = G_\alpha^{g_n,\mu_n}u$ and $g_n^{ij} = g_n^*(dx^i, dx^j)_q$. Further, for $d'+1 \le i \le d$, $1 \le j \le d'$,

$$\begin{aligned} &\left| \int_{\widetilde{M}} g_n^{ij}(X_i u_n)(X_j \pi_{g,\mu}v)\phi_k(q)\mu_n(dq)\right| \\ &\le \textit{const.}\ \max\{|g_n^*(dx^i, dx^j)_q| : q \in U^{(k)}\}/(K_n)^{1/2}. \end{aligned}$$

Hence, due to (2.8),

$$\lim_{n\to\infty} \sum_k \sum_{i=d'+1}^{d} \sum_{j=1}^{d'} \int_{\widetilde{M}} g_n^{ij}(X_i u_n)(X_j v)\phi_k(q)\mu_n(dq) = 0.$$

Thereby, it follows from (4.8), (4.9) and (B.1) that

$$\begin{aligned} &\lim_{n\to\infty} \mathcal{E}^{g_n,\mu_n}(G_\alpha^{g_n,\mu_n}u, \psi^* \circ \pi_{g,\mu}v) \\ &= \sum_k \sum_{i,j=1}^{d'} \int_{\widetilde{M}} g^*(\omega^i, \omega^j)(X_i G_\alpha u)(X_j \pi_{g,\mu}v)(q)\phi_k(q)\mu(dq) \\ &= \mathcal{E}^{g,\mu}(\pi_{g,\mu} \circ G_\alpha u, \pi_{g,\mu}v), \end{aligned}$$

for each $v \in H_1(\widetilde{M})$. Since

$$\mathcal{E}^{g_n,\mu_n}(G_\alpha^{g_n,\mu_n}u, \psi^* \circ \pi_{g,\mu}v) + \alpha(G_\alpha^{g_n,\mu_n}u, \psi^* \circ \pi_{g,\mu}v)_{\mu_n} = (u, \psi^* \circ \pi_{g,\mu}v)_{\mu_n},$$

and $(u, \psi^* f)_\mu = (\pi_{g,\mu}u, f)_{\psi_*\mu}$ for $u \in C(\widetilde{M})$ and $f \in C(M_g)$, this implies

$$\mathcal{E}^{g,\mu}(\pi_{g,\mu} \circ G_\alpha u, \pi_{g,\mu}v) + \alpha(\pi_{g,\mu} \circ G_\alpha u, \pi_{g,\mu}v)_{\psi_*\mu} = (\pi_{g,\mu}u, \pi_{g,\mu}v)_{\psi_*\mu}.$$

This proves $\pi_{g,\mu} \circ G_\alpha u = G_\alpha^{g,\mu} \circ \pi_{g,\mu}u$, or $G_\alpha u = \psi^* \circ \pi_{g,\mu} \circ G_\alpha u = \psi^* \circ G_\alpha^{g,\mu} \circ \pi_{g,\mu}u$. ∎

We next turn to the proof of Theorem 3, which is clear from Theorem 2 and the next lemma.

Lemma 4.2 *Let $(g,\mu) \in \mathcal{R}_r \times \mathcal{M}(\widetilde{M})$. Then the law of the process associate to the semigroup $T_t = \psi^* \circ T_t^{g,\mu} \circ \pi_{g,\mu}$, $t > 0$ is equal to that of $(\{q(t;q)\}_{t>0}, P)$, i.e., for each $0 < t_1 < t_2 < \cdots < t_N$ and $u_1, u_2, \cdots, u_N \in C(M)$,*

$$\begin{aligned} &E[u_1(q(t_1;q))u_2(q(t_2;q)) \cdots u_N(q(t_N;q))] \\ &= T_{t_N - t_{N-1}}(u_N T_{t_{N-1}-t_{N-2}}(u_{N-1} \cdots T_{t_2-t_1}(u_2 T_{t_1}u_1)))(q), \qquad q \in \widetilde{M}. \qquad (4.11) \end{aligned}$$

Proof We will show the assertion only for N=2. Also we denote as $p_i = \psi(q_i)$ and $p(t;p) = p_{g,\mu}(t;p)$. Then the equality (2.10) with our identification $(p(t;\psi(q)), \xi(t;p(t;\psi(q))))$ with $q(t;q) \in \widetilde{M}$ ensures

$$E[u(q(t;q))] = T_t^{g,\mu} \circ \pi_{g,\mu}u(\psi(q)), \qquad u \in C(M),\ t > 0.$$

Hence it follows that

$$\begin{aligned} &E[u_1(q(t_1;q))u_2(q(t_2;q))] \\ &= \int_M \int_M E[u_1((p_1,\xi(t_1,p_1)))u_2((p_2,\xi(t_2,p_2)));\\ &\qquad\qquad p(t_1;p)\in dp_1, p(t_2;p)\in dp_2] \\ &= T^{g,\mu}_{t_2-t_1}(\pi_{g,\mu}u_2 T^{g,\mu}_{t_1}\pi_{g,\mu}u_1)(q). \end{aligned}$$

This proves (4.11) with N=2.

The proof for a general $N \geq 3$ is similar. ∎

5. Proof of Theorem 4

In this section we will prove Theorem 4. After the proof of Theorem 2, the proof is rather a routine work. But we will give it here for completeness.

Proof of Theorem 4 In this proof we fix a $t > 0$, and denote as $T_t^{(n)} = T_t^{g_n,\mu_n}$, $S_t = T_t^{g,\mu}$ and $\nu = \psi_*\mu$. Let

$$\mathcal{L}_{\mu_n,k} = \{\Lambda : \Lambda \textit{ is a } (k+1)\textit{-dimensional subspace of } \ L_2(\mu_n)\},$$

$$\mathcal{L}_{\nu,k} = \{\Lambda : \Lambda \textit{ is a } (k+1)\textit{-dimensional subspace of } \ L_2(\nu)\}.$$

Then by means of the min-max principle

$$\lambda_k^{(n)} = \inf_{\Lambda\in\mathcal{L}_\mu} \sup_{u\in\Lambda\setminus\{0\}} \left(-\frac{1}{t}\log(\|T_t^{(n)}u\|_{\mu_n}/\|u\|_{\mu_n})\right),$$

$$\lambda_k = \inf_{\Lambda\in\mathcal{L}_{\nu,k}} \sup_{v\in\Lambda\setminus\{0\}} \left(-\frac{1}{t}\log(\|S_t v\|_\nu/\|v\|_\nu)\right).$$

We will first show that

$$\limsup_{n\to\infty} \lambda_k^{(n)} \leq \lambda_k. \tag{5.1}$$

For each $\delta > 0$, there exists a $\Lambda \in \mathcal{L}_{\nu,k}$ such that

$$\sup_{v\in\Lambda\setminus\{0\}} \left(-\frac{1}{t}\log(\|S_t v\|_\nu/\|v\|_\nu)\right) < \lambda_k + \delta.$$

On the other hand, since the finite dimensional set $\{v \in \Lambda : \|v\|_\nu = 1\}$ is compact in $L_2(\nu)$ and $\|v\|_\nu = \|\psi^* v\|_\mu$, we have from Theorem 2 that

$$\limsup_{n\to\infty} \sup_{v\in\Lambda\setminus\{0\}} \left(-\frac{1}{t}\log(\|T_t^{(n)}\circ\psi^* v\|_{\mu_n}/\|\psi^* v\|_{\mu_n})\right) \leq \sup_{v\in\Lambda\setminus\{0\}} \left(-\frac{1}{t}\log(\|S_t v\|_\nu/\|v\|_\nu)\right).$$

But

$$\lambda_k^{(n)} \leq \sup_{v\in\Lambda\setminus\{0\}} \left(-\frac{1}{t}\log(\|T_t^{(n)}\circ\psi^* v\|_{\mu_n}/\|\psi^* v\|_{\mu_n})\right)$$

by means of the min-max principle. Hence we arrive at (5.1).

We will next show the converse inequality. Combining the relation $T_t^{(n)}\varphi_k^{(n)} = \exp(-\lambda_k^{(n)}t)\varphi_k^{(n)}$ with (4.4), we have

$$\begin{aligned}\mathcal{E}^{(n)}(\varphi_k^{(n)},\varphi_k^{(n)}) &= \exp(2\lambda_k^{(n)}t)\mathcal{E}^{(n)}(T_t^{(n)}\varphi_k^{(n)},T_t^{(n)}\varphi_k^{(n)})\\ &\leq \exp(2\lambda_k^{(n)}t)/2t,\end{aligned}$$

where $\mathcal{E}^{(n)} = \mathcal{E}^{g_n,\mu_n}$. Thereby due to (5.1), the system $\{\varphi_k^{(n)}\}_{n\in\mathbf{N}}$ is bounded in $H_1(\tilde{M})$ for each $k \in \mathbf{Z}_+$. Since the set $\{\lambda_k^{(n)}\}_{n\in\mathbf{N}}$ is also bounded in $\mathbf{R}$, the diagonal method assures the existence of a system $\{\varphi_k\}_{k=0}^{\infty} \subset L_2(\tilde{\mu})$, a sequence $\{\lambda_k^{(0)}\}_{k=0}^{\infty} \subset \mathbf{R}_+$ and a subsequence $n_1 < n_2 < \cdots \to \infty$ such that

$$\lim_{m\to\infty} \|\varphi_k^{(n_m)} - \varphi_k\|_{\tilde{\mu}} = 0, \quad \lim_{m\to\infty} \lambda_k^{(n_m)} = \lambda_k^{(0)}, \quad k = 0,1,\cdots. \tag{5.2}$$

Clearly the system $\{\varphi_k\}_{k=0}^{\infty}$ is orthonormal in $L_2(\mu)$. Further, from Theorem 2 and the contraction property of $T_t^{(n)}$,

$$\begin{aligned}&\lim_{m\to\infty} \|T_t^{(n_m)}\varphi_k^{(n_m)} - \psi^* \circ S_t \circ \pi_{g,\mu}\varphi_k\|_{\tilde{\mu}} = 0,\\ &\lim_{m\to\infty} \|\exp(-\lambda_k^{(n_m)}t)\varphi_k^{(n_m)} - \exp(-\lambda_k^{(0)}t)\varphi_k\|_{\tilde{\mu}} = 0,\end{aligned} \tag{5.3}$$

so that $\exp(-\lambda_k t)\varphi_k = \psi^* \circ S_t \circ \pi_{g,\mu}\varphi_k$. This implies that $\varphi_k = \psi^* \circ \pi_{g,\mu}\varphi_k$, $\|\varphi_k\|_\mu = \|\pi_{g,\mu}\varphi_k\|_\nu$ and the system $\{\pi_{g,\mu}\varphi_k\}_{k=0}^{\infty}$ is also orthonormal in $L_2(\nu)$. Now let $\Lambda_k^{(n)} = l.s.[\varphi_0^{(n)},\varphi_1^{(n)},\cdots,\varphi_k^{(n)}]$. Then

$$\begin{aligned}\lambda_k^{(n)} &= \sup\{-\frac{1}{t}\log(\|T_t^{(n)}u\|_{\mu_n}/\|u\|_{\mu_n}) : u \in \Lambda_k^{(n)}\setminus\{0\}\}\\ &= \max_{0\leq i\leq k}\left(-\frac{1}{t}\log(\|T_t^{(n)}\varphi_i^{(n)}\|_{\mu_n}/\|\varphi_i^{(n)}\|_{\mu_n})\right).\end{aligned}$$

Hence, (5.3) and the min-max principle verify

$$\lim_{m\to\infty} \lambda_k^{(n_m)} = \max_{0\leq i\leq k}\left(-\frac{1}{t}\log(\|S_t \circ \pi_{g,\mu}\varphi_i\|_\nu/\|\pi_{g,\mu}\varphi_i\|_\nu)\right) \geq \lambda_k. \tag{5.4}$$

Now (5.1) and (5.4) ensure $\lim_{m\to\infty}\lambda_k^{(n_m)} = \lambda_k$. Since $\{\lambda_k\}$ is independent of the choice of a subsequence $\{n_m\}$, we obtain (2.12).

To show (2.13), note first that $\lambda_k = \lambda_k^{(0)}$, so that $\varphi_k \in E(\lambda_k)$. Then (5.3) implies $\lim_{m\to\infty} d_\mu(\varphi_k^{(n_m)},\psi^*(E(\lambda_k))) = 0$. Thus, repeating the similar argument to that in the above to take subsequences, we obtain (2.13). ∎

REFERENCES

[1] S. Albeverio, S. Kusuoka and L. Streit : Convergence of Dirichlet forms and associated Schrödinger operators, *J. Func. Analysis*, **68**(1986), 130-148.

[2] R. L. Bishop and B. O'neil : Manifolds of negative curvature, *Transactions Amer. Math. Soc.*, **145**(1969), 1-49.

[3] M. I. Freidlin and A. D. Wentzell : *Random Perturbation of Dynamical Systems*, Springer, 1984.
[4] K. Fukaya : Collapsing Riemannian manifolds and eigenvalues of Laplace operator, *Invent. Math.*, **87**(1987), 517-547.
[5] M. Fukushima : *Dirichlet Forms and Markov Processes*, Kodansha/North-Holland, 1980.
[6] M. Fukushima and Y. Oshima : On the skew product of symmetric diffusion processes, *Forum Math.*, **1**(1989), 103-142.
[7] P. B. Gilkey : *Invariance Theory, The Heat Equation, And the Atiya- Singer Index Theorem*, Publish or Perish, Inc., 1984.
[8] M. Gromov : *Structures métriques pour les variétés riemanniennes*, rédigé par J. Lafontaine et P. Pansu, Cedic-Nathan, Paris, 1981.
[9] N. Ikeda and S. Watanabe : The local structure of a class of diffusions and related problems, in *Proceedings of the Second Japan-USSR Symposium on Probability Theory*, Edited by G. Maruyama and Yu. V. Prohorov, Lect. Notes in Math. **330**, Springer, 1973, 124-169.
[10] N. Ikeda and S. Watanabe : *Stochastic Differential Equations and Diffusion Processes*, Kodansha/North-Holland, 1989.
[11] K. Itô and H. P. McKean, Jr. : *Diffusion Processes and their Sample Paths*, Springer, 1965.
[12] Y. Ogura : One-dimensional bi-generalized diffusion processes, *J. Math. Soc. Japan*, **41**(1989), 213-242.
[13] G. C. Papanicolaou, D. Stroock and S. R. S. Varadhan : Martingale approach to some limit theorems, *1976 Duke Turbulence Conference*, Duke Univ. Math. Series III, 1977.
[14] B. L. Reinhart : Foliated manifolds with bundle like metrics, *Ann. Math.*, **69**(1958), 119-132.
[15] F. W. Warner : *Foundations of Differentiable Manifolds and Lie Groups*, Springer, 1983.

Nobuyuki Ikeda
Department of Mathematics
Osaka University
Toyonaka, Osaka 560, Japan

Yukio Ogura
Department of Mathematics
Saga University
Saga 840, Japan

A Stochastic Criterion for Yang–Mills Connections

Seth Stafford

1. Introduction. The application of probability to differential geometry hinges on the intimate connection between the Laplace operator and Brownian motion. The geometric use of the Laplace operator can be thought of as progressing from the study of function theory on manifolds (harmonic functions), to Hodge theory (harmonic forms), to nonlinear variational problems such as harmonic maps (minimizing the energy $\int |df|^2 \mathrm{dVol}$ for maps into a nonlinear manifold) or Yang-Mills theory (essentially nonlinear Hodge theory).

So far, probabilistic techniques have been applied successfully in function theory – for example D. Sullivan's solution of the Dirichlet problem at infinity for manifolds of boundedly negative curvature (cf. [14], and antecedents in [7] and [11]); in Hodge theory – for example the work of K. D. Elworthy and S. Rosenberg on refining the classical vanishing and finiteness theorems of Bochner and Myers (cf. Rosenberg's paper in this volume and [3]); and in harmonic maps – for example W. Kendall's use of Γ-martingales in proving vanishing, uniqueness and (fine)-existence theorems (cf. [5], [6] and references therein, or [13]).

The remaining chapter of this thematic tour of geometry, namely Yang-Mills theory, has remained unexplored. In this note, we make a small step in this direction with a result which was directly inspired by work of M. Liao and M. Pinsky [9]. The result gives a criterion for a connection to be Yang-Mills (i.e. have harmonic curvature) in terms of the stochastic parallel transport of the connection. Unfortunately, this criterion is apparently unverifiable (apart from trivial cases), so much remains to be done if there is to be any meaningful exchange between probability theory and this chapter of geometry. The present result suggests that further progress may involve getting quantitative control of the stochastic holonomy of a connection – a problem raised earlier by K. D. Elworthy and S. Rosenberg in another geometric context (cf. [12], pp.296-7).

The remainder of this note proceeds essentially by adapting the notation and terminology of [9] to a more general context (sections 2 - 7), and adding a new twist in the *dénouement* (section 8). In particular, this new twist shows that a minor modification of the Corollary of [9] (p.210) strengthens the conclusion from 'constant scalar curvature' to 'harmonic curvature'.

2. The Setting. There are few good expositions of Yang-Mills theory, but the reader unacquainted with it may benefit from reading parts of the books by Gabor Tóth [15] and Jürgen Jost [4]. Those too well acquainted to need a review may wish to skip to section 3.

Research partially supported by a graduate stipend from the NSF.

Let M be an n-dimensional Riemannian manifold; G a compact Lie group – for definiteness we work with $G = SO(d)$ for some d, but one may easily generalize to other cases of interest; P a principal G-bundle over M, and V the R^d vector bundle associated to P by the usual matrix representation of $SO(d)$ on R^d. Thus P is an $(n + \frac{1}{2}d(d-1))$-dimensional manifold, G acts freely and properly on P on the right, defining a quotient manifold $M \cong P/G$. V is the quotient of $P \times R^d$ by the action of G defined as follows: on P do as before; on R^d rotate by g^{-1} regarded as a matrix in $SO(d)$. Thus the vector v at point p is carried to $g^{-1}(v)$ at point $p.g$.

Each T_pP has a distinguished subspace T_pG tangential to the G-orbit of p. A connection on P is a (G-equivariant) choice of subspace H_p, so that $T_pP \cong T_pG \oplus H_p$. If $\pi : P \longrightarrow M$ is the quotient map, then $\pi_* : H_p \longrightarrow T_{\pi(p)}M$ is an isomorphism, and π_*^{-1} gives a prescription for lifting tangent vectors on M to tangent vectors on P – and by (Stratonovich) integration, a prescription for lifting (Brownian) paths. In practice, H_p is defined as the kernel of a map $A : T_pP \longrightarrow T_pG$. Since $T_pG \cong Lie(G)$, A is a Lie-algebra-valued one-form on P.

The distribution H_p need not be involutive. That is, tracing out a small parallelogram on M and lifting to P may result in a non-closed path. Both endpoints always project to the same point of M, but may be displaced from one another by an element of G. This phenomenon is measured by the covariant exterior derivative $F = D_A(A)$ of the one-form A, also known as its curvature.

To make this less abstract, trivialize P over some small open set $U \subset M$. This amounts to choosing a map $\kappa : U \longrightarrow P$ such that $\pi \circ \kappa = id|_U$. The image $\kappa_*(T_mM) \subset T_{\kappa(m)}P$ is a complement to $T_{\kappa(m)}G$ in general different from H_p. As we move in direction ∂_μ on M, we move in direction $\kappa_*(\partial_\mu)$ on P. The connection tells us the 'vertical projection' $A(\kappa_*(\partial_\mu))$ of $\kappa_*(\partial_\mu)$ onto $T_{\kappa(m)}G$. Thus as we move in direction $\kappa_*(\partial_\mu)$, we find ourselves drifting in the vertical direction $A(\kappa_*(\partial_\mu))$. To stay horizontal on P, we subtract this component of the motion. If you carry the vector v as you move in direction $\kappa_*(\partial_\mu)$ on P, A tells you to keep v 'parallel' by canceling out the vertical component $A(\kappa_*(\partial_\mu))$. On M we get the equation: $\partial_\mu v - A(\kappa_*(\partial_\mu)) \cdot v = 0$. Since $G = SO(d)$, and acts by matrix multiplication on R^d, $A(\kappa_*(\partial_\mu))$ is just a skew-symmetric matrix, whose components are the symbols $[_\mu A_i^j]$ of A.

The curvature form F takes the 'infinitesimal parallelogram' with sides ∂_μ and ∂_ν and returns the direction of displacement along G as an element $F(\partial_\mu, \partial_\nu)$ of $Lie(G)$. The components of $F(\partial_\mu, \partial_\nu)$ as a matrix are written as $[_{\mu\nu}F_i^j]$.

Because of the way this connection on V is defined, it is compatible with an inner product on the fibers induced by the standard one on R^d – meaning that when we parallel translate a vector v in V along a path in M using A, its norm $|v|$ is defined and is constant along the path.

In theoretical physics, Yang-Mills theories generalize the gauge symmetry of Maxwell's equations – solutions are preserved by (circular) rotations of the electric and magnetic components of the field, to gauge symmetries of other force fields – solutions are preserved by non-abelian rotations of (> 2) component fields.

In modern mathematical language, the electro-magnetic 4-vector potential A is a connection form on a $U(1)$-bundle over space-time, and the electro-magnetic field is its curvature form. Maxwell's equations (*in vacuo*) then say that $D_A * F = 0$, where $*F$ denotes the Hodge dual of F. For general reasons this implies that F is harmonic in the sense of deRham-Hodge ($(D_A D_A^* + D_A^* D_A)F = 0$). $D_A * F$ can be interpreted as a measure of the electric current.

Yang-Mills theory replaces $U(1)$ with some other group, notably $SU(2)$ in the case of the weak nuclear force. Since $Lie(U(1)) = iR$, all of the forms in the last paragraph were Lie-algebra-valued without knowing it. In this formulation, the Yang-Mills equations are identical to Maxwell's equations – but with a non-abelian Lie-algebra in the background. For a fuller explanation of the physics, consult Drechsler and Meyer [2].

3. The Result. Let X_t be the Brownian motion on M generated by the Laplace-Beltrami operator Δ of M, and ξ_t the A-parallel transport of a vector ξ_0 in V_{X_0}. More formally, we lift X_t to V using A starting from $(\kappa(X_0), \xi_0)$, then record this process as (X_t, ξ_t) and regard it as an oddly defined process on $M \times R^d$.

Suppose we start the Brownian motion X_t at m, and wait until it first exits the ball $B(m, r)$ of radius r about m – call this time T_r. Choose a unit vector z in $T_m M$. Using normal coordinates centered at m, we can regard rz as a point of $\partial B(m, r)$. If we condition X_t to hit $\partial B(m, r)$ at the point rz, then we can ask what has happened to some vector ξ_0 in V_m as it is parallel translated from m to rz along the Brownian path. By adopting a suitable gauge, we can compare this vector ξ_{T_r} in V_{rz} to ξ_0 in V_m. In particular, we can define the vector:

$$E(\xi_{T_r} | X_{T_r} = rz)$$

and ask how this vector differs from ξ_0. Since the bundle P has structure group $SO(d)$, ξ_{T_r} can differ from ξ_0 only by a rotation. Hence this conditional expectation is defined and takes some value in the ball of radius $|\xi_0|$ in V_m. We adopt the following definition:

DEFINITION (SPT$_i$ CONDITION). *The connection A is said to satisfy the* **SPT$_i$** *condition at m if*

$$E(\xi_{T_r} | X_{T_r} = rz) = \xi_0 + O(r^i)$$

for every $\xi_0 \in V_m$ and unit vector z of $T_m M$.

In effect, translated vectors feel an average net rotation which vanishes to order i as r tends to 0.

The remainder of this note is dedicated to proving:

PROPOSITION (SPT$_4 \iff$ YM). *A connection A satisfies the SPT$_4$ property at m if and only if it is Yang-Mills at m – i.e. $D_A * F = 0$ at m.*

In fact, SPT$_3$ holds for any connection, so the Yang-Mills current $D_A * F$ makes the dominant contribution to the average net rotation of translated vectors.

4. Radial Gauge. To do any computations we must choose an appropriate coordinate system. For small r we can always trivialize V, so we will work on $B(m, r_0) \times R^d$. We use normal coordinates x^l about m and identify $B(m, r_0)$ with the ball of radius r_0 in $T_m M$. In these coordinates we have convenient bases $\{\partial_l\}$ $(= \frac{\partial}{\partial x^l})$ and $\{dx^l\}$ for TM and T^*M respectively.

Naturally, we can use Cartesian coordinates $\{y_i\}$ for the R^d fibers of $P \times R^d$. To get coordinates on the quotient V, we need to resolve the ambiguity caused by the G action on the fibers. Because $G = SO(d)$, the fibers of '$P/G \times R^d$' are only defined up to a rotation. We resolve this ambiguity by making a smooth choice of orthonormal basis for the fibers. This amounts to choosing a map like κ

defined in section 2. It is not hard to see that one has the freedom to ensure that $\kappa_*(T_mM) = H_{\kappa(m)}$ for any single point m. This choice has the effect of making the symbols $[_\mu A^j_i]$ all vanish at m. So we start with orthonormal bases $\{y_i\}$ ($= \frac{\partial}{\partial y^i}$) for the fibers and coordinates $\{x^l, y^i\}$ for $B(m, r_0) \times R^d$.

To facilitate the comparison of ξ_{T_r} and ξ_0, we will make a gauge transformation a^i_j which will take vectors from the basis $\{y_i\}$ to a new basis $\{Y_i\}$. Thus:

$$Y_i = a^j_i y_j = a^j_i \frac{\partial}{\partial y^j}$$

The gauge transformation a^j_i will satisfy the following gauge condition:

$$x^l \partial_l a^j_i - x^l [_l A^k_i] a^j_k = 0 \tag{4.1}$$

subject to the initial condition that $a^j_i(0) = \delta^j_i$. This equation says that the new basis $\{Y_i\}$ will be parallel along radial geodesics. Hence in this basis the components of ξ_{T_r} will match those of ξ_0 translated first along the Brownian path to rz, then back along the radial geodesic from rz to m – making a direct comparison of ξ_{T_r} and ξ_0 possible.

In this gauge, the connection A will become:

$$[_l \tilde{A}^h_i] = a^j_i [_l A^k_j] b^h_k + [\partial_l a^k_i] b^h_k \tag{4.2}$$

where $a^k_i b^j_k = \delta^j_i$. Note that, because $[_l A^k_j](m) = 0$, $[_l \tilde{A}^k_j](m) = 0$ also.

5. Infinitesimal Generator. We now derive the infinitesimal generator of the joint (X_t, ξ_t) process on $B(m, r_0) \times R^d$. The first step is to write down the relevant stochastic differential equations. Because X_t is the Brownian motion generated by the Laplace-Beltrami operator Δ rather than $\frac{1}{2}\Delta$, as is often used, we have the following s.d.e. for X_t:

$$dX^i_t = \sqrt{2}\, \sigma^i_l dW^l_t - (g^{jk}\Gamma^i_{jk})dt \tag{5.1}$$

where σ^i_l is a square root matrix of g^{jk}, W^l_t denote independent copies of a standard Wiener process, and we take $X^i_0 = 0$. On the other hand, the stochastic parallel transport of ξ_0 is generated by the following s.d.e.:

$$d\xi^i_t = [_l \tilde{A}^i_j] \xi^j_t \circ dX^l_t \tag{5.2}$$

where $\circ$ denotes Stratonovich 'multiplication'. In [9], this equation has a minus sign on the right-hand side. This is inconsistent with the choice of sign for $[_l A^k_i]$ used in formula (4.1), so we have changed it. This produces several sign discrepancies between the formulas here and in [9], but has no effect on the argument.

Using Itô's formula, we may derive the differential generator of the joint (X_t, ξ_t) process from equations 5.1 and 5.2. First, note that the differential generator of the X_t process by itself looks like:

$$\Delta = g^{jk}\partial_j \partial_k - g^{jk}\Gamma^i_{jk}\partial_i \tag{5.3}$$

The differential generator of (X_t, ξ_t) turns out to be:

(5.4) $$L = \Delta$$

(a) $$+ 2[{}_l\tilde{A}^i_k][{}_l\tilde{A}^j_h]\xi^k\xi^h \frac{\partial^2}{\partial\xi^i\partial\xi^j}$$

(b) $$+ 2[{}_\alpha\tilde{A}^i_k]\xi^k \frac{\partial^2}{\partial x^\alpha \partial\xi^i}$$

(c) $$+ \{\partial_l[{}_l\tilde{A}^i_j] + [{}_l\tilde{A}^i_k][{}_l\tilde{A}^k_j]\}\xi^j \frac{\partial}{\partial\xi^i}$$

6. Taylor Expansions. The next step in the Liao-Pinsky program is to expand this operator as:

$$L = \Delta_{-2} + \sum_{k=0}^{\infty} L_k$$

where L_k is an operator which takes a degree d polynomial in the x^i and ξ^j into a polynomial of degree $d + k$. We will then solve a series of Poisson equations involving the lowest order terms of this series. First, in order to expand L we will need to expand the coefficients of L, which in turn involves expanding the gauge transformations and the symbols of the connection.

The original gauge condition on a^j_i (formula 4.1) implies that its Taylor expansion is:

(6.1) $$a^j_i(x) = \delta^j_i - \tfrac{1}{2}x^p x^q \partial_p[{}_qA^j_i](0) - \tfrac{1}{6}x^p x^q x^r \partial_p\partial_q[{}_rA^j_i](0) + O(|x|^4)$$

while its inverse b^l_k will have the expansion:

(6.2) $$b^l_k(x) = \delta^l_k + \tfrac{1}{2}x^p x^q \partial_p[{}_qA^l_k](0) + \tfrac{1}{6}x^p x^q x^r \partial_p\partial_q[{}_rA^l_k](0) + O(|x|^4)$$

The next order of business is to expand $[{}_l\tilde{A}^i_j]$. Bearing in mind the fact that $[{}_lA^i_j]$ vanishes to first order at x = 0, we substitute into formula 4.2 to obtain:

(6.3) $$\begin{aligned}[{}_l\tilde{A}^i_j](x) &= \tfrac{1}{2}\left[\partial_p[{}_lA^i_j](0) - \partial_l[{}_pA^i_j](0)\right]x^p \\ &+ \tfrac{1}{3}\left[\partial_p\partial_q[{}_lA^i_j](0) - \partial_p\partial_l[{}_qA^i_j](0)\right]x^p x^q \\ &+ O(|x|^3)\end{aligned}$$

This formula further simplifies, when we use the definition of the curvature of the connection A:

(6.4) $$[{}_{ij}F^k_l] = \partial_i[{}_jA^k_l] - \partial_j[{}_iA^k_l] + [{}_iA^k_h][{}_jA^h_l] - [{}_jA^k_h][{}_iA^h_l]$$

which upon making the appropriate substitutions gives us:

(6.5) $$\begin{aligned}[{}_l\tilde{A}^i_j](x) &= \tfrac{1}{2}[{}_{pl}F^i_j](0)x^p \\ &+ \tfrac{1}{3}\left[\partial_p[{}_{ql}F^i_j](0)\right]x^p x^q \\ &+ O(|x|^3)\end{aligned}$$

Now we are equipped to compute the decomposition of L. Although Δ will in general have many non-zero terms (cf. [8]), we will not deal with them explicitly here. Instead, we will decompose only $L^* = L - \Delta$.

First observe that all of the terms of L (formula 5.4) which involve derivatives with respect to ξ^j are of degree at least -1. Hence $L_{-2} = \Delta_{-2}$. The only term of degree -1 will arise from the constant term of the expansion of 5.4(b), but since the symbols of the connection vanish at 0, this term will be zero. Hence $L_{-1} = \Delta_{-1}$, which in turn happens to be zero [8].

Terms of degree zero will arise from the constant terms of 5.4(a) and 5.4(c), and the first order term of 5.4(b). Term 5.4(a) vanishes to second order at 0, however, because each copy of the connection vanishes to first order there. Likewise, only the first half of 5.4(c) makes a zero order contribution since the second half vanishes to second order. From 6.5 we see that $\partial_l[{}_l\tilde{A}^i_j](0) = \frac{1}{2}[{}_{ll}F^i_j](0)$, but $[{}_{ij}F^l_k]$ is antisymmetric in i and j, as is clear from 6.4, so 5.4(c) actually makes no contribution. The first order term of 5.4(b), however, makes a non-zero contribution, and we obtain:

$$L_0 = \Delta_0 + \left[[{}_{p\alpha}F^i_k](0)\right] x^p \xi^k \frac{\partial^2}{\partial x^\alpha \partial \xi^i} \tag{6.6}$$

Contributions to L_1 will come from the second order term of 5.4(b) and the first order terms of 5.4(a) and 5.4(c). Again 5.4(a) and the second part of 5.4(c) make no contribution in first order. Assembling the non-zero terms we obtain:

$$\begin{aligned}
&(6.7) && L_1 = \Delta_1 \\
&(a) && \quad + \tfrac{2}{3}\left[\partial_q[{}_{p\alpha}F^i_k](0)\right] x^p x^q \xi^k \frac{\partial^2}{\partial x^\alpha \partial \xi^i} \\
&(b) && \quad + \tfrac{1}{3}\left[\partial_l[{}_{pl}F^i_j](0)\right] x^p \xi^j \frac{\partial}{\partial \xi^i}
\end{aligned}$$

These terms give us enough information for present purposes.

7. The Perturbation Argument. We want to compute certain conditional expectations at the first exit time T from a small ball. If the radius r of the ball were fixed, we would have the formula:

$$E[g(X_T)] = u(0) \tag{7.1}$$

where u solves the Dirichlet problem: $\Delta u = 0$ in the interior of the ball, and $u = g$ on the boundary sphere. When r is allowed to vary, this formula can be generalized (cf. [8], [9], [10]) to:

$$E[g(X_{T_r}/r)] = u(0) + r^2 u_0(0) + r^3 u_1(0) + O(r^4) \tag{7.2}$$

where u_i solve the Dirichlet problems:

$$\begin{aligned}
&(D1) && \Delta_{-2}u = 0 \quad \text{in } B_r; && u = g \quad \text{on } \partial B_r \\
&(D2) && \Delta_{-2}u_0 + \Delta_0 u = 0 \quad \text{in } B_r; && u_0 = 0 \quad \text{on } \partial B_r \\
&(D3) && \Delta_{-2}u_1 + \Delta_1 u = 0 \quad \text{in } B_r; && u_1 = 0 \quad \text{on } \partial B_r
\end{aligned}$$

A small further generalization (cf. [9]) gives us:

$$E[g(X_{T_r}/r, \xi_{T_r})] = w(0,\xi) + r^2 w_0(0,\xi) + r^3 w_1(0,\xi) + O(r^4) \tag{7.3}$$

for any fixed ξ, where w_i solve the Dirichlet problems:

(Da) $\Delta_{-2} w = 0$ in $B_r \times \{\xi\}$; $w = g$ on $\partial B_r \times \{\xi\}$

(Db) $\Delta_{-2} w_0 + L_0 w = 0$ in $B_r \times \{\xi\}$; $w_0 = 0$ on $\partial B_r \times \{\xi\}$

(Dc) $\Delta_{-2} w_1 + L_1 w = 0$ in $B_r \times \{\xi\}$; $w_1 = 0$ on $\partial B_r \times \{\xi\}$

To apply this formula, one needs to evaluate $w_i(0)$ somehow. To do this, we use a generalization of Pizetti's formula (cf. [8]) which states that if $\Delta_{-2} f + p = 0$ in B_r, $f = 0$ on ∂B_r, and $\Delta_{-2}^{(k)} p = 0$ for some k then:

$$f(0) = \sum_{h=1}^{k} \left(\frac{1}{2^h (h!) n(n+2)...(n+2h-2)}\right) \Delta_{-2}^{(h-1)} p(0) \tag{P}$$

Our next step will be to find suitable formulas for the $w_i(0)$. Note that we can divide the task for L_i into one for Δ_i and another for $L_i - \Delta_i$. Each $w_i(0)$ will have two parts corresponding to Δ_i and $L_i - \Delta_i$ respectively. Again, we will not deal with the Δ_i expressions, as they are given in [8]. We denote $L_i - \Delta_i$ by L_i^*.

To apply formula (P) to L_0^*, we must check that some power of Δ_{-2} applied to $L_0^* w$ vanishes. Computing, we find that:

$$\begin{aligned} \Delta_{-2}(L_0^* w) = {} & 2\left[[{}_{a\alpha}F_k^i](0)\right]\xi^k \frac{\partial^2}{\partial x^\alpha \partial \xi^i}\left(\frac{\partial w}{\partial x^a}\right) \\ & + \left[[{}_{p\alpha}F_k^i](0)\right] x^p \xi^k \frac{\partial^2}{\partial x^\alpha \partial \xi^i}(\Delta_{-2} w) \end{aligned}$$

and since $\Delta_{-2} w = 0$, the second term drops out. Applying Δ_{-2} to the first term above we get:

$$\Delta_{-2}^{(2)}(L_0^* w) = 2\left[[{}_{a\alpha}F_k^i](0)\right]\xi^k \frac{\partial^2}{\partial x^\alpha \partial \xi^i}\left(\frac{\partial}{\partial x^a}(\Delta_{-2} w)\right)$$

which is clearly zero. Forming the sum in (P) gives:

$$\begin{aligned} w_0(0) = {} & \text{terms involving } \Delta_0 \\ & + \frac{1}{4n(n+2)}\left[[{}_{a\alpha}F_k^i](0)\right]\xi^k \frac{\partial^2}{\partial x^\alpha \partial \xi^i}\left(\frac{\partial w}{\partial x^a}\right) \end{aligned}$$

The $\Delta_{-2}^{(0)}(L_0^* w)(0)$ term of (P) vanishes at $x^p = 0$, and is not shown. Moreover, the second term above is zero because the curvature is antisymmetric in a and α while the partial derivatives are symmetric in these indices. Thus $w_0(0)$ only depends on Δ.

Now we apply the same procedure to L_1^*. First we compute:

$$\begin{aligned} \Delta_{-2}(L_1^* w) = {} & + \tfrac{4}{3}\left[\partial_a[{}_{a\alpha}F_k^i](0)\right]\xi^k \frac{\partial^2 w}{\partial x^\alpha \partial \xi^i} \\ & + \tfrac{4}{3}\left[\partial_a[{}_{p\alpha}F_k^i](0)\right] x^p \xi^k \frac{\partial^2}{\partial x^\alpha \partial \xi^i}\left(\frac{\partial w}{\partial x^a}\right) \\ & + \tfrac{2}{3}\left[\partial_l[{}_{al}F_j^i](0)\right]\xi^j \frac{\partial^2 w}{\partial x^a \partial \xi^i} \end{aligned}$$

The third term has the same form as the first, if we substitute a for l, α for a, and k for j. These two terms then coalesce, leaving:

$$\begin{aligned}\Delta_{-2}(L_1^* w) = &+ \tfrac{2}{3}\left[\partial_a[{}_{a\alpha}F^i_k](0)\right]\xi^k \frac{\partial^2 w}{\partial x^\alpha \partial \xi^i} \\ &+ \tfrac{4}{3}\left[\partial_a[{}_{p\alpha}F^i_k](0)\right]x^p \xi^k \frac{\partial^2}{\partial x^\alpha \partial \xi^i}\left(\frac{\partial w}{\partial x^a}\right)\end{aligned}$$

Writing out $\Delta_{-2}^{(2)}(L_1^* w)$ produces three terms. Two of these will look like 'operator applied to $\Delta_{-2}w$', the other one will vanish due to the antisymmetry of the curvature, hence $\Delta_{-2}^{(2)}(L_1^* w)$ is zero. We next form the sum in (P), suppressing the $\Delta_{-2}^{(0)}(L_1^* w)(0)$ term since it is zero:

$$\begin{aligned} w_1(0) = &\quad \text{terms involving } \Delta_1 \\ &+ \frac{1}{12n(n+2)}\left[\partial_a[{}_{a\alpha}F^i_k](0)\right]\xi^k \frac{\partial^2 w}{\partial x^\alpha \partial \xi^i}(0)\end{aligned}$$

The second term of $\Delta_{-2}(L_1^* w)$ vanishes, so it is not shown.

When we assemble the sum in (7.3) we find that $w(0,\xi)$ and $w_0(0,\xi)$ do not depend on the ξ_t part of the (X_t, ξ_t) process. We can collect all of the terms attributable to Δ_k in one expression, and the one non-zero term attributable to L_k^* which we have computed in another leaving:

$$\begin{aligned} E[g(X_{T_r}/r, \xi_{T_r})] = &\ E[g(X_{T_r}/r, \xi_0)] \\ &+ r^3\left[\frac{1}{12n(n+2)}\left[\partial_a[{}_{a\alpha}F^i_k](0)\right]\xi_0^k \frac{\partial^2 w}{\partial x^\alpha \partial \xi^i}(0)\right] \\ &+ O(r^4)\end{aligned} \tag{7.4}$$

Because we are now evaluating the partial derivatives of w at the point 0, we can use a 'Poisson kernel' representation of this quantity:

$$\frac{\partial^2 w}{\partial x^\alpha \partial \xi^i}(0) = \int_{\partial B_r \times \{\xi\}} y^\alpha \frac{\partial g(y)}{\partial \xi^i} dI(y) \tag{7.5}$$

where dI denotes the exit distribution of $((X_t), \xi)$ on $\partial B_r \times \{\xi\}$, starting from $(0, \xi)$.

8. The Proof. We want to study the expectation of each component $\xi^i_{T_r}$ of ξ_{T_r} conditioned on the event that $X_{T_r} = rz \in \partial B_r$. To do this we apply formula 7.4 to a special choice of $g(x,\xi)$ defined as follows. First take a sequence of functions ψ_N on $\partial B(m, r_0)$ such that $\psi_N \Longrightarrow \delta_{r_0 z}$. Next extend these functions to $\phi_N(x)$ on $B(m, r_0) - \{0\}$ by $\phi_N(x) = \psi_N(\frac{r_0 x}{|x|})$. Finally set $g(x,\xi) = \phi_N(x)\xi^i$. From (7.4) and (7.5) we get:

$$\begin{aligned} E[\phi_N(X_{T_r}/r)\xi^i_{T_r}] = &\ E[\phi_N(X_{T_r}/r)]\xi^i_{T_r} \\ &+ r^3\left[\frac{1}{12n(n+2)}\left[\partial_a[{}_{a\alpha}F^i_k](0)\right]\xi_0^k \int_{\partial B_r \times \{\xi\}} y^\alpha \frac{\partial \phi_N(y)}{\partial \xi^i} dI(y)\right] \\ &+ O(r^4)\end{aligned} \tag{8.1}$$

In the limit as $N \longrightarrow \infty$, we obtain:

THEOREM.

$$E[\xi^i_{T_r}|X_{T_r} = rz] = \lim_{N\longrightarrow\infty} \frac{E[\phi_N(X_{T_r}/r)\xi^i_{T_r}]}{E[\varphi_N(X_{T_r}/r)]} \tag{8.2}$$

$$= \xi^i_0$$

$$+ r^3\left[\frac{1}{12n(n+2)}[\partial_a[_{a\alpha}F^i_k](0)]\xi^k_0 z^\alpha\right] \tag{$B_z(\xi)$}$$

$$+ O(r^4)$$

Note that B_z is antisymmetric in i and k, so $B_z(\xi)$ is orthogonal to ξ.

We are now prepared to prove the Proposition stated in section 3.

COROLLARY ($\mathrm{SPT}_4 \Longleftrightarrow \mathrm{YM}$).

PROOF: SPT_4 says that B_z must vanish. Literally, this means

$$\partial_a[_{a\alpha}F^i_k] = 0$$

for all i, k, and α. When M is flat, the Yang-Mills equations read:

$$(D_A * F)_\alpha = \partial_a[_{a\alpha}F^i_k] + [[_aA^i_k],[_{a\alpha}F^i_k]] = 0$$

where $[..,..]$ denotes the commutator of two matrices – recall that A and F are antisymmetric matrices in the indices i and k. Since A vanishes at m, the second term disappears, leaving a multiple of B_z. When M is not flat, $D_A * F$ gains an additional zero order term whose coefficient involves derivatives of the metric g^{ij} and $\sqrt{\det g}$. Because we are using normal coordinates centered at m, these derivatives vanish at m, and again we are left with nothing but B_z. •

In the case studied by Liao and Pinsky in [9], the bundle P is the orthonormal frame bundle of M, V is TM, and the connection A is the Levi-Civita connection from the metric on M. This connection is then used to transport vectors from a point m to small spheres about m. Their Corollary (p.210) imposes a condition like SPT_4 for every direction z, but only for those vectors in T_mM (read V_m) which 'point at' rz – in the sense that they are tangent to the geodesic joining m to rz. This hypothesis, with a little help from the Bianchi identities, allows them to conclude that M must have constant scalar curvature. In the Corollary above, a condition is imposed which must hold for every direction z (indexed by α) *and* every vector ξ_0 (indexed by i). This gives the stronger conclusion that M has harmonic curvature (cf. [1] for a discussion of manifolds with harmonic curvature).

REFERENCES

[1] J-P. Bourguignon, *Metrics with harmonic curvature*, in "Global Riemannian Geometry," Ellis Horwood/Wiley, New York, 1984, pp. 18-26.

[2] W. Drechsler & M. E. Meyer, "Fiber-bundle techniques in gauge theory," Springer-Verlag, New York, 1977.

[3] K. D. Elworthy & Steven Rosenberg, *Generalized Bochner theorems and the spectrum of complete manifolds*, Acta Appl. Math. **12** (1988), 1-33.

[4] Jürgen Jost, "Nonlinear methods in Riemannian and Kählerian geometry," Birkhäuser, Boston, 1988.

[5] Wilfred S. Kendall, *Martingales on manifolds*, in "Geometry of Random Motion," Contemporary Math. **73**, American Mathematical Society, Providence RI, 1987, pp. 121–157.

[6] ______, *Probability, convexity and harmonic maps with small image, I: uniqueness and fine existence*, Journal of the London Math. Soc. (1990) (to appear).

[7] Yu. I. Kifer, *Brownian motion and harmonic functions on manfolds of negative curvature*, Theory of Probability and its Applications **21** (1976), 81-95.

[8] Ming Liao, *Hitting distributions of small geodesic spheres*, Annals of Probability **16** (1988), 1039-1050.

[9] Ming Liao & Mark Pinsky, *Stochastic parallel translation for Riemannian Brownian motion conditioned to hit a fixed point of a sphere*, in "Geometry of Random Motion," Contemporary Math. **73**, American Mathematical Society, Providence RI, 1987, pp. 203–210.

[10] Mark Pinsky, *On non-Euclidean harmonic measure*, Ann. Inst. Henri Poincaré Sect. B (N.S.) **21** (1985), 39-46.

[11] J-J. Prat, *Étude asymptotique et convergence angulaire du mouvement brownien sur une variété à courbure negative*, C.R. Acad. Sci. Paris Ser A-B **280** (1975), 1539-1542.

[12] Steven Rosenberg, *Semigroup domination and vanishing theorems*, in "Geometry of Random Motion," Contemporary Math. **73**, American Mathematical Society, Providence RI, 1987, pp. 287–302.

[13] Seth Stafford, *A probabilistic proof of S-Y. Cheng's Liouville theorem*, Annals of Probability (1990) (to appear).

[14] Dennis Sullivan, *The Dirichlet problem at infinity for negatively curved manifolds*, Journal of Differential Geometry **18** (1983), 723–732.

[15] Gabor Tóth, "Harmonic and Minimal Maps," Ellis Horwood/Wiley, New York, 1984.

Seth Stafford
Department of Mathematics
White Hall
Cornell University
Ithaca, New York
14853

Part V
General Theory of Processes

Dirichlet Forms and Markov Fields—A Report on Recent Developments

Sergio Albeverio, Zhi Ming Ma, and Michael Röckner

Abstract

We give a survey of recent developments in the study of Dirichlet forms. We present a general approach to the problem of constructing a right process to a given Dirichlet form on a metrizable space, without assumption of regularity. We also discuss Dirichlet forms on Souslin spaces which lead to diffusions with infinite dimensional state space, satisfying stochastic differential equations. The latter processes are associated to certain homogeneous Markov fields, which we also briefly discuss.

Introduction

In this report we shall discuss some recent developments in the theory of Dirichlet forms and Markov fields. The theory of Dirichlet forms can be seen as a natural extension to multi-dimensional space of the general Feller theory of one dimensional (generalized) diffusions. It can also be seen as a natural L^2-space extension of classical and axiomatic potential theories, as formulated in a space of continuous functions (and their duals). The origins of the theory of Dirichlet forms are in classical potential theory, but Beurling-Deny's work was fundamental, as well as Fukushima-Silverstein's work, see [Fu1-4], [Si1-2] and references therein.

The main features of the theory is a fruitful and general interplay between (functional) analytic-potential theoretic tools and probabilistic tools, extending in the symmetric case (i.e. the case of time reflection invariant processes) the classical correspondence between Laplacian and Brownian motion to very general singular operators and processes. E.g. the processes associated with Dirichlet forms over finite dimensional spaces corresponding to second order elliptic (possibly degenerate, with singular coefficients) operators are diffusions satisfying (weakly) stochastic differential equations with singular drifts and diffusion coefficients.

The theory of Dirichlet forms and associated processes is however not limited to the case of finite dimensional state space, in fact it can be extended to the case of infinite dimensional state space.

The generality of the theory makes Dirichlet forms a most adequate tool to handle the kind of situations including singularities and/or infinite dimensionality which arise in the study of quantum mechanics resp. quantum field theory and in the study of multidimensional homogeneous Markov fields.

In the first part of this paper we shall report on some new developments in the general theory of Dirichlet forms. In the second part we discuss developments concerned with Dirichlet forms of the local type – leading to diffusions – on infinite dimensional spaces – and we shall illustrate their relations to recent developments in the study of homogeneous Markov fields. We also take the opportunity to mention briefly some new intrinsic developments in the theory of homogeneous Markov fields, in their connection with the fascinating problem of the mathematical construction of (relativistic) quantum fields.

1. Dirichlet forms

Let us recall briefly some basic definitions concerning Dirichlet forms.

Let $\mathcal{H}$ be a real Hilbert space. We call *symmetric form* on $\mathcal{H}$ a positive, symmetric, bilinear form $\mathcal{E}$ defined on a dense linear subset $D(\mathcal{E})$ of $\mathcal{H}$ (i.e. $\mathcal{E}$ maps $D(\mathcal{E}) \times D(\mathcal{E})$ bilinearly into $\mathbb{R}$ and $\mathcal{E}(u,v) = \mathcal{E}(v,u), \mathcal{E}(u,u) \geq 0 \quad \forall u,v \in D(\mathcal{E})$).

Let $\mathcal{E}_1$ be the symmetric form defined by $\mathcal{E}_1(u,v) = \mathcal{E}(u,v) + (u,v)$, with (u,v) the scalar product of $u,v \in D(\mathcal{E})$ in $\mathcal{H}$. $\mathcal{E}_1$ defines a scalar product in $D(\mathcal{E}_1)$. $\mathcal{E}$ is closed (resp. closable) if $D(\mathcal{E})$ is complete with respect to the topology defined by $\mathcal{E}_1$ (resp. $\mathcal{E}$ has a closed extension). One shows easily, see e.g. [ABR], that if $(\mathcal{E}^x, D(\mathcal{E}^x))$ are closed (resp. closable) symmetric forms, for all x in some ordered index set I, then the quadratic form $\mathcal{E}[u] \equiv \sup_x \mathcal{E}^x(u,u)$ with domain $D(\mathcal{E}) \equiv \{u \in \bigcap_x D(\mathcal{E}^x) | \sup \mathcal{E}^x(u,u) < \subset\}$ is again closed (resp. closable).

There is a 1 - 1 correspondence between symmetric forms $\mathcal{E}$ and positive self-adjoint operators A in $\mathcal{H}$ given by $\mathcal{E} \leftrightarrow A$ with

$$\mathcal{E}(u,v) = (A^{\frac{1}{2}}u, A^{\frac{1}{2}}v) \quad \forall u \in D(A^{\frac{1}{2}}) = D(\mathcal{E}). \tag{1.1}$$

Such operators A are in 1 - 1 correspondence with strongly continuous *contraction semigroups* $T_t, t > 0$, in the sense that A is the infinitesimal generator of such a semigroup $T_t := e^{-tA}$. Hence also symmetric forms $\mathcal{E}$ are in 1 - 1 correspondence with contraction semigroups.

Now let $\mathcal{H}$ be an L^2- space over a σ- finite measure space (E, B, μ) i.e.

$$\mathcal{H} = L^2(E, B, \mu) \equiv L^2(\mu).$$

A *Markov semigroup* $T_t, t > 0$, is a contraction semigroup s.t.

$$0 \leq u \leq 1 \ \mu - \text{ a.e. } \Rightarrow 0 \leq T_t u \leq 1 \mu \text{ - a.e.} \forall u \in L^2(\mu).$$

What are the symmetric forms which are in 1 - 1 correspondence with Markov semigroups? The answer is given precisely by the *Dirichlet forms*: a closed symmetric form on $\mathcal{H} = L^2(\mu)$ is called Dirichlet if it has the contraction property i.e. $u^\sharp \in D(\mathcal{E})$ and $\mathcal{E}(u^\sharp, u^\sharp) \leq \mathcal{E}(u, u) \forall u \in D(\mathcal{E})$, where $u^\sharp \equiv (u \vee 0) \wedge 1$

Remark There are many other equivalent formulations of this property, e.g. $\mathcal{E}(\varphi \circ u, \varphi \circ u) \leq \mathcal{E}(u, u) \quad \forall u \in D(\mathcal{E}), \quad \varphi$ being any Lipschitz continuous map on $\mathbb{R}$ with $\varphi(0) = 0$. A basic result (see [Fu1] : note however that the additional topological assumptions in [Fu1] are not needed here, cfr. [AR4] (Sect. 1)) states that the above relation (1.1) gives a 1 - 1 correspondence between Dirichlet forms and Markov semigroups.

Remark A direct characterization of generators A of Markov semigroups T_t, is A self-adjoint and

$$(Au, (u - 1) \vee 0) \geq 0 \quad \forall u \in D(A).$$

See [BH1-2].

We call *pre-Dirichlet forms* symmetric forms which are closable and such that their closure is a Dirichlet form.

Remark A useful criterium for a symmetric form to be a pre-Dirichlet form can be formulated in terms of a family of smooth modifications $\varphi_\mathcal{E}$ of the cut-off function φ defined by

$$\varphi(x) = (x \vee 0) \wedge 1, \ x \in \mathbb{R} \, ,$$

see [Fu1], [ABR].

Remark There are several structure theorems for Dirichlet forms: e.g. if E is locally compact separable and $D(\mathcal{E})$ contains the $C_0(E)$ functions then $\mathcal{E}$ decomposes in the sum of a diffusion, a killing and a jumping part, (Beurling - Deny's result, cfr. [Fu1]).

Extensions to cases where E is infinite dimensional can be given [ABR]. A representation of local Dirichlet forms (i.e. such that $\mathcal{E}(u,v) = 0$ if supp $u \cap$ supp $v = \emptyset$, supp u, supp v compacts) on locally compact separable spaces as "gradient forms" associated with certain Hilbert spaces is also possible [AHKRS] (which is based on an abstract representation in [Fu4]).
The question arises whether one can associate to a Dirichlet form, hence a Markov semigroup, some nice Markov process X_t on E, at least when E is some nice topological space.

If E is a locally compact separable space and $\mathcal{E}$ is a *regular Dirichlet form*, in the sense that the continuous function of compact support are dense in $C_0(E)$ w.r.t. the supremum and in $D(\mathcal{E})$ w.r.t. the $\mathcal{E}_1$- norm, then by Fukushima - Silverstein's theory (cfr. [Fu1]) one can indeed associate to $\mathcal{E}$ a *Hunt process* X_t s.t.

$$(T_t u)(x) = E^x(u(X_t))\mu \text{ a.e. } \quad x \in E\,,$$

for all $u \in L^2(\mu)$ (where T_t is the unique Markov semigroup corresponding to $\mathcal{E}$), see [Fu1] for details (in fact the association is proper in the sense of [Fu2] i.e. $x \longrightarrow E^x(u(X_t))$ for u in $L^2(\mu)$ is a quasi continuous modification of $T_t u$.

Also by Fukushima's theory, $u(X_t) - u(X_0)$ for $u \in D(\mathcal{E})$ has a Doob - Meyer type decomposition in a martingale part (continuous additive functional) and a zero energy part (continuous additive functional of zero energy). If $\mathcal{E}$ is local, then X_t is a diffusion, cfr.[Fu1]. A particularly interesting case is given by $E = \mathbb{R}^d, \mathcal{E}$ the closure of the pre-Dirichlet form

$$\mathcal{E}^0(u,v) = \frac{1}{2}\int \nabla u \cdot \nabla v d\mu, \quad u,v \in C_0^\infty(\mathbb{R}^d)$$

(μ a Radon measure assumed to be such that $\mathcal{E}^0$ is closable, see below for closability criteria). In this case under a weak additional condition on μ (namely $d\mu = \rho dx, \beta_i = \frac{\partial_i \rho^{\frac{1}{2}}}{\rho^{\frac{1}{2}}}$ in distributional sense is in $L^2(\mathbb{R}^d)$) one can show that X_t satisfies the stochastic differential equation

$$dX_t = \beta(X_t)dt + dw_t,$$

with β the vector with component β_i, see [AHKS], [Fu1-3].

The whole theory sketched here concerning associating a nice process to the Dirichlet form is based on two basic assumptions:

(a) regularity of $\mathcal{E}$,
(b) E locally compact separable.
In recent years extensions of the theory in the sense of dropping (a),(b) or both have been studied. We report here briefly on corresponding results. We shall now discuss separately two recent approaches to such extensions, starting with one concerned with removing (a).

2. Nonregular Dirichlet forms and associated processes

We shall report here mainly on a line of research pursued in recent years in the publications [AM1-9], [AMR]. The aim is to develop a theory of Dirichlet forms on general metrizable spaces, not necessarily locally compact, without the regularity assumption. Before we start describing our work on this approach, let us mention some related work. Fukushima in [Fu4] and Silverstein in [Si2] already remarked that a suitable compactification method permits to reduce a part of the analytic theory for Dirichlet forms on non locally compact spaces to the case of regular Dirichlet forms on locally compact spaces. However the construction of an associated process in the original space meets difficulties, the overcoming of which needs additional assumptions, cfr. [AKH3], [Ku], [AR3], [AR6,7] for the case of Dirichlet forms of diffusion type on locally convex spaces. Only recently [AR6], [Sch] for such forms, a general construction has been achieved, on which we report in sect.3. The analytic side and a functional calculus has been developed for general Dirichlet forms on σ- finite measurable spaces by Bouleau- Hirsch [BH1,2], who gave striking applications in particular to finite dimensional stochastic differential equations with Lipschitz coefficients.

A different point of view of starting from a given nice process and recovering the associated non necessarily regular Dirichlet form has been pursued by several authors, in particular [D1],[Fi]. We are going to present here an approach by which it is possible to associate "directly" a process to a given nonregular Dirichlet form, without going to a compactification. This will permit to cover in particular cases of Dirichlet forms over $\mathbb{R}^d$ which have no nontrivial continuous function in their domain, hence are certainly not regular. That such forms exist is illustrated by the following observation, see [AM1,2] for details.

Let $(\mathcal{E}, D(\mathcal{E}))$ be a regular Dirichlet form on $L^2(E;\mu)$, where (as in Fukushima's setting [Fu1]), E is locally compact separable metric and μ is a positive Radon measure with supp $\mu = E$. There is a μ- symmetric Hunt process X_t associated with $(\mathcal{E}, D(\mathcal{E}))$, see [Fu1]. To any positive continuous additive functional A (in the sense of [Fu1]) there is a unique positive Borel measure ν on E, called Revuz measure of A, s.t.

$$\lim_{t \downarrow 0} \frac{1}{t} E_{h \cdot \mu} \int_0^t f(X_s) dA_s = \int h(x)(f \cdot \mu)(dx),$$

for all Borel positive functions f and all γ- excessive (w.r.t. X_t) functions h, $\gamma \geq 0$ (h is a γ-excessive function w.r.t. X_t if $h \geq 0$ and $e^{-\gamma t} E_x h(X_t) \uparrow h(x)$ as $t \downarrow 0$, $\forall x \in X$.) Let $\mathcal{S}$ be the set of all Revuz measures, called "smooth measures". S is in 1 - 1 correspondence with the set of all positive continuous additive functionals. From [Fu1] one knows that a Borel measure μ on E is smooth iff it charges no set of capacity zero (the capacity being the one associated with $\mathcal{E}$) and μ is tight on the sense that there exists an increasing sequence of compact subsets K_n of E, s.t.$\mu(K_n) <\subset$, $\mu(E - \bigcup K_n) = 0$, and $\lim_{n \longrightarrow \infty}$ Cap $(K - K_n) = 0$ for any compact set K. It follows that $\mathcal{S}$ contains in particular all positive Radon measures which charge no set of capacity zero. It is also known that any measure $\nu \in \mathcal{S}$ can be approximated by Radon measures of finite energy integral (in the sense that there exists a sequence $\{F_n\}$ of increasing closed sets s.t. $\mu(X - \bigcup F_n) = 0$, lim cap $(K - F_n) = 0$ for any compact K and $\chi_{F_n} \mu \in \mathcal{S}_0 \quad \forall n$). It was shown in [AM2] that the following Proposition holds:

Proposition *To any subset B of E of zero capacity and any smooth measure ν s.t. supp $\nu \supset \bar{B}$ there exists at least one smooth measure $\tilde{\nu}$ which is equivalent to ν and $\tilde{\nu}(G) =\subset$ for all open sets G s.t. $G \cap B \neq \emptyset$.* ■

Let us call a measure μ nowhere Radon if $\mu(G) = \infty$ for any non-empty open set G. In spaces where each point is a set of capacity zero, it is then easy to construct many nowhere Radon smooth measures by the above Proposition, see [AM1,2]. E.g. if $E = \mathbb{R}^d, d \geq 2, \{x_j\}$ is a dense countable subset of $\mathbb{R}^d$ and $\mathcal{E}$ is the classical Dirichlet form $\frac{1}{2} \int |\nabla u|^2 dx$ associated with Brownian motion on $\mathbb{R}^d$, then for any sequence of real numbers $\{\alpha_j\}$, one can find a sequence of strictly positive numbers $\{c_j\}$ such that $\mu(dx) := V(x)dx$, with $V(x) = \sum c_j |x - x_j|^{-\alpha_j}$, is a smooth measure (see [AM1,2]). Obviously if $\alpha_j \geq d$, $\forall j \geq 1$, then μ is nowhere Radon.

On the other hand one can approximate (in the above sense,with F_n compacts) every smooth measure (also those which are nowhere Radon) by smooth measures (of finite energy) in Kato's class (in the sense of [AM1,2,3]; in the case where $\mathcal{E}$ is the Dirichlet form associated with the Brownian motion on $\mathbb{R}^d$, the measures absolutely continuous with respect to Lebesgue measure which are in Kato's class are precisely those which have densities in the Kato class of functions as defined in [AS]).

Remark If $\mathcal{E}$ is a Dirichlet form in a $L^2(E;m)$ space, with m σ- finite and

$$\mathcal{E}^\mu(u,v) \equiv \mathcal{E}(u,v) + \int uvd\mu \ , \ \forall u,v \in D(\mathcal{E}^\mu)$$

for μ a nowhere Radon smooth measure, with $D(\mathcal{E}^\mu) := D(\mathcal{E}) \cap L^2(E,\mu)$, then by [AM3] (Prop. 3.1), $(\mathcal{E}^\mu, D(\mathcal{E}^\mu))$ is a Dirichlet form (nonregular since it has no nontrivial continuous function in its domain). Hence it is easy to perturb a very nice regular Dirichlet form by a "killing term" making it a nonregular one.

Remark The perturbations $\mathcal{E}^\mu$ of regular Dirichlet forms $\mathcal{E}(u,v)$ on locally compact separable spaces by "killing terms" $\int uvdk$, with k not necessarily positive, has been studied in [AM3], where e.g. necessary and sufficient conditions for lower boundedness of $\mathcal{E}^\mu$ are given, as well as the description for cores of $D(\mathcal{E}^\mu)$ in terms of cores for $D(\mathcal{E})$.

What about the construction of processes associated with nonregular Dirichlet forms? We shall report here shortly on recent results in [AM5], [AM9]. It turns out that a 1 - 1 correspondence can be maintained between Dirichlet forms and a class of right processes called perfect processes (see below).

Let $(E, B(E))$ be a metrizable space and let μ be σ- finite positive Borel measure on E. A process $(X_t)_{t\geq 0}$ on E with life time ζ is called perfect if it is normal ($P_x(X_0 = x) = 1$ $\forall x \in E_\Delta \equiv E \cup \{\Delta\}, \Delta$ being the cemetery), right continuous with left limits up to ζ, $R_1 u$ defined by $R_1 u(x) \equiv E^x \left[\int_0^\zeta e^{-s} u(X_s) ds\right]$ has the (strengthened) fine continuity property

$$R_1 u(X_{t-})\chi_{\{t<\zeta\}} = R_1(X_t)_-\chi_{\{t<\zeta\}} \quad P_x \text{ a.s.,} \quad x \in E.$$

for all bounded measurable u.

One has the inclusions among different families of strong Markov processes: (Feller) $\subset$ (Hunt) $\subset$ (special standard) $\subset$ (perfect) $\subset$ (right) (right being here extended to non Radon spaces).

One calls a process μ- tight if there exists an increasing sequence $\{K_n\}$ of compact sets s.t.

$$P_x(\lim_{n\to\infty} \tau_{E-K_n} \geq \zeta) = 1 \quad \mu \text{ a. e. } x \in E$$

One calls a process μ- perfect if it is μ- tight and perfect. From a basic idea of [LR] one can prove that in polish spaces μ- perfect is equivalent with perfect. We need now to define the analytic correlates of such concepts, in terms of Dirichlet forms and associated capacities.

Let $(\mathcal{E}, D(\mathcal{E}))$ be a Dirichlet form on $L^2(m)$. Let F be closed subset of E. We define

$$D(\mathcal{E})_F \equiv \{f \in D(\mathcal{E}) | f = 0 \text{ m-a.e. on } E - F\} .$$

$D(\mathcal{E})_F$ is then a closed subset of $D(\mathcal{E})$. Let F_k be an increasing sequence of closed subsets of E. $\{F_n\}$ is called and *$\mathcal{E}$-nest* if $\bigcup_k \mathcal{D}(\mathcal{E})_{F_k}$ is $\mathcal{E}_1$-dense in $D(\mathcal{E})$. $B \subset E$ is called *$\mathcal{E}$-polar* if there exists an $\mathcal{E}$-nest $\{F_k\}$ s.t. $B \subset \bigcap_k (E - F_k)$.

A function f on E is called *$\mathcal{E}$-quasi-continuous* if there exists an $\mathcal{E}$-nest $\{F_k\}$ s.t. $f|F_k$ is continuous for all k.

The basic theorem which relates Dirichlet forms and the above m-perfect processes is the following

Theorem *Let $(\mathcal{E}, D(\mathcal{E}))$ be a Dirichlet form on $L^2(E,m)$, with m a σ-finite measure on a metrizable space E. Then the conditions* (i), (ii), (iii) *below are necessary and sufficient for the existence of an m-perfect process $(X_t)_{t\geq 0}$ associated with $(\mathcal{E}, D(\mathcal{E}))$:*

(i) *There exists an $\mathcal{E}$-nest $\{K_k\}$ consisting of compact sets K_k;*

(ii) *There exists an $\mathcal{E}_1$-dense subset $\mathcal{F}_0$ of $D(\mathcal{E})$ consisting of $\mathcal{E}$-quasi-continuous functions;*

(iii) *There exists a countable subset B_0 of $\mathcal{F}_0$ and an $\mathcal{E}$-polar subset N of E s.t. the σ-algebra generated by all u in B_0 contains $\mathcal{B}(E) \cap (E - N)$.*

Moreover, if an m-perfect process (X_t) is associated with $\mathcal{E}$, then it is always properly associated with $\mathcal{E}$ (in the sense that $P_t u$ is an $\mathcal{E}$-quasi-continuous version of $T_t u$, for all $n \in L^2(m)$, where P_t is the transition-semigroup associated with (X_t) and T_t is the Markov semigroup associated with $\mathcal{E}$).

Remarks

(i) If $\mathcal{E}$ is a local Dirichlet form then all conditions are satisfied. However the form need not be regular for the existence of a diffusion (or an m-perfect process). In fact in the example of the non regular form $(\mathcal{E}^\mu, \mathcal{F}^\mu)$ discussed above there is , by above theorem, an m-perfect process associated with $(\mathcal{E}^\mu, \mathcal{F}^\mu)$. This is a diffusion process if $\mathcal{E}$ is local.

(ii) A necessary and sufficient condition for (X_t) to be a Hunt process is the requirement that $\mathcal{F}_0$ consists of strictly $\mathcal{E}$-quasi-continuous functions u.

For the proofs and further work see [AM1-9] – [ABM].

3. Classical Dirichlet forms on infinite dimensional spaces

We shall report here shortly on some recent developments (years 1988-90) concerning the theory of Dirichlet forms on infinite dimensional spaces, which are of the classical "gradient square" type, and the associated diffusion processes. For more details we refer to the original references [AR1-7]. The aim is to extend to the infinite dimensional setting the study of finite dimensional Dirichlet forms of diffusions type, given on $C_0^\infty(\mathbb{R}^d)$ functions by $\mathcal{E}(u,v) = \frac{1}{2}\int \nabla u \cdot \nabla v d\mu$, for some σ-finite measure μ, as forms in $L^2(\mu)$ (let us remark that the extension to more general forms, of the type $\frac{1}{2}\int A\nabla u \cdot \nabla v d\mu$, with A some strictly positive operator on $\mathbb{R}^d$ is easy, both in finite and in infinite dimensions).

The setting is as follows. The state space $\mathbb{R}^d$ is replaced by a general locally convex Hausdorff topological vector space E, only required to have a Souslin structure i.e. to be the continuous image of a complete separable metric (i.e. polish) space. E.g. E can be any separable Banach space or a distributional space like $\mathcal{D}'(\mathbb{R}^d)$ or $\mathcal{S}'(\mathbb{R}^d)$ (which are even Lusin) (or,of course, a locally compact separable metrizable space, the latter being the finite dimensional version of the general case) (see e.g. [Schw] for other examples).

Let μ be a probability measure on E. We assume for simplicity that supp $\mu = E$ (but this can easily be relaxed, cfr. [AR1]). The role of the minimal domain $C_0^\infty(\mathbb{R}^d)$ in the finite dimensional situation is here taken by the algebra FC_b^∞ of function on E which are cylinder functions (i.e. finitely based) and are C^∞ on the base with all derivatives bounded. It is well known that FC_b^∞ is dense in $L^2(\mu)$ (see [AR4] Rem. 3.1).

Let $k \in E - \{0\}$ and consider the form $\mathcal{E}_k^0(u,v) \equiv \frac{1}{2}\int \frac{\partial u}{\partial k}\frac{\partial v}{\partial k} d\mu$, $u, v \in FC_b^\infty$ on $L^2(E;\mu) \equiv L^2(\mu)$ ($\frac{\partial}{\partial k}$ is the derivative in the direction k). We call $\mathcal{E}_k^0$ the *pre-Dirichlet form in the direction k given by μ*.

Let us assume there is in E a densely and continuously embedded separable Hilbert space $\mathcal{H}$, such that the dualisation between E and its topological dual E' coincides with the scalar product in $\mathcal{H}$ when restricted to $\mathcal{H}$.

Let K be a complete orthonormal set in $\mathcal{H}$. Define $\mathcal{E}^0 \equiv \sum_{k\in K} \mathcal{E}_k^0$, we call this form with domain FC_b^∞, the *classical pre-Dirichlet form given by μ*. It can also be written as

$$\mathcal{E}^0(u,v) = \frac{1}{2}\int <\nabla u, \nabla v>_{\mathcal{H}} d\mu \; ,$$

with $<,>_{\mathcal{H}}$ the scalar product in $\mathcal{H}$, where ∇ is the gradient operator i.e.

$$<\nabla u(z), k>_{\mathcal{H}} = \frac{\partial}{\partial k} u(z)$$

for all $z \in E$.

When is $\mathcal{E}^0$ closable?

Certainly when all $\mathcal{E}_k^0$ are closable, cfr. [AR4] (Th. 1.2). For the closability of $\mathcal{E}_k^0$ necessary and sufficient conditions have been established in [AR1]. The finding of these conditions relies on the observation that a necessary and sufficient condition can be found when E is one-dimensional, coupled with a desintegration theorem for μ (hence for $L^2(\mu)$ and $\mathcal{E}_k^0$), which holds since E is Souslin (desintegration is , probabilistically, building of regular conditional probability measures). The desintegration procedure is as follows: We can write, using Hahn-Banach's theorem, E as the direct sum of $k\mathbb{R}$ and a supplementary space $E_{(k)}$, in such a way that any $z \in E$ decomposes as $z = ks \oplus x$, with $s \in \mathbb{R}$, $x \in E_{(k)}$ (of course $E_{(k)}$ is, in general, not unique, it is unique e.g. if E is a Hilbert space and $E_{(k)} \perp k$). Then by desintegration we can write

$$\mu(dz) = \rho_k(x, ds)\nu_k(dx) ,$$

for some probability measure ν_k on $E_{(k)}$ and some kernel $\rho_k(x, ds)$ on $E_{(k)} \times \mathcal{B}(\mathbb{R})$ ($\mathcal{B}(\mathbb{R})$ being the Borel subsets of $\mathbb{R}$) (so that $\rho_k(x, \cdot)$ is the conditional measure to μ given $x \in E_{(k)}$).

Let us now call condition **(H)**the following condition: $\rho_k(x, ds)$ is absolutely continuous with respect to ds, with density $\tilde{\rho}_k(x, s)$ for ν_k a.e. x, and $\tilde{\rho}_k(x, s) = 0$ for Lebesgue a.e. s on the singular set for $\tilde{\rho}_k(x, \cdot)$, i.e. the largest open subset of $\mathbb{R}$ s.t. $\frac{1}{\tilde{\rho}_k(x, \cdot)}$ is locally integrable. Note that **(H)**is satisfied e.g. if $s \longrightarrow \tilde{\rho}_k(x, s)$ is continuous (or even lower semicontinuous).

The following basic theorem was proven in [AR1]:

Theorem 3.1 *$\mathcal{E}_k^0$ is closable if and only if $\rho_k(x, ds)$ satisfies condition* **(H)***for ν_k a.e. x.* ■

Remark

(a) This theorem solves a conjecture of Fukushima, in the case $\dim E < \infty$.

(b) The assumptions are satisfied in applications, see Sect. 3

(c) there exist corresponding necessary and sufficient conditions for the closability of the related form

$$\frac{1}{2} \int \frac{\partial u}{\partial k} \frac{\partial v}{dk} d\bar{\mu} \ , u, \ v \in FC_b^\infty \text{ in } L^2(\mu) \ ,$$

with $\bar{\mu}$ not necessarily coinciding with μ.

The above theorem immediately yields a sufficient condition for closability of $\mathcal{E}^0$. (The sum of closable forms being closable). More generally we have, see [AR4]:

Theorem 3.2 *Suppose each $\mathcal{E}_k^0$ is closable. Let $A : E \longrightarrow \mathcal{B}(H)$ ($\equiv$ space of bounded linear operators on $\mathcal{H}$), be strongly measurable, symmetrical, such that $\|A(\cdot)\| \in L^1(\mu)$ (with $\|\cdot\|$ the operator norm) and such that $\exists \sigma \in L^1(\mu), \sigma > 0$, σ lower semicontinuous (e.g.) with $A(z) - \sigma(z)\mathbb{1}_{\mathcal{H}}$ positive definite. Then*

$$\mathcal{E}_A^0(u, v) \equiv \frac{1}{2}\int < A\nabla u,\ \nabla v >_{\mathcal{H}} d\mu,\ u, v \in FC_b^\infty$$

is closable (in $L^2(\mu)$). Its closure $\mathcal{E}_A$ is a Dirichlet form on $L^2(\mu)$. ∎

For $A = \mathbb{1}_{\mathcal{H}}$ (the identity operator in $\mathcal{H}$) we have $\mathcal{E}_A^0 = \mathcal{E}^0$. We call $\mathcal{E} = \overline{\mathcal{E}^0}$ the closure of $\mathcal{E}^0$: it is the *lassical Dirichlet form given by* μ (the contraction property characteristic of Dirichlet forms is easily verified).

Remark For $E = \mathbb{R}^d$, $\mathcal{E}$ is the Dirichlet form to distorted Brownian motion given by μ, see [EKS], [AHKS], [Fu2].

We shall now discuss the construction of a diffusion process associated with the Dirichlet form $\mathcal{E}$. Since $\mathcal{E}$ is a Dirichlet form there is a concept of capacity associated with it. In the following a property is said to hold quasi everywhere if it holds except for a capacity zero set. The following theorem holds:

Theorem 3.3 *Suppose that one of the following conditions holds:*
(i) *E is a separable Banach space*
(ii) *E it is the dual of a nuclear space (e.g. $E = \mathcal{S}'(\mathbb{R}^d)$) such that $\int_E |< k, z >| \mu(dz) < \infty \forall k \in E'$.*
Then there exists a diffusion process $(X_t, P_z, z \in E, t \in \mathbb{R}_+)$ (i.e. a strong Markov process with continuous paths) properly associated with $\mathcal{E}$, in the sense that

$$E^z u(X_t) = T_t u(z)$$

for μ-a.e. $z \in E$ and all bounded measurable u. T_t is the (symmetric) Markov semigroup uniquely associated with $\mathcal{E}$. Moreover the left hand side is a quasi continuous version of the right hand side, and the capacity associated with $\mathcal{E}$ is always tight.

Proof In case (a) the proof follows from the recent work of Schmuland [Sch], extending previous works [Ku], [AR3]. Case (ii) was proved in [AR3]. The key idea is to construct a certain Gelfand-Naimark compactification $\hat{E}$ of E. Then one shows that $\mathcal{E}$ extends to a regular local form $\hat{\mathcal{E}}$ on $\hat{E}$, and that if the diffusion process $\hat{X}_t$ associated with it (by Fukushima's theory) is started in $z \in E$, for quasi every z, it will not leave E, so that it has a modification X_t running in E. In case (ii) the tightness of the capacity was proven in [AR3]; in case (i) it follows from a general result in [LR].

Remark The idea of compactifying E is already present in [AHK3]. However before Schmuland further assumptions were needed, to show that the process $\hat{X}_t$ on the extension can be restricted to have E as an invariant set. Schmuland uses a particular compactification. For further discussions in this connection see also [So1-3].

It also follows from the compactification method described above that Fukushima's decomposition proved in the locally compact case [Fu1] (extensions of Doob-Meyer and Ito's decompositions) extends to our infinite dimensional situation:

Any $u \in D(\mathcal{E})$ has a quasi-continuous version $\tilde{u}$ and one has, P_z a.e., except for a set of capacity zero:

$$u(X_t) - u(X_0) = M_t^{[u]} + N_t^{[u]} \ \forall t \geq 0 ,$$

where $M_t^{[u]}$ is a martingale additive functional of finite energy and $N_t^{[u]}$ is a continuous additive functional of zero energy, see [AR6] for details.

Remark

(a) For $u \in D(H)$ then

$$N_t^{[u]} = - \int_0^t (Hu)(X_s)ds$$

(b) As mentioned in the proof, the capacity associated with $\mathcal{E}$ is tight in the sense that there exist compact subsets K_n of E s.t. $\lim_{n\longrightarrow\infty} \text{cap}(E - K_n) = 0$, see [AR6].

Can one show that the process X_t satisfies a stochastic equation (in the weak sense, somewhat in the spirit of the martingale problem à la Stroock-Varadhan)?

The answer is yes, under a slightly stronger assumption ([AR6], see also [AR7]):

Theorem 3.4 *If for any $k \in K$ the integration by parts formula holds with $\beta_k \in L^2(\mu)$, i.e.*

$$\int u\beta_k d\mu \equiv - \int \frac{\partial u}{\partial k} d\mu \quad \forall u \in FC_b^\infty ,$$

the linear functions $< k, \cdot >$ over E are in $L^2(\mu)$, and E, H are such that there exists a Brownian motion W on E over $\mathcal{H}$, then

$$< k, X_t > = < k, z > + < k, N_t > + < k, W_t > ,$$

for quasi-every $z \in E$, P_z a.s. where $(N_t)_{t \geq 0}$ is a continuous process on E satisfying

$$< k, N_t > = \frac{1}{2} \int_0^t \beta_k(X_s) ds \ \forall k \in K .$$

If there exists a Hilbert space $\mathcal{H}_0$ s.t. E' is densely injected in $\mathcal{H}_0$, with Hilbert-Schmidt injection, and a constant $c > 0$ such that $\|\beta_k\|_\mu^2 \leq c\|k\|_{\mathcal{H}_0}^2 \forall k \in K$, $\| \ \|_\mu$ resp. $\| \ \|_{\mathcal{H}_0}$ being the $L^2(\mu)$- resp. $\mathcal{H}_0$-norms, then there exists a measurable map $\beta : E \longrightarrow E$ ("drift vector") s.t. $< k, \beta > = \beta_k \mu$ a.e.

In this case

$$dX_t = \beta(X_t)dt + dw_t$$

holds P_z a.s. on E, for quasi every initial condition in E, in the weak sense of stochastic equations. ■

Remark There are other questions that can be discussed. E.g. are there other Dirichlet forms, besides $\mathcal{E}$, extending $\mathcal{E}^0$? There is a natural order in the family of such forms . $\mathcal{E}$ is clearly the element with minimal domain. A characterization of the form $\mathcal{E}$ having maximal domain, extending $\mathcal{E}^0$, has been obtained in [AK], [AKR]. Criteria when $\mathcal{E} = \mathcal{E}^{\max}$ have been given in the case of $\mathbb{R}^d$ [Wi], [Tak3], in the case of Wiener space [Tak1,2], and in the case of the forms associated with quantum fields in [AHK1] (Gaussian case) and [RZ] (non-Gaussian case with space cut-off).

Other structure results on $\mathcal{E}$ are given in [AK].

4. Markov fields as infinite dimensional processes, and associated Dirichlet forms

We shall discuss some examples of homogeneous Markov fields over $\mathbb{R}^d$ (mainly $d = 2$), for two purposes. One is for their intrinsic interest: it is not obvious that there should exist non-Gaussian (generalized) homogeneous random fields which are not independent at every point and yet are Markov, in fact, although the fields themselves had been known for over 15 years, the proof of the Markov property of some of them is rather recent (due to work of Albeverio, Høegh-Krohn and Zegarlinski [AHKZ]). The other reason for considering homogeneous Markov fields is that they provide interesting examples of Dirichlet forms of the type discussed in Sect. 3.

Let us shortly describe the construction of homogeneous Markov fields. The basic role (comparable to the one of Brownian motion of ordinary processes) is played by the so-called *Nelson's free field* on $\mathbb{R}^d$. This is, as a probability measure, the standard normal distribution $N(0;(-\Delta_d + m^2)^{-1}) \equiv \mu_N$, with mean zero and covariance the fundamental solution of $-\Delta_d + m^2$, $-\Delta_d$ being the Laplacian on $\mathbb{R}^d$ and $m > 0$ a constant. μ_N can be realized (by Minlos-Sazonov Theorem) as a probability measure on $\mathcal{S}'(\mathbb{R}^d)$.

In fact it is not difficult to see that the coordinate process X can be extended to $< \rho, X >$, where ρ is any Borel measure of finite energy i.e.

$$\int d\rho(x)d\rho(y)(-\Delta_d + m^2)^{-1}(x,y) < \infty$$

$(((-\Delta_d + m^2)^{-1}(x,y)$ being Newton's kernel).

Basic properties of (X, μ_N) are:

(1) Euclidean property i.e. homogeneity (stationarity) with respect to the proper Euclidean group $SO(d) \odot \mathbb{R}^d$ and reflections with respect to hyperplanes (full Euclidean group).

(2) Sharp global Markov property i.e. for any Lebesgue measurable $G \subset \mathbb{R}^d$, denoting by $\sigma(G)$ the σ-algebra associated with G (i.e. the σ-algebra generated by all $< \rho, X >$ with ρ as above and supp $\rho \subset G$), we have: that $\sigma(\bar{G})$ is independent of $\sigma(\mathbb{R}^d - G)$ given $\sigma(\partial G)$. The Markov property for more special regions was first proven by Wong, Molchan, Nelson. In the present form it is due to Röckner [R1,2], see also [Roz], [D2].

It is well known that the only Gaussian (generalized) random fields with (1), (2) are Gaussian white noise and Nelson's free field. The existence of non Gaussian, (generalized) random fields μ which are not independent at every point and satisfy (1), (2) was postulated by Nelson in '71 and answered positively in an increasing number of models for $d = 2$ in '79 [AHK5], '84 [Gie], [Z] and '88 [AHKZ]. μ is constructed as weak limit as $\Lambda \uparrow \mathbb{R}^2$, for $\Lambda \subset \mathbb{R}^2$ bounded measurable, of probability measures μ_Λ associated with Λ. μ_Λ is defined by $d\mu_\Lambda \equiv e^{-V_\Lambda} d\mu_N (\int e^{-V_\Lambda} d\mu_N)^{-1}$, with $V_\Lambda \sigma(\Lambda)$-measurable and additive (i.e. $\Lambda_1 \cap \Lambda_2 = \emptyset \longrightarrow V_{\Lambda_1 \cup \Lambda_2} = V_{\Lambda_1} + V_{\Lambda_2}$ a.s.) ("additive functional of μ_N").

Heuristically $V_\Lambda(X)$ has the form

$$V_\Lambda(X) = \text{``}\int_\Lambda v(X((x))dx\text{''} \ ,$$

with a real valued function v on $\mathbb{R}$, of exponential, trigonometric resp. polynomial type. Mathematically V_Λ has to be constructed via a suitably regularization procedure $X \longrightarrow X_\varepsilon$, $v \longrightarrow v_\varepsilon$, $\varepsilon \downarrow 0$, see e.g. [AHK4] and references therein. The resulting measure μ and associated coordinate process X are called *field with interaction* v, it constitutes the so-called $v(\varphi)_2$-model (for v a polynomial we have the $P(\varphi)_2$-model). In the above cases (X, μ) is proven to be full Euclidean invariant and (sharp global) Markov.

Remark The proof is quite involved and is carried through in the framework of the theory of local specifications. It uses in particular the solution of a Dirichlet problem with distributional boundary values (as studied by Albeverio/Høegh-Krohn, Dobrushin/Minlos and Röckner), the establishment of FKG-order structure on associated Gibbs measures, uniqueness results for Gibbs measures, uniform continuity of the specifications. We refer to [AHKZ] (and references therein) for details.

We shall now associate to (X, μ) Dirichlet forms, in fact three types of them.

(1) Let $E = \mathcal{S}'(\mathbb{R}^{d-1})$, $\mathcal{H} = L^2(\mathbb{R}^{d-1})$ and take the measure μ of Sect. 3 to be one of the above measures μ.

Let us consider the splitting $x = (t, \overrightarrow{x}) \in \mathbb{R} \times \mathbb{R}^{d-1}$ of $\mathbb{R}^d$, $d \geq 1$. Since the random field X is such that $< \rho, X > = < \delta_t \otimes f, X >$ for $f \in \mathcal{S}(\mathbb{R}^{d-1})$ makes sense, and X is sharp Markov with respect to the hyperplane $\{t = 0\}$ in $\mathbb{R}^d$,

$$t \longrightarrow < \delta_t \otimes f, X > \equiv < f, X_t > \equiv \int f(\overrightarrow{x}) X_t(\overrightarrow{x})) d\overrightarrow{x}$$

(in the sense of distributions) can be looked upon as a Markov process X_t with values in $\mathcal{S}'(\mathbb{R}^{d-1})$, evaluated at f. This Markov process is symmetric (by the fact that the random field X is reflection symmetric), hence its transition semigroup e^{-tH}, $t \geq 0$ is symmetric in $L^2(\mu^0)$, where μ^0 is the restriction of μ to $\sigma(\{t = 0\})$. We can consider the associated Dirichlet form $(H^{\frac{1}{2}}u, H^{\frac{1}{2}}v)$, $u, v \in D(H^{\frac{1}{2}}) \subset L^2(\mu^0)$. As we shall see below, this form is an extension of the classical Dirichlet form given by μ^0.

Remark By analytic continuation $t \longrightarrow it$ of $E(e^{i<t,X_t>})$ one gets relativistic functions yielding relativistic fields, the quantization of the solution fields of the classical non-linear hyperbolic equation $\Box\varphi + m^2\varphi + v'(\varphi) = 0$.

(2) Let $E, \mathcal{H}, \mu$ be as above. One shows, cfr. [AR1], that the classical Dirichlet form $\mathcal{E}$ given by μ^0 is well defined. $\mathcal{E}(u, v)$ coincides with

$(H^{\frac{1}{2}}u, H^{\frac{1}{2}}v)$ and $\mathcal{E}^0(u,v)$ for all $u, v \in FC_b^\infty$. In the case of v of exponential type it can be shown [AK] that $(H^{\frac{1}{2}}u, H^{\frac{1}{2}}v)$ coincides with the maximal Dirichlet form $\mathcal{E}^{\max}$ extending $\mathcal{E}^0$, the pre-Dirichlet form given by μ^0. μ is such that all properties as in Sect. 3 hold. To $\mathcal{E}$ one can associate a diffusion satisfies the stochastic differential equation given in Sect. 3, with a drift vector β which depends on v. For $v \equiv 0$ one has $\beta(X_t) = -\sqrt{-\Delta_x + m^2}X_t(\vec{x}), \vec{x} \in \mathbb{R}^{d-1}$.

Remark Hölder's continuity properties for the paths of X_t have been established, for certain v, in [AHK2], [Ha], [Can2], [Car], [R3], [K].

(3) Let $E = \mathcal{S}'(\mathbb{R}^d)$, $H = L^2(\mathbb{R}^d)$, μ as above. The classical Dirichlet $\mathcal{E}_\mu$ form given by μ is well defined, μ has all properties required by Sect. 3, so that one gets a diffusion process $X_\tau, \tau \geq 0$ associated with $\mathcal{E}_\mu$, satisfying the stochastic differential equation given in Sect. 3. The drift is given by $\beta(X_\tau) =: v'(X_\tau)$: (with : denoting Wick-ordering, see e.g. [AR6]).

X_τ has invariant measure μ, by construction, and it is a realization of the stochastic process of stochastic quantization (cfr. [JoLaM], [BoCM].

Remarks

(1) The fact that $(X_t)_{t\geq 0}$ constructed as described here using Dirichlet forms indeed coincides with the process in [JoLaM] has been proved recently in [RZ].

(2) It has been proven in [AHiPRS1,2] that the Dirichlet forms 2), 3) are given by generalized positive functionals in the sense of white noise calculus. In particular this provides other cores, besides FC_b^∞, for the forms.

(3) Recently the extension of the above construction of (X, μ) to Riemannian manifolds has been discussed [AZ1], convergent simplicial approximation has also been given, see also [AZ2]. It would be interesting to look at the simplicial approximation of the associated Dirichlet form. (Dirichlet forms on finite dimensional manifolds have been considered e.g. in [Fu3], [ABR].

(4) Some of the above ideas have been extended to gauge fields and Higgs fields, see [AHKK1,2] and references therein.

(5) Recently construction of Markov fields over $\mathbb{R}^4$ which are homogeneous with respect to the proper Euclidean group has been achieved [AHKI], [AIK]. This solves the existence problem for proper Euclidean Markov fields over the physical space-time $\mathbb{R}^4$. The fields are shortly described as follows:

The role of X is here played by a distributional-valued stochastic vector field $A = (A_\mu)$, $\mu = 0, 1, 2, 3$ with distribution s.t.

$$E\left(e^{i<f,A>}\right) = e^{-\int_{\mathbb{R}^4} \psi(S*f)dx} ,$$

$f = (f_\mu) \in \mathcal{S}(\mathbb{R}^4)$, with $< f, \mu >= \sum f_\mu A_\mu$, ψ a certain positive definite real function on $\mathbb{R}^4$ consisting of two parts, a Gaussian part ψ_G and a Poisson-like part ψ_P. S is a kernel described below.

A solves a system of first order partial differential equation, most conveniently expressed by realizing $\mathbb{R}^4$ as the (linear structure of the) field quaternions $\mathbb{H}$: A is then identified with a quaternionic valued function ove the quaternions, the equation is $\partial A = F$, with ∂ the Cauchy-Riemann quaternionic operator and F a suitable 4-components (Gaussian + Poisson) white noise over $\mathbb{R}^4 \simeq \mathbb{H}$. $S = (S_\mu)$ is the kernel $S(x) = \frac{x}{\pi|x|^4}$. For $\psi_P = 0$ A is the free Euclidean free electromagnetic potential field. It is possible to associate to it a Dirichlet form on a certain L^2-space.

For $\psi_P \neq 0$, A can still be proven to be global Markov, see [Iw], [AIK] and an analytic continuation of moments of A to "relativistic functions" can be achieved, [AIK]. It would be interesting to see in which sense non symmetric Dirichlet forms still give the time evolution in this model.

Acknowledgements

It is a great pleasure for two of us (S.A. and M.R.) to thank Professor Mark Pinsky for very kind invitations to a most enjoyable and stimulating conference. We gratefully acknowledge very interesting discussions with Masatoshi Fukushima, Wolfhard Hansen, Terry Lyons, Byron Schmuland and all friends with whom we coworked on the papers given in the references. The partial support by an EEC-project and BiBoS is also gratefully acknowledged.

References

[A] S. Albeverio, *Some new developments concerning Dirichlet forms, Markov fields and quantum fields,* SFB–Preprint, pp.250–259 in "IXth International Congress on Mathematical Physics", July 1988, Swansea '88, Edts., B. Simon, A. Truman, I.M. Davis, Adam Hilger, Bristol and New York (1989)

[ABR] S. Albeverio, J. Brasche, M. Röckner, *Dirichlet forms and generalized Schrödinger operators*, pp. 1-42 in Proc. Sønderborg Conf. "Schrödinger operators", Edts. H. Holden, A. Jensen, Lect. Notes Phys. 345, Springer, Berlin (1989)

[AFHKL1] S. Albeverio, J.E. Fenstad, R. Høegh-Krohn, T. Lindstrøm, *Non standard methods in stochastic analysis and mathematical physics*, Academic Press (1986) (translation into Russian by Akad. Nauk. is in preparation).

[AHiPRS1] S. Albeverio, T. Hida, J. Potthoff, M. Röckner, L. Streit, *Dirichlet forms in terms of white noise analysis I - Construction and QFT examples*, Rev. Math. Phys. **1**, 291-312 (1990).

[AHiPRS2] S. Albeverio, T. Hida, J. Potthoff, M. Röckner, L. Streit, *Dirichlet forms in terms of white noise analysis II - Closability and Diffusion Processes*, Rev. Math. Phys. **1**, 313-323 (1990).

[AHK1] S. Albeverio, R. Høegh-Krohn, *Quasi invariant measures, symmetric diffusion processes and quantum fields*, pp.11–59 in "Proceedings of the International Colloquium on Mathematical Methods of Quantum Field Theory", Editions du CNRS. 1976, (Colloques Internationaux du Centre National de la Recherche Scientifique, No. 248).

[AHK2] S. Albeverio, R. Høegh-Krohn, *Dirichlet forms and diffusion processes on rigged Hilbert spaces*, Zeitschrift für Wahrscheinlichkeitstheorie und verwandte Gebiete **40**, 1–57 (1977)

[AHK3] S. Albeverio, R. Høegh-Krohn, *Hunt processes and analytic potential theory on rigged Hilbert spaces*, Ann. Inst. H. Poincaré (Probability Theory) **B13**, 269–291 (1977)

[AHK4] S. Albeverio, R. Høegh-Krohn, *Diffusion Fields, Quantum Fields and Fields with Values in Groups*, pp.1–98, "Stochastic Analysis and Applications", Adv. in Probability, Edt. M. Pinsky, M. Dekker, New York (1984)

[AHK5] S. Albeverio, R. Høegh-Krohn, *Uniqueness and the global Markov property for Euclidean fields, the case of trigonometric interactions*, Commun. Math. Phys. **68** 95-128 (1979)

[AHKHK1] S. Albeverio, R. Høegh-Krohn, H. Holden, T. Kolsrud, *A covariant Feynman–Kac formula for unitary bundles over Euclidean space*, pp. 1-12 in Proc. Trento Conference "Stochastic Partial Differential Equations and Applications II", Edts. G. Da Prato, L. Tubaro, Lect. Notes Maths., 1390 Springer (1988).

[AHKI] S. Albeverio, R. Høegh-Krohn, K. Iwata, *Covariant Markovian random fields in four space-time dimensions with nonlinear electromagnetic interaction*, pp.69–83 in "Applications of Self-Adjoint Extensions in Quantum Physics", in Proc. Dubna Conf. 1987, Edts. P. Exner, P. Seba, Lec. Notes in Physics **324**, Springer, Berlin (1989)

[AHKRS] S. Albeverio, R. Høegh-Krohn, M. Röckner, U. Spönemann, *On the structure of symmetric diffusion processes*, in preparation

[AHKS] S. Albeverio, R. Høegh-Krohn, L. Streit, *Energy forms, Hamiltonians and distorted Brownian paths*, Journal of Mathematical Physics **18**, 907–917 (1977),

[AHKZ] S. Albeverio, R. Høegh-Krohn, B. Zegarlinski, *Uniqueness and global Markov property for Euclidean fields: The case of general polynomial interactions*, Commun. Math. Phys. **123**, 377–424 (1989)

[AIK] S. Albeverio, K. Iwata, T. Kolsrud, *Random fields as solutions of the inhomogeneous quaternionic Cauchy-Riemann equation I. Invariance and Analytic Continuation*, SFB 237 - Preprint (1989), to appear in Commun. Math. Phys.; and paper in preparation

[AK1] S. Albeverio, S. Kusuoka, *Maximality of infinite dimensional Dirichlet forms and R. Høegh-Krohn's model of quantum fields*, to appear in Memorial Volume for Raphael Høegh-Krohn.

[AKR] S. Albeverio, S. Kusuoka, M. Röckner, *On partial integration in infinite dimensional space and applications to Dirichlet forms*, SFB 237 - Preprint to appear in J. London Math. Soc. (1989).

[AM1] S. Albeverio, Zhi Ming Ma, *Nowhere Radon smooth measures, perturbations of Dirichlet forms and singular quadratic forms*, pp. 3-45 in Proc. Bad Honnef Conf.1988, ed. N. Christopeit, K. Helmes, M. Kohlmann Lect. Notes Control and Inform. Sciences 126, Springer, Berlin 1989).

[AM2] S. Albeverio, Zhi Ming Ma, *Additive functionals, nowhere Radon and Kato class smooth measure associated with Dirichlet forms*, SFB 237 - Preprint.

[AM3] S. Albeverio, Zhi Ming Ma, *Perturbation of Dirichlet forms - lower semiboundedness, closability and form cores*, SFB 237 - Preprint, to appear in J. Funct. Anal. (1990).

[AM4] S. Albeverio, Zhi Ming Ma, *A note on quasicontinuous kernels representing quasi- linear positive maps*, BiBoS - Preprint 1990, subm. to Forum Math.

[AM5] S. Albeverio, Zhi Ming Ma, *A general correspondence between Dirichlet forms and right processes*, SFB 237 - Preprint 1990

[AM6] S. Albeverio, Zhi Ming Ma, *Local property for Dirichlet forms on general metrizable spaces*, in preparation

[AM7] S. Albeverio, Zhi Ming Ma, *Diffusion processes associated with singular Dirichlet forms*, to appear in Proc. Lisboa Conf., Ed. A.B. Cruzeiro, 1990

[AM8] S. Albeverio, Zhi Ming Ma, *Characterization of Dirichlet spaces associated with symmetric Hunt processes*, in preparation

[AM9] S. Albeverio, Zhi Ming Ma, *Necessary and sufficient conditions for the existence of m-perfect processes associated with Dirichlet forms*, BiBoS-Preprint 1990

[AMR] S. Albeverio, Zhi Ming Ma, M. Röckner, in preparation

[AR1] S. Albeverio, M. Röckner, *Classical Dirichlet forms on topological vector spaces—closability and a Cameron–Martin formula*, J. Funct. Anal., **88**, 395-436 (1990).

[AR2] S. Albeverio, M. Röckner, *Dirichlet forms, quantum fields and stochastic quantisation*, pp.1–21 in "Stochastic Analysis, path integration and dynamics", Emanations from "Summer Stochastics", Warwick 1987, Edts. D. Elworthy, J. C. Zambrini, Pitman Res. Notes, Longman, Harlow (1989)

[AR3] S. Albeverio, M. Röckner, *Classical Dirichlet forms on topological vector spaces - the construction of the associated diffusion process*, Prob. Theory and Rel. Fields, **83**, 405-434 (1989).

[AR4] S. Albeverio, M. Röckner, *New developments in theory and applications of Dirichlet forms*, to appear in "Stochastic Processes, Physics and Geometry", Proc. 2nd Int. Conf. Ascona - Locarno - Como 1988, Ed. S. Albeverio, G. Casati, U. Cattaneo, D. Merlini, R. Moresi, World Scient. (1990).

[AR5] S. Albeverio, M. Röckner, *On Dirichlet forms on topological vector spaces: Existence and maximality*, pp. 14-31 in Proc. Bad Honnef '88 Conf., Edts. N. Christopeit, K. Helmes, M. Kohlmann, Lect. Notes Inform. Control 126, Springer Verlag (1989)

[AR6] S. Albeverio, M. Röckner, *Stochastic differential equations in infinite dimension: solutions via Dirichlet forms*, SFB 237 - Preprint.

[AR7] S. Albeverio, M. Röckner, *Infinite dimensional diffusions connected with positive generalized white noise functionals*, Edinburgh Preprint (1990), to appear in Proc. Bielefeld Conference "White Noise Analysis", Ed. T. Hida, H. H. Kuo, J. Potthoff, L. Streit

[AS] M. Aizenman, B. Simon: *Brownian motion and Harnack's inequality for Schrödinger operators*, Comm. Pure Appl. Math. **35**, 209-271 (1982)

[AZ1] S. Albeverio, B. Zegarlinski, *Construction of convergent simplicial approximations of quantum fields on Riemannian manifolds*, SFB 237 - Preprint (1989), to appear in Commun. Math. Phys..

[AZ2] S. Albeverio, B. Zegarlinski, *Some stochastic techniques in quantization, new developments in Markov fields and quantum fields*, SFB 237 - Preprint (1990), to appear in Proc. Cargèse 89, " Stochastic Quantization", Edts. P. H. Damgaard, H. Hüffel, Plenum Press (1990)

[BH1] N. Bouleau, F. Hirsch, *Propriétés d'absolue continuité dans les espaces de Dirichlet et application aux équations différentielles stochastiques*, Séminaire de Probabilité, no. XX.

[BH2] N. Bouleau, F. Hirsch, *Formes de Dirichlet générales et densité des variables aléatoires réelles sur l'espace de Wiener*, J. Funct. Anal. **69**, 229-259 (1986)

[BKo] Yu. M. Berezanski, Yu. G. Kondratjev *Spectral methods in infinite dimensional analysis* 1988, Ak. Nauk (russ.) (transl. Kluwer, to appear)

[BoCMi] V.S. Borkar, R.T. Chari, S.K. Mitter, *Stochastic quantization of field theory in finite and infinite Volume*, J. Funt. Anal. **81**, 184–206 (1988)

[Can] J.T. Cannon, *Continuous single paths in quantum field theory*, Comm. Math. Phys. **35**, 215–233 (1974)

[Car] R. Carmona, *Measurable norms and some Banachspace valued Gaussian processes*, Duke Math. J. **44**, 109–127 (1977)

[D1] E.B. Dynkin, *Green's and Dirichlet spaces for a symmetric Markov transition function*, Lecture Notes of the LMS, 1982

[D2] E.B. Dynkin, *Markov processes and random fields*, Bull. AMS **3**, 975–999 (1980)

[DeMe] C. Dellacherie, P.A. Meyer: *Probabilités et potentiel*, Ch. XII–XVI, Hermann, Paris (1987)

[DrYM2] , B. Driver, YM2, *Continuum expectations, lattice convergence and lassos*, Commun. Math. Phys. **123**, 575–616 (1989)

[EUSh] H. Ezawa, J.R. Klander, L.A. Shepp, *Vestigial effects of singular potentials in diffusion theory and quantum mechanics*, J. Math. Phys. **16**, 783-799 (1975)

[Fa] Fan, R. *On absolute continuity of symmetric diffusion processes on Banach spaces*, Beijing Prepr. 1989, to appear in Acta Mat. Appl. Sinica

[Fi1] P.J. Fitzsimmons, *Markov processes and nonsymmetric Dirichlet forms without regularity*, to appear in J. Funct. Anal. (1988)

[Fi2] P.J. Fitzsimmons, *Time changes of symmetric Markov processes and a Feynman-Kac formula*, to appear in J. Theoretical Prob.

[FiG] P.J. Fitzsimmons, R.K. Getoor, *On the potential theory of symmetric Markov processes*, Math. Annalen., **281**, 495-512 (1988)

[Fu1] M. Fukushima, *Dirichlet forms and Markov processes*, Amsterdam - Oxford - New York: North-Holland (1980)

[Fu2] M. Fukushima, *On a stochastic calculus related to Dirichlet forms and distorted Brownian motion*, Physical Reports **77**, 255-262 (1981)

[Fu3] M. Fukushima, *Basic properties of Brownian motion and a capacity on the Wiener space*, J. Math. Soc. Japan **36**, 161-175 (1984)

[Fu4] M. Fukushima, *Regular representation of Dirichlet spaces*, Trans. Am. Math. Soc. **155**, 455-473 (1971)

[Fu5] M. Fukushima, *Energy forms and diffusion processes*, pp. 65-98 in "Mathematics + Physics", Vol. 1, Ed. L. Streit, World Scient., Singapore (1985)

[Gie] R. Gielerak, *Verification of the global Markov property in some class of strongly coupled exponential interactions*, J. Math. Phys. **24**, 347–355 (1983)

[GlJ] J. Glimm, A. Jaffe, *Quantum physics: A functional integral point of view*, New York - Heidelberg - Berlin: Springer (1981)

[GrKiSen] , L. Gross, Ch. King, A. Sengupta, *Two dimensional Yang–Mills theory via stochastic differential equations*, Ann. Phys.

[Ha] Z. Haba, *Hölder continuity of sample paths in Euclidean field theory*, Commun. Math. Phys. **69**, 247-256 (1979)

[Iw] K. Iwata, *On linear maps preserving Markov properties and applications to multicomponent generalized random fields*, PhD Thesis, Bochum (1990)

[Jo–LaM] G. Jona–Lasinio, P.K. Mitter, *On the stochastic quantization of field theory*, Comm. Math. Phys. **101**, 409–436 (1985)

[K] T. Kolsrud, *Gaussian random fields, infinite dimensional Ornstein–Uhlenbeck prosses, and symmetric Markov processes*, Acta Appl. Math. **12**, 237–263 (1988)

[Ku] S. Kusuoka, *Dirichlet forms and diffusion processes on Banach space*, J. Fac. Science Univ. Tokyo, Sec. 1A **29**, 79-95 (1982)

[Lj] Y. LeJan, *Quasi-continuous functions and Hunt processes*, J. Math. Soc. Japan, **35**, 37-42

[LR] T. Lyons, M. Röckner, *A note on tightness of capacities associated with Dirichlet forms*, Edinburgh Preprint (1990)

[Pa] D. Pantic, *Stochastic calculus on distroted Brownian motion*, J. Math. Phys. **39**, 207–210 (1988)

[Pac] P. Paclet, *Espaces de Dirichlet et capacités fonctionelles sur triplet de Hilbert-Schmidt*, Séminare P. Krée, no. 5 (1978)

[PR] J. Potthoff, M. Röckner, *On the contraction property of infinite dimensional Dirichlet forms*, Preprint Edinburgh (1989), to appear in J. Funct. Anal.

[R1] M. Röckner, *Generalized Markov fields and Dirichlet forms*, Acta Appl. Math. **3**, 285-311 (1985)

[R2] M. Röckner, *A Dirichlet problem for distributions and the construction of specifications for Gaussian generalized random fields*, Mem. An. Math. Soc. **54** (1985)

[R3] M. Röckner, *Taraces of harmonic functions and a new path space for the free quantum field*, J. Funct. Anal.

[RZ] M. Röckner, T. Zhang, *On uniqueness of generalized Schrödinger operators*, Preprint Edinburgh (May 1990). Publication in preparation.

[Roz] Yu. Rozanov, *Markov Random Fields*, Springer, Berlin (1982)

[Sch] B. Schmuland, *An alternative compactification for classical Dirichlet forms and topological vector spaces*, Vancouver Preprint (1989)

[Schw] L. Schwarz, *Radon measures on arbitrary topological spaces and cylindrical measures*, Tata, Oxford (1973)

[Sha] M.J. Sharpe, *General theory of Markov processes*, Academic Press, New York (1988)

[Sil] M.L. Silverstein, *Symmetric Markov processes.* Lecture Notes in Math. **426**, Berlin - Heidelberg - New York, Springer (1974)

[Si2] M.L. Silverstein, *Boundary theory for symmetric Markov processes*, Lect. Notes Maths. **516**, Springer, Berlin (1976)

[Sim] B. Simon, *The $P(\Phi)_2$ Euclidean (quantum) field theory*, Princeton, Princeton University Press (1974)

[So1] S. Song, *The closability of classical Dirichlet forms on infinite dimensional spaces and the Wiener measure*, Acad. Sinica Preprint, Beijing

[So2] S. Song, *Admissible vectors and their associated Dirichlet forms*, Acad. Sinica Preprint, Beijing (1990)

[So3] S. Song, *An infinite dimensional analogue of the Albeverio–Röckner's theorem on Fukushima's conjecture*, Acad. Sinica Preprint, Beijing (1989)

[Tak1] M. Takeda, *On the uniqueness of the Markovian self–adjoint extension*, pp. 319–325 in "Stochastic Processes – Mathematics and Physics II"=, Edts. S. Albeverio, Ph. Blanchard, L. Streit, Lect. Notes Maths. **1250**, Springer, Berlin (1987)

[Tak2] M. Takeda, *On the uniqueness of Markovian self-adjoint extensions of diffusion operators on infinite dimensional spaces*, Osaka J. Math. **22**, 733-742 (1985)

[Tak3] M. Takeda, *The maximum markovian self–adjoint extensions of generalized Schrödinger operators*, Preprint (1990)

[Tak4] M. Takeda, *On the uniqueness of Markovian self-adjoint extensions of diffusion operators on infinite dimensional spaces*, Osaka J. Math. **22**, 733-742 (1985)

[Tam] H. Tamura, *Nonlinear electromagnetic fields confine charges*, Kanazawa Preprint (1989)

[Wi] N. Wielens, *On the essential self-adjointness of gneralized Schrödinger operators*, J. Funct. Anal. **61**, 98-115 (1985)

[Z] B. Zegarlinski, *Uniqueness and the global Markov property for Euclidean fields: the case of general exponential interaction*, Comm. Math. Phys. **96**, 195–221 (1984)

Sergio Albeverio
Fakultät für Mathematik,
Ruhr-Universität,
D 4630 Bochum (FRG)

BiBoS Research Center,
D 4800 Bielefeld (FRG)

SFB 237 -Essen, Bochum,
Düsseldorf;
CERFIM, Locarno

Zhi Ming Ma
BiBoS Research Center,
D 4800 Bielefeld (FRG)
On leave of absence from
Inst. of Appl. Mathematics
Academia Sinica, Beijing

Michael Röckner
Dept. of Mathematics
Univ. of Edinburgh,
Edinburgh EH9 3JZ
Scotland

Itô's Formula for Stochastic Integration in Banach Spaces

J.K. Brooks and N. Dinculeanu

Introduction

In this paper we shall establish Itô's formula for stochastic processes taking their values in a Banach space. This formula is indispensible in the study of stochastic differential equations.

The real case was first established for Brownian motion by Itô. The full extension to the general case resulted form the works of Kunita, Watanabe, Meyer and Doleans–Dade.

The infinite dimensional case was first considered by Kunita for Hilbert–valued martingales after he established a stochastic integration theory for Hilbert spaces [5], and the final stronger version in Hilbert spaces was given by Métivier and Métivier and Pellaumail, after Métivier introduced the tensor quadratic variation (see [7] for the treatment in Hilbert space). Kussmaul [6] presented a different approach in this setting.

A number of technical difficulties emerge when Itô's formula is considered for Banach–valued processes, and one truly appreciates the geometry that the Hilbert space setting provides in stochastic integration, after considering the general case. Obviously, one difficulty encountered in the Banach case is the construction of the stochastic integral. Recently, a viable theory in this setting has been developed by the authors in [2]. This theory stemmed from the general functional analytic theory of integration in Banach spaces in 1976 [1].

In [2], a stochastic integral is established in Banach spaces, and the space of integrable processes constitutes a Lebesgue space, that is, a space in which the Lebesgue and Vitali convergence theorems hold—necessary tools for interchanging limits in order to derive Itô's formula. This necessary ingredient was missing from previous attempts to establish Itô's formula in this setting (e.g. [4]).

Another hurdle is the existence of the tensor quadratic variation $[[X]]$. While not all bounded, predictable Banach–valued processes are necessarily integrable (unless a further assumption is made concerning the relative weak compactness of the family of measures induced by the stochastic measure I_X), it turns out, fortunately, that the caglad process X_{s-} is locally integrable, if X is locally summable; the desired tensor quadratic variation $[[X]]_t$ exists, and is defined by means of stochastic integrals as follows:

$$[[X]]_t = X_t^{\otimes 2} - \int_{[0,t]} (X_{s-} \otimes dX_s + dX_s \otimes X_{s-}).$$

The tensor quadratic variation $[[X]]_t$ does not necessarily have finite variation, unless the quadratic variation $[X]$ exists. In the Hilbert space case, $[X]$ is defined by the usual equality

$$[X]_t = |X_t|^2 - 2 \int_{[0,t]} \langle X_{s-}, dX_s \rangle,$$

where $\langle \cdot, \cdot \rangle$ is the inner product. In the absence of the inner product, we have to postulate the existence of $[X]_t$, defined pointwise by

$$[X]_t = X_0 + \lim_n \sum_{k \geq 0} |X_{v(n,k+1)\wedge t} - X_{v(n,k)\wedge t}|^2,$$

where the $v(n,k)$ are stopping times satisfying certain conditions (see secction 3).

The processes X for which Itô's formula holds are those which have finite quadratic variation and are locally summable relative to certain bilinear mappings or processes having paths of finite variation. In the Hilbert space case these conditions are automatically satisfied for any semimartingale X. It would be interesting to find conditions which would insure the existence of the quadratic variation in the Banach space case.

Even in the Hilbert space case, our approach enables us to prove the Itô formula for more general processes than semimartingales. We remark that the proof of Itô's formula presented in this paper fills some lacunas in the proof given for the Hilbert space case in [7].

In section 1, the general theory of integration is presented, and the application to stochastic integration is given in section 2. The vector quadratic variation and quadratic variation are defined in sections 4 and 5 respectively, and some properties are derived in section 6. The Itô formula is proved in section 7.

1. Integration in Banach spaces

In this section we shall sketch the general theory of integration in Banach spaces. This material is based on [1].

Let E, F, G, D be Banach spaces with norm denoted by $|\cdot|$. The dual of any Banach space M is denoted by M^*, and the unit ball of M by M_1.

We shall assume that $E \subset L(F,G)$ isometrically and we shall denote by $B : E \times F \to G$ the bilinear mapping defined by $B(x,y) = x(y)$ for $x \in E$ and $y \in F$. If there is no confusion possible, we shall write $x(y) = xy = yx$. We have $|x| = \sup\{|xy| : y \in F_1\}$ for every $x \in E$.

Conversely, any bilinear mapping $B : E \times F \to G$ satisfying $|x| = \sup\{|B(x,y)| : y \in F_1\}$, for $x \in E$, realizes a linear isometry $x \to B(x,\cdot)$ of E into $L(F,G)$.

Usual examples of embedding $E \subset L(F,G)$ are $E = L(\mathbb{R}, E)$, $E \subset L(E^*, R)$, $E \subset L(F, E \hat{\otimes}_\pi F)$; E a real Hilbert space and $E = L(E, \mathbb{R})$; $\mathbb{R} \subset L(F,F)$ for any Banach space F.

Let S be a nonempty set, $\sum$ a σ-algebra of subsets of S and $m : \sum \to E \subset L(F,G)$ a σ-additive measure. The semivariation $\tilde{m}_{F,G}$ of m, relative to the pair (F,G), is defined for every $A \in \sum$ by

$$\tilde{m}_{F,G}(A) = \sup \left| \sum m(A_i) x_i \right|,$$

where the supremum is taken over all finite families (A_i) of disjoint sets from $\sum$ with union A, and all finite families (x_i) from F_1. The semivariation $\tilde{m}_{F,G}(A)$ is also denoted by $\text{svar}_{F,G} m(A)$. If we consider $E = L(\mathbb{R}, E)$, the semivariation $\tilde{m}_{\mathbb{R},E}$ is denoted simply by $\tilde{m}$. In this case $\tilde{m}$ is finite on $\sum$. However, in general, $\tilde{m}_{F,G}$ may be infinite. If we consider also $E \subset L(E^*, I\!R)$, then we have

$$\tilde{m}_{I\!R,E} \leq \tilde{m}_{F,G} \leq \tilde{m}_{E^*,I\!R} = |m|,$$

where $|m|$ is the variation of m, defined for every set $A \in \sum$ by

$$|m|(A) = \sup \sum |m(A_i)|,$$

where the supremum is taken over all finite families (A_i) from $\sum$ with union A.

The semivariation $\tilde{m}_{F,G}$ satisfies the following equality:

$$\tilde{m}_{F,G} = \sup\{|m_z| : z \in G_1^*\},$$

where $|m_z|$ is the total variation of the measure $m_z : \sum \to F^*$ defined for $A \in \sum$ and $x \in F$ by

$$\langle x, m_z(A)\rangle = \langle m(A)x, z\rangle.$$

We shall denote by $m_{F,G}$ the set of positive measures $\{|m_z| : z \in G_1^*\}$.

In case $F = I\!R$, then $\tilde{m}_{I\!R,E}$ is finite; hence each measure m_z has finite variation. Moreover, the set of measures $m_{I\!R,E}$ is uniformly σ–additive.

In general, if the semivariation $\tilde{m}_{F,G}$ is finite, the measures m_z have finite variation, and the set $m_{F,G}$ consists of finite σ–additive measures, but $m_{F,G}$ is not necessarily uniformly σ–additive.

In the sequel, we shall assume that m has finite semivariation $\tilde{m}_{F,G}$.

If D is any Banach space, we denote by $\mathcal{F}_D(m_{F,G})$ the vector space of functions $f : S \to D$ belonging to the intersection

$$\bigcap\{L_D^1(|m_z|) : z \in G_1^*\},$$

and such that

$$\tilde{m}_{F,G}(f) := \sup\{\int |f| d|m_z| : z \in G_1^*\} < \infty.$$

Then $\tilde{m}_{F,G}(\cdot)$ is a seminorm and $\mathcal{F}_D(m_{F,G})$ is complete for this seminorm. The space $\mathcal{F}_D(m_{F,G})$ contains all bounded measurable functions. (We define $f : S \to D$ to be measurable if there exists a sequence of D–valued $\sum$–simple functions such that the subset of S for which the sequence does not converge to f is $|m_z|$–negligible for every $z \in G_1^*$).

The closure $\mathcal{F}_D(B, m_{F,G})$ of the set B of bounded measurable functions is a "Lebesgue space" in the sense that it is complete and the Vitali and Lebesgue convergence theorems are valid for convergence of functions in $\tilde{m}_{F,G}$–measure, and in particular for uniform convergence of functions; but not necessarily for pointwise convergence. Consequently, if $\mathcal{R}$ is a ring generating $\sum$, then the $\mathcal{R}$–step functions are not necessarily dense in $\mathcal{F}_D(B, m_{F,G})$. If the set of measures $m_{F,G}$ is uniformly σ–additive (for example, if $F = I\!R$), then an Egorov–type theorem can be proved, and the Vitali and Lebesgue theorems hold for pointwise convergence (see [1] for details). Also, if $m_{F,G}$ is uniformly σ–additive, and if $\mathcal{R}$ is a ring generating the σ–algebra $\sum$, then the $\mathcal{R}$–step functions are dense in $\mathcal{F}_D(B, m_{F,G})$.

If $D = F$, we denote $\mathcal{F}_{F,G}(m) = \mathcal{F}_F(m_{F,G})$. This particular case is very important since we can define an integral $\int f dm \in G^{**}$, for $f \in \mathcal{F}_{F,G}(m)$ as follows: since $f \in L^1_F(|m_z|)$ for every $z \in G_1^*$, the mapping $z \to \int f dm_z$ is an element of G^{**}, which we shall denote by $\int f dm$:

$$\langle z, \int f dm \rangle = \int f dm_z, \quad \text{for } z \in G^*.$$

Under certain conditions we have $\int f dm \in G$. This is true, first for step functions f, and then, by continuity of the integral, for any limit in $\mathcal{F}_{F,G}(m)$ of step functions.

More generally, let $\mathcal{C} \subset \mathcal{F}_{F,G}(m)$ be a subspace and denote by $\mathcal{F}_{F,G}(\mathcal{C}, m)$, the closure of $\mathcal{C}$ in $\mathcal{F}_{F,G}(m)$. If $\int f dm \in G$ for every $f \in \mathcal{C}$, then the same is true for $f \in \mathcal{F}_{F,G}(\mathcal{C}, m)$. If the set of measures $m_{F,G}$ is uniformly σ–additive (for example, if $F = I\!R$), then $\int f dm \in G$ for any f in the closure $\mathcal{F}_{F,G}(B, m)$ of the bounded measurable functions.

2. The Stochastic Integral

In this section we shall apply the results of section 1 to define the stochastic integral. Detailed proofs will be given elsewhere.

Let E, F and G be Banach spaces such that $E \subset L(F,G)$ and denote by $B : E \times F \to G$ the bilinear mapping defined by $B(x,y) = y(x) = xy = yx$, for $x \in E$ and $y \in F$. Let $(\Omega, \mathcal{F}, P)$ be a probability space with a filtration $(\mathcal{F}_t)_{t\geq 0}$ satisfying the usual conditions (the terminology of [3] will be used throughout this paper). Let $\mathcal{R}$ be the ring of subsets of $\mathbb{R}_+ \times \Omega$ generated by the predictable rectangles $[0_A] = \{0\} \times A$, with $A \in \mathcal{F}_0$ and $(s,t] \times A$, with $A \in \mathcal{F}_s$. The σ–algebra $\mathcal{P}$ of predictable subsets of $\mathbb{R}_+ \times \Omega$ is generated by $\mathcal{R}$.

Let $X : \mathbb{R}_+ \times \Omega \to E \subset L(F,G)$ be a cadlag, adapted process, let $1 \leq p < \infty$, and assume that $X_t \in L^p_E = L^p_E(P)$, for $t \geq 0$. We define the stochastic measure I_X on the class of predictable rectangles by

$$I_X([0_A]) = 1_A X_0 \text{ and } I_X((s,t] \times A) = 1_A(X_t - X_s)$$

and then extend I_X to an additive measure $I_X : \mathcal{R} \to L^p_E$. Occasionally we shall omit the index X and write I for I_X.

The isometric embedding $E \subset L(F,G)$ implies that $L^p_E \subset L(F, L^p_G)$ isometrically. We say that X is *p–summable* relative to the pair (F,G), or relative to the bilinear mapping B, if I_X has a σ–additive extension to the σ–algebra $\mathcal{P}$ with values in L^p_E (still denoted by I_X or I), and if I has finite semivariation relative to the pair (F, L^p_G). If $p = 1$, we say simply that X is *summable* relative to (F,G). Note that if $r < s$, then for any Banach space M, we have $L^s_M \subset L^r_M$ and $\|f\|_r \leq \|f\|_s$, for $f \in L^s_M$. Thus if X is s–summable relative to (F,G), then it is r–summable relative to (F,G) and $\text{svar}_{F,L^r_G} I \leq \text{svar}_{F,L^s_G} I$. In particular, if X is p–summable, then it is summable and $\text{svar}_{F,L^1_G} I \leq \text{svar}_{F,L^p_G} I$. Moreover, if the set of measures I_{F,L^p_G} is uniformly σ–additive, then so is I_{F,L^1_G}.

In view of the fact that summability plays a central role in this theory, we present the following characterization of summability, which is rather unexpected since it yields σ–additivity of a set function from its boundedness.

2.1 THEOREM. *Assume E does not contain a copy of c_0. (This is the case, for example, if E is weakly sequentially complete, in particular if E is reflexive).*

Then X is p–summable relative to (F,G) if and only if I_X has bounded semivariation on the ring $\mathcal{R}$, relative to (F, L^p_G).

If X is p–summable relative to (F, G), then

$$X_\infty := \lim_{t\to\infty} X_t \text{ exists in } L_E^p,$$

and for any stopping time T, we have $X_T \in L_E^p$; for any stochastic interval $(S, T]$, we have $I_X((S,T]) = X_T - X_S$.

Assume now that X is p–summable relative to (F, G). Then we can apply the results of section 1 with $S = \mathbb{R}_+ \times \Omega$, $\sum = \mathcal{P}$, E and G replaced by L_E^p and L_G^p respectively, and $m = I_X$. The summability condition means that $m = I_X$ is σ–additive and the semivariation $\tilde{m}_{F,L_G^p}$ is finite. The corresponding space $\mathcal{F}_F(m_{F,L_G^p})$ will be denoted by $\mathcal{F}_{F,G}(X)$, and consists of predictable processes $H : \mathbb{R}_+ \times \Omega \to F$, such that $\tilde{m}_{F,L_G^p}(H) < \infty$. If $H \in \mathcal{F}_{F,G}(X)$, the integral $\int H dI_X$ belongs to the bidual of $L_G^p(P)$. If $H \in \mathcal{F}_{F,G}(X)$, then for every $t \geq 0$, we have $H1_{[0,t]} \in \mathcal{F}_{F,G}(X)$; we denote $\int_{[0,t]} H dI_X = \int 1_{[0,t]} H dI_X$. In this way we obtain a process $(\int_{[0,t]} H dI_X)_{t\geq 0}$ with values in the bidual of L_G^p.

We shall be interested in processes $H \in \mathcal{F}_{F,G}(X)$ having the following property:

(1) $\int_{[0,t]} H dI_X \in L_G^p$, for every $t \geq 0$.

We shall follow the usual practice to identify functions in L_G^p which are equal a.s., and we shall denote by the same symbol $\int_{[0,t]} H dI_X$ the equivalence class in L_E^p as well as any representative of this class.

Furthermore, we shall be interested in those processes $H \in \mathcal{F}_{F,G}(X)$ satisfying condition (1) above and the following:

(2) the L_G^p–valued process $(\int_{[0,t]} H dI_X)_{t\geq 0}$ has a cadlag version, that is, from each equivalence class $\int_{[0,t]} H dI_X$ in L_G^p we can choose a representative, denoted by $(H \cdot X)_t$ such that the process $H \cdot X = ((H \cdot X)_t)_{t\geq 0}$ is cadlag.

Such a cadlag version $H \cdot X$, if it exists, is called the *stochastic integral of H with respect to X*, and this process is also denoted by $\int H dX$ or $\int H_s dX_s$:

$$(H \cdot X)_t = \left(\int H dX\right)_t = \int_{[0,t]} H dI_X.$$

We remark that the stochastic integral is defined up to an indistinguishable process.

We now wish to study spaces of processes H for which the stochastic integral $H{\cdot}X$ is defined. We shall introduce the following notation: if $\mathcal{C}$ is any linear subspace of $\mathcal{F}_{F,G}(X)$, we denote by $\mathcal{F}_{F,G}(\mathcal{C},X)$ the closure of $\mathcal{C}$ in $\mathcal{F}_{F,G}(X)$. By the continuity of the integral, if $\int H dI_X \in L^p_G(P)$ for every $H \in \mathcal{C}$, then the same is true for any $H \in \mathcal{F}_{F,G}(\mathcal{C},X)$.

The smallest space of interest is the space $\mathcal{C}_s$ of *simple* processes of the form

$$H = y_0 1_{\{0\}\times A_0} + \sum\nolimits_{1\le i\le n} y_i 1_{(t_i,t_{i+1}]\times A_i},$$

where $0 = t_1 < t_2 < \cdots < t_{n+1}$, and for $0 \le i \le n$, we have $y_i \in F$ and $A_i \in \mathcal{F}_{t_i}$. This is the space of step processes over the ring $\mathcal{R}$ generated by the predictable rectangles. The integral of such a process with respect to I_X is defined pathwise by

$$\int H dI_X = y_0 1_{A_0} X_0 + \sum\nolimits_{1\le i\le n} y_i 1_{A_i}(X_{t_{i+1}} - X_{t_i})$$

and belongs to L^p_G. The stochastic integral of such a process is defined, and satisfies

$$(H\cdot X)_t = \int_{[0,t]} H dI_X = y_0 1_{A_0} X_0 + \sum\nolimits_{1\le i\le n} y_i 1_{A_i}(X_{t_{i+1}\wedge t} - X_{t_i\wedge t}).$$

If the set of measures I_{F,L^p_G} is uniformly σ–additive (which is the case if $F = \mathbb{R}$), then the closure $\mathcal{F}_{F,G}(\mathcal{C}_s,X)$ of $\mathcal{C}_s$ contains the space of all bounded predictable processes, and the stochastic integral $H\cdot X$ is defined for all $H \in \mathcal{F}_{F,G}(\mathcal{C}_s,X)$.

However, if I_{F,L^p_G} is not uniformly σ–additive, then $\mathcal{F}_{F,G}(\mathcal{C}_s,X)$ might not contain even the predictable step processes of the form $H = \sum_{1\le i\le n} y_i 1_{E_i}$, with $y_i \in F$ and $E_i \in \mathcal{P}$. In this case we must look for a larger space $\mathcal{C}$ in which $\mathcal{C}_s$ is not necessarily dense in the topology of $\mathcal{F}_{F,G}(X)$, but for which the stochastic integral is still defined. Such a space is the set $\mathcal{E}$ of F–valued *elementary* processes of the form

$$H = H_0 1_{\{0\}} + \sum\nolimits_{1\le i\le n} H_i 1_{(T_i,T_{i+1}]},$$

where $0 \le T_1 \le T_2 \le \cdots \le T_{n+1}$ is an increasing set of stopping times and for each $i = 0,1,\dots,n$, H_i is an F–valued, bounded, $\mathcal{F}_{T_i}$–measurable random variable. For such an elementary process, the integral $\int H dI_X$ can be defined pathwise by

$$\int H dI_X = H_0 X_0 + \sum_{1 \leq i \leq n} H_i (X_{T_{i+1}} - X_{T_i});$$

hence the stochastic integral $H \cdot X$ is defined and

$$(H \cdot X)_t = H_0 X_0 + \sum_{1 \leq i \leq n} H_i (X_{T_{i+1} \wedge t} - X_{T_i \wedge t}).$$

As a result, the stochastic integral $H \cdot X$ is defined for all processes H in the closure $\mathcal{F}_{F,G}(\mathcal{E}, X)$ of $\mathcal{E}$. This follows from a more general theorem.

2.2 THEOREM. *Assume X is p–summable relative to (F, G). Let $\mathcal{C} \subset \mathcal{F}_{F,G}(X)$ be a subspace and assume that every $H \in \mathcal{C}$ has the following two properties:*

(1) *The stochastic integral $H \cdot X$ is defined;*

(2) $(H \cdot X)_v = \int 1_{[0,v]} H dI_X$ a.s. *for every stopping time v. (Condition (2) follows from (1) if the set of measures I_{F, L^p_G} is uniformly σ–additive; in particular if $F = \mathbb{R}$).*

Then every $H \in \mathcal{F}_{F,G}(\mathcal{C}, X)$ has the above two properties.

Moreover, if $H^n \to H$ in $\mathcal{F}_{F,G}(\mathcal{C}, X)$, then there is a subsequence (n_r) such that, a.s., $(H^{n_r} \cdot X)_t \to (H \cdot X)_t$, as $r \to \infty$, uniformly on any compact time interval.

All the integrable processes of interest for this paper belong to the space $\mathcal{F}_{F,G}(\mathcal{E}, X)$. For this reason, in the sequel, we shall restrict ourselves to the space $\mathcal{F}_{F,G}(\mathcal{E}, X)$ and we shall denote this space by $L^1_{F,G}(X)$. The processes H of $L^1_{F,G}(X)$ will be called *integrable processes* with respect to X. The space $L^1_{F,G}(X)$ is complete for the seminorm $\widetilde{I}_{F, L^p_G}$, and the space $\mathcal{E}$ of elementary processes is dense in $L^1_{F,G}(X)$. For every $H \in L^1_{F,G}(X)$, the stochastic integral $H \cdot X$ is defined; if $H^n \to H$ in $L^1_{F,G}(X)$, then there is a subsequence (n_r) such that $(H^{n_r} \cdot X)_t \to (H \cdot X)_t$ a.s., uniformly on compact time intervals.

2.3 THEOREM. *(Lebesgue) Assume X is p–summable relative to (F, G) and let $(H^n)_{n \geq 0}$ be a sequence from $L^1_{F,G}(X)$ and $H \in \mathcal{F}_{F,G}(X)$. Assume that $|H^n| \leq |H^0|$, for each n, and that either*

(a) $H^n \to H$ *in* $\widetilde{I}_{F,L^p_G}$*-measure;*

or

(b) $H^n \to H$ *uniformly on* $\mathbb{R}_+ \times \Omega$*;*

or

(c) $H^n \to H$ *pointwise, and the set of measures* I_{F,L^p_G} *is uniformly* σ*–additive (this is the case if* $F = \mathbb{R}$*).*

Then $H \in L^1_{F,G}(X)$ *and* $H^n \to H$ *in* $L^1_{F,G}(X)$*. Moreover, there is a subsequence* (n_r) *such that a.s.,* $(H^{n_r} \cdot X)_t \to (H \cdot X)_t$*, uniformly on compact time intervals.*

To prove the Itô formula for more general processes, we must consider locally summable processes. We say that a process $X : \mathbb{R}_+ \times \Omega \to E \subset L(F,G)$ is *locally p–summable* relative to (F,G) if there exists an increasing sequence $T_n \nearrow \infty$ of stopping times such that each X^{T_n} is p–summable relative to (F,G). A locally 1–summable process is called, simply, locally summable. Notice that a locally p–summable process is locally summable.

A predictable process $H : \mathbb{R}_+ \times \Omega \to F$ is said to be *locally integrable* with respect to a process X, which is *locally p–summable* relative to (F,G), if there is an increasing sequence $T_n \nearrow \infty$ of stopping times, such that for each n, the process X^{T_n} is p–summable relative to (F,G) and $1_{[0,T_n]}H$ is integrable with respect to X^{T_n}, that is, $1_{[0,T_n]}H \in L^1_{F,G}(X^{T_n})$. In this case, the stochastic integral $(1_{[0,T_n]}H) \cdot X^{T_n}$ is defined for each n, and for $n < m$, we have, a.s.,

$$\left((1_{[0,T_m]}H) \cdot X^{T_m}\right)^{T_n} = (1_{[0,T_n]}H) \cdot X^{T_n}.$$

Then we define the stochastic integral $H \cdot X$ of H with respect to X, up to an evanescent set, by the equality

$$H \cdot X = \lim_n (1_{[0,T_n]}H) \cdot X^{T_n}, \text{ a.s.}$$

The stochastic integral is also denoted by $\int H dX$.

REMARK: Let X be a cadlag, adapted, E–valued process with paths of finite variation on each compact time interval. Such a process is called, simply, a process with ***finite variation***. The variation process $|X|$ of X is defined for every $t \geq 0$ by

$$|X|_t = |X_0| + \mathrm{Var}_{[0,t]} X.$$

The variation $|X|$ is cadlag and adapted. We say X has ***bounded variation*** if

$$|X|_\infty = \sup_{t\geq 0} |X|_t < \infty.$$

X is said to have integrable variation if $E(|X|_\infty) < \infty$. We say a predictable process $H : \mathbb{R}_+ \times \Omega \to F$ is integrable with respect to a process X with ***bounded variation*** X if $\int_{[0,\infty)} |H_s(w)| d|X|_s(w) < \infty$. Finally, H is said to be ***locally integrable*** with respect to a process X with ***finite variation***, if for every $t \geq 0$ and $w \in \Omega$ we have $\int_{[0,t]} |H_s(w)| d|X|_s(w) < \infty$. In this case the Stieltjes integral $\int_{[0,t]} H_s(w) dX_s(w)$ is defined for each $t \geq 0$ and $w \in \Omega$, and the process $(\int_{[0,t]} H_s(w) dX_s(w))_{t\geq 0}$ has values in G, is cadlag and adapted, and has finite variation.

If X is locally p–summable and has finite variation, and if the stochastic integral $H \cdot X$ is defined, then $\int_{[0,t]} |H_s(w)| d|X|_s(w) < \infty$ for each $t \geq 0$ and $w \in \Omega$, and $H \cdot X$ can be defined pathwise by the equality

$$(H \cdot X)_t(w) = \int_{[0,t]} H_s(w) dX_s(w).$$

We use this equality to extend the definition of the stochastic integral $H \cdot X$, for any predictable, F–valued process H which is locally integrable with respect to a process X with finite variation.

In view of the above remark, we make the following general definition:

2.4 DEFINITION. *An E–valued process Z is said to be semi locally p–summable relative to the pair (F, G), if it is of the form $Z = X + Y$, where X is locally p–summable relative to (F, G) and Y is an E–valued, cadlag, adapted process with finite variation.*

An F–valued predictable process H is said to be locally integrable with respect to a semi locally p–summable process $Z = X + Y$ with decomposition as above, if both stochastic integrals $H \cdot X$ and $H \cdot Y$ are defined. In this case we define the stochastic integral $H \cdot Z$ by the equality

$$H \cdot Z = H \cdot X + H \cdot Y.$$

The definition of local integrability of H and of the stochastic integral $H \cdot Z$ is independent of the decomposition $Z = X + Y$.

The above definition is an extension of the classical concept of semimartingale in the scalar case. But even in the Hilbert space case, the concept of semi local summability is more general than that of semimartingale.

An example of F–valued processes, which are locally integrable with respect to any semi locally p–summable process relative to (F, G) is that of the σ–*elementary* processes of the form

$$H = H_0 1_{\{0\}} + \sum\nolimits_{1 \leq n \leq \infty} H_n 1_{(T_n, T_{n+1}]},$$

where $T_n \nearrow \infty$ is an increasing sequence of stopping times, H_0 is an $\mathcal{F}_0$–measurable random variable, and for each n, H_n is an $\mathcal{F}_{T_n}$–measurable random variable. We do not assume that H_0 or H_n are bounded

For such a σ–elementary process H, its stochastic integral with respect to a semi locally p–summable process X relative to (F, G) can be computed pathwise, a.s., by the equality

$$(H \cdot X)_t = H_0 X_0 + \sum\nolimits_{i \leq n \leq \infty} H_n (X_{T_{n+1} \wedge t} - X_{T_n \wedge t}).$$

Note that for each w, the above series is a finite sum.

The Lebesgue theorem extends for locally integrable processes.

2.5 THEOREM. *(Lebesgue) Assume that X is locally p–summable relative to (F, G) and let $(H^n)_{n \geq 0}$ be a sequence of F–valued processes which are locally integrable with respect to X, such that $|H^n| \leq |H^0|$, for each n. Let $H : \mathbb{R}_+ \times \Omega$ - F be a predictable process and assume that either*

(a) $H^n \to H$ locally uniformly (i.e. there is a sequence $T_i \nearrow \infty$ of stopping times such that $H^n \to H$ uniformly on each $[0, T_i]$);

or

(b) $H^n \to H$ pointwise and the set of measures I_{F, L^p_G} is uniformly σ–additive on $\mathcal{P} \cap [0, T_i]$ (this is the case when $F = \mathbb{R}$).

Then $(H^n \cdot X)_t \to (H \cdot X)_t$ in probability for each $t \geq 0$, and there is a subsequence (n_r) such that, a.s., $(H^{n_r} \cdot X)_t \to (H \cdot X)_t$ uniformly on compact time intervals.

The Lebesgue theorem for processes with finite variation is the usual Lebesgue theorem for the Stieltjes integral, pathwise.

2.6 COROLLARY. *Any caglad process $H : \mathbb{R}_+ \times \Omega \to F$ is locally integrable with respect to any semi locally p–summable process X relative to (F, G).*

This corollary follows from the following observation.

Let $a_n \downarrow 0$ and for each n, consider the stopping times $v(n,0) = 0$, and for $k \geq 0$,

$$v(n, k+1) = \inf\{t : |H_{t+} - H_{v(n,k)+}| > a_n\} \wedge (v(n,k) + a_n).$$

For each n, consider the σ–elementary process

$$H^n = \sum\nolimits_{k\geq 0} H_{v(n,k)+} 1_{(v(n,k),v(n,k+1)]}.$$

Then $H^n \to H$ uniformly, and we can apply the Lebesgue theorem.

We now state some properties.

2.7 THEOREM. *If X is semi locally p–summable relative to (F, G) and T is a stopping time, then X^T is semi locally p-summable relative to (F, G) and we have $X^T = 1_{[0,T]} \cdot X$.*

Moreover, if H is an F–valued process which is locally integrable with respect to X, then H is locally integrable with respect to X^T, $1_{[0,T]}H$ is locally integrable with respect to X, and we have

$$(H \cdot X)^T = H \cdot X^T = (1_{[0,T]}H) \cdot X.$$

2.8 THEOREM. *Let $X : \mathbb{R}_+ \times \Omega \to E$ be a semi locally p–summable process relative to $(\mathbb{R}, E)$ and let H be a real valued process, locally integrable with respect to X. Then $H \cdot X$ is semi locally p–summable relative to $(\mathbb{R}, E)$.*

If $E \subset L(F, G)$ and if X is semi locally p–summable relative to (F, G), then so is $H \cdot X$.

Assume that X is semi locally p–summable relative to (F, G). Then an F–valued predictable process K is locally integrable with repsect to $H \cdot X$ if *and only if KH is locally integrable with respect to X, and in this case we have*

$$K \cdot (H \cdot X) = (KH) \cdot X$$

2.9 **THEOREM.** *If H is an F–valued process, locally integrable with respect to a locally p–summable process X relative to (F, G), then the G–valued process $H \cdot X$ is locally p–summable relative to $(\mathbb{R}, G)$ in each of the following cases:*

(1) $1_A H$ is locally integrable with respect to X, for every $A \in \mathcal{P}$;

(2) E does not contain a copy of c_0;

(3) The set of measures I_{F, L^p_G} is locally uniformly σ–addtive (this is the case if $F = \mathbb{R}$).

Assume now X is semi locally p–summable relative to (F, G), H is locally integrable with respect to X and $H \cdot X$ is semi locally p–summable relative to $(\mathbb{R}, G)$. If K is a real valued predictable process such that KH is locally integrable with respect to X, then K is locally integrable with respect to $H \cdot X$, and we have

$$K \cdot (H \cdot X) = (KH) \cdot X.$$

2.10 **THEOREM.** *Let $X : \mathbb{R}_+ \times \Omega \to E \subset L(F, G)$ be a semi locally p–summable process relative to (F, G) and let H be an F–valued process, locally integrable with respect to X. Assume that either*

(1) The set of measures I_{F, L^p_G} is locally uniformly σ–additive (in case X is locally p–summable)

or

(2) H is caglad.

Then

$$\Delta(H \cdot X) = H \Delta X.$$

The case of Hilbert-valued square integrable martingales

Previously, the stochastic integral was defined and studied only for Hilbert-valued square integrable martingales (see [5], [6], and [7]). The square integrable martingales with values in a Hilbert space are 2-summable, and our approach yields a stochastic integral for a larger class of predictable processes than previously considered.

Let E and G be Hilbert spaces, and suppose F is a Banach space such that $E \subset L(F,G)$. Then $L^2_E \subset L(F, L^2_G)$. Let $M : I\!R_+ \times \Omega \to E$ be a square integrable martingale. Consider the measure $I_M : \mathcal{R} \to L^2_E$ corresponding to M. Then I_M has finite semivariation on the ring $\mathcal{R}$, relative to the pair (F, L^2_G). This follows from the fact that if A and B are disjoint sets from $\mathcal{R}$, then $I_M(A)$ and $I_M(B)$ are orthogonal in L^2_E. Since E does not contain a copy of c_0, it follows from Theorem 2.1 that M is 2-summable, relative to (F,G). Moreover, the semivariation relative to the pair (F, L^2_G) satisfies

$$\mathrm{svar}_{F,L^2_G} I_M(A) = \|I_M(A)\|_{L^2_E}, \quad A \in \mathcal{P}.$$

In particular, taking $F = I\!R$ and $G = E$, we have

$$\mathrm{svar}_{I\!R,L^2_E} I_M(A) = \|I_M(A)\|_{L^2_E} = \mathrm{svar}_{F,L^2_G} I_M(A).$$

Hence the semivariation is independent of the spaces F and G. Since the set of measures $(I_M)_{I\!R,L^2_E}$ is uniformly σ-addtive, it follows that the set of measures $(I_M)_{F,L^2_G}$ is also uniformly σ-additive. This also follows from the fact that $A_n \downarrow \emptyset$ implies $I_M(A_n) \to 0$ in L^2_E.

In particular, if M is a real valued square integrable martingale and if we embed $I\!R \subset L(E,E)$, then we have

$$\mathrm{svar}_{E,L^2_E} I_M(A) = \mathrm{svar}_{I\!R,L^2_{I\!R}} I_M(A) = \|I_M(A)\|_{L^2_{I\!R}},$$

for $A \in \mathcal{P}$.

Since $L^2_E \subset L^1_E$, we can regard the measure I_M as taking values in $L^1_E \subset L(F, L^1_G)$. We deduce that M is also 1-summable, and

$$\mathrm{svar}_{F,L^1_G} I_M \leq \mathrm{svar}_{F,L^2_G} I_M;$$

therefore, the set of measures $(I_M)_{F,L^1_G}$ is uniformly σ-additive. It follows that the spaces $L^1_{F,G}(m_{F,L^2_G})$ and $L^1_{F,G}(m_{F,L^1_G})$, where $m = I_M$, contain all F-valued, bounded predicatble processes, and that the space $\mathcal{C}_s$ of F-valued simple processes is dense in these spaces. As a consequence, the stochastic integral $H{\cdot}X$ is defined for every process $H \in L^1_{F,G}(m_{F,L^1_G})$ and $(H \cdot X)_t \in L^1_G$ for every $t \geq 0$.

If $H \in L^1_{F,G}(m_{F,L^2_G})$, then $(H \cdot X)_t \in L^2_G$, for $t \geq 0$, and $H \cdot X$ is a square integrable martingale, and we have

$$\|(H{\cdot}X)_\infty\|^2_{L^2_G} = E\left(\int_{[0,\infty)} |H|^2 d\langle M,M\rangle\right),$$

where $\langle M,M\rangle$ is the sharp bracket of M. This realizes the classical isometry between the spaces $L^1_{F,G}(m_{F,L^2_G})$ and $L^2_E(\langle M,M\rangle)$ since $m_{F,L^2_G}(H) = \|(H{\cdot}X)_\infty\|^2_{L^2_G}$, if $H \in L^1_{F,G}(m_{F,L^2_G})$. However, if $H \in L^1_{F,G}(m_{F,L^1_G})$, then $H{\cdot}X$ is defined, but is no longer a square integrable martingale. Thus our approach yields a larger class, than before, of processes H for which the stochastic integral $H \cdot X$ can be defined.

If X is a locally square integrable martingale, then it is locally 2–summable relative to (F, G), therefore it is also locally summable relative to (F, G).

REMARK: In the above considerations, we assumed that both E and G are Hilbert spaces. If G is not a Hilbert space, then an E-valued square integrable martingale is not necessarily summable, since $\mathrm{svar}(I_M)_{F,L^1_G}$ may be infinite. This is the case with the example considered by Yor [8], where X is the real Brownian motion $(B_t)_{t\geq 0}$, $F = G = \ell^1$. Yor gives an example of an ℓ^1-valued σ-elementary process which is not locally integrable with respect to B. Since σ-elementary processes are locally integrable with respect to any locally summable process, it follows that the Brownian motion does not have local finite semivariation relative to the pair $(\ell^1, L^1_{\ell^1})$.

3. Preliminary results

For the remainder of the paper, we shall assume that $E \subset L(F, G)$ and $X : \mathbb{R}_+ \times \Omega \to E$ is a cadlag, adapted process with $X_t \in L^1_E(P)$, for $t \geq 0$. In the sequel we shall state and prove the results concerning only local summability. The results concerning local p-summabililty will then follow as corollaries.

We shall reserve the notation $(v(n,k))_{k\geq 0}$, $n = 1,2,\ldots$ for a family of stopping times satisfying the following three conditions:

(i) for each n, we have $v(n,0) = 0$, and $v(n,k) \nearrow \infty$, as $k \to \infty$;

(ii) $\lim_n \sup_k (v(n,k+1) - v(n,k)) = 0$;

(iii) there is a sequence $a_n \downarrow 0$ such that for $t \in [v(n,k), v(n,k+1))$, we have

$$|X_t - X_{v(n,k)}| \leq a_n.$$

Given $a_n \downarrow 0$, an example of such a family of stopping times is:

$v(n,0) = 0,$

and

$$v(n,k+1) = \inf\{t > v(n,k) : |X_t - X_{v(n,k)}| > a_n\} \wedge (v(n,k) + a_n).$$

These stopping times will play an important role in the development of Itô's formula.

If for each n, we denote

$$X^n = \sum\nolimits_{k\geq 0} X_{v(n,k)} 1_{(v(n,k),v(n,k+1)]}.$$

From condition (ii) we deduce that $X^n \to X_-$ pointwise. From condition (iii), we deduce that

$$|X_{t-} - X_{v(n,k)}| \leq a_n, \text{ for } t \in (v(n,k), v(n,k+1)],$$

and this shows that $X^n \to X_-$ uniformly on $\mathbb{R}_+ \times \Omega$. We use the standard convention that $Y_{0-} = 0$, for any process Y.

3.1 THEOREM. *Assume that X is locally summable relative to (F,G) and let $H : \mathbb{R}_+ \times \Omega \to F$ be a cadlag, adapted process; therefore, H_- is locally integrable with respect to X. Assume further that either*

(a) the set of measures $(I_X)_{F,L^1_G}$ is locally uniformly σ-additive (this is the case, for example, if $F = \mathbb{R}$);

or

(b) there is a sequence $b_n \downarrow 0$ such that for $t \in [v(n,k), v(n,k+1))$ we have

$$|H_t - H_{v(n,k)}| \le b_n.$$

Then for every $t \ge 0$, we have

$$\begin{aligned}\int_{[0,t]} H_{s-}dX_s &= \lim\text{prob} \sum_{k\ge 0} H_{v(n,k)\wedge t}(X_{v(n,k+1)\wedge t} - X_{v(n,k)\wedge t}) \\ &= \lim\text{prob} \sum_{k\ge 0} H_{v(n,k)} \left(X_{v(n,k+1)} - X_{v(n,k)}\right) 1_{\{v(n,k)\le t\}}.\end{aligned}$$

Moreover, there is a subsequence (n_r) such that a.s., the first limit is uniform on compact time intervals, and the second limit is pointwise.

PROOF: For each n, define

$$H^n = \sum_{k\ge 0} H_{v(n,k)} 1_{(v(n,k),v(n,k+1)]}.$$

The process H^n is σ–elementary, hence locally integrable with respect to X and

$$\int_{[0,t]} H^n_s dX_s = \sum_{k\ge 0} H_{v(n,k)\wedge t}(X_{v(n,k+1)\wedge t} - X_{v(n,k)\wedge t}).$$

Condition (ii) of the stopping times $v(n,k)$ implies that $H^n \to H_-$ pointwise. If, in addition, H satisfies condition (b) above, then $H^n \to H_-$ uniformly on $\mathbb{R}_+ \times \Omega$.

Let $M_i \nearrow \infty$, and for each i, define the stopping time

$$T_i = \inf\{t : |H_t| > M_i\}.$$

Then $T_i \nearrow \infty$, and on $[0, T_i)$, we have $|H_t| \le M_i$; therefore $|H^n_t| \le M_i$, for each n. Since H^n is left continuous, we deduce that $|H^n_t| \le M_i$ on $[0, T_i]$ for each n. The process

$$\Phi = \sum_{i\ge 0} M_{i+1} 1_{(T_i, T_{i+1}]}$$

is σ–elementary; if we choose $x \in F$, with $|x| = 1$, then $H^0 = \Phi x$ is locally integrable with respect to X and for each n, we have $|H^n| \le |H^0|$. Under either of the conditions (a) or (b) above, we can apply the Lebesgue theorem 2.5 and deduce that

$$\begin{aligned}\int_{[0,t]} H_{s-}dX_s &= \lim\text{prob} \int_{[0,t]} H^n_s dX_s \\ &= \lim\text{prob} \sum_{k\geq 0} H_{v(n,k)\wedge t}\left(X_{v(n,k+1)\wedge t} - X_{v(n,k)\wedge t}\right),\end{aligned}$$

and that there exists a subsequence (n_r) such that a.s., the convergence is uniform on any compact time interval.

The equality involving the second limit follows from the fact that if $t \in [v(n,j), v(n,j+1))$, then

$$\begin{aligned}&\lim_n \sum_{k\geq 0} \left[H_{v(n,k)}\left(X_{v(n,k+1)} - X_{v(n,k)}\right) - H_{v(n,k)\wedge t}\left(X_{v(n,k+1)\wedge t} - X_{v(n,k)\wedge t}\right)\right] \\ &= \lim_n \left[H_{v(n,j)}\left(X_{v(n,j+1)} - X_{v(n,j)}\right) - H_{v(n,j)}\left(X_t - X_{v(n,j)}\right)\right] = 0.\end{aligned}$$

REMARKS:

(1) In the proof of Theorem 3.1, we did not use condition (iii) of the stopping times.

(2) If there are no conditions imposed on the family $v(n,k)$ of stopping times, other than conditions (i) and (ii), we can always choose $v(n,k)$ to satisfy conditions (i)–(iii) and condition (a) above. In fact we can take $v(n,0) = 0$, and

$$\begin{aligned}v(n,k+1) = \inf\{t > v(n,k) : |X_t - X_{v(n,k)}| > a_n \text{ or } |H_t - H_{v(n,k)}| > b_n\}\wedge \\ \wedge\left(v(n,k) + a_n + b_n\right).\end{aligned}$$

But, in some cases, for example, when the existence of the quadratic variation $[X]$ is assumed (see section 5), then, in general, the stopping times $v(n,k)$ cannot be replaced. They can be replaced if E is a Hilbert space, and X is summable relative to the inner product (see section 5).

(3) Since $H_{0-} = 0$, we have $\int_{(0,t]} H_{s-}dX_s = \int_{[0,t]} H_{s-}dX_s$.

The following theorem is the analog of Theorem 3.1 for processes of finite variation.

3.2 THEOREM. *Assume that X has finite variation and let $H : I\!R_+ \times \Omega \to F$ be a cadlag, adapted process. Then H_- is locally integrable with respect to X*

and for every $t \geq 0$ we have

$$\int_{[0,t]} H_{s-} dX_s = \lim_n \sum_{k\geq 0} H_{v(n,k)\wedge t} \left(X_{v(n,k+1)\wedge t} - X_{v(n,k)\wedge t}\right)$$
$$= \lim_n \sum_{k\geq 0} H_{v(n,k)} \left(X_{v(n,k+1)} - X_{v(n,k)}\right) 1_{\{v(n,k)\leq t\}}.$$

pointwise. The first limit is uniform on compact time intervals if H satisfies the following condition:

$(b(w))$ for each $w \in \Omega$, there is a sequence $b_n \downarrow 0$ (depending on w) such that for $t \in [v(n,k), v(n,k+1))$ we have $|H_t - H_{v(n,k)}| \leq b_n$.

PROOF: Using the notations in the proof of Theorem 3.1, we have $H^n \to H_-$ pointwise. Let $t \geq 0$. Since H is cadlag, it is pathwise bounded on $[0,t]$ by a constant M (depending on w and t); also $|H^n| \leq M$ on $[0,t]$ for each n. We can then apply the Lebesgue theorem pathwise for the Stieltjes integral and deduce that, pathwise,

$$\int_{[0,t]} H_{s-} dX_s = \lim_n \int_{[0,t]} H_s^n dX_s.$$

If condition $b(w)$ is satisfied, then $H^n \to H_-$ uniformly for each w. Let $[0,a]$ be a compact interval. If $t \in [0,a]$, then

$$\left| \int_{[0,t]} H_s^n dX_s - \int_{[0,t]} H_{s-} dX_s \right| \leq \sup_{s\in[0,a]} |H_s^n - H_{s-}| \, |X_a| \to 0,$$

which implies that

$$\int_{[0,t]} H_{s-} dX_s = \lim_n \int_{[0,t]} H_s^n dX_s$$

uniformly on $[0,a]$. This proves the conclusion concerning the first limit in the statement. The part concerning the second limit is proved as in the proof of Theorem 3.1.

3.3 THEOREM. *Assume that X is locally summable relative to (F,G) and suppose $h : E \to F$ is a continuous function. The $h(X_-)$ is locally integrable with respect to X. Assume further that either*

(a) The set of measures $(I_X)_{F,L_G^1}$ is uniformly σ-additive (for example, when $F = \mathbb{R}$);

or

(b′) h is uniformly continuous and bounded on bounded subsets of E.

Then for every $t > 0$, we have

$$\int_{(0,t]} h(X_{s-})dX_s = \operatorname{lim\,prob} \sum\nolimits_{k\geq 0} h\left(X_{v(n,k)\wedge t}\right)\left(X_{v(n,k+1)\wedge t} - X_{v(n,k)\wedge t}\right)$$
$$= \operatorname{lim\,prob} \sum\nolimits_{k\geq 0} h\left(X_{v(n,k)}\right)\left(X_{v(n,k+1)} - X_{v(n,k)}\right) 1_{\{v(n,k)\leq t\}}$$

and there is a subsequence (n_r) such that, a.s., the first limit is uniform on any compact time interval, and the second limit is pointwise.

PROOF: The process $H = h(X_t)$ is F–valued, cadlag and adapted. We note that $H_{t-} = h(X_{t-})$ for $t > 0$, but not necessarily for $t = 0$. Condition (a) above is identical to condition (a) of Theorem 3.1, therefore the conclusion follows in this case. Now assume condition (b′).

Denote

$$X^n = \sum\nolimits_{k\geq 0} X_{v(n,k)} 1_{(v(n,k),v(n,k+1)]}$$

and

$$H^n = \sum\nolimits_{k\geq 0} H_{v(n,k)} 1_{(v(n,k),v(n,k+1)]}.$$

We notice that for $t > 0$ we have $H_t^n = h(X_t^n)$. By condition (iii) of the stopping times $v(n,k)$, we have $X_t^n \to X_{t-}$ uniformly on $\mathbb{R}_+ \times \Omega$. Since h is continuous, we see that $H_t^n \to H_{t-}$ pointwise. We shall prove that $H_t^n \to H_{t-}$ locally uniformly. Let $M_i \nearrow \infty$ and for each i, define the stopping time $T_i = \inf\{t : |X_t| > M_i\}$. Then $T_i \nearrow \infty$ and $1_{[0,T_i)}|X| \leq M_i$; thus $1_{[0,T_i)}|X_-| \leq M_i$. It follows that $1_{[0,T_i)}|X^n| \leq M_i$, and by left continuity of X^n, also $1_{[0,T_i]}|X^n| \leq M_i$ for each n and i. Since $1_{[0,T_i]}X^n \to 1_{[0,T_i]}X_-$ uniformly, and since h is uniformly continuous on the bounded set $B_i = \{|x| \leq M_i\}$, we see that

$$1_{[0,T_i]}h(X^n) \to 1_{[0,T_i]}h(X_-) \text{ uniformly.}$$

Since $H_0^n = H_{0-} = 0$, we deduce that $1_{[0,T_i]}H^n \to 1_{[0,T_i]}H_-$ uniformly.

Note that h is bounded on B_i; set $N_i = \sup\{|h(x)| : |x| \leq M_i\}$. Then $1_{[0,T_i]}|H^n| \leq N_i$ for all n. The σ–elementary process $H^0 = \sum_{i\geq 1} xN_i 1_{(T_i,T_{i+1}]}$, where $x \in F$ and $|x| = 1$, is locally integrable with respect to X, and we have $|H^n| \leq |H^0|$ for all n. By the Lebesgue Theorem 2.5, we see that (since $H_0^n = H_{0-} = 0$),

$$\int_{[0,t]} H^n dX \to \int_{[0,t]} H_- dX$$

in probability, and uniformly on compact time intervals for a certain subsequence (n_r). The conclusion follows by using the equalities $1_{[0,t]}H^n = 1_{(0,t]}h(X^n)$ and $1_{[0,t]}H_- = 1_{(0,t]}h(X_-)$.

This proves the part of the theorem concerning the first limit. The part concerning the second limit follows from the fact that the difference of the sums involved in the two limits converges to zero a.s., as a consequence of condition (ii) of the stopping times $v(n,k)$, and the cadlag property of H.

REMARK: Under assumption (a), we do not need condition (iii) of the stopping times $v(n,k)$.

3.4 COROLLARY. *Assume that $F = L(E,G)$; hence $E \subset L(F,G)$. Assume X is locally summable relative to (F,G). Let $f : E \to G$ be a function of class C^1. Then $f' : E \to L(E,G) = F$ is continuous and $f'(X_-)$ is locally integrable with respect to X. Assume further that either*

(a) the set of measures $(I_X)_{F,L^1_G}$ is uniformly σ–additive (for example, if $F = \mathbb{R}$);

or

(b″) the second derivative $f'' : E \to L(E\hat{\otimes}_\pi E, G)$ is bounded on bounded subsets of E.

Then, for every $t > 0$, we have

$$\begin{aligned}\int_{(0,t]} f'(X_{t-})dX_s &= \lim_n \sum_{k\geq 0} f'\left(X_{v(n,k)\wedge t}\right)\left(X_{v(n,k+1)\wedge t} - X_{v(n,k)\wedge t}\right)\\ &= \lim_n \sum_{k\geq 0} f'\left(X_{v(n,k)}\right)\left(X_{v(n,k+1)} - X_{v(n,k)}\right) 1_{\{v(n,k)\leq t\}}\end{aligned}$$

in probability, and there exists a subsequence (n_r) such that a.s. the first limit is uniform on compact time intervals and the second limit is pointwise.

PROOF: Let $B \subset E$ be a bounded set and let $S = \{|x| < r\}$ be an open sphere containing B. Let $M = \sup\{|f''(x)| : |x| < r\} < \infty$. We can apply the mean value theorem and deduce that

$$|f'(x) - f'(y)| \leq M|x - y|, \text{ for } x, y \in S.$$

It follows that f' is uniformly continuous on S. Also, we see that f' is bounded on S. Since the function $h = f'$ satisfies condition (b′) of Theorem 3.3, the conclusion follows.

REMARK: Under assumption (a) we do not use condition (iii) of the stopping times.

The following theorem and corollary are the analog of Theorem 3.3 and Corollary 3.4 respectively, for processes of finite variation.

3.5 THEOREM. *Assume X has finite variation and let $h : E \to F$ be a continuous function. Then $h(X_-)$ is locally integrable with respect to X and for every $t > 0$ we have*

$$\begin{aligned}\int_{(0,t]} h(X_{s-})dX_s &= \lim_n \sum_{k\geq 0} h\left(X_{v(n,k)\wedge t}\right)\left(X_{v(n,k+1)\wedge t} - X_{v(n,k)\wedge t}\right)\\ &= \lim_n \sum_{k\geq 0} h\left(X_{v(n,k)}\right)\left(X_{v(n,k+1)} - X_{v(n,k)}\right) 1_{\{v(n,k)\leq t\}}\end{aligned}$$

pointwise. If h is uniformly continuous and bounded on bounded subsets of E, then the first limit is uniform on bounded time intervals.

PROOF: If we set $H_t = h(X_t)$, then H is cadlag, adapted and the first part of the theorem follows from Theorem 3.2, using $1_{[0,t]}H_- = 1_{(0,t]}h(X_-)$. Assume now that h is uniformly continuous and bounded on bounded subsets of E. Fix $w \in \Omega$ and $a > 0$; we shall prove that the first limit is uniform for $t \in (0, a]$. We shall use the notations of Theorem 3.3. For a given sequence $M_i \nearrow \infty$, we can find a sequence of stopping times $T_i \nearrow \infty$ such that $1_{[0,T_i]}|X^n| \leq M_i$ and

$$1_{[0,T_i]}X_t^n \to 1_{[0,T_i]}X_{t-} \text{ uniformly.}$$

Since h is uniformly continuous on the set $B_i = \{|x| \leq M_i\}$, we see that

$$1_{[0,T_i]}h(X_t^n) \to 1_{[0,T_i]}h(X_{t-}) \text{ uniformly.}$$

Taking i such that $a < T_i(w)$, we deduce that $H_t^n \to H_{t-}$ uniformly on $[0, a]$. Let

$$N = \sup\{|h(x)| : |x \leq M_i\} < \infty.$$

Then $|H_t^n| \leq N$ for all n. We can then apply the Lebesgue theorem for the Stieltjes integral and conclude that for every $t \in (0, a]$ we have

$$\left| \int_{(0,t]} h(X_{s-})dX_s - \int_{(0,t]} h(X_s^n)dX_s \right|$$
$$\leq \sup_{s\in[0,a]} |H_{s-} - H_s^n| \, |X|_a \to 0$$

as $n \to \infty$, which proves the second conclusion of the theorem.

3.6 COROLLARY. *Assume X has finite variation and let $f : E \to G$ be a function of class C^2 such that the second derivative $f'' : E \to L(E \hat{\otimes}_\pi E, G)$ is bounded on bounded subsets of E. Then $f'(X_-)$ is locally integrable with respect to X on $[0,t]$ for every $t > 0$ and we have*

$$\int_{(0,t]} f'(X_{s-})dX_s = \lim_n \sum\nolimits_{k\geq 0} f'\left(X_{v(n,k)\wedge t}\right)\left(X_{v(n,k+1)\wedge t} - X_{v(n,k)\wedge t}\right)$$
$$= \lim_n \sum\nolimits_{k\geq 0} f'\left(X_{v(n,k)}\right)\left(X_{v(n,k+1)} - X_{v(n,k)}\right) 1_{\{v(n,k)\leq t\}}$$

pointwise, and the first limit is uniform on bounded time intervals.

The proof is the same as that of Corollary 3.4.

4. The vector quadratic variation

In this section, E and D are Banach spaces and $B : E \times E \to D$ is a continuous bilinear mapping, denoted by $B(x, y) = xy$, such that $|x| = \sup\{|xy| : |y| \leq 1\}$ for every $x \in E$. We denote by $B' : E \times E \to D$ the bilinear mapping defined by $B'(x, y) = B(y, x) = yx$, for $x, y \in E$. We write $x^2 = x \cdot x$. Important examples of such bilinear mappings are:

(1) The tensor product $B(x, y) = x \otimes y$ from $E \times E$ into $E \hat{\otimes}_\pi E$. We write $x^{\otimes 2} = x \otimes x$.

(2) The inner product $B(x, y) = \langle x, y \rangle$, if E is a Hilbert space.

If $B : E \times E \to D$ is a bilinear mapping as above, we can embed isometrically $E \subset L(E, D)$. We say a process $X : \mathbb{R}_+ \times \Omega \to E$ is locally summable or semi locally summable relative to the bilinear mapping B, if we regard X as taking values in $L(E, D)$ and X is locally summable or semi locally summable relative to (E, D).

If X is semi locally summable relative to both bilinear mappings B and B', then we can integrate E–valued processes H and the stochastic integrals $\int H \cdot dX$ and $\int dX \cdot H$ have values in D. In particular, if X is semi locally summable relative to both B and B', then X_- is locally integrable with respect to X for both B and B' and the stochastic integrals $\int_{[0,t]} X_{s-} \cdot dX_s$ and $\int_{[0,t]} dX_s \cdot X_{s-}$ are defined and have values in D.

We remark that if B is the tensor product, or the inner product, then local summability relative to B implies local summability relative to B', since in this case

$$\left|\sum B(x_i, y_i)\right| = \left|\sum B(y_i, x_i)\right| = \left|\sum B'(x_i, y_i)\right|;$$

as a result, the semivariations of I_X relative to B and B' are equal.

We now define a process which is fundamental to the establishment of Itô's formula.

4.1 Definition. *Assume that $X : \mathbb{R}_+ \times \Omega \to E$ is semi locally summable relative to B and B'. The D–valued, cadlag, adapted process $[[X]]^B$ defined by*

$$[[X]]^B_t = X_t^2 - \int_{[0,t]} (X_{s-} \cdot dX_s + dX_s \cdot X_{s-})$$

is called the vector quadratic variation (or vector square bracket) of the process X with respect to B.

We note that $[[X]]^B_0 = X_0^2$.

If $B(x, y) = x \otimes y$ and $D = E \hat{\otimes}_\pi E$, we denote $[[X]] = [[X]]^B$, and call $[[X]]$ the tensor quadratic variation of X. Hence, if X is semi locally summable relative to the tensor product mapping $(x, y) \to x \otimes y$, then

$$[[X]]_t = X_t^{\otimes 2} - \int_{[0,t]} (X_{s-} \otimes dX_s + dX_s \otimes X_{s-}).$$

If E is a Hilbert space and X is semi locally summable relative to the inner product B, we denote $[X] = [[X]]^B$, and call $[X]$ the *scalar quadratic variation* of X. In this case we have

$$[X]_t = |X_t|^2 - 2 \int_{[0,t]} X_{s-} dX_s.$$

We shall present, in the next section, an extension of $[X]$, when E is a general Banach space.

Remark: If E is a Hilbert space and X is a semimartingale, then both $[[X]]$ and $[X]$ can be defined since in this case X is semi locally summable relative to the corresponding bilinear maps.

4.2 **Theorem.** *Assume $X : \mathbb{R}_+ \times \Omega \to E$ is semi locally summable relative to B and B'. For each $t \geq 0$, we have*

$$[[X]]_t^B = X_0^2 + \lim_n \text{prob} \sum_{k\geq 0} \left(X_{v(n,k+1)\wedge t} - X_{v(n,k)\wedge t}\right)^2$$
$$= X_0^2 + \lim_n \text{prob} \sum_{k\geq 0} \left(X_{v(n,k+1)} - X_{v(n,k)}\right)^2 1_{\{v(n,k)\leq t\}}.$$

If X is locally summable, then there is a subsequence such that the first limit is a.s., uniform on compact time intervals, and the second limit exists pointwise a.s.

If X has finite variation, the first limit is a.s. uniform on compact time intervals and the second limit is pointwise a.s.

Proof: We use theorems 3.1 and 3.2 with H and H^n replaced by X and X^n, defined by

$$X^n = \sum_{k\geq 0} X_{v(n,k)} 1_{(v(n,k),v(n,k+1)]}.$$

We have $X^n \to X_-$ pointwise uniformly, by condition (iii) of the stopping times $v(n,k)$. The processes X^n and X_- are locally integrable with respect to X. If X is locally summable, then, by Theorem 3.1

$$\int_{[0,t]} X_s^n dX_s \to \int_{[0,t]} X_{s-} dX_s \text{ and } \int_{[0,t]} dX_s \cdot X_s^n \to \int_{[0,t]} dX_s \cdot X_{s-}$$

in probability, and uniformly on compact time intervals for a certain subsequence. For each $t \geq 0$, we have

$$\sum_{k\geq 0} \left(X_{v(n,k+1)\wedge t} - X_{v(n,k)\wedge t}\right)^2 = \sum_{k\geq 0} \left(X^2_{v(n,k+1)\wedge t} - X^2_{v(n,k)\wedge t}\right)$$
$$- \sum_{k\geq 0} X_{v(n,k)\wedge t} \cdot \left(X_{v(n,k+1)\wedge t} - X_{v(n,k)\wedge t}\right)$$
$$- \sum_{k\geq 0} \left(X_{v(n,k+1)\wedge t} - X_{v(n,k)\wedge t}\right) \cdot X_{v(n,k)\wedge t}$$

and this converges in probability, as $n \to \infty$, to

$$X_t^2 - X_0^2 - \int_{[0,t]} (X_{s-} \cdot dX_s + dX_s \cdot X_{s-}) = [[X]]_t^B - X_0^2.$$

If X has finite variation, we apply Theorem 3.2, with H replaced by X, using the fact that condition (iii) of the stopping times $v(n,k)$ implies condition b(w) of Theorem 3.2.

REMARKS:

(1) As a rule, either of the sums involved in the convergence to the vector quadratic variation $[[X]]$ will be denoted by W^n:

$$[[X]]_t = X_0^2 + \lim \operatorname{prob} W_t^n$$

(2) If the set of measures $(I_X)_{E,L_D^1}$ is uniformly σ–additive, we can require that the stopping times satisfy only conditions (i) and (ii) of section 3.

(3) We shall see in the following sections that if X is semi locally summable relative to B and B' and has "finite quadratic variation," then $[[X]]^B$ has finite variation.

(4) If E is a Hilbert space and X is semi locally summable relative to the inner product, then $[X]$ is an increasing, positive process, and we have

$$[X]_t = X_0^2 + \lim_n \operatorname{prob} \sum_{k\geq 0} \left|X_{v(n,k+1)\wedge t} - X_{v(n,k)\wedge t}\right|^2.$$

5. The quadratic variation

In this section $X : \mathbb{R}_+ \times \Omega \to E$ is a cadlag, adapted process and $B(x,y) = xy$ is a bilinear mapping of $E \times E$ into D, with $|x| = \sup\{|xy| : |y| \leq 1\}$, for $x \in E$. In order to ensure that $[[X]]$ has finite variation paths, we need to examine the quadratic variation $[X]$.

5.1 DEFINITION. *We say that X has finite quadratic variation if there exists a family of stopping times $(v(n,k))_{n,k}$ (cf. section 3), such that*

$$\lim_n \sum_{k\geq 0} \left|X_{v(n,k+1)\wedge t} - X_{v(n,k)\wedge t}\right|^2$$

exists and is finite a.s. for every $t \geq 0$. The process

$$[X]_t = |X_0|^2 + \lim_n \sum_{k\geq 0} \left|X_{v(n,k+1)\wedge t} - X_{v(n,k)\wedge t}\right|^2$$

is called the quadratic variation of X.

Note that $[X]_0 = |X_0|^2$ and $[X]$ is positive, increasing and adapted. Moreover,

$$[X]_t = |X_0|^2 + \lim_n \sum_{k\geq 0} \left|X_{v(n,k+1)} - X_{v(n,k)}\right|^2 1_{\{v(n,k)\leq t\}}.$$

In fact, for $v(n,k) \leq t < v(n,k+1)$, the difference of the two sums is $|X_t - X_{v(n,k)}|^2 - |X_{v(n,k+1)} - X_{v(n,k)}|^2$, and since X is right continuous, this has limit zero, as $n \to \infty$. In the above equality one can replace $1_{\{v(n,k)\leq t\}}$ by $1_{\{v(n,k)<t\}}$ if $t > 0$.

REMARKS:

(1) As a rule, either of the sums involved in the definition of the quadratic variation $[X]$ will be denoted by V^n, hence $[X]_t = |X_0|^2 + \lim_n V_t^n$.

(2) If E is a Hilbert space and X is a semimartingale, then X has finite quadratic variation $[X]$, which is equal to the vector quadratic variation $[[X]]^B$, relative to the inner product B.

Next we state the following important property concerning the variation of $[[X]]^B$.

5.2 PROPOSITION. *Assume that* X *is semi locally summable relative to* B *and* B' *and suppose that* X *has finite quadratic variation. Then* $[[X]]^B$ *has finite variation and*

$$\left|[[X]]_t^B - [[X]]_s^B\right| \leq [X]_t - [X]_s, \text{ for } s \leq t.$$

PROOF: For $0 \leq s < t$, we have

$$\begin{aligned} &\left| \sum_{k\geq 0} \left(X_{v(n,k+1)} - X_{v(n,k)}\right)^2 1_{\{s<v(n,k)\leq t\}} \right| \\ &\leq \sum_{k\geq 0} \left|X_{v(n,k+1)} - X_{v(n,k)}\right|^2 1_{\{s<v(n,k)\leq t\}}. \end{aligned}$$

If we pass to the limit along a convenient subsequence, we obtain the desired inequality.

REMARK: When $[[X]]^B$ has finite variation and takes values in D, one can integrate pathwise, with respect to $[[X]]^B$, processes H with values in F if $F \subset L(D,G)$. Although $[[X]]_t^B$ has an integral representation, we cannot use this formula in evaluating $\int H_s d[[X]]_s^B$, since we do not know if the processes $\int X_{s-} \cdot dX_s$ and $\int dX_s \cdot X_{s-}$ have local finite variation or if they are locally summable relative to the bilinear mapping $D \times F \to G$.

We shall use the following convergence result in studying integration with respect to $[[X]]^B$ in Theorem 5.4

5.3 PROPOSITION. *Let $(x, y) \to xy$ be a continuous bilinear mapping from $E \times F$ into G. Suppose $h_n, h : [a, b] \to E$ are functions with equally bounded variations such that $\lim_n h_n = h$ pointwise. Let $g : [a, b] \to F$ be a caglad function.*

Then

$$\int_{[a,s]} g dh^n \to \int_{[a,s]} g dh, \text{ for } s \in [a, b].$$

Moreover, the convergence is uniform if $h_n \to h$ uniformly.

PROOF: Choose a number V larger than $\sup_n V_{[a,b]}(h_n)$ and $V_{[a,b]}(h)$. Let $\epsilon > 0$. Obtain a partition $a = t_0 < t_1 < \cdots < t_{p+1} = b$ such that the oscillation of g on each interval $(t_i, t_{i+1}]$, $0 \leq i \leq p$, is smaller than $\epsilon/4V$. To do this, set $g(t) = g(a)$ for $t < a$ and $g(t) = g(b)$ for $t > b$; for each $t \in [a, b]$, obtain an open interval (a_t, b_t) containing t such that g has oscillation less than $\epsilon/4V$ on $(a_t, t]$, by left continuity of g, and oscillation less than $\epsilon/4V$ on $(t, b_t]$, since right limits exist. Use the compactness of $[a, b]$ to obtain a covering (a_{t_i}, b_{t_i}), $i = 0, 1, \ldots, p+1$ of $[a, b]$. We then intersect $(a_{t_i}, t_i] \cap (a, t_i]$ and $(t_i, b_{t_i}] \cap (t_i, b]$.

We adopt the convention that $h^n(a-) = h(a-) = 0$. If $s \in [a, b]$, then

$$\begin{aligned} I_n := &\int_{[a,s]} gd(h^n - h) \\ = &\int_{\{a\}} gd(h^n - h) + \sum\nolimits_{0 \leq i \leq p} \int_{(t_i, t_{i+1}] \cap [a,s]} [g(t) - g(t_i)] dh^n(t) \\ &- \sum\nolimits_{0 \leq i \leq p} \int_{(t_i, t_{i+1}] \cap [a,s]} [g(t) - g(t_i)] dh(t) \\ &+ \sum\nolimits_{0 \leq i \leq p} \int_{(t_i, t_{i+1}] \cap [a,s]} g(t_i) d(h^n - h)(t). \end{aligned}$$

Then

$$\left| \int_{\{a\}} gd(h^n - h) \right| \leq |g(a)| \, |h^n(a) - h(a)| < \epsilon/4$$

for $n \geq n_0$, for a suitable n_0.

Also,

$$\sum_{0\leq i\leq p}\left|\int_{(t_i,t_{i+1}]\cap[a,s]}[g(t)-g(t_i)]dh^n\right|$$
$$\leq \sum_{0\leq i\leq p}\int_{(t_i,t_{i+1}]}|g(t)-g(t_i)|d|h^n| < \epsilon/4$$

and similarly

$$\sum_{0\leq i\leq p}\left|\int_{(t_i,t_{i+1}]\cap[a,s]}[g(t)-g(t_i)]dh\right| < \epsilon/4.$$

Finally,

$$\left|\int_{(t_i,t_{i+1}]\cap[a,s]} g(t_i)d(h^n-h)\right|$$

$$\leq \|g\|_\infty\,|h^n(t_{i+1}\wedge s)-h(t_{i+1}\wedge s)-(h^n(t_i\wedge s)-h(t_i\wedge s))|$$
$$< \epsilon/4V(p+1),$$

for n sufficiently large, for $0\leq i\leq p$. Note that $\|g\|_\infty<\infty$, since g is caglad.

Thus, for n large enough, $|I_n|<\epsilon$, that is, $I_n\to 0$, which proves the first conclusion. If $h_n\to h$ uniformly, then $I_n\to 0$ uniformly on $[a,b]$.

We now apply the above to study the integral with respect to $[[X]]^B$.

5.4 THEOREM. *Assume that X is semi locally summable relative to the bilinear maps B and B' and suppose X has finite quadratic variation $[X]$.*

Let $F = L(D,G)$ and let H be an F–valued cadlag, adapted process. Then $[[X]]^B$ has finite variation, hence the Stieltjes integral $\int_{[0,t]}H_{s-}d[[X]]^B_s$ is defined pathwise. Let $(v(m,k))$ be a family of stopping times used to define the quadratic variation $[X]$. Assume that for each $w\in\Omega$, there is a sequence $b_n(w)\downarrow 0$ such that for $t\in[v(n,k),v(n,k+1))$, we have

(b″) $|H_t(w)-H_{v(n,k)}(w)|\leq b_n(w)$.

Then there is a subsequence $\{n\}$ of $\{m\}$ such that a.s.

$$\int_{[0,t]}H_{s-}d[[X]]^B_s = \lim_n\sum_{k\geq 0}H_{v(n,k)}\left(X_{v(n,k+1)\wedge t}-X_{v(n,k)\wedge t}\right)^2$$
$$= \lim_n\sum_{k\geq 0}H_{v(n,k)}\left(X_{v(n,k+1)}-X_{v(n,k)}\right)^2 1_{\{v(n,k)\leq t\}}.$$

Proof: Denote

$$H^m = \sum\nolimits_{k\geq 0} H_{v(m,k)} 1_{(v(m,k),v(m,k+1)]}.$$

Then $H^m \to H_-$ pathwise, uniformly on compact time intervals by condition (b″) above.

Denote

$$W_t^m = \sum\nolimits_{k\geq 0} \left(X_{v(m,k+1)} - X_{v(m,k)}\right)^2 1_{\{v(m,k)\leq t\}}.$$

This process is simple, constant on $[v(m,k), v(m,k+1))$ and has jumps at $v(m,k)$ equal to $(X_{v(m,k+1)} - X_{v(m,k)})^2$. Let $|W^m|$ denote the variation of the process W^m defined by

$$|W^m|_t = |W_0^m| + \mathrm{Var}_{[a,t]} W^m.$$

Then $|W^m|_0 = |W_0^m|$ and for every t we have

$$\begin{aligned} |W^m|_t &= \sum\nolimits_{k\geq 0} \left|\Delta W^m_{v(m,k)}\right| 1_{\{v(m,k)\leq t\}} \\ &\leq \sum\nolimits_{k\geq 0} \left|X_{v(m,k+1)} - X_{v(m,k)}\right|^2 1_{\{v(m,k)\leq t\}} := V_t^m. \end{aligned}$$

Now let $m \to \infty$. We have $\lim_m V_t^m = [X]_t - |X_0|^2$ and $\lim_m \mathrm{prob}\, W_t^m = [[X]]_t^B - X_0^2$. Choose a subsequence $v(n,k)$ of $v(m,k)$ such that $\lim_n W_t^n = [[X]]_t^B - X_0^2$ a.s. by Theorem 4.2. Since the sequence (V^n) is bounded a.s. on $[0,t]$, the sequence of variations $(|W^n|)$ is also bounded a.s. on $[0,t]$. By proposition 5.3, we see that a.s.,

$$\int_{[0,t]} H_{s-} dW_s^n \to \int_{[0,t]} H_{s-} d\left([[X]]_t^B - X_0^2\right) = \int_{[0,t]} H_{s-} d[[X]]_s^B.$$

On the other hand

$$\int_{[0,t]} (H_s^n - H_{s-})\, dW_s^n \to 0 \text{ a.s.}$$

In fact, let $\epsilon > 0$. Then for almost all w, there is an n_0 such that for $n \geq n_0$ and $s \in [0,t]$, we have $|H_s^n(w) - H_{s-}(w)| < \epsilon$; therefore

$$\left|\int_{[0,t]} (H_s^n(w) - H_{s-}(w))\, dW_s^n(w)\right| \leq \epsilon |W^n|_t(w) \leq \epsilon V_t^n(w).$$

It follows that

$$\limsup_n \left| \int_{[0,t]} (H_s^n(w) - H_{s-}(w))\, dW_s^n(w) \right| \le \epsilon [X]_t(w) \text{ a.s.}$$

Thus

$$\lim_n \int_{[0,t]} (H_s^n(w) - H_{s-}(w))\, dW_s^n(w) = 0 \text{ a.s.}$$

Now observe that

$$\begin{aligned} \int_{[0,t]} H_s^n dW_s^n &= \sum_{k\ge 0} H_{v(n,k)} \Delta W_{v(n,k)}^n 1_{\{v(n,k)\le t\}} \\ &= \sum_{k\ge 0} H_{v(n,k)} \left(X_{v(n,k+1)} - X_{v(n,k)}\right)^2 1_{\{v(n,k)\le t\}} \end{aligned}$$

and

$$\begin{aligned} &\int_{[0,t]} H_s^n dW_s^n - \sum_{k\ge 0} H_{v(n,k)} \left(X_{v(n,k+1)\wedge t} - X_{v(n,k)\wedge t}\right)^2 \\ &\quad = H_{v(n,k)} \left[\left(X_{v(n,k+1)} - X_{v(n,k)}\right)^2 - \left(X_t - X_{v(n,k)}\right)^2 \right] 1_{[v(n,k),v(n,k+1))}(t), \end{aligned}$$

and this last term approaches zero as $n \to \infty$. This completes the proof.

REMARK: Assume E is a Hilbert space and the bilinear mapping B is the inner product. In this case $[X] = [[X]]^B$, and by Theorem 4.2 any family of stopping times satisfying conditions (i)–(iii) of section 3.1 can be used in the definition of $[X]$. Then we can take $v(n,k)$ to satisfy condition (b) of Theorem 3.1, which implies condition (b″) of the above theorem.

5.5 COROLLARY. *Assume X is semi locally summable relative to the tensor product and has finite quadratic variation $[X]$, and let $f : E \to G$ be a function of class C^2, such that f'' is uniformly continuous on bounded subsets of E. Then the Stieltjes integral $\int_{(0,t]} f''(X_{s-}) d[[X]]_s$ is defined pathwise and there is a family of stopping times (which can be used in the definition of $[X]$) such that a.s.*

$$\begin{aligned} \int_{(0,t]} f''(X_{s-}) d[[X]]_s &= \lim_n \sum_{k\ge 0} f''\left(X_{v(n,k)}\right) \left(X_{v(n,k+1)\wedge t} - X_{v(n,k)\wedge t}\right)^{\otimes 2} \\ &= \lim_n \sum_{k\ge 0} f''\left(X_{v(n,k)}\right) \left(X_{v(n,k+1)} - X_{v(n,k)}\right)^{\otimes 2} 1_{\{v(n,k)\le t\}}. \end{aligned}$$

PROOF: If we set $F = L(E \hat{\otimes}_\pi E, G)$, then $f'' : E \to F$. Let $H_t = f''(X_t)$, for $t \ge 0$. Then H satisfies condition (b″) of Theorem 5.4. In fact, there is a sequence $T_i \nearrow \infty$ of stopping times such that $|1_{[0,T_i)}X| \le i$. Let $w \in \Omega$ and $t_0 \ge 0$. Let i be such that $t_0 < T_i(w)$. Let $b_n(w) \downarrow 0$. Since f'' is uniformly continuous on the bounded set $\{x : |x| \le i\} \subset E$, there is a $\delta > 0$ such that for $x, y \in E$, each bounded by i, with $|x-y| < \delta$, we have $|f''(x) - f''(y)| < b_n(w)$. Since $a_n \downarrow 0$, there is a $j(n)$, depending on w, such that $a_{j(n)} < \delta$. For any $t \le t_0$, let k be such that

$$t \in [v(j(n),k)(w), v(j(n),k+1)(w)).$$

Then by condition (iii) of section 3, we have

$$|X_t(w) - X_{v(j(n),k)}(w)| \le a_{j(n)} < \delta;$$

therefore

$$|f''(X_t(w)) - f''((X_{v(j(n),k)}(w))| < b_n(w).$$

Hence H_t satisfies condition (b$''$) of Theorem 5.4 for the family of stopping times $v(j(n),k)$. We can then apply Theorem 5.4 and deduce the corollary, using the fact that $1_{[0,t]}H_- = 1_{(0,t]}f''(X_-)$.

6. The process of jumps

In this section we shall develop further results about the jumps of X which will enable us to prove Itô's formula.

The first is a simple but useful lemma.

6.1 LEMMA. *Let $X : \mathbb{R}_+ \times \Omega \to E$ be cadlag and adapted. Let $(v(n,k))$ be the usual family of stopping times defined in section 3 satisfying conditions (i)–(iii).*

Then for every n and w we have

$$\{s : |\Delta X_s| > 2a_n\} \subset \{v(n,k) : k \ge 0\}.$$

PROOF: Assume $|\Delta X_s| > 2a_n$. If $s = 0$, then $s = v(n,0)$. If $s > 0$ and $v(n,k) < s \le v(n,k+1)$, then $s = v(n,k-1)$, since, otherwise,

$|X_s - X_{v(n,k)}| \leq a_n$, and hence $|X_s - X_{s'}| \leq 2a_n$ for $v(n,k) < s' < s$; therefore, letting $s' \nearrow s$, we obtain $|\Delta X_s| \leq 2a_n$, a contradiction.

6.2 THEOREM. *Assume that X is E–valued, cadlag, adapted, and has finite quadratic variation $[X]$, and let $(v(n,k))$ be a family of stopping times defining $[X]$.*

Then

(1) the family $\{|\Delta X_s|^2 : s \leq t\}$ is summable for each t;

(2) the process of jumps defined for any $t \geq 0$ by

$$J_t := \sum_{s\leq t} |\Delta X_s|^2$$

is increasing, cadlag, adapted and satisfies

$$\sum_{s\leq t} |\Delta X_s|^2 \leq [X]_t;$$

(3)

$$\begin{aligned}\sum_{s\leq t} |\Delta X_s|^2 &= \lim_n \sum_{k\geq 0} |\Delta X_{v(n,k)}|^2 1_{\{v(n,k)\leq t\}} \\ &= \lim_n \sum_{k\geq 0} |\Delta X_{v(n,k)}|^2 1_{\{|\Delta X_{v(n,k)}|>2a_n\}} 1_{\{v(n,k)\leq t\}}\end{aligned}$$

uniformly for t on any compact time interval.

(4) Let $V_t^n = \sum_{k\geq 0} |X_{v(n,k+1)\wedge t} - X_{v(n,k)\wedge t}|^2$ and assume that for every w there exists a subsequence (n_i), which may depend on w, such that $V_t^{n_i}(w)$ converges uniformly to $[X]_t(w) - |X_0(w)|^2$ on compact time intervals. Then $[X]$ is cadlag and the process

$$[X]_t^c(w) = [X]_t(w) - \sum_{s\leq t} |\Delta X_s|^2(w)$$

is continuous.

REMARK: The assumption in (4) is satisfied, for example, if E is a Hilbert space and X is locally summable relative to the inner product, by Theorem 4.2

PROOF: Since X is cadlag and by using condition (ii) for the stopping times $v(n,k)$, it follows that for $s > 0$, we have

$$\Delta X_s = \lim_n \sum_{k\geq 0} \left(X_{v(n,k+1)} - X_{v(n,k)}\right) 1_{[v(n,k),v(n,k+1))}(s).$$

Let $0 < s_1 < s_2 < \cdots < s_j \le t$, and choose $(v(n,k))$ to be a defining sequence for $[X]$. Then

$$\begin{aligned}
&\sum_{1\le i\le j} |\Delta X_{s_i}|^2 \\
&= \lim_n \sum_{1\le i\le j} \sum_{k\ge 0} \left|X_{v(n,k+1)} - X_{v(n,k)}\right|^2 1_{[v(n,k),v(n,k+1))}(s_i) \\
&\le \lim_n \sum_{k\ge 0} \left|X_{v(n,k+1)} - X_{v(n,k)}\right|^2 1_{\{v(n,k)\le t\}} = [X]_t - |X_0|^2.
\end{aligned}$$

It follows that $\sum_{0<s\le t} |\Delta X_s|^2 \le [X]_t - |X_0|^2$, hence $\sum_{0\le s\le t} |\Delta X_s|^2 \le [X]_t$, which proves (1).

It is obvious that the process of jumps J is increasing and cadlag. The fact that J is adapted will follow from (3).

To prove (3), we note that, by the above Lemma 6.1, we have

$$\begin{aligned}
&\sum_{s\le t} |\Delta X_s|^2 - \sum_{k\ge 0} |\Delta X_{v(n,k)}|^2 1_{\{v(n,k)\le t\}} \\
&\quad \le \sum_{s\le t} |\Delta X_s|^2 - \sum_{k\ge 0} |\Delta X_{v(n,k)}|^2 1_{\{|\Delta X_{v(n,k)}|>2a_n\}} 1_{\{v(n,k)\le t\}} \\
&\quad \le \sum_{s\le t} |\Delta X_s|^2 - \sum_{s\le t} |\Delta X_s|^2 1_{\{|\Delta X_s|>2a_n\}}
\end{aligned}$$

and the last term converges to zero, as $n \to \infty$. But the last term is equal to $\sum_{s\le t} |\Delta X_s|^2 1_{\{|\Delta X_s|\le 2a_n\}}$, which is increasing as a function of t; therefore, it converges to zero, as $n \to \infty$, uniformly on compact time intervals.

Now we shall prove (4). For $t = 0$, we have $[X]_0 - J_0 = |X_0|^2 - |\Delta X_0|^2 = 0$, hence $[X] - J$ is continuous at 0. Now fix w; we shall show that $[X] - J$ is continuous on $(0, +\infty)$. By the assumption in (4), there is a subsequence (n_i) such that $V_t^{n_i}(w) \to [X]_t(w) - |X_0(w)|^2$ uniformly on compact time intervals. To simplify notation, assume $n_i = n$. Each V^n is cadlag, since for fixed t, V_t^n is a finite sum; as a consequence, $[X]$ is cadlag. For $t \in (v(n,j), v(n,j+1)]$, the jump of V^n at t is

$$\Delta V_t^n = \left|X_t - X_{v(n,j)}\right|^2 - \left|X_{t-} - X_{v(n,j)}\right|^2,$$

therefore $|\Delta V_t^n| \le 2a_n^2$. Denote

$$J_t^n = \sum_{k\ge 0} \left|\Delta X_{v(n,k)}\right|^2 1_{\{v(n,k)\le t\}}.$$

Then by (3), $J_t^n \to J_t$, uniformly on compact time intervals. The jumps of J_t^n are at $t = v(n,j+1)$, and

$$\Delta J_{v(n,j+1)}^n = \left|\Delta X_{v(n,j+1)}\right|^2$$

For $t = v(n, j+1)$ we have

$$|\Delta(V_t^n - J_t^n)| = \left|X_{v(n,j+1)} - X_{v(n,j)}\right|^2 - \left|X_{v(n,j+1)-} - X_{v(n,j)}\right|^2$$
$$- \left|\Delta X_{v(n,j+1)}\right|^2 \le 2\left|\Delta X_{v(n,j+1)}\right| \left|X_{v(n,j+1)-} - X_{v(n,j)}\right|$$
$$\le 2 \sup_{s\le t} |\Delta X_s| \, a_n.$$

Since $\sum_{s\le t} |\Delta X_s|^2 < \infty$, we have $\sup_{s\le t} |\Delta X_s| < \infty$, therefore $\lim_n (\Delta V_t^n - \Delta J_t^n) = 0$, uniformly on compact time intervals.

If $t \ne v(n, j+1)$, then for any n and j, we have $\Delta J_t^n = 0$, hence

$$|\Delta(V_t^n - J_t^n)| = |\Delta V_t^n| \le 2a_n^2.$$

As a result, in either case, for $t > 0$, we have $\Delta(V_t^n - J_t^n) = 0$ uniformly on bounded time intervals. Since $\lim_n J_t^n = J_t$ uniformly on compact time intervals, we can interchange limits and conclude that for $t > 0$,

$$\Delta([X]_t - J_t) = \lim_{s\nearrow t} \{([X]_t - [X]_s) - (J_t - J_s)\}$$
$$= \lim_{s\nearrow t} \lim_n \{(V_t^n - V_s^n) - (J_t^n - J_s^n)\}$$
$$= \lim_n \Delta(V_t^n - J_t^n) = 0.$$

Thus $[X] - J$ is continuous at $t > 0$. This completes the proof.

We now turn to the regularity of $[[X]]^B$.

6.3 THEOREM. *Assume X is E-valued, cadlag, adapted and has finite quadratic variation $[X]$. Let $(v(n,k))$ be a family of stopping times defining $[X]$.*

Then

(1) the family $\{(X_s)^2 : s \le t\}$ is summable for each t;

(2) the process of vector valued jumps

$$vJ_t := \sum_{s\le t} (\Delta X_s)^2$$

is cadlag, adapted and has finite variation $|vJ|$; in addition, we have

$$|vJ|_t \le \sum_{s\le t} |\Delta X_s|^2;$$

(3) We have

$$\sum_{s\le t} (\Delta X_s)^2 = \lim_n \sum_{k\ge 0} (\Delta X_{v(n,k)})^2 1_{\{v(n,k)\le t\}}$$
$$= \lim_n \sum_{k\ge 0} (\Delta X_{v(n,k)})^2 1_{\{|\Delta X_{v(n,k)}| > 2a_n\}} 1_{\{v(n,k)\le t\}}$$

uniformly for t in any compact time interval.

Now let $B : E \times E \to D$ be a continuous bilinear map such that $|x| = \sup\{|xy| : |y| \leq 1\}$ for $x \in E$.

(4) Assume that X is semi locally summable relative to B and B'. Then the process

$$[[X]]_t^c = [[X]]_t^B - \sum_{s\leq t}(\Delta X_s)^2$$

is continuous.

(5) Let $H : \mathbb{R}_+ \times \Omega \to L(D,G)$ be a cadlag, adapted process. The family $\{H_{s-}(\Delta X_s)^2 : s \leq t\}$ is summable and we have

$$\begin{aligned} J_t(H,X) :&= \sum_{0<s\leq t} H_{s-}(\Delta X_s)^2 \\ &= \lim_n \sum_{k>0} H_{v(n,k)-}(\Delta X_{v(n,k)})^2 1_{\{v(n,k)\leq t\}} \\ &= \lim_n \sum_{k>0} H_{v(n,k)-}(\Delta X_{v(n,k)})^2 1_{\{|\Delta X_{v(n,k)}|>2a_n\}} 1_{\{v(n,k)\leq t\}} \end{aligned}$$

uniformly on compact time intervals. The process $J_t(H,X)$ is adapted and right continuous.

Proof: (1) By Theorem 6.2 it follows that $\{(\Delta X_s)^2 : s \leq t\}$ is absolutely summable.

(2) The process of jumps $J_t = \sum_{s\leq t} |\Delta X_s|^2$ is increasing and finite. For $t' < t$ we have

$$\begin{aligned} |vJ_t - vJ_{t'}| &= \left|\sum_{t'<s\leq t'} (\Delta X_s)^2\right| \\ &\leq \sum_{t'<s\leq t'} |\Delta X_s|^2 = J_t - J_{t'}. \end{aligned}$$

It follows that the process vJ has finite variation

$$|vJ|_t := |vJ_0| + \mathrm{Var}_{[0,t]} vJ;$$

and $|vJ|_t \leq J_t$. The fact that vJ is cadlag and adapted follows from (3) and from the fact that the limits in (3) of the cadlag adapted processes are uniform on compact time intervals.

(3) Note that, by Lemma 6.1,

$$\begin{aligned} &\left|\sum_{s\leq t}(\Delta X_s)^2 - \sum_{k\geq 0} \left(\Delta X_{v(n,k)}\right)^2 1_{\{v(n,k)\leq t\}}\right| \\ &\qquad \leq \sum_{s\leq t} |\Delta X_s|^2 1_{\{|\Delta X_s|\leq 2a_n\}}, \end{aligned}$$

and

$$\left|\sum_{s\leq t}(\Delta X_s)^2 - \sum_{k\geq 0}\left(\Delta X_{v(n,k)}\right)^2 1_{\{|\Delta X_{v(n,k)}|>2a_n\}}1_{\{v(n,k)\leq t\}}\right|$$
$$\leq \sum_{s\leq t}|\Delta X_s|^2\, 1_{\{|\Delta X_s|\leq 2a_n\}}.$$

Since the last term converges to zero, as $n \to \infty$, uniformly on any compact time interval, assertion (3) follows.

(4) Denote

$$W_t^n = \sum_{k\geq 0}\left(X_{v(n,k+1)\wedge t} - X_{v(n,k)\wedge t}\right)^2,$$

and

$$vJ_t^n = \sum_{k\geq 0}\left(\Delta X_{v(n,k)}\right)^2 1_{\{v(n,k)\leq t\}}.$$

Note that $\Delta W_t^n = (X_t - X_{v(n,j)})^2 - (X_{t-} - X_{v(n,j)})^2$. If $t = 0$, then $[[X]]_0^B - vJ_0 = X_0^2 - (\Delta X_0)^2 = 0$, hence $[[X]]^B - vJ$ is continuous at 0. If $t > 0$ and $t \in (v(n,j), v(n,j+1)]$, the jumps of vJ_t^n are at $t = v(n,j+1)$ and $\Delta vJ_t^n = (\Delta X_{v(n,j+1)})^2$; for this value of t, we have

$$\begin{aligned}
&|\Delta\, (W_t^n - vJ_t^n)|\\
&= \left|\left(X_{v(n,j+1)} - X_{v(n,j)}\right)^2 - \left(X_{v(n,j+1)-} - X_{v(n,j)}\right)^2 - \left(\Delta X_{v(n,j+1)}\right)^2\right|\\
&= \left|\Delta X_{v(n,j+1)}\left(X_{v(n,j+1)-} - X_{v(n,j)}\right) + \left(X_{v(n,j+1)-} - X_{v(n,j)}\right)\Delta X_{v(n,j+1)}\right|\\
&\leq 2\left|\Delta X_{v(n,j+1)}\right|\left|X_{v(n,j+1)-} - X_{v(n,j)}\right|\\
&\leq 2\sup_{s\leq t}|\Delta X_s|\, a_n,
\end{aligned}$$

where $a_n \downarrow 0$ is the sequence used in condition (iii) of the stopping times $v(n,k)$.

In this case, $\Delta(W_t^n - vJ_t^n) \to 0$, as $n \to \infty$, uniformly on compact time intervals.

If $t \in (v(n,j), v(n,j+1))$, then $\Delta vJ_t^n = 0$ and

$$|\Delta\,(W_t^n - vJ_t^n)| \leq \left|X_t - X_{v(n,j)}\right|^2 + \left|X_{t-} - X_{v(n,j)}\right| \leq 2a_n^2.$$

Again, $\Delta(W_t^n - vJ_t^n) \to 0$, as $n \to \infty$, uniformly on compact time intervals.

By Theorem 4.2 and assertion (3), there exists a subsequence (n_r) such that $\lim_r(W_t^{n_r} - vJ_t^{n_r}) = [[X]]_t^B - \sum_{s\leq t}(\Delta X_s)^2$ uniformly on compact time intervals. If we interchange limits, it follows that $[[X]]_t^B - \sum_{s\leq t}(\Delta X_s)^2$ is a continuous process.

(5) Since H is cadlag, for fixed t and w, the function $s \to H_s(w)$ is bounded on the interval $[0,t]$ by a constant $M_t(w)$.

If we denote

$$J_t = \sum_{s\leq t} |\Delta X_s|^2,$$

then

$$\sum_{0<s\leq t} \left|H_{s-}(\Delta X_s)^2\right| \leq M_t J_t,$$

hence $J_t(H,X)$ is defined. By Lemma 6.1 we have

$$\left|\sum_{0<s\leq t} H_{s-}(\Delta X_s)^2 - \sum_{k>0} H_{v(n,k)-}\left(\Delta X_{v(n,k)}\right)^2 1_{\{v(n,k)\leq t\}}\right|$$
$$\leq M_t \sum_{s\leq t} |\Delta X_s|^2 1_{\{|\Delta X_s|\leq 2a_n\}}$$

and

$$\sum_{0<s\leq t} H_{s-}(\Delta X_s)^2 - \sum_{k>0} H_{v(n,k)-}\left(\Delta X_{v(n,k)}\right)^2 1_{\{|\Delta X_{v(n,k)}|>2a_n\}} 1_{\{v(n,k)\leq t\}}$$
$$\leq M_t \sum_{s\leq t} |\Delta X_s|^2 1_{\{|\Delta X_s|\leq 2a_n\}},$$

and the last term converges to 0, as $n \to \infty$, uniformly on compact time intervals. Assertion (5) follows. From the equalities of assertion (5) we also deduce that $J_t(H,X)$ is adapted. This process is cadlag since J_t is cadlag and for $s < t$ we have

$$|J_t(H,X) - J_s(H,X)| \leq M_t(J_t - J_s).$$

This completes the proof.

We shall use the following corollary of assertion (5) in establishing Itô's formula.

6.4 COROLLARY. *Assume X is E-valued, cadlag, adapted and has finite quadratic variation $[X]$. Let $(v(n,k))$ be a family of stopping times defining $[X]$. Let $f : E \to G$ be a function of class C^2; hence $f'' : E \to L(E \hat{\otimes}_\pi E, G)$.*

Then the family $\{f''(X_{s-})(\Delta X_s)^2 : s \leq t\}$ is summable and

$$\sum_{0<s\leq t} f''(X_{s-})(\Delta X_s)^{\otimes 2} = \lim_n \sum_{k>0} f''\left(X_{v(n,k)-}\right)\left(\Delta X_{v(n,k)}\right)^{\otimes 2} 1_{\{v(n,k)\leq t\}}$$

uniformly on compact time intervals.

We take $H_t = f''(X_t)$ and $D = E\hat{\otimes}_\pi E$ in assertion (5) of Theorem 6.3. We notice that the equality remains true if we allow $0 \leq s \leq t$ in the summation of the left–hand side, and $k \geq 0$ in the summation of the right–hand side.

We state Taylor's formula for reference.

6.5 PROPOSITION. *Let $f : E \to G$ be a function of class C^2, with f'' uniformly continuous on bounded subsets of E. Then there exists a function $R : E \times E \to L(E \hat{\otimes}_\pi E, G)$, with $R(x,x) = 0$ for $x \in E$, such that $\lim_{y\to x} R(y,x) = 0$ uniformly with respect to x, belonging to any given bounded set, and such that for $x, y \in E$, we have*

$$f(y) = f(x) + f'(x)(y-x) + \tfrac{1}{2}f''(x)(y-x)^{\otimes 2} + R(y,x)(y-x)^{\otimes 2}.$$

For the proof, we use the Taylor formula

$$f(y) = f(x) + f'(x)(y-x) + \int_{[0,1]} (1-s)f''(x+s(y-x))ds(y-x)^{\otimes 2}$$

and let $R(x,y) = \int_{[0,1]}(1-s)(f''(x+s(y-x)) - f''(x))ds$.

6.6 PROPOSITION. *Suppose X is E–valued and has finite quadratic variation $[X]$. Let $f : E \to G$ be a function of class C^2 such that f'' is bounded on bounded subsets of E.*

Then for every t, we have

$$\sum_{s\leq t} |f(X_s) - f(X_{s-}) - f'(X_{s-})\Delta X_s| < \infty.$$

The process

$$Q_t := \sum_{s\leq t} (f(X_s) - f(X_{s-}) - f'(X_{s-})\Delta X_s)$$

is adapted, cadlag, and has finite variation.

PROOF: Since X is cadlag, for fixed w and t, $X_s(w)$ is bounded in norm on $[0,t]$ by a constant a depending on w and t; hence f'' is bounded in norm by a constant c on the set $\{x : |x| \leq a\}$. We then have

$$|X_{t-} + u(X_t - X_{t-})| = |X_{t-}(1-u) + X_t u| \leq a,$$

therefore

$$|f''(X_{t-} + u(X_t - X_{t-}))| \leq c,$$

consequently

$$\left|\int_{[0,1]} (1-u) f''(X_{t-} + u(X_t - X_{t-}))du\right|$$
$$\leq \int_{[0,1]} (1-u)\,|f''(X_{t-} + u(X_t - X_{t-}))|\,du \leq \frac{c}{2}.$$

If we use Taylor's formula for f, it follows that

$$|f(X_t) - f(X_{t-}) - f'(X_{t-})(X_t - X_{t-})|$$
$$\leq \left(\frac{c}{2}\right) |(X_t - X_{t-})^{\otimes 2}| = \left(\frac{c}{2}\right) |\Delta X_t|^2.$$

Since X has finite quadratic variation, the summability follows, hence Q_t is defined and we have

$$|Q_t - Q_{t'}| \leq \left(\frac{c}{2}\right)(J_t - J_{t'}),$$

for $t' < t$. This proves that Q_t is cadlag and of finite variation.

To show that Q_t is adapted, note that the jumps of $f(X_t)$ and X are contained in a countable union of disjoint graphs of stopping times T_n. Hence

$$Q_t = \sum\nolimits_{1 \leq n} f\left(X_{T_n \wedge t}\right) - f\left(X_{(T_n \wedge t)-}\right) - f'\left(X_{(T_n \wedge t)-}\right) \Delta X_{T_n \wedge t},$$

and it follows that Q_t is adapted.

7. Itô's formula

Theorem. *Let $X : \mathbb{R}_+ \times \Omega \to E$ be semi locally summable with respect to the bilinear mappings $B_1(x, y) = x \otimes y$ of $E \times E$ into $E \hat{\otimes}_\pi E$ and $B_2(x, y) = u(x)$ of $E \times L(E, G)$ into G (that is, we regard X, in this instance, as taking values in $L(L(E, G), G)$). Assume that X has finite quadratic variation $[X]$.*

Let $f : E \to G$ be a function of class C^2 such that $f'' : E \to L(E \hat{\otimes}_\pi E, G)$ is uniformly continuous and bounded on bounded subsets of E. Then for every

$t > 0$, *we have, a.s.,*

$$\begin{aligned} f(X_t) &= f(X_0) + \int_{(0,t]} f'(X_{s-})dX_s + \tfrac{1}{2}\int_{(0,t]} f''(X_{s-})d[[X]]_s \\ &+ \sum\nolimits_{0<s\le t} \left[f(X_s) - f(X_{s-}) - f'(X_{s-})\Delta X_s - \tfrac{1}{2}f''(X_{s-})(\Delta X_s)^{\otimes 2}\right] \\ &= f(X_0) + \int_{(0,t]} f'(X_{s-})dX_s + \tfrac{1}{2}\int_{(0,t]} f''(X_{s-})d[[X]]^c_s \\ &+ \sum\nolimits_{0<s\le t} \left[f(X_s) - f(X_{s-}) - f'(X_{s-})\Delta X_s\right]. \end{aligned}$$

Proof: All the processes in the formula are defined. In fact, $f' : E \to L(E, G)$ is continuous, hence $f'(X_{s-})$ is caglad, therefore it is locally integrable with respect to X, which is semi locally summable relative to B_2; consequently, the first integral is defined. Next, $f''(X_{s-})$ is caglad, therefore pathwise integrable, on every compact time interval, with respect to the process $[[X]]$, which is of finite variation; consequently the second integral is also defined. The sum in the formula is defined and is cadlag and adapted by Corollary 6.4 and Proposition 6.6. Now we have to establish the equality.

Choose a sequence $a_n \downarrow 0$ and the corresponding family $(v(n,k))$ of stopping times defining the quadratic variation $[X]$ (see Definition 5.1).

Fix $t > 0$. By Taylor's formula, for each n we have

$$\begin{aligned} f(X_t) - f(X_0) &= \sum\nolimits_{k\ge 0} \left[f\left(X_{v(n,k+1)\wedge t}\right) - f\left(X_{v(n,k)\wedge t}\right)\right] \\ (1) \qquad &= \sum\nolimits_{k\ge 0} f'\left(X_{v(n,k)\wedge t}\right)\left(X_{v(n,k+1)\wedge t} - X_{v(n,k)\wedge t}\right) \\ (2) \qquad &+ \tfrac{1}{2}\sum\nolimits_{k\ge 0} f''\left(X_{v(n,k)\wedge t}\right)\left(X_{v(n,k+1)\wedge t} - X_{v(n,k)\wedge t}\right)^{\otimes 2} \\ (3) \qquad &+ \sum\nolimits_{k\ge 0} R_{n,k}\left(X_{v(n,k+1)\wedge t} - X_{v(n,k)\wedge t}\right)^{\otimes 2}, \end{aligned}$$

where we set $R_{n,k} = R(X_{v(n,k+1)\wedge t}, X_{v(n,k)\wedge t})$. Note that $R_{n,k} = R_{n,k}1_{\{v(n,k)<t\}}$. By Corollary 3.4, taking a subsequence if necessary, we see that the sum (1) converges pointwise to $\int_{(0,t]} f'(X_{s-})dX_s$. By Corollary 5.5, again by taking a subsequence if necessary, we deduce that the sum (2) converges pointwise to $\frac{1}{2}\int_{(0,t]} f''(X_{s-})d[[X]]_s$.

We must prove that $\lim_n \sum_{k\ge 0} R_{n,k}\left(X_{v(n,k+1)\wedge t} - X_{v(n,k)\wedge t}\right)^{\otimes 2}$ is equal, pointwise, to the sum

$$\sum\nolimits_{0<s\le t} \left[f(X_s) - f(X_{s-}) - f'(X_{s-})\Delta X_s - \tfrac{1}{2}f''(X_{s-})(\Delta X_s)^{\otimes 2}\right].$$

The above sum can be written as

$$\sum_{0<s\le t} R(X_s, X_{s-})(\Delta X_s)^{\otimes 2}.$$

We shall assume, at first, that X is bounded, and let $B = \{x : |x| < a\}$ be an open ball in E containing the range of X.

For $v(n,k) < t$, we can write

$$\begin{aligned}
R_{n,k}\left(X_{v(n,k+1)\wedge t} - X_{v(n,k)\wedge t}\right)^{\otimes 2} &= R_{n,k}\left(X_{v(n,k+1)\wedge t} - X_{v(n,k)}\right)^{\otimes 2} \\
&= f\left(X_{v(n,k+1)\wedge t}\right) - f\left(X_{v(n,k)}\right) - f'\left(X_{v(n,k)}\right)\left(X_{v(n,k+1)\wedge t} - X_{v(n,k)}\right) \\
&\quad - \tfrac{1}{2} f''\left(X_{v(n,k)}\right)\left(X_{v(n,k+1)\wedge t} - X_{v(n,k)}\right)^{\otimes 2} \\
&= v_{n,k} + w_{n,k} + y_{n,k},
\end{aligned}$$

where

$$\begin{aligned}
v_{n,k} &= f\left(X_{v(n,k+1)\wedge t}\right) - f\left(X_{v(n,k+1)\wedge t-}\right) - f'\left(X_{v(n,k)}\right)\Delta X_{v(n,k+1)\wedge t} \\
&\quad - \tfrac{1}{2} f''\left(X_{v(n,k)}\right)\left(\Delta X_{v(n,k+1)\wedge t}\right)^{\otimes 2};
\end{aligned}$$

$$\begin{aligned}
w_{n,k} &= f\left(X_{v(n,k+1)\wedge t-}\right) \\
&\quad - f\left(X_{v(n,k)}\right) - f'\left(X_{v(n,k)}\right)\left(X_{v(n,k+1)\wedge t-} - X_{v(n,k)}\right) \\
&\quad - \tfrac{1}{2} f''\left(X_{v(n,k)}\right)\left(X_{v(n,k+1)\wedge t-} - X_{v(n,k)}\right)^{\otimes 2} \\
&= R\left(X_{v(n,k+1)\wedge t-}, X_{v(n,k)}\right)\left(X_{v(n,k+1)\wedge t-} - X_{v(n,k)}\right)^{\otimes 2};
\end{aligned}$$

$$\begin{aligned}
y_{n,k} &= \tfrac{1}{2} f''\left(X_{v(n,k)}\right)\left[\left(\Delta X_{v(n,k+1)\wedge t}\right) \otimes \left(X_{v(n,k+1)\wedge t-} - X_{v(n,k)}\right)\right. \\
&\quad \left. + \left(X_{v(n,k+1)\wedge t-} - X_{v(n,k)}\right) \otimes \Delta X_{v(n,k+1)\wedge t}\right].
\end{aligned}$$

We shall first prove

(a) $\lim_n \sum_{k\ge 0} |w_{n,k}| 1_{\{v(n,k)<t\}} = 0$

uniformly on compact time intervals.

By Proposition 6.5, given an $\epsilon > 0$, there is a $\delta > 0$ such that $|R(x,y)| < \epsilon$ whenever $|x - y| < \delta$ and $x, y \in B$. Choose n_0 such that $a_n < \delta$ for $n \ge n_0$. Then by condition (iii) of the stopping times $v(n,k)$, we have for $v(n,k) < t$, $|X_{v(n,k+1)\wedge t-} - X_{v(n,k)}| < \delta$, hence

$$\left|R\left(X_{v(n,k+1)\wedge t-}, X_{v(n,k)}\right)\right| < \epsilon,$$

for $n \geq n_0$. Thus for any $t > 0$ and $n \geq n_0$, we have

$$\sum_{k\geq 0} |w_{n,k}| 1_{\{v(n,k)<t\}} \leq \epsilon \sum_{k\geq 0} \left|X_{v(n,k+1)\wedge t-} - X_{v(n,k)}\right|^2 1_{\{v(n,k)<t\}}$$
$$\leq 2\epsilon \left(\sum_{k\geq 0} \left|\Delta X_{v(n,k+1)\wedge t}\right|^2 1_{\{v(n,k)<t\}} + \sum_{k\geq 0} \left|X_{v(n,k+1)\wedge t} - X_{v(n,k)\wedge t}\right|^2\right).$$

As a result, for every $t > 0$, we have

$$\limsup_n \sum_{k\geq 0} |w_{n,k}| \, 1_{\{v(n,k)<t\}}$$
$$\leq 2\epsilon \left(\sum_{s\leq t} |\Delta X_s|^2 + [X]_t - [X]_0\right),$$

and the last term converges to zero, as $\epsilon \downarrow 0$, uniformly on compact time intervals. This proves (a).

Next we choose a sequence $b_n \downarrow 0$ such that $a_n \leq b_n^2$ and $2a_n \leq b_n$ (e.g. $b_n = 2\sqrt{a_n}$). Define the following subsets of Ω:

$$A_{n,k} = \left\{\left|\Delta X_{v(n,k+1)}\right| > b_n\right\}$$

and

$$B_{n,k} = \left\{\left|\Delta X_{v(n,k+1)}\right| \leq b_n\right\} = \Omega - A_{n,k}.$$

We note that if $w \in B_{n,k}$, then by Lemma 6.1, for every $s > 0$, we have $|\Delta X_s(w)| \leq b_n$.

Next we prove

(b) $\lim_n \sum_{k\geq 0} |y_{n,k}| 1_{A_{n,k}} 1_{\{v(n,k)<t\}} = 0$,

uniformly on compact time intervals.

To see this, let $|f''|$ have bound M on B. Note that on $A_{n,k}$, we have, for $t > v(n,k)$,

$$\left|X_{v(n,k+1)\wedge t-} - X_{v(n,k)}\right| \leq a_n \leq b_n^2 < b_n \left|\Delta X_{v(n,k+1)}\right|;$$

therefore

$$\left|\Delta X_{v(n,k+1)\wedge t} \otimes \left(X_{v(n,k+1)\wedge t-} - X_{v(n,k)}\right)\right| \leq b_n \left|\Delta X_{v(n,k+1)}\right|^2$$

and

$$\left|\left(X_{v(n,k+1)\wedge t-} - X_{v(n,k)}\right) \otimes \Delta X_{v(n,k+1)\wedge t}\right| \leq b_n \left|\Delta X_{v(n,k+1)}\right|^2.$$

It follows that

$$\sum_{k\geq 0}|y_{n,k}|1_{A_{n,k}}1_{\{v(n,k)<t\}} \leq \sum_{k\geq 0} Mb_n \left|\Delta X_{v(n,k+1)\wedge t}\right|^2$$
$$\leq Mb_n \sum_{s\leq t} |\Delta X_s|^2 ,$$

and the last term tends to 0, as $n \to \infty$, uniformly on compact time intervals, which proves (b).

Next we prove that for $t \geq 0$, we have

(c) $\lim_n \sum_{k\geq 0} R(X_{v(n,k+1)\wedge t}, X_{v(n,k)\wedge t})(X_{v(n,k+1)\wedge t} - X_{v(n,k)\wedge t})^{\otimes 2} 1_{B_{n,k}} = 0$

and

(d) $\lim_n \sum_{k\geq 0} R\left(X_{v(n,k+1)}, X_{v(n,k)}\right) \left(X_{v(n,k+1)} - X_{v(n,k)}\right)^{\otimes 2} 1_{B_{n,k}}$
$\cdot 1_{\{v(n,k)\leq t\}} = 0.$

To prove this, let $\epsilon > 0$, and as before, choose a $\delta > 0$ such that for x and y in B, with $|x - y| < \delta$, we have $|R(x, y)| < \epsilon$. Choose n_0 so that $a_{n_0} + b_{n_0} < \delta$. Then on $B_{n,k}$, with $n \geq n_0$, we have

$$\left|X_{v(n,k+1)} - X_{v(n,k)}\right| \leq b_n + a_n < \delta,$$

which implies that

$$\left|R\left(X_{v(n,k+1)}, X_{v(n,k)}\right)\right| 1_{B_{n,k}} \leq \epsilon.$$

Let M_n denote

$$\left|\sum_{k\geq 0} R\left(X_{v(n,k+1)}, X_{v(n,k)}\right) \left(X_{v(n,k+1)} - X_{v(n,k)}\right)^{\otimes 2} 1_{B_{n,k}} 1_{\{v(n,k)\leq t\}}\right|,$$

and denote

$$V_t^n = \sum_{k\geq 0} \left|X_{v(n,k+1)} - X_{v(n,k)}\right|^2 1_{\{v(n,k)\leq t\}}.$$

Then $M_n \leq \epsilon V_t^n$. Since $V_t^n \to [X]_t - [X]_0$, we deduce $\lim_n M_n = 0$. This proves (d).

To prove (c), fix $t \geq 0$. Then for every n, there is a j such that $v(n, j) \leq t < v(n, j + 1)$. The difference between the two sums in (c) and (d) is equal to

$$A_n = R\left(X_t, X_{v(n,j)}\right) \left(X_t - X_{v(n,j)}\right)^{\otimes 2}$$
$$- R\left(X_{v(n,j+1)}, X_{v(n,j)}\right) \left(X_{v(n,j+1)} - X_{v(n,j)}\right)^{\otimes 2}$$

on $B_{n,j}$. As $n \to \infty$, we have $v(n,j) \nearrow t$ and $v(n,j+1) \searrow t$, therefore $\lim A_n = 0$, which means that the two limits in (c) and (d) are equal. This proves (c).

We shall now prove that for every $t > 0$, we have

$$\begin{aligned}
\text{(e)}\quad & \lim_n \sum_{k\geq 0} v_{n,k} 1_{A_{n,k}} 1_{\{v(n,k)<t\}} \\
& = \lim_n \sum_{k\geq 0} \Big[f(X_{v(n,k+1)}) - f(X_{v(n,k+1)-}) - f'(X_{v(n,k)}) \Delta X_{v(n,k+1)} \\
& - \tfrac{1}{2} f''(X_{v(n,k)}) (\Delta X_{v(n,k+1)})^{\otimes 2} \Big] 1_{A_{n,k}} 1_{\{v(n,k)<t\}} \\
& = \sum_{0<s\leq t} R(X_s, X_{s-}) (\Delta X_s)^{\otimes 2} \\
& = \sum_{0<s\leq t} \Big[f(X_s) - f(X_{s-}) - f'(X_{s-}) \Delta X_s - \tfrac{1}{2} f''(X_{s-}) (\Delta X_s)^{\otimes 2} \Big],
\end{aligned}$$

and

$$\text{(f)}\quad \lim_n \sum_{k\geq 0} v_{n,k} 1_{\{v(n,k)<t\}} = \sum_{0<s\leq t} R(X_s, X_{s-}) (\Delta X_s)^{\otimes 2}.$$

To prove this, we define the following functions, for $t > 0$.

$$\begin{aligned}
v'_{n,k} & = f\left(X_{v(n,k+1)\wedge t}\right) - f\left(X_{v(n,k+1)\wedge t-}\right) \\
& - f'\left(X_{v(n,k+1)\wedge t-}\right) \Delta X_{v(n,k+1)\wedge t} \\
& - \tfrac{1}{2} f''\left(X_{v(n,k)\wedge t-}\right) \left(\Delta X_{v(n,k+1)\wedge t}\right)^{\otimes 2} \\
& = R\left(X_{v(n,k+1)\wedge t}, X_{v(n,k)\wedge t-}\right) \left(\Delta X_{v(n,k+1)\wedge t}\right)^{\otimes 2}; \\
v''_{n,k} & = \left[f'\left(X_{v(n,k+1)\wedge t-}\right) - f'\left(X_{v(n,k)}\right)\right] \Delta X_{v(n,k+1)\wedge t}; \\
v'''_{n,k} & = \tfrac{1}{2} \left[f''\left(X_{v(n,k+1)\wedge t-}\right) - f''\left(X_{v(n,k)}\right)\right] \left(\Delta X_{v(n,k+1)\wedge t}\right)^{\otimes 2}.
\end{aligned}$$

Thus $v_{n,k} = v'_{n,k} + v''_{n,k} + v'''_{n,k}$.

In order to prove (e) and (f), we shall prove first (g), (h) and (i) below:

$$\begin{aligned}
\text{(g)}\quad & \lim_n \sum_{k\geq 0} v'_{n,k} 1_{A_{n,k}} 1_{\{v(n,k)<t\}} = \sum_{0<s\leq t} R(X_s, X_{s-}) (\Delta X_s)^{\otimes 2} \\
& = \lim_n \sum_{k\geq 0} R(X_{v(n,k+1)}, X_{v(n,k+1)-}) (\Delta X_{v(n,k+1)})^{\otimes 2} 1_{A_{n,k}} 1_{\{v(n,k)\leq t\}};
\end{aligned}$$

$$\text{(h)}\quad \lim_n \sum_{k\geq 0} v''_{n,k} 1_{A_{n,k}} 1_{\{v(n,k)<t\}} = 0;$$

and

$$\text{(i)}\quad \lim_n \sum_{k\geq 0} v'''_{n,k} 1_{A_{n,k}} 1_{\{v(n,k)<t\}} = \lim_n \sum_{k\geq 0} v'''_{n,k} 1_{\{v(n,k)<t\}} = 0$$

uniformly on compact time intervals.

To prove (g), note that, since $2a_n < b_n$, by Lemma 6.1, we have

$$\{s > 0 : |\Delta X_s| > b_n\} \subset \{s > 0 : |\Delta X_s| > 2a_n\} \subset \{v(n, k+1) : k \geq 0\}.$$

Then

$$\begin{aligned} &\left|\sum\nolimits_{0<s\leq t} R(X_s, X_{s-})(\Delta X_s)^{\otimes 2} - \sum\nolimits_{k\geq 0} v'_{n,k} 1_{A_{n,k}} 1_{\{v(n,k)<t\}}\right| \\ &= \Big|\sum\nolimits_{0<s\leq t} R(X_s, X_{s-})(\Delta X_s)^{\otimes 2} \\ &- \sum\nolimits_{k\geq 0} R\left(X_{v(n,k+1)\wedge t}, X_{v(n,k+1)\wedge t-}\right)\left(\Delta X_{v(n,k+1)\wedge t}\right)^{\otimes 2} 1_{A_{n,k}}\Big| \\ &\leq \sum\nolimits_{0<s\leq t} |R(X_s, X_{s-})|\,|\Delta X_s|^2\, 1_{B_{n,k}} \end{aligned}$$

and

$$\begin{aligned} &\Big|\sum\nolimits_{0<s\leq t} R(X_s, X_{s-})(\Delta X_s)^{\otimes 2} \\ &- \sum\nolimits_{k\geq 0} R\left(X_{v(n,k+1)}, X_{v(n,k+1)-}\right)\left(\Delta X_{v(n,k+1)}\right)^{\otimes 2} 1_{A_{n,k}} 1_{\{v(n,k)<t\}}\Big| \\ &\leq \sum\nolimits_{s\leq t} |R(X_s, X_{s-})|\,|\Delta X_s|^2\, 1_{B_{n,k}}. \end{aligned}$$

Let $\epsilon > 0$ and take $\delta > 0$ such that if $x, y \in B$, with $|x - y| < \delta$, then $|R(x, y)| < \epsilon$. Choose n_0 so that $b_{n_0} < \delta$.

For $n \geq n_0$ and $s > 0$, we have, on $B_{n,k}$,

$$|X_s - X_{s-}| = |\Delta X_s| \leq b_n < \delta,$$

therefore $|R(X_s, X_{s-})| < \epsilon$; consequently

$$\sum\nolimits_{0<s\leq t} |R(X_s, X_{s-})|\,|\Delta X_s|^2\, 1_{B_{n,k}} \leq \epsilon \sum\nolimits_{0<s\leq t} |\,|\Delta X_s|^2\, 1_{B_{n,k}},$$

and the last term tends to 0 as $\epsilon \downarrow 0$.

It follows that

$$\lim_n \sum\nolimits_{k\geq 0} v'_{n,k} 1_{A_{n,k}} 1_{\{v(n,k)<t\}} = \sum\nolimits_{0<s\leq t} R(X_s, X_{s-})(\Delta X_s)^{\otimes 2}$$

uniformly on compact time intervals, that is, the first equality of (g) holds.

In a similar fashion, one can show that

$$\begin{aligned} &\lim_n \sum\nolimits_{k\geq 0} R\left(X_{v(n,k+1)}, X_{v(n,k+1)-}\right)\left(\Delta X_{v(n,k+1)}\right)^{\otimes 2} 1_{A_{n,k}} 1_{\{v(n,k)\leq t\}} \\ &= \sum\nolimits_{0<s\leq t} R(X_s, X_{s-})(\Delta X_s)^{\otimes 2} \end{aligned}$$

uniformly on compact time intervals. This establishes (g).

To prove (h), let M again be the bound of $|f''|$ on B. Use the mean value theorem to deduce

$$\begin{aligned}
&\sum_{k\geq 0} v''_{n,k} 1_{A_{n,k}} 1_{\{v(n,k)<t\}} \\
&\qquad \leq M \sum_{k\geq 0} \left| X_{v(n,k+1)\wedge t-} - X_{v(n,k)} \right| \left| \Delta X_{v(n,k+1)\wedge t} \right| 1_{A_{n,k}} 1_{\{v(n,k)<t\}} \\
&\qquad \leq M a_n \sum_{k\geq 0} \left| X_{v(n,k+1)\wedge t} \right| 1_{A_{n,k}} 1_{\{v(n,k)<t\}} \\
&\qquad \leq M b_n \sum_{k\geq 0} b_n \left| X_{v(n,k+1)\wedge t} \right| 1_{A_{n,k}} 1_{\{v(n,k)<t\}} \\
&\qquad \leq M b_n \sum_{s\leq t} |\Delta X_s|^2 ,
\end{aligned}$$

and the last sum tends to 0, as $n \to \infty$, uniformly on compact time intervals. This proves (h).

To prove (i), we use the uniform continuity of f'' on B. Let $\epsilon > 0$ and take $\delta > 0$ such that $x, y \in B$, with $|x - y| < \delta$, implies $|f''(x) - f''(y)| < \epsilon$. Take n_0 such that for $n \geq n_0$ we have $a_n < \delta$; therefore, if $v(n,k) < t$, then

$$\left| X_{v(n,k+1)\wedge t-} - X_{v(n,k)} \right| \leq a_n < \delta;$$

consequently

$$\left| f''\left(X_{v(n,k+1)\wedge t-} \right) - f''\left(X_{v(n,k)} \right) \right| < \epsilon.$$

It follows that for $n \geq n_0$, we have

$$\begin{aligned}
&\sum_{k\geq 0} v'''_{n,k} 1_{A_{n,k}} 1_{\{v(n,k)<t\}} \\
&\leq \sum_{k\geq 0} \left| f''\left(X_{v(n,k+1)\wedge t-} \right) - f''\left(X_{v(n,k)\wedge t} \right) \right| \left| \Delta X_{v(n,k+1)\wedge t} \right|^2 1_{A_{n,k}} 1_{\{v(n,k)<t\}} \\
&\leq \epsilon \sum_{k\geq 0} \left| \Delta X_{v(n,k+1)\wedge t} \right|^2 1_{A_{n,k}} 1_{\{v(n,k)<t\}} \leq \epsilon \sum_{s\leq t} |\Delta X_s|^2 ,
\end{aligned}$$

and similarly

$$\sum_{k\geq 0} v'''_{n,k} 1_{\{v(n,k)<t\}} \leq \epsilon \sum_{s\leq t} |\Delta X_s|^2 .$$

This proves (i).

Since $v_{n,k} = v'_{n,k} + v''_{n,k} + v'''_{n,k}$, from (g), (h) and (i), we deduce that

$$\lim_n \sum_{k\geq 0} v_{n,k} 1_{A_{n,k}} 1_{\{v(n,k)<t\}} = \sum_{0<s\leq t} R(X_s, X_{s-}) (\Delta X_s)^{\otimes 2} ,$$

which proves (e). From (a), (b) and (e), we deduce that

$$\begin{aligned}
&\lim_n \sum_{k\geq 0} R_{n,,k} \left(X_{v(n,k+1)\wedge t} - X_{v(n,k)\wedge t} \right)^{\otimes 2} 1_{A_{n,k}} \\
&\lim_n \sum_{k\geq 0} (v_{n,k} + w_{n,k} + y_{n,k}) 1_{A_{n,k}} 1_{\{v(n,k)<t\}} \\
&= \sum_{0<s\leq t} R(X_s, X_{s-}) (\Delta X_s)^{\otimes 2} .
\end{aligned}$$

Since, by (c), we have

$$\lim_n \sum_{k\geq 0} R_{n,k}\left(X_{v(n,k+1)\wedge t} - X_{v(n,k)\wedge t}\right)^{\otimes 2} 1_{B_{n,k}} = 0,$$

we deduce finally that

$$\lim_n \sum_{k\geq 0} R_{n,k}\left(X_{v(n,k+1)\wedge t} - X_{v(n,k)}\right)^{\otimes 2} = \sum_{0<s\leq t} R(X_s, X_{s-})(\Delta X_s)^{\otimes 2},$$

and this proves the theorem when X is bounded. We note that to establish the above equality, we only used the fact that X has finite quadratic varition.

For the case X is unbounded, obtain a sequence $T_i \nearrow \infty$ of stopping times such that $X^{T_i -}$ is bounded. We note that $X^{T_i -}$ still has finite quadratic variation and $[X^{T_i -}]_t = [X]_t^{T_i -}$; therefore, for each $X^{T_i -}$, the above equality is true. For a given w and $t > 0$, choose i such that $T_i(w) > t$. Then for every $0 < s \leq t$, we have $X_s^{T_i -} = X_s$, and the equality is valid for $X_t(w)$. This concludes the proof.

REFERENCES

1. J.K Brooks and N. Dinculeanu, *Lebesgue-type spaces for vector integration, linear operators, weak completeness and weak compactness*, J. Math. Anal. Appl. **54** (1976), 348–389.
2. ———, *Stochastic integration in Banach space*, Advances in Math. (in press).
3. C. Dellacherie and P. Meyer, "Probabilities and Potential," North–Holland, (1978), (1980).
4. B. Gravereaux and J. Pellaumail, *Formule de Itô pour des processus à valeurs dans des espaces de Banach*, Ann. Inst. H. Poincaré, **10** (1974), 339–422.
5. H. Kunita, *Stochastic integrals based on martingales taking their values in Hilbert spaces*, Nagoya Math. J. **38** (1970), 41–52.
6. A.U. Kussmaul, Regularität and Stochastische Integration von Semimartingalen mit Werten in einem Banachraum, Dissertation, Stuttgart, 1978.
7. M. Métivier, Semimartingales, Walter de Gruyter, Berlin, 1982.
8. M. Yor, *Sur les intégrales stochastiques à valeurs dans un espace de Banach*, Ann. Inst. H. Poincaré, vol. X, n° 1 (1974), 31–36.

Department of Mathematics
University of Florida
Gainesville, Florida 32611
USA

Regular, Stationary Gaussian Processes with the Markov Property of Infinite Order, and Diffusive Filtrations

Frank B. Knight

INTRODUCTION: This paper records the substance of a talk at the Conference on Diffusion Processes, and is not to be regarded as a complete research work, although it is as complete as the author can make it. The purpose of the talk was partly to point out the presence of certain difficulties, which are only partially resolved. The main subject goes back to a paper of N. Levinson and H. P. McKean [**9**, (1964)], in which they showed that there exist regular stationary real Gaussian processes with an infinite-dimensional germ-Markov property. In the same paper, they indicated the existence of such processes which are, in addition, Markov of order $d = \infty$, that is, Markov relative to all the finite order derivatives. Subsequent papers by H. Dym and H. P. McKean, [**3**, **4**], as well as their book [**5**], indicated the same fact but gave no concrete example. Meanwhile, in our paper [**7**] the germ-Markov property was studied further in a general setting, but no serious example of a Markov process of infinite order was found. It is trivial, indeed, to write examples having analytic path functions. But for these, analytic continuation shows that the process is completely deterministic given the derivatives at a single time, hence such examples are degenerate. It is to be expected that there exists a large class of processes whose paths that are in C_∞ but nowhere analytic, and having the desired Markov property. But prior to the present work no explicit example seems to have been given.

It turns out, indeed, that it is not difficult to write such an example in the spectral domain by following the prescription of [**9**] and [**5**]. However, this is still not explicit enough to satisfy the intuitive and realistic requirements for such an example. In this regard we may quote from page 9 of [**5**]: "It is a further source of disappointment (or, if you will, a challenge) that the answers obtained to simple statistical questions are seldom expressed in such a way as to make plain their statistical content." In fact, the spectral form

$$X_t = \int e^{it\lambda} h(\lambda) W(d\lambda); \quad W(d\lambda) = \text{ complex Gaussian white noise, } \quad W(d\lambda) = \bar{W}(-d\lambda), \quad E(W(d\lambda))^2 = 0, \tag{0.1}$$

is unsatisfactory in this situation for at least 3 reasons. First, the process $W(d\lambda)$ cannot be "observed" without knowing the future of X. Second, the representation shows little about the paths of X_t, and one cannot see intuitively whether X_t has or has not any Markov property. Third, in the cases in point it seems impossible to obtain the inverse Fourier transform of h explicitly, without which the Kolmogorov-Wiener prediction theory of X_t cannot be implemented.

Accordingly, the main objective of this paper is to give an example of such a Markov process of infinite order in the "time domain," where no Fourier inversion is necessary in order to proceed further. It must be added, however, that we are not 100% successful in finding one. Our example is a complex (or 2-dimensional) Gaussian process, where as the treatment in **[5]** is for real processes. We have not been able to express such a real example in the time domain, for reasons to be indicated.

In either case, the next question to ask in the present context is whether such a process, regarded as having its values in the countable product space with one coordinate for each derivative, is a true diffusion process. Here we take as definition of diffusion the standard one of probabilistic potential theory and the theory of Markov processes. Namely, a diffusion may equivalently be defined either as a Hunt process with continuous paths, or as a "right" process with continuous paths (inhomogeneous diffusion may be treated by the usual "space-time" device). As our examples are all Borel measurable, we may presume a Borel state space and transition function. However, it does not seem easy to verify the strong Markov property for such examples. The Gaussian character is of no help at stopping times, and the infinite-dimensional transition function may be intractable. Instead, we have to be content with showing the diffusive character at a more primitive level–that of the probability filtration.

We propose to call a probability filtration "diffusive" if it is equivalent to the natural filtration generated by some true diffusion. With this as definition, it is possible to show that such examples are indeed diffusive. Moreover, it turns out (and this is the main result of Section 2) that a probability filtration $(\mathcal{F}_t, P)$ is diffusive (under broad conditions) if and only if all right-continuous, $\mathcal{F}_{t+}$–martingales have continuous paths. Thus the diffusive filtrations are already rather familiar in the context of the general theory of processes, where they are sometimes called "continuous" filtrations. Since they are clearly not continuous in a literal sense, it would seem better to call them now diffusive, especially since a true diffusion may be used to aid in their analysis.

Section 1. Examples of Regular, Stationary, Gaussian Markov Processes of Infinite Order

Let X_t, $-\infty < t < \infty$, be a real or complex valued stochastic process with continuous path functions, and for each t let $\mathcal{F}_t^- = \sigma(X_s,\ s \le t)$ and $\mathcal{F}_t^+ = \sigma(X_s,\ s \ge t)$ denote the σ–fields of "past" and "future" at time t. We invoke the following terminology.

DEFINITION 1.1: For $0 < d \le \infty$, we call X_t a Markov process of order d if, for every t, $\mathcal{F}_t^-$ and $\mathcal{F}_t^+$ are conditionally independent given $\sigma(\frac{d^k}{dt^k}x_t;\ 0 \le k < d)$, where the derivatives of order $d-1$ are assumed to be continuous quadratic mean.

It follows by a familiar criterion of Kolmogorov that we may assume that the paths of $\frac{d^k}{dt^k}X_t$ are continuous for $k < d-1$, so that in case $d = \infty$ the paths of all derivatives of X_t are continuous provided that X_t is chosen to be separable (see [**10**, Sec. 35, B. Corollary]).

If X_t is assumed to be mean 0-Gaussian, and if it is regular, i.e. $\mathcal{F}_{-\infty}^-(= \cap \mathcal{F}_t^-) \equiv (\phi, \Omega)$ so that there is no deterministic component, then for $d < \infty$ the following is well-known ([**5**, **4.7**], [**6**]).

THEOREM 1.2. *A real-valued, regular, stationary Gaussian process (mean 0) is Markov of order $d < \infty$ if and only if it has a representation (0.1) with $h^{-1} = p(-i\lambda)$, where p is a real polynomial of degree d having all of its roots in the lower half-plane.*

Accordingly, a reasonable idea for finding a Markov process of order $d = \infty$ might seem to be an iteration of such a representation, namely to consider the processes

$$X_t^{(n)} = \int e^{it\lambda} p(-i\lambda)^{-n} W(d\lambda) / \left(\int |p(-i\lambda)|^{-2n} d\lambda\right)^{1/2}$$

as $n \to \infty$. However, it is not hard to see that the spectral density then concentrates on the set of values of λ for which $|p(-i\lambda)|$ is minimal, which is a finite set (symmetric about 0, since X_t is real). Hence in the limit X_t becomes analytic in t, and only a degenerate Markov property can be obtained in this way. Other limit procedures may be considered, but only by knowing in advance an exact expression for the desired limit $h(\lambda)$ does it seem an easy task to obtain both regularity and a Markov property of infinite order.

A direct prescription for $h(\lambda)$ may be sorted out from [**9**] as follows. According to Section 7, $|h|^2$ should be of Hardy class H^{2+}, and h itself should be an outer function. In this case, if h is the reciprocal of an entire function of minimal exponential type, then using Sections 6 and 7 of [**9**] it follows that X_t of (0.1) is a germ-Markov process. This being the case, we seek the further condition under which the germ is generated by $\sigma(\frac{d^k}{dt^k}X_t;\ 0 \le k)$. In Section 10 is found the sufficient condition that all of the roots of h^{-1} lie in the sector $-\frac{3\pi}{4} \le \theta \le -\frac{\pi}{4}$. Therefore, we need only construct a function h having all of these properties. We may state

PROPOSITION 1.3. *The function $h(z) = \prod_{n=1}^{\infty}(1 - \frac{iz}{n^2})^{-1}$ satisfies the above conditions. Hence for this $h(z)$ the process X_t of (0.1) is a regular, stationary, real Gaussian process and it is Markov of infinite order (and not of finite order).*

PROOF: The details are not difficult. The product $h^{-1}(z) = \prod_{n=1}^{\infty}(1 - \frac{iz}{n^2})$ converges uniformly on bounded sets (this is "self-evident" in the language of [**5**, 4.10, Example 2, Proof], but we would like to add that it is equally self-evident that the above Example 2 is written with a misprint in which Π is replaced by $\sum$). Indeed, we have $\sum_{n=N}^{\infty} \ln^+ |1 - \frac{iz}{n^2}| = \sum_N^{\infty} \ln(1 + \frac{|z|^2}{n^4}) < \frac{|z|^2}{N^4}$ when N is so large that $\frac{|z|^2}{N^4} < 1$ on the set in question. Moreover, for $|z| = R \to \infty$, we have

$$R^{-1}\sum_1^{\infty} \ell n^+(1 + \frac{R^2}{n^4}) \le R^{-1} \le R^{-1}(1 + \frac{R^2}{n^4}) + R\sum_{\sqrt{R}}^{\infty}\frac{1}{n^4}$$

$$\le R^{-1/2}\ell n^+(1 + R^2) + \frac{2}{3}RR^{-3/2} \longrightarrow 0,$$

implying that $h^{-1}(z)$ is of minimal exponential type. Since its zeros are on the negative imaginary axis, it only remains to see that h is an outer function of class H^{2+}.

A single factor satisfies, for $b > 0$,

$$\int |1 - \frac{i(x+ib)}{n^2}|^{-2} dx = \int ((1+\frac{b}{n^2})^2 + \frac{x^2}{n^4})^{-1} dx < \int (1+\frac{x^2}{n^4})^{-1} dx < \infty,$$

and this provides an upper bound for $\int |h(x+ib)|^2 dx$, implying that $h \in H^{2+}$. This implies that $\int (1+x^2)^{-1} \ln |h(x)| dx > -\infty$ (ibid, 2.6), so that by Szegö's Alternative the process X_t is regular. Finally, to show that h is an outer function, it suffices (ibid, 2.7) to show that

$$\ln |h(i)| = \pi^{-1} \int (1+x^2)^{-1} \ln |h(x)| dx.$$

Equivalently, for a single factor it suffices to note that $\ln |1 - iz| (= \ln |z+i|)$ is a harmonic function in the upper half-plane, and as such satisfies the Poisson representation

$$\ln |z+i| = \frac{b}{\pi} \int \frac{\ln |x+i|}{(a-x)^2 + b^2} dx; \quad z = a + ib.$$

The proof is completed by setting $z = i$.

As mentioned in the introduction, such an example seems intractable. We now seek an example in the time domain. Let us introduce the kernels

$$K_1(t) = \exp(-(\frac{1}{t} + t)); \quad K_2(t) = e^{-t} \exp(-e^{-t}); \quad t > 0.$$

Letting $W(d\lambda)$ now denote a real white noise, derived from a two-sided Wiener process $W(t)$ with $W(0) = 0$, $-\infty < t < \infty$, we set $X_t = \int_{-\infty}^t K_1(t-s) W(ds)$. It is easy to see (for instance using Ito's formula) that integration by parts is justified and we have for

each t

$$X_t = \int_{-\infty}^t K_1'(t-s) W(s) ds, \quad P\text{–a.s.} \tag{1.1}$$

Since this is well-defined for all t, P–a.s., we adopt (1.1) as our pathwise definition of X_t (discarding a P-nullset). Furthermore, the usual iterated log estimate of $W(t)$ as

$t \to \pm\infty$ shows that, by dominated convergence of the differential quotients, (1.1) may be differentiated arbitrarily often under the integral. Since $\frac{d^k}{dt^k} K_1(0) = 0$ for all k, we have for all t

$$X^{(k)}(t) = \int_{-\infty}^t K_1^{(k+1)}(t-s) W(s) ds, \quad P\text{–a.s.} \tag{1.2}$$

Thus the paths of $X(t)$ are C_∞, but they are nowhere analytic because it is easy to see that $X(t)$ is not predictable from its derivatives in any interval (this would make it free of $W(ds)$ in the interval). Moreover, for X_t we have $\mathcal{F}^-_{-\infty} \equiv (\phi, \Omega)$ by Kolmogorov $0-1$ Law, since this is contained in the corresponding σ-field of $W(t)$. Thus X_t is a regular, stationary, Gaussian process with C_∞-paths (we are indebted to a remark of Professor J. L. Doob for this step). However, we do not yet have any Markov property.

Next, we set

$$Y_t = \int_{-\infty}^{t} K_2(t-s)X_s ds, \quad -\infty < t < \infty \tag{1.3}$$

which is well-defined for all t, P–a.s., and gives a pathwise definition of a second regular, stationary, Gaussian process. The same argument as before justifies differentiation of Y_t under the integral, but since $\frac{d^k}{dt^k}K_2(0) \neq 0$ for most k, extra terms appear.

LEMMA 1.4. *For $k > 0$, the derivative $Y_t^{(k)}$ has the form*

$$Y_t^{(k)} = D_k(X_t) + \sum_{j \le k} c_{j,k} \int^t e^{-j(t-s)} K_2(t-s) X_s ds$$

where D_k is a linear differential operator of degree $k-1$, and $c_{j,k}$ are real constants with $c_{k,k} = 1$.

PROOF: For $k = 1$ we have directly

$$Y_t' = e^{-1}X_t + \int_{-\infty}^{t} (-1 + e^{-(t-s)})K_2(t-s)X_s ds.$$

Assuming the result for k, we have

$$Y_t^{(k+1)} = (D\ D_k(X_t) + \sum_{j \le k} c_{j,k} e^{-1} X_t)$$
$$+ \sum_{j \le k} c_{j,k} \int_{-\infty}^{t} (-(j+1) + e^{-(t-s)}) e^{-j(t-s)} K_2(t-s) X_s ds,$$

which has the required form since $1 = c_{k,k} = c_{k+1,k+1}$.

Therefore, for any continuous function f on $[0,\infty)$ with $\lim_{t\to\infty} f(t) = 0$, by the Stone–Weierstrass theorem there are constants $N(\epsilon)$ and $d_n(\epsilon)$, $0 \le n \le N(\epsilon)$,

such that
$|\sum_{n=0}^{N(\epsilon)} d_n(\sum_{j\leq n} c_{j,k}e^{-jt}) - f(t)| < \epsilon$, and thus

$$\sum_{n=0}^{N(\epsilon)} d_n Y_t^{(n)} = D^{(N(\epsilon))}(X_t) + \int_{-\infty}^{t} f(t-s)K_2(t-s)X_s ds + \mathcal{E}(\epsilon), \tag{1.4}$$

where $D^{(N(\epsilon))}$ is a linear differential operator, and
$E(\mathcal{E}(\epsilon))^2 \leq \epsilon^2 E(\int_{-\infty}^{0} K_2(s)|X_s|ds)^2$, which tends to 0 as $\epsilon \to 0$. We now set

$$Z_t = X_t + iY_t, \tag{1.5}$$

so that Z_t is a regular, stationary, complex Gaussian process with C_∞-paths. Then $\sigma(Z_t^{(k)}) = \sigma(X_t^{(k)}) \vee \sigma(Y_t^{(k)})$, so that for every ϵ, $D^{(N(\epsilon))}(X_t)$ in (1.4) is measurable over $\sigma(Z_t^{(k)}, 0 \leq k)$, and therefore so is $\int_{-\infty}^{t} f(t-s)K_2(t-s)X_s ds + \mathcal{E}(\epsilon)$. Then letting $\epsilon \to 0$, and then letting f approximate an indicator $I_{[0,u]}$, it is easy to see that $\int_{t-u}^{t} K_2(t-s)X_s ds$ is measurable over $\sigma(Z_t^{(k)}, 0 \leq k)$, and so its derivative in u is as well. Therefore

PROPOSITION 1.5. *The process Z_t is Markov of order $d = \infty$ (and not of smaller order). In fact, $\sigma(Z_t^{(k)}, 0 \leq k) \equiv \sigma(Z_s, s \leq t)$.*

PROOF: We have only to note that $\sigma(Z_s, s \leq t) \subset \sigma(X_s, s \leq t) \subset \sigma(Z_t^{(k)}, 0 \leq k)$, as just shown.

A natural question at this point would be whether a real process Z_t may be written down having the same combination of properties. For this, it would be necessary to find a substitute $K_2^*(t)$ for $K_2(t)$ having all derivatives 0 at the origin, but such that $(K_2^{*(k)}(t), 0 \leq k)$ span $L^2(e^{-t}dt)$, $0 \leq t$. Considerable effort has led us to doubt the existence of such a kernel. The corresponding prescription for a spectral density $h(z)$ in (0.1) is also known from [**9**, *p.* 103]. It requires that, in addition to the properties needed for Proposition 1.3, it be possible to express $\frac{\bar{h}}{h}$ as the ratio of two inner functions. We also have no idea how to exhibit such an h.

Meanwhile, two remarks are in order concerning Z_t of Proposition 1.5. First, the same construction works in case $W(s)$ in (1.1) is replaced by some non-Gaussian process with homogeneous independent increments, mean 0, and finite variance.

Of course Z_t will no longer be Gaussian, but it will still have C_∞ paths and the Markov property of infinite order, just as before. (We can justify the initial integration by parts as an L^2–equivalence). Secondly, in the Gaussian case one should ask whether Z_t is also Markovian of infinite order in the wide sense, in the language of Doob [**2**, II, §3]. This would mean that the conditional expectation of Z_{t+u} given $\sigma(Z_s,\ s \leq t)$, is a linear function of $(Z_t^{(k)},\ 0 \leq k)$, and in the real case it would hold automatically. Unfortunately, in the complex case it is not automatic, and evidently fails to hold for our example. In short, we need the "real part" $\sigma(X_t^{(k)};\ 0 \leq k)$ separately from $\sigma(Y_t^{(k)},\ 0 \leq k)$, but even using complex coefficients the operation of taking the real part of $Z_t^{(k)}$ is nonlinear.

Section 2. Diffusive Filtrations

It is known (see [**5**, §4.5]) that the filtration generated by $X_t = \int_{-\infty}^t K_1(t-s)W(ds)$, where $W(ds)$ is real Gaussian white noise, is the same up to P-nullsets as that of $W(ds)$ if and only if the Fourier transform of K_1 is an outer function, but this may be difficult to ascertain (we refrained from using a K_1 (such as $t^{-3/2}e^{-1/t}$) which is known to have an outer transform, especially to illustrate this point). In any case, however, if we factor $|h|^2$ with an outer function h, and then solve for a K_1^* as its inverse Fourier transform, we obtain for this K_1^* (possibly $K_1^* \neq K_1$) a new representation with a white noise which does generate $\sigma(X_s,\ s \leq t)$. Since $W(t+u) - W(t)$ for any t is a true diffusion starting at 0, it is plausible (but perhaps not yet obvious) that $\sigma(X_s,\ s \leq t)$ is equivalent for all t to the filtration generated by a single diffusion process $(V_t,\ -\infty < t < \infty)$. Here we mean that V_t is a strong Markov process with continuous paths, relative to a single Borel transition function on a Borel state space (it turns out that we could assume without loss of generality that the state space is even a compact metric space). Correspondingly, we introduce

DEFINITION 2.1: A probability filtration $(\mathcal{F}_t, P)$, $-\infty \leq a < t < b \leq \infty$, is called "diffusive" on (a, b) if it is equivalent (up to P-nullsets) to the natural filtration of a diffusion V_t, $a < t < b$.

The main result concerning diffusive filtrations is the following.

THEOREM 2.1. *If $\mathcal{F}_\infty = \vee_t \mathcal{F}_t$ is such that $L^2(\mathcal{F}_\infty, P)$ is separable, then $(\mathcal{F}_t, P)$*

is diffusive on (a,b) if and only if every right-continuous, $\mathcal{F}_{t+}$–martingale is continuous, P–a.s. on (a,b).

REMARKS: It is not hard to see that when $L^2(\mathcal{F}_\infty, P)$ is separable this is a local property in t, in such a way that $(\mathcal{F}_t, P)$ is diffusive on $(-\infty,\infty)$ if and only if it is diffusive on (a,b) for every $-\infty < a < b < \infty$. Indeed, any martingale on $[a,b]$ may be prolonged to one on $(-\infty,\infty)$ by using its value at a for $t \le a$, and its value at b for $t \ge b$, and continuity on (a,b) is equivalent to continuity on each $[a+\epsilon,\ b-\epsilon]$, $\epsilon > 0$. In view of K Ito's stochastic integral representation of martingales of $\sigma(W(s),\ 0 \le s)$, it looks obvious that all martingales of $\sigma(W(ds),\ s \le t)$ should be continuous. However, there is a detail missing—namely, to reduce matters to $(W(s),\ s \le t)$ we have to fix a value, say $W(0) = 0$, and then $W(s)$, $s \le 0$ is not a Wiener process, but a Brownian bridge.

An easier approach is to consider any martingale of $\mathcal{F}_t = \sigma\{W(ds);\ s \le t\}$ conditionally on $\mathcal{F}_{t_1}$. For $t > t_1$ it is $P - a.s.$ a martingale under the conditional probabilities, and we can set $W(t_1) = 0$ to reduce it to Ito's representation for $t > t_1$. Therefore, it is continuous for $t > t_1$ conditionally on $\mathcal{F}_{t_1}$, which obviously implies that it is unconditionally continuous. Consequently, we see that the filtrations of all regular, stationary, real Gaussian processes are diffusive, and this applies in particular to our examples X_t and Z_t.

Turning to the proof of Theorem 2.1, suppose first that $(\mathcal{F}_t, P)$ is the natural filtration of a diffusion process X_t, with resolvent operators $R_\lambda f$, $f \in b(\mathcal{E})$. According to a result of Kunita and Watanabe, extended by Meyer (see [**1**, XV, **19**]), the martingales of the form

$$(2.1) \qquad R_\lambda f(X_t) - R_\lambda f(X_a) - \int_a^t \lambda R_\lambda f(X_s) - f(X_s) ds \quad f \in b(\mathcal{E}),$$

are dense in the space of all square-integrable martingales in $a \le t < b$, and we may restrict to a countable set of f which are dense under bounded pointwise convergence (say, simple functions based on the rationals). Thus it is enough to show that each martingale (2.1) is continuous. (Note that we do not have $R_\lambda f$ continuous, except perhaps in the Ray topology where it is not obvious that X_t itself is still continuous). The proof now rests in showing that the left limits $\lim_{s \to t-} R_\lambda f(X_s)$ are the same as $R_\lambda f(X_{t-})$ for all t, or (what is the same here) equal $R_\lambda f(X_t)$. This fact seems to depend on the previsible section theorem. The proof (from [1]) is that since both left limits are previsible processes, it suffices to prove them equal at an arbitrary previsible stopping time $T < \infty$ (of the usual augmented filtration of $\mathcal{F}_t$). Then there are stopping times $T_n < T$, $\lim_{n\to\infty} T_n = T$, such that $\lim_{n\to\infty} R_\lambda f(T_n) = \lim_{s\to T-} R_\lambda f(s) = \lim_{n\to\infty} E(\int_0^\infty e^{-\lambda s} X_{T_n+s} ds|\mathcal{F}_{T_n}) = E(\int_0^\infty e^{-\lambda s} f(X_{T+s})|\mathcal{F}_{T-})$. But since $R_\lambda f(X_T) = E(\int_0^\infty e^{-\lambda s} f(X_{T+s}) ds|\mathcal{F}_T)$, and is also measurable over $\mathcal{F}_{T-}$ along with $X_T (= X_{T-})$, we see that $R_\lambda f(X_T) = E(\int_0^\infty e^{-\lambda s} f(X_{T+s}) ds|\mathcal{F}_{T-})$, as required. Here we used, of course, both Hunt's Lemma and the strong Markov property of X_t.

Suppose, finally, that all of the right-continuous, $\mathcal{F}_{t+}$–martingales are continuous on (a, b), and that $L^2(\mathcal{F}_\infty, P)$ is separable. Let (X_n) be a sequence which is dense in $L^2(\mathcal{F}_\infty, P)$, and consider the sequence-valued process $(E(X_n|\mathcal{F}_{t+}))$. It is not hard to see that the natural filtration of this process is equivalent to $\mathcal{F}_{t+}$. Indeed, for $X \in \mathcal{F}_{t+}$, let $X_{n_j} \to X$ in L^2. Then by Jensen's Inequality, $E(X_{n_j}|\mathcal{F}_{t+}) \to E(X|\mathcal{F}_{t+}) = X$ in L^2, and a subsequence converges P–a.s. Hence X is in the generated σ–field at time t (up to P–nullsets).

We note that the process $(E(X_n|\mathcal{F}_{t+}))$ has continuous paths, although of course it is not a diffusion. However, let Z_t^P be the prediction process of $(E(X_n|\mathcal{F}_{t+}))$, as constructed in [8, Definition 1.6]. We know from [ibid, Theorem 1.17] that Z_t^P is a realization of a Borel right process (this construction and result go through without change even for $-\infty < t < \infty$). It remains to show that Z_t^P is continuous, and hence a diffusion. Here, in fact, continuity is defined in terms of the measures Z_{t-}^P, hence in purely measure-theoretic terms, and so is more or less free of the choice of a topology. In particular, we are free to use the Ray topology of [ibid, p. 25], which is generated by the Ray cone containing $\{R_\lambda g; \lambda > 0, g \in C^+\}$ and closed under both R_λ, $\lambda > 0$, and formation of minima. Now it is enough, since the cone contains a countable uniformly dense subset, to show that for every f in the cone, $f(Z_t^P)$ is continuous in t, P–a.s. For $f = R_\lambda g$, this follows by continuity of the martingales (2.1), as assumed. Moreover, the set of bounded measurable f for which $f(Z_t^P)$ is continuous is closed under R_λ (for the same reason as before), and it is obviously closed under formation of minima. Therefore, by the same transfinite construction as for the cone itself, $f(Z_t^P)$ is P–a.s. continuous for each f in the Ray cone. This means that Z_t^P is a diffusion in the Ray topology. Finally, by [ibid, Theorem 1.9], Z_t^P generates the same filtration as $(E(X_n|\mathcal{F}_{t+}))$, and since this is continuous in t up to P–nullsets, it is equivalent to $\mathcal{F}_t$.

References

[1]. C. Dellacherie and P.–A. Meyer, *Probabilités et Potentiel, Chap. XII–XVI*, Hermann, Paris.

[2]. J. L. Doob, "Stochastic Processes," Wiley and Sons, New York, 1953.

[3]. H. Dym and H. P. McKean, *Extrapolation and interpolation of stationary Gaussian processes*, Ann. Math. Stat. **41** (1970b), 1817–1844.

[4]. H. Dym and H. P. McKean, *Application of de Branges spaces of integral functions to the prediction of stationary Gaussian processes, Ill.*, J. Math. **14** (1970a), 299–343.

[5]. H. Dym and H. P. McKean, "Gaussian Processes, Function Theory, and the Inverse Spectral Problem," Academic Press, New York, 1976.

[6]. T. Hida, *Canonical representations of Gaussian processes and their applications*, Mem. Coll. Sci. Univ. Kyoto **33** (1960), 109–155.

[7]. F. B. Knight, *Prediction processes and an autonomous germ-Markov property*, The Annals of Probability **7** (1979), 385–405.

[8]. F. B. Knight, *Essays on the Prediction Process*, Inst. of Math. Statistics Monographs and Lecture Notes Series **1** (1981).

[9]. N. Levinson and H. P. McKean, *Weighted trigonometrical approximation on the line with application to the germ field of a stationary Gaussian noise,*, Acta Math. **112** (1964), 99–143.

[10]. M. Loeve, "Probability Theory, 3rd Ed.," D. Van Nostrand Co. Inc., Princeton, 1963.

Professor Frank B. Knight
Department of Mathematics
University of Illinois
1409 West Green Street
Urbana, Illinois 61801
U.S.A.

Topics in the Theory of Noncausal Stochastic Integral Equations

Shigeyoshi Ogawa

1. We are going to show some recent results as well as their applications in the theory of noncausal stochastic integral equations, to be more precise, we will select the following two topics; we will discuss in the paragraph 2, (A) the Cauchy problem of noncausal type and in the paragraph 3, (B) the integral equation of Fredholm type.

For the convenience of discussions, we prepare some terminology and notations which will be of frequent use in what follows ; We fix a probability space $(\Omega,\mathcal{F},P)$ and a real process $\{Z(t,\omega),t\geq 0\}$ which can be in all cases the standard Brownian motion $\{B(t,\omega),t\geq 0\}$. We understand by the random function (or process) $f(t,\omega)$ $(0\leq t\leq 1)$ the real function, which is measurable in (t,ω) with respect to the σ -field $\mathcal{B}_{[0,1]}\times\mathcal{F}$ and satisfy the condition, $P\{\int_0^1 f(t,\omega)^2dt<\infty\}=1$. We will suppose nothing more about the measurability in ω of the random functions unless the following special mentions are given; we say that a random function $f(t,\omega)$ is *causal* (or, *anti-causal*) when it is adapted to the family of σ-fields, $\mathcal{F}_{t,Z}=\sigma\{Z(s,\omega):s\leq t\}$ (or, $\mathcal{F}^+_{t,Z}=\sigma\{Z(1,\omega)-Z(s,\omega):s\geq t\}$, respectively).
The stochastic integral of noncausal type $\int_0^1 f(t,\omega)d_\phi Z(t)$ of a random function $f(t,\omega)$, with respect to the process $Z(\cdot)$ and a c.o.n.s. $\{\phi_n\}$ (a complete orthonormal system) in the space $L^2(0,1)$, is defined as being the limit in probability of the sequence of Stieltjes integrals,

$$\int_0^1 f(t,\omega)d_\phi Z(t)=p-\lim_{n\to\infty}\int_0^1 f(t,\omega)dZ^\phi_n(t),$$

where, $Z^\phi_n(t,\omega)=Z(0,\omega)+\sum_{k=1}^{n}\int_0^t\phi_k(t)dt\int_0^1\phi_k(s)dZ(s)$ and the integral

$\int_0^1\phi_k(s)dZ(s)$ is supposed to be well defined.

We will often say that the function $f(t,\omega)$ is ϕ-integrable if it is integrable with respect to the basis $\{\phi_n\}$.

For the case that $Z(\cdot)$ is the Brownian motion $\{B(t,\omega),t\geq 0\}$, we have a fundamental result due to the author which we repeat here, as Theorem 0, to make this article self-contained ; We call such random function $f(t,\omega)$ a quasi martingale if its differential (in t) df admits the decomposition as follows, $df(t,\omega) = df_1(t,\omega) + f_2(t,\omega)d^0B(t)$, where $f_1(t,\omega)$ is a random function (causal or not) almost all sample of which are of bounded variation and the last term stands for the differential of the function represented by the Ito-integral $\int f_2 d^0 B$ of a causal function $f_2(t,\omega)$.

Theorem 0, (OGAWA [1]). *Let $\{\phi_n\}$ be a c.o.n.s. in the space $L^2(0,1)$. Then every quasi martingale $f(t,\omega)$ is ϕ-integrable if and only if the basis $\{\phi_n\}$ is regular in the sense that;*

$$\sup_n \|u_n\|_{L^2} < +\infty \quad \textit{where} \quad u_n(t) = \sum_{k=1}^{n} \phi_k(t) \int_0^t \phi_k(s)ds.$$

In this case, the integral $\int f d_\phi B$ coincides with the symmetric (or Stratonovich) integral which we will denote by $\int f dB$.

(*Remark*) (i) When a random function $f(t,\omega)$ is integrable with respect to any c.o.n.s., then the function is said to be universally integrable (or, *u-integrable* for short). It is shown in OGAWA [1] that a quasi martingale $f(t,\omega)$ is u-integrable if the function $f_2(t,\omega)$ in the decomposition above is again a quasi martingale.

(ii) Recently, Zakai and Nualart published an article ([7]) in which they discussed the relation between the Stratonovich integral and the noncausal integral in this very limited sense (i.e., the integral in the sense of *u-integrability*) and they referred to the latter as "ogawa integral". As for the relation between the noncausal integral in its original form ([1],[5]) and the symmetric integrals, we refer to the article of the author [1].

It is worthwhile to notice that, if we take the system of Haar functions $\{H_{n,i}\}$ for the basis,

$$\begin{cases} H_{0,0}(t) = 1 \\ H_{n,i}(t) = \sqrt{2}^{(n-1)} \left[\mathbf{1}_{[2\cdot 2^{-n}i,\ 2^{-n}(2i+1))}(t) - \mathbf{1}_{[2^{-n}(2i+1),\ 2^{-n}(2i+2))}(t) \right] \\ \qquad\qquad\qquad\qquad (0 \le i \le 2^{n-1}-1, \quad n \ge 1), \end{cases}$$

then the process Z_n^H defined in the following form is the "Cauchy polygon approximation" of $Z(t,\omega)$;

$$Z_n^H(t,\omega) = Z(0,\omega) + \sum_{k=0}^{n} \sum_{i=0}^{2^{k-1}-1} \int_0^t H_{k,i}(s)ds \cdot \int_0^1 H_{k,i}(s)dZ(s).$$

In this case, the corresponding Stieltjes integral $\int f dZ_n^H$ is the Riemann sum as follows (see, [6]);

$$\int_0^1 f(t,\omega)dZ_n^H = \sum_{i=0}^{2^n-1}\left[\frac{1}{t_{i+1}^n - t_i^n}\int_{t_i^n}^{t_{i+1}^n} f(s,\omega)ds\right]\cdot\{Z(t_{i+1}^n) - Z(t_i^n)\}$$

where, $t_i^n = 2^{-n}i$.

2. Given an arbitrary random variable $\xi(\omega)$ we are concerned with the question of the existence and the uniqueness of solutions of the following noncausal initial value problem,

$$X(t,\omega) - \xi(\omega) = \int_0^t a(s,X(s),\omega)ds + \int_0^t b(s,X(s),\omega)d_\phi B(s). \tag{1}$$

where $a(t,x,\omega)$ and $b(t,x,\omega)$ $(t \geq 0, x \in R^1)$ are some random functions, causal in (t,ω) with respect to the Brownian motion for each fixed $x \in R^1$.

This problem was studied by the author in 1984 ([2]) in slightly restricted circumstances and now we are to study the same problem under more general assumptions as follows;

(A,1) The basis $\{\phi_n\}$ is regular.

(A,2) Almost all sample functions of the $a(t,x,\omega)$, $b(t,x,\omega)$ are sufficiently smooth in x (of C^3-class with bounded derivatives, for example).

In order to construct a solution, we consider the unique causal solution, say $\tilde{X}(t,\omega;x)$, of the following symmetric (or Stratonovich) integral equation,

$$\tilde{X}(t,\omega;x) - x = \int_0^t a(s,\tilde{X}(s,\omega;x),\omega)ds + \int_0^t b(s,\tilde{X}(s,\omega;x),\omega)dB(s). \tag{2}$$

Since almost all sample of the $\tilde{X}$ are continuous in x by virtue of the assumption (A,2), we can substitute the random variable $\xi(\omega)$ into x to get the new random function, $\hat{X}(t,\omega) = \tilde{X}(t,\omega;\xi(\omega))$, for which we have the following result,

Theorem 1; *(i) The function $\hat{X}$ is a solution of (1). (ii) For any function $F(t,x)$, which is differentiable in t and of C^4-class in x, we have the following equality,*

$$\begin{aligned} &F(t,\hat{X}(t)) - F(0,\xi(\omega)) \\ &= \int_0^t \{F'_t + F'_x \cdot a\}(s,\hat{X}(s))ds + \int_0^t (F'_x \cdot b)(s,\hat{X}(s))d_\phi B(s), \end{aligned} \tag{3}$$

where F'_t, F'_x *are the partial derivatives of F with respect to the variables* t, x .

(*Proof*) The statements can be verified by the arguments, similar to those given in the proofs of Theorem 1 in [2] and of Theorem in [3],

(i) ; Let

$$X_n^\phi(t,\omega;x) = x + \int_0^t a(s,\tilde{X}(s,\omega;x),\omega)ds + \int_0^t b(s,\tilde{X}(s,\omega;x),\omega)dB_n^\phi(s). \tag{4}$$

Then we have, for each t fixed, the equality; $\lim_{n\to\infty} X_n^\phi(t,\omega;x) = \tilde{X}(t,\omega;x)$ (in probability) by virtue of Theorem 0.

We claim that this convergence is uniform in x on any compact set. In fact, we see from the assumptions (A,1), (A,2) that the functions $X_n^\phi(t,\omega;x)$ and $\tilde{X}(t,\omega;x)$ are differentiable in x and their derivatives satisfy the following equations,

$$\tilde{X}'(t,\omega;x) - 1 = \int_0^t a'(s,\tilde{X}(s,\omega;x),\omega)\tilde{X}'(s,\omega;x)ds + \int_0^t b'(s,\tilde{X}(s,\omega;x),\omega)\tilde{X}'(s,\omega;x)dB(s) \tag{5}$$

$$\frac{\partial}{\partial x}X_n^\phi(t,\omega;x) - 1 = \int_0^t a'(s,\tilde{X}(s,\omega;x),\omega)\tilde{X}'(s,\omega;x)ds + \int_0^t b'(s,\tilde{X}(s,\omega;x),\omega)\tilde{X}'(s,\omega;x)dB_n^\phi(s) \tag{6}$$

where the symbol ' stands for the partial differentiation $\dfrac{\partial}{\partial x}$.

From (5) and (6) we get, for an arbitrary positive A, the inequality as follows,

$$\begin{aligned}
&\sup_{|x|\le A} |\tilde{X}(t,\omega;x) - X_n^\phi(t,\omega;x)| \\
&= \sup_{|x|\le A} \Big| \int_{-A}^{A} \frac{\partial}{\partial x}\{\tilde{X}(t,\omega;x) - X_n^\phi(t,\omega;x)\}dx + \{\tilde{X}(t,\omega;-A) - X_n^\phi(t,\omega;-A)\}\Big| \\
&\le \sqrt{2A}\,[\int_{-A}^{A}\{\frac{\partial}{\partial x}(\tilde{X}(t,\omega;x) - X_n^\phi(t,\omega;x)\}^2dx]^{1/2} + |\tilde{X}(t,\omega;-A) - X_n^\phi(t,\omega;-A)| \\
&= \sqrt{2A}\,[\int_{-A}^{A}\{\int_0^t b'(s,\tilde{X}(s,\omega;x),\omega)\tilde{X}'(s,\omega;x)(dB_n^\phi - dB)\}^2dx]^{1/2} \\
&\qquad + |\tilde{X}(t,\omega;-A) - X_n^\phi(t,\omega;-A)|
\end{aligned}$$

On the other hand, it is not difficult to verify (see, Proposition 1 of [1]) the following,

$$\lim_{n\to\infty}\int_{-A}^{A}\{\int_0^t b'(s,\tilde{X}(s,\omega;x),\omega)\tilde{X}'(s,\omega;x)(dB_n^\phi - dB)\}^2 dx = 0 \quad \text{(in probability)},$$

which implies that; $\lim_{n\to\infty}\sup_{|x|\le A}|X_n^\phi(t,\omega;x) - \tilde{X}(t,\omega;x)| = 0$ for each t. Hence, by applying the usual method of truncation, we confirm that ; $\lim_{n\to\infty} X_n^\phi(t,\omega;\xi(\omega)) = \hat{X}(t,\omega)$, (in P). Replacing the x by $\xi(\omega)$ in (4) and letting $n\to\infty$ on both sides of the equation, we obtain the conclusion (i).

(ii) For the causal solution $\tilde{X}$ of the equation (2), we have;

$$F(t,\tilde{X}(t,\omega;x))$$
$$= F(0,x) + \int_0^t \{F_t' + F_x'\cdot a\}(s,\tilde{X}(s,\omega;x))ds + \int_0^t (F_x'\cdot b)(s,\tilde{X}(s,\omega;x))dB(s).$$

Let

$$F_n(t,x) = F(0,x) + \int_0^t \{F_t' + F_x'\cdot a\}(s,\tilde{X}(s,\omega;x))ds + \int_0^t (F_x'\cdot b)(s,\tilde{X}(s,\omega;x))dB_n^\phi(s). \tag{7}$$

Following the argument given in the first part we will get the conclusion (ii), namely; we verify that for each fixed t, the sequence $\{F_n(t,x)\}$ converges (in P) to $F(t,\tilde{X}(t,\omega;x))$, uniformly in x on any compact set and consequently we get the relation; $\lim_{n\to\infty} F_n(t,\hat{X}(t)) = F(t,\hat{X}(t))$ (in P).

On the other hand, we have from (6) the equality,

$$F_n(t,\hat{X}(t)) = F(0,\xi(\omega)) + \int_0^t \{F_t' + F_x'\cdot a\}(s,\hat{X}(s),\omega)ds + \int_0^t (F_x'\cdot b)(s,\hat{X}(s),\omega)dB_n^\phi(s). \tag{8}$$

Letting $n\to\infty$ on both sides of (8), we see from (7) and from the definition itself of the noncausal integral the validity of the conclusion. *q.e.d.*

As we have seen in the discussions above our approach to the problem is of somewhat qualitative nature in the sense that we need not use explicitly the tools from the infinite dimensional calculus, such as the Wiener-Ito decomposition of the homogeneous chaos. The next example shows the efficiency of such qualitative theory.

Example A.1 (*"go-return problem" in SDE*) Let us take the system of Haar functions $\{H_{n,i}\}$ for the basis $\{\phi_n\}$. Let $\tilde{X}(t,\omega;x)$ be the causal solution of (2) and let $Y(t,\omega;y)$ be the unique anti-causal solution of the following symmetric integral equation,

$$(9)\quad Y(t,\omega;y)-y=\int_1^t a(s,Y(s),\omega)ds+\int_1^t b(s,Y(s),\omega)d_H B(s),\ (0\le t\le 1).$$

Then we can consider the function, $Y(t,\omega;\tilde{X}(1,\omega;x))$ (which is a noncausal solution of the equation (9) for the case that $y=\tilde{X}(1,\omega;x)$) and it is not difficult to see that this function Y satisfies the equality ; $Y(0,\omega;\tilde{X}(1,\omega;x))=x$, (P-a.s. $\forall x\in R^1$).

We notice that the equality above always holds in the case of the ordinary differential equation whenever the uniqueness of solution is assured, but in the case of stochastic differential equations the problem itself had not been considered meaningful until the noncausal stochastic calculus was introduced.

(*Proof of the equality*) Let $X_n(t,x)$, $Y_n(t,y)$ $(0\le t\le 1)$ be the solutions of the following equations,

$$X_n(t,x)-x=\int_0^t a(s,X_n(s,x),\omega)ds+\int_0^t b(s,X_n(s,x),\omega)dB_n^H(s).$$

$$Y_n(t,y)-y=\int_1^t a(s,Y_n(s,y),\omega)ds+\int_1^t b(s,Y_n(s,y),\omega)dB_n^H(s).$$

Notice that, $Y_n(0,X_n(1,x))=x$ (P-a.s.) for any x.

On the other hand, the coefficients a, b being sufficiently smooth, we see that the sequences $\{X_n(1,x)\}$, $\{Y_n(0,y)\}$ converge in probability, as $n\to\infty$, to the $\tilde{X}(1,\omega;x)$ and $Y(0,y,\omega)$ (respectively) uniformly in x, y on any compact sets. The conclusion follows from these and the inequality below,

$$|Y(0,\omega;\tilde{X}(1,\omega;x))-x|\le|Y(0,\omega;\tilde{X}(1,\omega;x))-Y(0,\omega;X_n(1,x))|$$
$$+|Y(0,\omega;X_n(1,x))-Y_n(0,X_n(1,x))|.\quad q.e.d.$$

As for the question of uniqueness of the solution, we have not yet obtained a final result, but we have an interesting example which suggests us the existence of a close connection between the uniqueness property and the validity of the Ito-formula of noncausal type; Let us call the random function $f(t,\omega)$ a *chain* if it satisfies the following differential form,

$$(10)\qquad df(t,\omega)=\alpha(t,\omega)dt+\beta(t,\omega)d_\phi B(t)\ ,$$

where $\alpha(\cdot)$ and $\beta(\cdot)$ are some random functions.

We say that a chain has the *C-property* provided that the following Ito-formula of noncausal type holds for any smooth function $F(t,x)$;

$$dF(f(t)) = \{F'_t(f) + F'_x(f) \cdot \alpha(t)\}dt + F'_x(f) \cdot \beta(t)d_\phi B.$$

We have seen that the solution $\hat{X}$ of (1), just constructed above, has the C-property. The following example shows that, in the particular case mentioned below, the $\hat{X}(\cdot)$ is the unique solution in the class of all functions having the C-property.

Example A,2; Let the coefficient b in (1) be positive for all (t,x). For the simplicity, we suppose that $b(t, x, \omega)$ is independent of ω. Then the solution of (1) having C-property is unique.

(*Proof*) Let $F(t, x) = \int_0^x dy/b(t, y)$ and let X be an arbitrary solution of (1) having the C-property. Then we have,

$$F(t, X(t)) - F(0, \xi(\omega)) = \int_0^t \{F'_t + a/b(t, X(t))\}dt + \int_0^t d_\phi B.$$

Since $F(t, x)$ is sufficiently smooth and is strictly monotone in x, it is clear that the last equation admits only one solution. *q.e.d.*.

Stimulated by the Example A,2, we conceive the following,

Conjecture ; *That the solution of (1) is unique is equivalent to the fact that every solution has the C-property.*

3. One of the most typical subjects in the noncausal stochastic calculus is the stochastic integral equation of Fredholm type,

$$X(t, \omega) = f(t, \omega) + \alpha \int_0^1 L(t, s, \omega)X(s)ds + \beta \int_0^1 K(t, s, \omega)X(s)d_\phi Z(s), \tag{11}$$

where L and K $(0 \leq t, s \leq 1)$ are some random kernels such that almost all sample functions belong to the class $L^2([0, 1] \times [0, 1])$.

In the article of 1986 ([4]) the author studied this subject in connection with the boundary value problems of stochastic differential equations and he showed the existence and the uniqueness of solutions (see Theorem 1 in [4]), under some reasonable assumptions on the choice of the fundamental pair $(Z, \{\phi_n\})$ and on the sample regularity of the kernels K and L. These assumptions ((H,1)-(H,3) in [4]) are in fact sufficiently reasonable in the sense that the results can apply to the stochastic boundary value problems as follows;

$$\begin{cases} [(d/dt)p(t, \omega)d/dt + q(t, \omega)]X(t) = X(t)dB(t, \omega)/dt + h(t, \omega) \\ X(0) = \xi_0(\omega), \;\; X(1) = \xi_1(\omega) \end{cases} \tag{12}$$

where ξ_0 and ξ_1 are arbitrary random variables.

But the discussions given there were extensively based on the one dimensionality of the parameter t and are hardly applicable to the multi-dimensional cases. Hence, as a first step toward the general theory, we are interested in studying the equation (11) without assuming the sample regularities (such as, the piecewise smoothness) of the kernels. We have not yet obtained the general results on this direction except for some special cases as follows;

Example B ; (13) $$X(t) = f(t,\omega) + \alpha \int_0^1 K(t,s,\omega)X(s)d_\phi B(s) .$$

We will finish this report by giving a result concerning the existence and the uniqueness of solutions of this equation;

Let $S(l^2)$ be the set of all random functions $f(t,\omega)$ for which there exist a pair (ψ,ϵ) of an orthonormal basis $\{\psi_n(t)\}$ and an l^2-sequence $\{\epsilon_n\}$ satisfying the following conditions;

(S-1) the integral, $f_n = \int_0^1 f(t,\omega)\psi_n(t)d_\phi B(t)$, exists $(\forall n)$ and

(S-2) $\lim_{m\to\infty} \sum_{n=1}^{\infty} [\epsilon_n\{f_n - \int_0^1 f(t,\omega)\psi_n(t)dB_m^\phi(t)\}]^2 = 0$, in probability,

(S-3) $\sum_{m,n}^{\infty} {\epsilon_m}^2{\epsilon_n}^2 \int_0^1 \psi_m(t)^2\psi_n(t)^2 dt < \infty.$

We will indicate this property by saying that the $S(l^2)$-function admits the $\{\epsilon_n\}$ as the weight. Notice that $S(l^2) \supset L^2(0,1)$.

Here are the assumptions that we pose on the functions $f(\cdot)$ and $K(t,s)$;

(B,1) $f(\cdot) \in S(l^2)$,

(B,2) With an appropriate weight $\{\epsilon_n\}$ for $f(\cdot,\omega)$, it holds the condition;

$$\sum_{m,n} (k_{m,n}/\epsilon_m\epsilon_n)^2 < \infty, \text{ where } k_{m,n} = \int_0^1\int_0^1 K(t,s,\omega)\psi_m(t)\psi_n(s)dtds.$$

Proposition. *Under the assumptions (B,1)-(B,2) the equation (13) has the unique $S(l^2)$-solution for almost all α .*

(*Proof*) We are going to show that there exists an $S(l^2)$- solution X admitting the $\{\epsilon_n\}$ in (B,1) as the weight sequence and that the solution X is unique in the class $S(l^2)$.

Let X be an $S(l^2)$-solution of (8) admitting the $\{\epsilon_n\}$ as the weight, then we see that;

$$(14)\qquad X(t) = f(t,\omega) + \sum_{m,n} \epsilon_m k'_{m,n}\psi_m(t)\hat{x}_n\ ,$$

$$\text{where } k'_{m,n} = k_{m,n}/\epsilon_m\epsilon_n,\quad \hat{x}_n = \epsilon_n\int_0^1 X(t)\psi_n(t)d_\phi B(t).$$

Let $\gamma_{m,n}(\omega) = \int_0^1 \psi_m(t)\psi_n(t)dB(t)$ then, we get from (14) the next relation

$$(15)\qquad \hat{x}_l = \hat{f}_l + \alpha\sum_{m,n}\hat{\gamma}_{l,m}k'_{m,n}\hat{x}_n,\qquad \text{where } \hat{\gamma}_{l,m} = \epsilon_l\gamma_{l,m}\epsilon_m,\ \hat{f}_n = \epsilon_n f_n.$$

We notice that the random sequence $\hat{K}_{l,n} = \sum_m \hat{\gamma}_{l,m}k'_{m,n}$ satisfies the condition, $P[\sum_{l,n}\hat{K}^2_{l,n} < +\infty] = 1.$

Now we introduce the new functions as follows,

$$(16)\qquad \begin{cases} \hat{f}(t,\omega) = \sum_n \hat{f}_n\psi_n(t),\ \hat{X}(t) = \sum_n \hat{x}_n\psi_n(t) \\ \hat{K}(t,s,\omega) = \sum_{m,n}\hat{K}_{m,n}\psi_m(t)\psi_n(s) \end{cases}$$

Then from the relation (15) we see that the $\hat{X}(\cdot)$ is a solution of the following random integral equation,

$$(17)\qquad \hat{X}(t) = \hat{f}(t) + \alpha\int_0^1 \hat{K}(t,s,\omega)\hat{X}(s)ds.$$

Conversely, by tracing the above process in the inverse direction we see that any L^2-solution $\hat{X}$ of (17) can be transformed into an $S(l^2)$-class solution of (13) and that this correspondence is one-to-one and onto. Thus the problem is reduced to the question of existence and uniqueness of L^2-solution of the ordinary random integral equation (17). Since the kernel $\hat{K}(t,s,\omega)$ is of Hilbert-Schmidt type (P-a.s.), we get the conclusion. *q.e.d.*.

REFERENCES

[1] OGAWA,S. ; The stochastic integral of noncausal type as an extension of the symmetric integrals. Japan J.Appl.Math., vol.2, No.1 (1985) pp.229-240.

[2] OGAWA,S. ; Sur la question d'éxistence de solutions d'une équation differentielle stochastique du type noncausal. J.Math. of Kyoto Univ., 24-1 (1984). pp.699-704.

[3] OGAWA,S. and T.SEKIGUTI ; On the Ito formula of noncausal type. Proc.Japan Acad., vol.60, Ser.A, No.7 (1984), pp.249-251.

[4] OGAWA,S. ; On the stochastic integral equation of Fredholm type. In: Waves and Patterns etd.by T.Nishida et al., (1986) Kinokuniya and North-Holland, pp.597-605.

[5] OGAWA,S. ; Sur le produit direct du bruit blanc par lûi-même. C.R.Acad.Sc.Paris,t.288, Série A (1979), pp.359-362.

[6] OGAWA,S. ; Une remarque sur l'approximation de l'integrale stochastique du type noncausal par une suite des integrales de Stieltjes. Tôhoku Math.J.,vol.36,no.1,(1984),pp.41-48.

[7] NUALART,D and M.ZAKAI ; On the relation between the Stratonovich and Ogawa integrals, Ann. Probability, vol.17, No.4, (1989) pp.1536-1540

OGAWA Shigeyoshi

Kyoto Institute of Technology
Kyoto-shi, Sakyo-ku Matsugasaki
Kyoto 606, JAPAN

Orlicz and Luxemburg Norm Inequalities

S.D. Jacka

§1. Introduction

The Orlicz and Luxemburg norms associated with a Young function Φ are defined in Krasnoselski and Ruticki (1961). We denote (a slight adaptation of) the Luxemburg norm by $\|\cdot\|_\Phi$:

$$\|X\|_\Phi = \inf\left\{\mu > 0 : \mathbf{E}\,\Phi\left[\frac{|X|}{\mu}\right] \leq \Phi(1)\right\},$$

and (a slight adaptation of) the Orlicz norm by $|||\cdot|||_\Phi$:

$$|||X|||_\Phi = \inf\{\mu > 0 : \sup_{\lambda \geq 0}(\mu\lambda - \mathbf{E}\,\Phi(\lambda|X|)) \geq \Psi(1)\}.$$

where Ψ is the convex conjugate of Φ.

The Orlicz space $L^\Phi(\Omega, \mathcal{F}, \mathbf{P})$ is given by:

$$L^\Phi \equiv \{X : \|X\|_\Phi < \infty\} = \{X : |||X|||_\Phi < \infty\},$$

the second equality holding because the two norms are equivalent.

The main result of this note is the following theorem.

Theorem 1. *Given a pair of non-negative processes X and Y, adapted to the filtered probability space $(\Omega, \mathcal{F}, (\mathcal{F}_t;\, t \geq 0), \mathbf{P})$,* and a collection $\mathcal{T}$ of $\mathcal{F}$-measurable, non-negative random variables (random times), define:*

$$\sigma_\Phi \equiv \sigma_\Phi(X, Y, \mathcal{T}) = \sup_{T \in \mathcal{T}} \frac{\mathbf{E}\,X_T}{|||Y_T|||_\Phi}.$$

* Measurable in the sense of Dellacherie and Meyer (1978) [p. 84, definition 3].

and

$$\Sigma_\Phi \equiv \Sigma_\Phi(X,Y,\mathcal{T}) = \sup_{T\in\mathcal{T}} \frac{\mathbf{E}X_T}{\|Y_T\|_\Phi}.$$

Then, defining, for each $\mu > 0$:

$$L^*(\mu) = \sup_{\lambda>0} \sup_{T\in\mathcal{T}(\lambda)} \mathbf{E}\Big(\frac{\lambda}{\mu}X_T - \Phi(\lambda Y_T)\Big)$$

where $\mathcal{T}(\lambda) = \{T \in \mathcal{T} : \mathbf{E}\Phi(\lambda Y_T) < \infty\}$

(i) $\sigma_\Phi = \sup\{\mu > 0 : L^*(\mu) \geq \Psi(1)\}$ *where* Ψ *is the conjugate of* Φ, *and*

(ii) $\Sigma_\Phi = \inf_{\mu>0}(\mu L^*(\mu) + \mu\Phi(1))$.

§2. Preliminary Remarks and Notation

First we review some of the properties of the Luxemburg and Orlicz norms.

Fix $X \in \mathcal{F}$ and define for each μ, $\lambda > 0$

$$L(\mu,\lambda) = \mu\lambda - \mathbf{E}\Phi(\lambda|X|),$$
$$\Lambda_1(\mu) = \sup_{\lambda>0} L(\mu,\lambda).$$

Note that, since, for each fixed λ, L is linear in μ, Λ_1 is a convex function of μ.

Now either

$$\begin{aligned} &\forall\lambda > 0 \;\; \mathbf{E}\Phi(\lambda|X|) = \infty \\ &\Leftrightarrow \|X\|_\Phi = \infty \\ &\Leftrightarrow \Lambda_1 \equiv -\infty \end{aligned} \tag{2.1}$$

or

$$\begin{aligned} &\exists\lambda > 0 \;\text{ such that }\; \mathbf{E}\,\Phi(\lambda|X|) < \infty \\ &\Leftrightarrow \|X\|_\Phi < \infty \\ &\Leftrightarrow \Lambda_1 \geq 0 \end{aligned} \tag{2.2}$$

Furthermore, if $\|X\|_\Phi < \infty$ either

$$\|X\|_\Phi = 0$$
$$\Leftrightarrow$$
$$\Lambda_1(\mu) = \infty \; \forall\mu > 0,$$

or

$$
\begin{aligned}
& 0 < \|X\|_\Phi < \infty \\
& \Leftrightarrow \\
(2.3) \qquad & 0 \le \Lambda_1(\mu) < \infty \; \forall \mu > 0 \\
& \Leftrightarrow \\
& \Lambda_1 \text{ is a continuous function of } \mu \\
& \text{with } \Lambda_1(0) = 0 \;\; \Lambda_1(\mu) \to \infty \text{ as } \mu \to \infty.
\end{aligned}
$$

To prove (2.2) observe that since Φ is positive, increasing and convex $\Phi(cx) \ge c\Phi(x)$ $(c \ge 1)$ so that, given $\lambda > 0$ such that $\mathbf{E}\Phi(\lambda|X|) = c < \infty$, set $\lambda' = \lambda(\frac{\Phi(1)}{c} \wedge 1)$ then

$$
\begin{aligned}
\mathbf{E}\,\Phi(\lambda'|X|) &\le \left(\frac{\Phi(1)}{c} \wedge 1\right) \mathbf{E}\,\Phi(\lambda|X|) \\
&\le \Phi(1)
\end{aligned}
$$

so that $\|X\|_\Phi \le \frac{1}{\lambda'}$; the rest is obvious.

To prove (2.3) observe that, since $\|X\|_\Phi > 0$, $\exists x > 0 : \mathbf{P}(|X| \ge x) = \varepsilon > 0$: thus $\mathbf{E}\,\Phi(\lambda|X|) \ge \varepsilon\Phi(\lambda x)$ so

$$
\begin{aligned}
L(\mu, \lambda) &\le \mu\lambda - \varepsilon\Phi(\lambda x) \\
&= \varepsilon\left(\frac{\mu}{\varepsilon x}(\lambda x) - \Phi(\lambda x)\right)
\end{aligned}
$$

so that

$$
\begin{aligned}
\sup_{\lambda > 0} L(\mu, \lambda) &\le \varepsilon \sup_{t > 0}\left(\frac{\mu}{\varepsilon x} t - \Phi(t)\right) \\
&= \varepsilon\Psi\left(\frac{\mu}{\varepsilon x}\right) < \infty
\end{aligned}
$$

where Ψ is the conjugate of Φ. Furthermore, since

$$
\|X\|_\Phi < \infty, \;\; \lim_{\lambda \to 0} \mathbf{E}\,\Phi(\lambda|X|) = 0
$$

so $\Lambda_1(\mu) \ge 0$. The rest is obvious. Now define

$$
\begin{aligned}
\Lambda_2(\mu) &= \mu\left(\mathbf{E}\,\Phi\left(\frac{|X|}{\mu}\right) - \Phi(1)\right) \\
&\equiv \mathbf{E}\, f(|X|, \mu) \\
\text{where} \quad f(x, \mu) &= \mu\Phi\left(\frac{x}{\mu}\right) - \mu\Phi(1).
\end{aligned}
$$

Note that, since Φ is a convex function on $\mathbb{R}_+$, so is $f(x,\cdot)$ for each fixed $x \geq 0$. We first prove

(2.4) Λ_2 is a decreasing convex function on $\mathbb{R}_+$.

The fact that Λ_2 is convex follows immediately from the convexity of $f(x,\cdot)$. The fact that it is decreasing follows from the fact that f is:

$$\begin{aligned} f(x, c\mu) &= c\mu\,\Phi\left(\frac{x}{c\mu}\right) - c\mu\,\Phi(1) \\ &\leq \mu\,\Phi\left(\frac{x}{\mu}\right) - c\mu\,\Phi(1) \quad (\text{for } c \geq 1) \\ &\leq f(x,\mu). \end{aligned}$$

Now either

$$\|X\|_\Phi = \infty$$
$$\Leftrightarrow$$
$$\Lambda_2(\mu) = \infty\ \forall \mu > 0$$

or

$$\|X\|_\Phi < \infty$$
$$\Leftrightarrow$$
$$\exists \mu > 0 : \Lambda(\mu) < \infty.$$

Furthermore if $\|X\|_\Phi < \infty$, either

$$\|X\|_\Phi = 0$$
$$\Leftrightarrow$$
$$\Lambda_2(\mu) < 0\ \forall \mu > 0,$$

or

$$0 < \|X\|_\Phi < \infty$$
$$\Rightarrow$$
$$\Lambda_2(\mu) \to -\infty \text{ as } \mu \to \infty \text{ and } \Lambda_2 \text{ is continuous on } C^o \text{ where } C = \{x: \Lambda_2(\mu) < \infty\}$$

The proof is obvious. We may conclude therefore that

$$0 < \|X\|_\Phi < \infty \Leftrightarrow 0 < |||X|||_\Phi < \infty \Leftrightarrow \Lambda_1(|||X|||_\Phi) = \Psi(1)$$

whilst if $0 < \|X\|_\Phi < \infty$

$$\begin{aligned} &\text{either} \quad \|X\|_\Phi = \inf\left\{\mu > 0 : \mathbf{E}\,\Phi\left(\frac{|X|}{\mu}\right) < \infty\right\} \\ &\text{or} \quad \Lambda_2(\|X\|_\Phi) = 0. \end{aligned}$$

It is well known that L^Φ and L^Ψ are dual spaces with the usual definition of the Luxemburg norm, denoted by $\|\cdot\|_\Phi^{Lux}$, at least if Ψ doesn't grow too fast (see, for example, the appendix on Orlicz spaces in Dellacherie and Meyer (1982)).

It is almost immediately obvious that definition (1.3) gives an equivalent norm since, for $c \geq 1 : \Phi(\frac{x}{c}) \leq \frac{1}{c}\Phi(x)$ so that

$$(1 \wedge \Phi(1))\|X\|_\Phi \leq \|X\|_\Phi^{Lux} \leq (1 \vee \Phi(1))\|X\|_\Phi$$

whilst the fact that $\|\cdot\|_\Phi$ is indeeed a norm is established in exactly the same way as for $\|\cdot\|_\Phi^{Lux}$; indeed we have the following

Lemma 2. *Given a Young function Φ, define*

$$c \equiv c_\Phi = \Phi^{-1}(1), \quad \text{and}$$
$$\hat{\Phi}(x) \equiv \Phi(cx)$$

then

(i) $\|X\|_{\hat{\Phi}} = c\|X\|_\Phi^{Lux}$

whilst, defining

$$c' = \Phi(1)$$
$$\bar{\Phi}(x) \equiv \Phi(x)/c'$$

(ii) $\|X\|_\Phi = \|X\|_{\bar{\Phi}}^{Lux}$.

Proof.

(i)
$$\|X\|_{\hat{\Phi}} = \inf\left\{\mu > 0 : \mathbf{E}\,\Phi\left(\frac{c|X|}{\mu}\right) \leq 1\right\}$$
$$= c\,\|X\|_\Phi^{Lux}$$

(ii)
$$\|X\|_{\bar{\Phi}}^{Lux} = \inf\left\{\mu > 0 : \mathbf{E}\,\Phi\left(\frac{|X|}{\mu}\right)/c' \leq 1\right\}$$
$$= \inf\left\{\mu > 0 : \mathbf{E}\,\Phi\left(\frac{|X|}{\mu}\right) \leq \Phi(1)\right\} \qquad \square$$

The advantage of using definition (1.3) for the Luxemburg norm is, of course, that for any constant x, $\|x\|_\Phi = |x|$.

We know that if we change the scale of Φ we still have a Young function: we may produce a whole class of equivalent norms on L^Φ:

Lemma 3. *For any t with $\Phi(t) > 0$ define*

$$\begin{aligned} \Phi_t(x) &= \Phi(tx), \quad \text{and} \\ {}^t\|X\|_\Phi &\equiv \|X\|_{\Phi_t} \end{aligned} \tag{2.5}$$

then if $\Phi(s), \Phi(t) > 0$:

$$(t/s)\left(1 \wedge \frac{\Phi(s)}{\Phi(t)}\right) {}^s\|X\|_\Phi \leq {}^t\|X\|_\Phi \leq (t/s)\left(1 \vee \frac{\Phi(s)}{\Phi(t)}\right) {}^s\|X\|_\Phi \tag{2.6}$$

Proof. Since Φ is increasing it is clearly sufficient to establish that

$$(t/s)\, {}^s\|X\|_\Phi \leq {}^t\|X\|_\Phi \leq \frac{t\Phi(s)}{s\Phi(t)}\, {}^s\|X\|_\Phi \tag{2.7}$$

for $s \geq t$.

From (1.3) and (2.5):

$${}^t\|X\|_\Phi = \inf\left\{\mu > 0 : \mathbf{E}\,\Phi\left(\frac{t|X|}{\mu}\right) \leq \Phi(t)\right\}$$

but $\Phi(s)/\Phi(t) \geq 1$ so

$$\begin{aligned} \mathbf{E}\,\Phi(t|X|/(\mu t\Phi(s)/s\Phi(t))) &= \mathbf{E}\,\Phi(s|X|/(\mu\Phi(s)/\Phi(t))) \\ &\leq (\Phi(t)/\Phi(s))\mathbf{E}\,\Phi(s|X|/\mu), \end{aligned}$$

establishing the right-hand inequality in (2.7). Conversely,

$$\mathbf{E}\,\Phi(t|X|/((t/s)\mu)) = \mathbf{E}\,\Phi(s|X|/\mu)$$

establishing the left-hand inequality. □

We are now at a suitable point to establish that, whilst $|||\cdot|||_\Phi$ is not, as yet, proven to be a norm, it is equivalent to $\|\cdot\|_\Phi$, which is

Lemma 4. *Suppose $\Phi(s), \Phi(t) > 0$ then, if Ψ is the conjugate of Φ, and $\Psi(1/t) > 0$:*

$$(t/s)\left(\Phi(s) \wedge \Psi\left(\frac{1}{t}\right)\right) {}^s\|X\|_\Phi \leq {}^t|||X|||_\Phi \leq (t/s)\left(\Phi(s) + \Psi\left(\frac{1}{t}\right)\right) {}^s\|X\|_\Phi$$

where

$${}^t|||\cdot|||_\Phi = |||\cdot|||_{\Phi_t}. \tag{2.8}$$

Proof. Assume ${}^{s}\|X\|_{\Phi} > 0$ and

$$\mu < \frac{t}{s}\left(\Phi(s) \wedge \Psi\left(\frac{1}{t}\right)\right) {}^{s}\|X\|_{\Phi};$$

and look at

$$\Lambda(\mu, \lambda) = \left(\lambda - \mathbf{E}\,\Phi\left(\frac{t\lambda|X|}{\mu}\right)\right). \tag{2.9}$$

Since $\mu < \frac{t}{s}\Phi(s)\,{}^{s}\|X\|_{\Phi}$, for $\lambda \geq \Phi(s) \wedge \Psi(\frac{1}{t})$

$$\Lambda(\mu, \lambda) \leq \lambda - \frac{\lambda}{\Phi(s) \wedge \psi(1/t)}\mathbf{E}\,\Phi\left(\frac{s|X|}{{}^{s}\|X\|_{\Phi}}\right) \leq 0.$$

Moreover, if $\lambda < \Phi(s) \wedge \Psi(\frac{1}{t})$

$$\Lambda(\mu, \lambda) < \Psi\left(\frac{1}{t}\right)$$

establishing the left-hand inequality in (2.8). Conversely, setting

$$\begin{aligned}
\lambda &= \left(\Phi(s) + \Psi\left(\frac{1}{t}\right)\right)(t/s)\sup_{\lambda \geq 0}(\mu\lambda - \mathbf{E}\,\Phi(t\lambda|X|)) \\
&= \sup_{\lambda \geq 0} \Lambda(\mu, \lambda) \\
&= \sup_{\lambda \geq 0}\left(\frac{s}{t}\lambda - \mathbf{E}\,\Phi\left(\frac{s\lambda|X|}{\mu}\right)\right) \\
&\geq \frac{s}{t}\lambda - \mathbf{E}\,\Phi\left(\frac{s\lambda|X|}{\mu}\right) = \rho(\mu)
\end{aligned}$$

and, setting $\bar{\mu} = \lambda\,{}^{s}\|X\|_{\Phi}$, $\rho(\bar{\mu}) = \Psi(\frac{1}{t})$, establishing the right-hand inequality in (2.8). $\square$

Corollary 5. *If Φ has conjugate Ψ with $\Psi(1) > 0$ then*

$$|||X|||_{\Phi} \leq (1 + \Psi(1))\,\|X\|_{\Phi}^{Lux}$$

Proof. From Lemma 2,

$$\|X\|_{\Phi}^{Lux} = \frac{1}{c}\,{}^{c}\|X\|_{\Phi},$$

where $c = \Phi^{-1}(1)$; setting $s = c$ and $t = 1$ in (2.8) we see that

$$\begin{aligned}|||X|||_\Phi &\le \frac{1}{c}(\Phi(c) + \Psi(1))\,{}^{c}\|X\|_\Phi \\ &= (1 + \Psi(1))\,\|X\|_\Phi^{Lux} \qquad \square\end{aligned}$$

§3. The Orlicz Norm

We're now nearly in a position to prove the following theorem.

Theorem 6. *The Luxemburg and Orlicz norms are dual, that is* $(L^\Phi, \|\cdot\|_\Phi)$ *and* $(E^\Psi, |||\cdot|||_\Psi)$ *are dual spaces where* Ψ *is the conjugate of* Φ *and* E^Ψ *is the closure of* L^∞ *in* L^Ψ *(see Dellacherie and Meyer* (1982), *p.* 168).

Most of the work required to prove Theorem 6 has been done for us; the duality of L^Φ and L^Ψ is already established so that, essentially, all we have to do is to prove a generalized version of Holder's inequality and show that equality is attained for any non-trivial r.v. $Y \in L^\Psi$, for some X.

To avoid confusion we denote, in this chapter, the conjugate of Φ by $\tilde{\Phi}$ rather than Ψ.

Proof of Theorem 6. We first show that

$$|\mathbf{E}XY| \le \mathbf{E}|XY| \le {}^{t}\|X\|_\Phi\,{}^{1/t}|||Y|||_{\tilde{\Phi}}. \tag{3.1}$$

We shall in fact assume that $|XY|$ is integrable and show that

$$\frac{\mathbf{E}|XY|}{{}^{t}\|X\|_\Phi} \le {}^{1/t}|||Y|||_{\tilde{\Phi}}. \tag{3.2}$$

Recall that $\tilde{\Phi}(x) = \sup_{t\ge 0}(tx - \Phi(t))$: now

$$\begin{aligned}\sup_{\lambda\ge 0}\left(\frac{\lambda\mathbf{E}|XY|}{{}^{t}\|X\|_\Phi} - \mathbf{E}\,\tilde{\Phi}\left(\frac{\lambda|Y|}{t}\right)\right) &\le \mathbf{E}\left(\sup_{\lambda\ge 0}\frac{\lambda|X|\,|Y|}{{}^{t}\|X\|_\Phi} - \tilde{\Phi}\left(\frac{\lambda|Y|}{t}\right)\right) \\ &= \mathbf{E}\left(\sup_{\lambda\ge 0}\frac{t|X|}{{}^{t}\|X\|_\Phi}\lambda - \tilde{\Phi}(\lambda)\right) \\ &= \mathbf{E}\,\Phi\left(\frac{t|X|}{{}^{t}\|X\|_\Phi}\right) \le \Phi(t)\end{aligned}$$

establishing (3.2) (since $(\tilde{\phi})_{1/t} = (\tilde{\Phi}_t))$. Conversely given $Y \in E^{\tilde{\Phi}}(\Omega, \mathcal{F}, \mathbf{P})$ with $\mathbf{P}(Y = 0) < 1$, define

$$Y_n = -n \vee (Y \wedge n)$$

and let $\Psi \equiv \Psi_m^\varepsilon$ be a Young function whose restriction to $[0, m]$ is C^1 with $\Psi'_+(0) = 0$ and with $|\Psi - \tilde{\Phi}| \leq \varepsilon$: note that it is possible to find such a Ψ for any m, $\varepsilon > 0$ since $\tilde{\Phi}$ has increasing left and right derivatives, and that $|\tilde{\Psi} - \Phi| \leq \varepsilon$. Let ψ be the derivative of Ψ and note that

$$\begin{aligned}\tilde{\Psi}(\psi(x)) &= \sup_{t\geq 0} t\psi(x) - \Psi(t) \\ &= x\psi(x) - \Psi(x),\end{aligned}$$

at least for $x : \Psi'_+(x) \leq m$. Now given $Y \in L^{\tilde{\Phi}}$, we see Y, $(Y_n; n \geq 1) \in L^{\Psi}$. Define

$$\begin{aligned} X_n &\equiv X_n(c, K) \\ &= c\, sgn(Y_n)\psi(K|Y_n|) : \\ {}^t\|X\|_{\tilde{\Psi}} &= \inf\left\{\mu \geq 0 : \mathbf{E}\,\tilde{\Psi}\left(\frac{ct}{\mu}\psi(K|Y_n|)\right) \leq \tilde{\Psi}(t)\right\}. \end{aligned} \tag{3.3}$$

Note that, since Ψ is continuous and Y_n is bounded ($|Y_n| \leq n$),

$$\sup_{\lambda\geq 0} f(\lambda) \equiv \sup_{\lambda\geq 0}({}^{1/t}|||Y_n|||_{\Psi}\lambda - \mathbf{E}\,\Psi(\frac{\lambda|Y_n|}{t})) \tag{3.4}$$

is attained, at λ say, and must equal $\tilde{\Psi}(t)$.

Moreover since for any λ, $\frac{\lambda|Y_n|}{t}$ is bounded, we may differentiate inside the expectation in (3.4) to see that λ must satisfy:

$${}^{1/t}|||Y_n|||_{\Psi} = \mathbf{E}\left(\frac{|Y_n|}{t}\psi\left(\frac{\lambda|Y_n|}{t}\right)\right), \tag{3.5}$$

at least for m sufficiently large. Substituting the expression (3.5) in (3.4) we see that

$$\begin{aligned}\tilde{\Psi}(t) = f(\lambda) &= \mathbf{E}\left[\frac{\bar{\lambda}|Y_n|}{t}\psi\left(\frac{\bar{\lambda}|Y_n|}{t}\right) - \Psi\left(\frac{\lambda|Y_n|}{t}\right)\right] \\ &= \mathbf{E}\,\tilde{\Psi}\left(\psi\left(\frac{\lambda|Y_n|}{t}\right)\right).\end{aligned}$$

Thus, setting $c = \frac{1}{t}$ and $K = \lambda/t$ in (3.3) we see that $\mathbf{E}\,\tilde{\Psi}(t|X_n|) = \tilde{\Psi}(t)$ and so ${}^t\|X_n\|_{\tilde{\Psi}} = 1$. Now

$$\mathbf{E}\,X_nY_n = \mathbf{E}\frac{|Y_n|}{t}\psi\left(\frac{\lambda|Y_n|}{t}\right)$$

which, from (3.5), is ${}^{1/t}|||Y_n|||_{\Psi}$. Thus, from (3.2) applied to $(\tilde{\Psi}, \Psi)$ (with $X = X_n$ $Y = Y_n$) we see that

$$\sup_{X:\|X_n\|_{\tilde{\Psi}}\leq 1} \mathbf{E}\,XY = {}^{1/t}|||Y_n|||_{\Psi}.$$

Equation (1.5) is then established by letting first ε tend to zero and then letting n tend to infinity, since

$$Y \in E^{\tilde{\Phi}},\ \lim_{n\to\infty}\lim_{\varepsilon\to 0}|||Y_n|||_{\Psi} = |||Y|||_{\tilde{\Phi}}.$$

Equation (1.6) follows by Lemma 2 and the definition of $|||\cdot|||_{\tilde{\Phi}}$ (equation (1.4)). □

Remarks.

1. Note that we have actually established that ${}^{1/t}|||\cdot|||_{\tilde{\Phi}}$ is the dual norm to ${}^t\|\cdot\|_{\Phi}$.
2. We have therefore shown that

$$\mathbf{E}|XY| \leq \inf_{t>0} {}^t\|X\|_{\Phi}\, {}^{1/t}|||Y|||_{\tilde{\Phi}}.$$

3. The result remains valid if we are working with a σ-finite measure m, rather than $\mathbf{P}$, with the revised definition:

$$|||f|||_{\Phi} = \inf\left\{\mu > 0 : \sup_{\lambda\geq 0}\int(\mu\lambda - \Phi(\lambda|f(w)|))dm(w) \geq \tilde{\Phi}(1)\right\}.$$

§4. Some Applications

The norm $|||\cdot|||_{\Phi}$ has certain advantages over the norm $\|\cdot\|_{\Phi}$ — certain calculations are much easier because of the Lagrangian form of the definition.

Note that we revert, in this section, to denoting the conjugate of Φ by Ψ.

The first application is the obvious

Lemma 7. *For any t with $\Psi(\frac{1}{t}) > 0$*

$$\frac{\mathbf{E}|X|}{{}^t|||Y|||_\Phi} = r(X,Y) \stackrel{\text{def}}{=} \inf\left\{\mu > 0 : \sup_{\lambda \in L(\Phi,Y)} \mathbf{E}\left(\frac{\lambda|X|}{t\mu} - \Phi(\lambda|Y|)\right) \leq \Psi\left(\frac{1}{t}\right)\right\}$$

where

$$L(\Phi, Y) = \{\lambda > 0 : \mathbf{E}\,\Phi(\lambda|Y|) < \infty\}.$$

Proof. By definition, and the continuity of Φ; assuming $X \in L^1$, $Y \in L^\Phi$, ${}^t|||Y|||_\Phi > 0$

$$\begin{aligned}\Psi\left(\frac{1}{t}\right) &= \sup_{\lambda \geq 0} \mathbf{E}[\lambda|X| - \Phi(tr(X,Y)\lambda|Y|)] \\ &= \sup_{\lambda \geq 0} \mathbf{E}\left[\frac{\lambda|X|}{r(X,Y)} - \Phi(t\lambda|Y|)\right]\end{aligned}$$

thus we see that if $X \in L^1$,

$$0 < {}^t|||Y|||_\Phi < \infty \Rightarrow \frac{\mathbf{E}|X|}{{}^t|||Y|||_\Phi} = r(X,Y).$$

For $X \notin L^1$ and $Y \in L^\Phi$ the result holds trivially (with $\inf\emptyset = \infty$). If $X \in L^1$ and $Y \notin L^\Phi$ the result holds trivially with $r(X,Y) = 0$ whilst if $X \in L^1$ and $Y = 0$ a.s. (i.e. if ${}^t|||Y|||_\Phi = 0$) either

$$X = 0 \text{ a.s.} \quad \Rightarrow r(X,Y) = 0$$

or

$$\mathbf{P}(|X| > 0) > 0 \Rightarrow r(X,Y) = \infty.$$

Finally if $X \notin L^1$, $Y \notin L^\Phi$ the result holds trivially with $r(X,Y) = 0$ at least if $\frac{\infty}{\infty}$ is taken to be 0. □

Lemma 8. *For any t with $\Phi(t) > 0$*

$$\frac{\mathbf{E}|X|}{{}^t\|Y\|_\Phi} = s(X,Y) \stackrel{\text{def}}{=} \inf_{\mu>0}\{\mu(\Lambda_t(\mu) + \Phi(t))\}$$

where

$$\Lambda_t(\mu) = \sup_{\lambda \in L(Y,\Phi)} \mathbf{E}\left(\frac{\lambda|X|}{t\mu} - \Phi(\lambda|Y|)\right)$$

and $L(Y,\Phi)$ is as defined in Lemma 7.

Proof. Assume first that $X \in L^1$ and note that if $\|Y\|_\Phi < \infty$ then setting $\lambda = \frac{t}{{}^t\|Y\|_\Phi + \varepsilon}$ we see that

$$\mu(\Lambda_t(\mu) + \Phi(t)) \geq \frac{\mathbf{E}X}{{}^t\|Y\|_\Phi + \varepsilon}$$

so letting $\varepsilon \to 0$ we see that $s(X,Y) \geq \mathbf{E}|X|/{}^t\|Y\|_\Phi$.

Conversely, if $0 < \|Y\|_\Phi = k < \infty$ there are three possibilities:

(i) $\mathbf{E}|Y|\phi_-(t|Y|/k) = \infty$;

or

(ii) $\mathbf{E}Y\phi_-(t|Y|/k) = K < \infty$ and $\mathbf{E}\,\Phi\left(\frac{t|Y|}{k}\right) = \Phi(t)$;

or

(iii) $\mathbf{E}Y\phi_-(T|Y|/k) = K < \infty$, $\mathbf{E}\,\Phi\left(\frac{t|Y|}{k}\right) < \Phi(t)$

and

$$\mathbf{E}Y\phi_-(t|Y|\lambda) = \infty\ \forall\lambda > 1/k,$$

where ϕ_- is the left derivative of Φ.

By considering the left derivative with respect to λ of

$$\mathbf{E}\frac{\lambda|X|}{\mu t} - \Phi(\lambda|Y|)$$

we see that: in case (i)

$$\Lambda_t(\varepsilon) = \sup_{\lambda \in L(Y,\Phi)} \mathbf{E}\left(\frac{\lambda|X|}{t\varepsilon} - \Phi(\lambda|Y|)\right)$$

is attained at a $\lambda < t/k$, $\forall\varepsilon > 0$ so

$$\begin{aligned} s(X,Y) &\leq \sup_{0<\lambda<t/k} \frac{\lambda\mathbf{E}|X|}{t} + \varepsilon\Phi(t) \\ &= \frac{\mathbf{E}|X|}{k} + \varepsilon\Phi(t) \end{aligned}$$

and letting $\varepsilon \to 0$ we obtain the desired result; in case (iii) the same argument holds; in case (ii) setting $\hat{\mu} = \mathbf{E}|X|/K$ we see that $\Lambda_t(\hat{\mu})$ is attained at $\lambda = t/k$ and so

$$
\begin{aligned}
s(X,Y) &\le \hat{\mu}(\Lambda_t(\hat{\mu}) + \Phi(t)) \\
&= \mathbf{E}\frac{|X|}{k} + \hat{\mu}\left(\Phi(t) - \mathbf{E}\,\Phi\left(\frac{t|Y|}{k}\right)\right) \\
&= \mathbf{E}\frac{|X|}{k}
\end{aligned}
$$

as required.

If $\mathbf{E}|X| > 0$ and ${}^t\|Y\|_\Phi = 0$ then clearly $s \equiv \infty$. If $X \notin L^1$ and $Y \in L^\Phi$ then, again $s \equiv \infty$ whilst if ${}^t\|Y\|_\Phi = \infty$ $s \equiv 0$ and if $\mathbf{E}|X| = 0$ again $s \equiv 0$. □

Lemma 9. Suppose X and Y are non-negative processes on a filtered probability space $(\Omega, \mathcal{F}, (\mathcal{F}_t), \mathbf{P})$. Define $\mathcal{L} = \{r.v.s.\ L \ge 0 : L \in \mathcal{F}\}$ then

$$\sup_{L\in\mathcal{L}} \sup_{F\in\mathcal{F}} \frac{\mathbf{E}\,X_L 1_F}{{}^x|||Y_L 1_F|||_\Phi} = {}_xC_\Phi(X,Y)$$

and

$$\sup_{L\in\mathcal{L}} \sup_{F\in\mathcal{F}} \frac{\mathbf{E}\,X_L 1_F}{\|Y_L 1_F\|_\Phi} = {}_xc_\Phi(X,Y)$$

where

$$(4.1)\quad {}_xC_\Phi(X,Y) \equiv \inf\left\{\mu > 0 : \sup_{\lambda\ge 0} \mathbf{E} \sup_{t\ge 0}\left(\frac{\lambda X_t}{\mu x} - \Phi(\lambda Y_t)\right)^+ \le \Psi\left(\frac{1}{x}\right)\right\}$$

and

$$(4.2)\qquad {}_xc_\Phi = \inf_{\mu>0}\left\{\mu\left(\sup_{\lambda\ge 0} \mathbf{E} \sup_{t\ge 0}\left(\frac{\lambda X_t}{\mu x} - \lambda Y_t\right)^+ + \Phi(x)\right\}$$

and $\frac{0}{0}$ and $\frac{\infty}{\infty}$ are defined as 0.

Proof. Apply Lemmas 7 and 8 simultaneously to all the pairs $(X_L 1_F, Y_L 1_F)$ and observe that for an arbitrary process (Z_t)

$$\sup_{L\in\mathcal{L}} \sup_{F\in\mathcal{F}} \mathbf{E}\,Z_L 1_F = \mathbf{E} \sup_{t\ge 0}(Z_t^+).$$

The interchange of $\inf_{\mu>0}$ and $\sup_{L\in\mathcal{L}}$ is justified by the fact that the inf is (essentially) attained.

We also have the following corollaries of Lemmas 7 and 8.

Theorem 10. *Let $\mathcal{T}$ be any collection of non-negative, $\mathcal{F}$-measurable r.v.s; X and Y non-negative measurable processes then define for each $\mu > 0$:*

$$\mathcal{T}(\mu,\Phi) = \{T \in \mathcal{T} : \mathbf{E}\,\Phi(\mu Y_T) < \infty\}$$
$$\mathcal{T}(\Phi) = \{T \in \mathcal{T} : Y_T \in L^{\Phi}(\Omega,\mathcal{F},\mathbf{P}\}$$

then

$$\mathcal{T}(\Phi) = \bigcup_{\mu>0} \mathcal{T}(\mu,\Phi)$$

and, defining

$$\sigma_\Phi \equiv \sigma_\Phi(X,Y,\mathcal{T}) = \sup_{T\in\mathcal{T}(\Phi)} \frac{\mathbf{E}\,X_T}{|||Y_T|||_\Phi},$$

and

$$\Sigma_\Phi \equiv \Sigma_\Phi(X,Y,\mathcal{T}) = \sup_{T\in\mathcal{T}(\Phi)} \frac{\mathbf{E}\,X_T}{||Y_T||_\Phi},$$

then

$$\begin{aligned} \sigma_\Phi &= \inf\{\mu > 0 : \sup_{\lambda\geq 0}\ \sup_{T\in\mathcal{T}(\lambda,\Phi)} \mathbf{E}\,(\lambda X_T/\mu - \Phi(\lambda Y_T)) \leq \Psi(1)\} \\ &= \inf\{\mu > 0 : \sup_{T\in\mathcal{T}(\Phi)}\ \sup_{\lambda\geq 0 : T\in\mathcal{T}(\lambda,\Phi)} \mathbf{E}\,((\lambda X_T)/\mu - \Phi(\lambda Y_T)) \leq \Psi(1)\} \end{aligned} \tag{4.3}$$

and

$$\Sigma_\Phi = \inf_{\mu>0} \mu\{(\sup_{\lambda\geq 0}\ \sup_{T\in\mathcal{T}(\lambda,\Phi)} \mathbf{E}\,(\lambda X_T/\mu - \Phi(\lambda Y_T)) + \Phi(1)\}. \tag{4.4}$$

Remarks. 1. The assumption that X and Y are measurable processes implies that X_T and Y_T are r.v.s.

2. Of course, measurability is the same as saying that X and Y are progressively measurable with respect to some filtration (of $\mathcal{F}$).

Corollary 11. *Suppose X, Y and $\mathcal{T}$ are as above then the best constant C_q appearing in*

$$\mathbf{E}\,X_T \leq C_q ||Y_T||_q\ \forall T \in \mathcal{T} \qquad (q > 1)$$

is $\sigma_q \equiv \sigma_{x^q}(X,Y,\mathcal{T})$ (appearing in (4.2)).

$$= p^{1/p}\, q^{1/q} (\sup_{\lambda\geq 0}\ \sup_{T\in\mathcal{T}(x^q)} \mathbf{E}[\lambda X_T - (\lambda Y_T)^q])^{1/p}$$

where $\frac{1}{p}+\frac{1}{q}=1$.

Proof. Simply apply Theorem 10 to the function $\Phi \equiv x^q$, observing that $|||\cdot|||_\Phi = \|\cdot\|_q$, the L^q-norm, to obtain the first expression. The second expression follows immediately from the fact that $\Psi \equiv x^p/(pq^{p-1})$ and $\mathcal{T}(\lambda, x^q) \equiv \mathcal{T}(x^q)$. $\square$

The following is an improved version of Theorem (2.2.1) of Barlow et al. (1986):

Theorem 12. *Suppose that* X *and* Y *are as in Lemma* 9 *and, in addition,* X *and* Y *are optional processes: define*

$$\mathcal{T} = \{\text{stopping times } T \text{ w.r.t the filtration } (\mathcal{F}_t)\}$$

and, for any $L \in \mathcal{L}$; $x, y > 0$

$$\begin{aligned} p_L(x,y) &= \mathbf{P}(X_L \geq x; Y_L \leq y) \\ p(x,y) &= \sup_{T\in\mathcal{T}} P_T(x,y) \\ I(c,\mu) &= \int_0^\infty p(\mu ca, c\Phi^{-1}(a))da \\ I(\mu) &= \sup_{c\geq 0} I(c,\mu) \end{aligned}$$

then the following are equivalent:

(i) $\exists x$ *with* $\Psi\left(\frac{1}{x}\right) > 0$ *and* ${}_xC_\Phi(X,Y) < \infty$

(ii) $\forall x$ *such that* $\Psi\left(\frac{1}{x}\right) > 0$: ${}_xC_\Phi(X,Y) < \infty$

(iii) $\exists \mu > 0$ *with* $I(\mu) < \infty$

(iv) $\forall x$ *such that*

$$\Psi\left(\frac{1}{x}\right) > 0 : \exists \mu(x) > 0 \text{ with } I(\mu x) \leq \Psi\left(\frac{1}{x}\right) \tag{4.5}$$

where ${}_xC_\Phi$ *is as given in* (4.1).

Moreover, if we fix Y, *and define for any optional process* X:

$$\begin{aligned} {}^x\|X\|_{\sigma(\Phi),Y} &=_x C_\Phi(|X|,Y), \\ {}^x\|X\|_{\Sigma(\Phi),Y} &=_x C_\Phi(|X|,Y), \\ {}^x\|X\|_{\pi(\Phi),Y} &= \inf\left\{\mu > 0 : I(\mu x) \leq \Psi\left(\frac{1}{x}\right)\right\}, \\ {}^x\|X\|_{\Pi(\Phi),Y} &= \inf_{\mu>0} \mu(I(\mu x) + \Phi(x)) \end{aligned}$$

then, defining

$$S_\Phi(Y) = \left\{ optional\ \ X :^x \|X\|_{\sigma(\Phi),Y} < \infty \text{ for some } x \text{ s.t. } \Psi\left(\frac{1}{x}\right) > 0 \right\},$$

(v) *the collection*

$$\left\{ {}^x\|\cdot\|_{\sigma(\Phi),Y} : \Psi\left(\frac{1}{x}\right) > 0 \right\} \cup \{ {}^x\|\cdot\|_{\Sigma\Phi,Y} : \Phi(x) > 0 \}$$

are equivalent norms on $S_\Phi(Y)$, *and*

(vi)

$$\tag{4.6} \begin{aligned} \frac{1}{2}{}^x\|X\|_{\Pi(\Phi),Y} &\le {}^x\|X\|_{\Sigma(\Phi),Y} \le {}^x\|X\|_{\Pi(\Phi),Y} \\ \frac{1}{2}{}^x\|X\|_{\pi(\phi),Y} &\le \|X\|_{\sigma(\Phi),Y} \le {}^x\|X\|_{\pi(\Phi),Y} \end{aligned}$$

Proof. We prove first that ${}^x\|\cdot\|_{\sigma(\Phi),Y}$ are norms: this is immediate on applying (4.1) and Lemma 8. Now we know from Lemmas 3 and 4 that the ${}^x|||\cdot|||_\Phi$ are all equivalent ($x : \Psi(\frac{1}{x}) > 0$), hence we see that (i) $\Leftrightarrow$ (ii) and establish (v).

We now establish that for $K \ge 1$

$$\tag{4.7} I(K\mu) \le \frac{1}{K} I(\mu).$$

This is obvious since

$$\begin{aligned} I(K\mu) &= \sup_{c\ge 0} \int p(K\mu ca, c\Phi^{-1}(a))da \\ &= \sup_{c\ge 0} \frac{1}{K} \int_0^\infty p(\mu cu, c\Phi^{-1}(u/K))du \text{ (substituting } u = Ka) \\ &\le \frac{1}{K} I(\mu) \text{ (since } p(x,y) \text{ increases in } y \text{ and } \phi^{-1} \text{ is increasing).} \end{aligned}$$

Then (iii) $\Leftrightarrow$ (iv), since taking $\bar\mu : I(\bar\mu) < \infty$, and setting

$$\hat\mu = \bar\mu \max(1, I(\bar\mu)/\Psi(1/x))$$

we see from (4.7) that $I(\hat\mu) \le \psi(1/x)$.

We shall now prove that (4.6) holds which will also establish that (ii) $\Leftrightarrow$ (iv).

Recall first that, from the optional section theorem (Dellacherie and Meyer (1978)):

$$\sup_L \mathsf{P}(X_L \geq x, Y_L < y) = \sup_T \mathsf{P}(X_T \geq x, Y_T < y). \tag{4.8}$$

Claim: If X and Y are non-negative r.v.s. then

$$E[X-Y]^+ = \int_0^\infty \mathsf{P}(X \geq x, Y < x)dx. \tag{4.9}$$

This is clear using Fubini's theorem since

$$\int_0^\infty \mathsf{P}(X \geq x, Y < x)dx = \mathbf{E}\int_0^\infty I_{(Y<x\leq X)}dx = \mathbf{E}(X-Y)^+.$$

To prove the right-hand inequality in (4.6) define

$$(Z_t) \equiv (Z_t^{\lambda,\mu}) \equiv \left(\frac{\lambda X_t}{\mu x} - \Phi(\lambda Y_t)\right)^+,$$

and observe that

$$\begin{aligned}
\mathbf{E}\sup_{t\geq 0} Z_t &= \sup_L \mathbf{E}\, Z_L \\
&= \sup_L \mathbf{E}\left(\frac{\lambda X_L}{\mu x} - \Phi(\lambda Y_L)\right)^+ \\
&= \sup_L \int_0^\infty \mathsf{P}\left(\frac{\lambda X_L}{\mu x} \geq t, \Phi(\lambda Y_L) < t\right)dt \qquad \text{(from (4.9))} \\
&\leq \int_0^\infty \sup_L \mathsf{P}\left(\frac{\lambda X_L}{\mu x} \geq t, \Phi(\lambda Y_L) < t\right)dt \\
&= \int_0^\infty \sup_L \mathsf{P}\left(X_L \geq \frac{\mu x t}{\lambda}, Y_L < \frac{\Phi^{-1}(t)}{\lambda}\right)dt \\
&= \int_0^\infty p\left(\frac{\mu x t}{\lambda}, \frac{\Phi^{-1}(t)}{\lambda}\right)dt \qquad \text{(from (4.8))}
\end{aligned}$$

Thus $\sup_{\lambda\geq 0} \mathbf{E}\sup_{t\geq 0} Z_t^{\lambda,\mu} \leq I(\mu)$ from which the RHS of (4.6) follows on applying (4.3) and (4.5). To prove the left-hand inequalities in (4.6) observe that (setting $Z^* = \sup_{t\geq 0} Z_t$)

$$\begin{aligned}
\mathbf{E}Z^* &= \int_0^\infty \mathsf{P}(Z^* \geq a)da \\
&= \int_0^\infty \sup_T \mathsf{P}(Z_T \geq a)da \qquad \text{(from the section theorem),}
\end{aligned}$$

where T runs through the set of stopping times $\mathcal{T}$. Now

$$\begin{aligned}\mathbf{P}(Z_T \geq a) &= \mathbf{P}\left(\frac{\lambda X_T}{\mu x} - \Phi(\lambda Y_T) \geq a\right)\\ &\geq \mathbf{P}\left(\frac{\lambda X_T}{\mu x} \geq 2a, \Phi(\lambda Y_T) < a\right)\\ &= p_T\left(\frac{2a\mu x}{\lambda}, \frac{\Phi^{-1}(a)}{\lambda}\right),\end{aligned}$$

Thus

$$\mathbf{E}Z^* \geq \int_0^\infty p\left(\frac{2\mu x}{\lambda}a, \frac{\Phi^{-1}(a)}{\lambda}\right) d\lambda$$

and so $\sup\limits_{\lambda} \mathbf{E} \sup\limits_{t\geq 0}(Z_t^{\lambda,\mu}) \geq I(2\mu)$ from which we may immediately deduce the LHS of (4.6). □

§5. Weak L^Ψ Spaces

The weak L^Ψ space was introduced in Barlow et al. (1986):

$$\begin{aligned}L_\Psi^W(\Omega, \mathcal{F}, \mathbf{P}) &= \left\{X \in \mathcal{F} : \exists c > 0 \text{ with } \mathbf{E}|X1_F| \leq c|||1_F|||_\Phi \ \forall F \in \mathcal{F}\right\}\\ &= \left\{X \in \mathcal{F} : \exists c > 0 \text{ with } \mathbf{E}|X1_F| \leq c||1_F||_\Phi \ \forall F \in \mathcal{F}\right\}\end{aligned}$$

the norms (on the equivalence classes of $\mathbf{P}$ a.s equal r.v.s) are, of course,

$$|||X|||_\Psi^W = \inf\{c > 0 : \mathbf{E}|X1_F| \leq c|||1_F|||_\Phi\}$$

and

$$||X||_\Psi^W = \inf\{c > 0 : \mathbf{E}|X1_F| \leq c||1_F||_\Phi\}.$$

From Lemma 7 we know that

$$|||X|||_\Psi^W = \inf\{\mu > 0 : \sup_{\lambda\geq 0} \sup_{F\in\mathcal{F}} \mathbf{E}(\lambda|X|1_F - \Phi(\mu\lambda 1_F)) \leq \Psi(1)\}$$

and

$$||X||_\Psi^W = \inf_{\mu>0}\left(\mu\left(\sup_{\lambda\geq 0} \sup_{F\in\mathcal{F}} \mathbf{E}\left(\frac{\lambda|X|1_F}{\mu} - \Phi(\lambda 1_F)) + \Phi(1)\right)\right)\right)$$

(we, of course, assume that $\Psi(1) > 0$). We may also introduce, the weak H^Ψ spaces:

$$H_\Psi^W = \{\text{processes } X : \mathbf{E}|X_L|1_F \leq c|||1_F|||_\Phi \ \forall L \in \mathcal{L}, F \in \mathcal{F}, \text{ for some } c\}$$

with the obvious norms

$$(5.1)\quad |||X|||^W_{H^\Psi} = \inf\{\mu > 0 : \sup_{\lambda\geq 0} \mathbf{E} \sup_{t\geq 0}(\lambda|X_t| - \Phi(\mu\lambda))^+ \leq \Psi(1)\}$$

$$(5.2)\quad = \inf\{\mu > 0 : \sup_{\lambda\geq 0} \mathbf{E}(\lambda X^* - \Phi(\mu\lambda))^+ \leq \Psi(1)\},$$

with the usual notation: $X^* = \sup_t |X_t|$, so that $|||X|||^W_{H^\Psi} = |||X^*|||^W_\Psi$ and $||X||^W_{H^\Psi} = ||X^*||^W_\Psi$. We may define a weaker L^Ψ-space:

$$\tilde{L}^W_\Psi = \left\{X \in \mathcal{F} : \exists\mu > 0 : \sup_{x\geq 0} \Psi(x)\mathbf{P}\left(\frac{|X|}{\mu} \geq x\right) < \infty\right\}$$

with the natural pseudo-norm:

$$||X||^{\tilde{W}}_\Psi = \inf\left\{\mu > 0 : \sup_{x\geq 0} \Psi(x)\mathbf{P}\left(\frac{|X|}{\mu} \geq x\right) \leq \Psi(1)\right\}$$

$$= \inf\left\{\mu > 0 : \sup_{x\geq 0} x\,\mathbf{P}\left(\Psi\left(\frac{|X|}{\mu}\right) \geq x\right) \leq \Psi(1)\right\}.$$

The question: 'when does $L^W_\Psi = \tilde{L}^W_\Psi$?' was settled in Barlow et al. (1986):

Lemma 13. (Lemma 1.6.1 of Barlow et al. (1986)): *The following assertions are equivalent:*

(i) $\exists K, \lambda\!\!\!-, C > 0$ *such that:*

$$(5.3)\quad \int_\lambda^\infty \frac{dt}{\Psi(Kt)} \leq \frac{C\lambda}{\Psi(\lambda)} \quad \forall\lambda \geq \lambda\!\!\!-.$$

(ii) *On any probability space* $L^W_\Psi = \tilde{L}^W_\Psi$.

We may then prove the following:

Lemma 14. *Suppose* Ψ *satisfies* (5.3) *then, defining*

$$(5.4)\quad \alpha \equiv \alpha(\Psi) = \Phi^{-1}(\Psi(\lambda\!\!\!-))\left[1 + \int_1^\infty \frac{dt}{\Psi(t)}\right]/\Psi(1)$$

and

$$(5.5)\quad \beta \equiv \beta = 2\max(K(1 \vee 2C), \alpha),$$

$$\|X\|_{\Psi}^{\tilde{W}} \le |||X|||_{\Psi}^{W} \le \beta \|X\|_{\Psi}^{\tilde{W}}. \tag{5.6}$$

We shall need the following lemma.

Lemma 15. (Lemma 1.1.1 of Barlow et al. (1986)). *For any $a > 0$,*

$$\Psi(a/\Phi^{-1}(a)) \le a \le \Psi(2a/\Phi^{-1}(a)),$$

or equivalently,

$$\frac{1}{2\Psi^{-1}(a)} \le \frac{\Phi^{-1}(a)}{a} \le \frac{1}{\Psi^{-1}(a)}.$$

Proof of Lemma 14. To prove the left-hand side of (5.6) set $\gamma = |||X|||_{\Psi}^{W}$ then $\forall F \in \mathcal{F}$:

$$E|X|1_F \le Y|||1_F|||_{\Phi}. \tag{5.7}$$

Observe that

$$\begin{aligned} |||1_F|||_{\Phi} &= \inf\{\mu > 0 : \sup_{\lambda \ge 0}(\mu\lambda - \Phi(\lambda)\mathbf{P}(F)) \ge \Psi(1)\} \\ &= \inf\{\mu > 0 : \mathbf{P}(F)\,\Psi(\mu/\mathbf{P}(F)) \ge \Psi(1)\} \\ &= \mathbf{P}(F)\,\Psi^{-1}(\Psi(1)/\mathbf{P}(F)). \end{aligned}$$

Thus, from (5.7), setting $F_x = 1_{(|X| \ge x)}$, $p(x) = \mathbf{E}\, F_x$;

$$\begin{aligned} xp(x) &\le \mathbf{E}|X|1_{F_x} \\ &\le \gamma p(x)\,\Psi^{-1}(\Psi(1)/p(x)), \qquad \text{(from (5.7))} \end{aligned}$$

and, setting $x = \gamma u$

$$\gamma u \le \gamma\, \Psi^{-1}(\Psi(1)/p(\gamma u))$$

so $\Psi(u)p(\gamma u) \le \Psi(1)\ \forall u \ge 0$.

To prove the right-hand side of (5.6); set

$$\gamma = |||X|||_{\Psi}^{\tilde{W}}$$

and observe that if $m \ge 1$

$$\int_{\lambda}^{\infty} \frac{dt}{\Psi(mKt)} \le \frac{1}{m} \int_{\lambda}^{\infty} \frac{dt}{\Psi(Kt)}.$$

Thus for all $\lambda > \bar{\lambda}$, $m > (1 \vee 2C)$

$$\int_\lambda^\infty \frac{dt}{\Psi(mKt)} < \frac{1}{2}\frac{\lambda}{\Psi(\lambda)}. \qquad \text{(from (5.3))}$$

Thus, from Lemma 15, for $\lambda > \Phi^{-1}(\Psi(\bar{\lambda}))$

$$\frac{2\Phi(\lambda)}{\lambda} \geq \Psi^{-1}(\Phi(\lambda)) > \bar{\lambda} \tag{5.8}$$

and so

$$\begin{aligned}
\mathbf{E}\left[\left(\frac{\lambda|X|}{\gamma mK} - 2\Phi(\lambda)\right)^+\right] &= \lambda\int_{2\Phi(\lambda)/\lambda}^\infty \mathbf{P}\left(\frac{|X|}{\gamma mK} \geq t\right)dt \\
&\leq \lambda\int_{2\Phi(\lambda)/\lambda}^\infty \frac{\Psi(1)dt}{\Psi(mKt)} \\
&< \frac{\Psi(1)\Phi(\lambda)}{\Psi(2\Phi(\lambda)/\lambda)} \qquad \text{(from (5.8))} \\
&\leq \Psi(1).
\end{aligned}$$

Moreover, for $\lambda \leq \Phi^{-1}(\Psi(\lambda))$,

$$\begin{aligned}
\mathbf{E}\left[\left(\frac{\lambda|X|}{\gamma mK} - 2\Phi(\lambda)\right)^+\right] &\leq \frac{\Phi^{-1}(\Psi(\lambda))}{mK}\mathbf{E}\frac{|X|}{\gamma} \\
&= \frac{\Phi^{-1}(\Psi(\lambda))}{mK}\int_0^\infty \mathbf{P}\left(\frac{|X|}{\gamma} \geq t\right)dt \\
&\leq \frac{\Phi^{-1}(\Psi(\lambda))}{mK}\int_0^\infty \left(1 \wedge \frac{1}{\Psi(t)}\right)dt \\
&\leq \Psi(1)\,\alpha/mK.
\end{aligned}$$

Thus for $\ell > \beta/2$

$$\sup_{\lambda\geq 0}\mathbf{E}\left[\left(\frac{\lambda|X|}{\gamma\ell} - 2\Phi(\lambda)\right)^+\right] < \Psi(1)$$

and so

$$\begin{aligned}
&\sup_{\lambda\geq 0}\mathbf{E}\left[\left(\frac{\lambda|X|}{\gamma\ell} - \Phi(2\lambda)\right)^+\right] \\
&= \sup_{\lambda\geq 0}\mathbf{E}\left[\left[\frac{\lambda|X|}{2\gamma\ell} - \Phi(\lambda)\right]^+\right] \\
&< \Psi(1). \qquad \square
\end{aligned}$$

Note: It is simple to prove that if Ψ is co-moderate (i.e. Φ is moderate) then (5.3) holds:

Lemma 16. *Suppose Φ is moderate with exponent r (note $r > 1$); that is $\forall K > 1$, $x \in \mathbb{R}_+$*

$$\Phi(Kx) \le K^r \Phi(x) \tag{5.9}$$

then

$$\int_\lambda^\infty \frac{dt}{\Psi(t)} \le \frac{(r-1)\lambda}{\Psi(\lambda)} \quad \forall \lambda > 0$$

so that 5.3 is satisfied for any $\lambda > 0$; $K = 1$ and $C = (r-1)$.

Proof. Observe that $\int_\lambda^\infty \frac{dt}{\Psi(t)} = \lambda \int_1^\infty \frac{du}{\Psi(\lambda u)}$. Now

$$\begin{aligned}\Psi(\lambda u) &= \sup_t (t\,\lambda\, u - \Phi(t)) \\ &= \sup_s (\lambda\, s\, y\, u - \Phi(sy)).\end{aligned}$$

So if we let $y = u^{1/r-1}$ then

$$\begin{aligned}\Psi(\lambda u) &\ge \sup_s (\lambda\, s\, u^{\tilde r} - u^{\tilde r}\,\Phi(s)) \qquad \text{(from (5.9))} \\ &= u^{\tilde r}\,\Psi(\lambda) \qquad \text{(where } \tilde r \text{ is the conjugate of } r\text{)}.\end{aligned}$$

Thus

$$\begin{aligned}\int_\lambda^\infty \frac{dt}{\Psi(t)} &\le \frac{\lambda}{\Psi(\lambda)} \int_1^\infty \frac{du}{u^{\tilde r}} \\ &= \frac{1}{\tilde r - 1} \frac{\lambda}{\Psi(\lambda)} = (r-1)\,\lambda/\Psi(\lambda). \quad \square\end{aligned}$$

We conclude with a weak version of Doob's inequality for submartingales:

Theorem 17. *Suppose Ψ satisfies condition (5.3) then for all uniformly integrable, cadlag, non-negative submartingales (M_t):*

$$\left\|\sup_t M_t\right\|_\Psi^W \le \beta_\Psi \|M_\infty\|_\Psi^W .$$

Proof. By standard embedding techniques we may reduce to the case where M is a discrete-time submartingale and hence Doob's inequality holds:

$$x\,\mathbf{P}(M^* \ge x) \le \mathbf{E}\, M_\infty I(M^* \ge x), \tag{5.10}$$

where $M^* = \sup_t M_t$. Now from the definition of $||M_\infty||_\Psi^W$, the RHS of (5.10) is dominated by

$$||M_\infty||_\Psi^W \, |||I(M^* \geq x)|||_\Phi = ||M_\infty||_\Psi^W \, p_x \, \Psi^{-1}(\Psi(1)/p_x)$$

where $p_x = \mathbf{P}(M^* \geq x)$. Thus $x\, p_x \leq ||M_\infty||_\Psi^W p_x \, \Psi^{-1}(\Psi(1)/p_x)$ so

$$p_x \, \Psi\left(\frac{x}{||M_\infty||_\Psi^W}\right) \leq \Psi(1).$$

We deduce that $||M^*||_\Psi^{\tilde{W}} \leq ||M_\infty||_\Psi^W$ and so the result follows from (5.6). □

REFERENCES

[1] M.T. Barlow, S.D. Jacka and M. Yor, *Inequalities for a pair of processes stopped at a random time,* Proc. London Math. Soc. **52** (1986), 142–172.

[2] C. Dellacherie and P.A. Meyer, *Probabilities and Potential A,* North Holland, Amsterdam–New York–Oxford, 1978.

[3] C. Dellacherie and P.A. Meyer, *Probabilities and Potential B,* North Holland, Amsterdam–New York–Oxford, 1982.

[4] M.A. Krasnoselski and Y.B. Ruticki, *Convex functions and Orlicz spaces,* Noordhoff, Groningen, 1961.

Department of Statistics
University of Warwick
Coventry
CV4 7AL
United Kingdom

Part VI
Large Deviations

Large Deviations by the Asymptotic Value Method

Wlodzimierz Bryc

0. Introduction. In this paper we present an approach to large deviations which is based on the converse to Varadhan's "asymptotic integral theorem". We call it the asymptotic value approach to the large deviation principle. The method puts less emphasis on rate functions and on the underlying probability theory; in particular, "changes of measure" are not used in our proofs.The asymptotic value approach follows a pattern analogous to the weak convergence of measures: limits over suitable continuous functions replace lower and upper bounds for probabilities. We need to find limits $L(F) = \lim_{V\to\infty} L_V(F)$ for some bounded continuous functions F, knowing the limits $L(f)=\lim_{V\to\infty} L_V(f)$ for much simpler (e. g. linear) functions f. An inspection of the proofs of the theorems shows that few properties of the asymptotic value mapping $f\to L(f)$ are really used: without any reference to probability theory, properties expressed by (3.1)—(3.4) below are responsible for a (non—standard) rate function representation of the asymptotic value; to get the standard rate function representation, we use $L(f\vee g)=L(f)\vee L(g)$ and compactness of the "state space" (or good enough approximations by compact sets, such as the one in the conclusion of lemma L.4.1 below). This might broaden the scope of the large deviation method in applications to those asymptotic problems, where there is no evident probabilistic representation behind the formulas analyzed, compare e. g. van den Berg, Lewis & Pulé [3].

Throughout this paper we assume that the asymptotic value $L(.)$ arises in a probabilistic context, and is given by (1.3) below. We shall show that effective and useful criteria for the large deviation principle follow naturally from the asymptotic value approach. These theorems, when accompanied by theorem 3.1 of de Acosta [8], give short proofs of some non—trivial large deviation principles; the lower—upper bounds pattern of proof is replaced here by the following two steps: verification of "exponential tightness" (see the definition below), and showing that the limit (1.3) exists for a large enough class of functions.

Some aspects of the traditional approach to large deviations were not retained in the paper. We didn't attempt to establish criteria for large deviations "uniform with respect to a starting point", or to separate conditions responsible for upper bounds from those responsible for lower bounds. Both of those aspects of large deviations are very well understood in the context of Markov chains, c. f. de Acosta [10].

Section 1 contains statements of the main results. Theorems T.1.1 and T.1.2 show that if $\mathcal{L}(F)$ exists for all bounded continuous functions F, then it has a rate function representation (1.4) or (1.9) . We also state two related criteria for the large deviation principle. Theorem T.1.3 is our basic asymptotic value criterion; theorem T.1.4 extends results known in the literature, see the commentary preceding corollary C.2.2. Section 2 gives applications of the general theorems in specific situations. Several corollaries of theorems T.1.3 and T.1.4 are stated. Examples (all well known) at the end of section 2 show how the method can be applied and also illustrate the convenience of having several related criteria. In section 3, the non—standard rate function representation T.1.1 is proved. Section 4 contains the proof of T.1.2. In section 5, useful auxiliary results are obtained. In section 6, the main large deviation criteria T.1.3 and T.1.4 are proved. The method of proof is new and is based on theorem T.1.2 .

Other authors used variants of the asymptotic value method less explicitly or less generally, see Baldi [2], Comets [6], Dawson & Gärtner [7], Ellis [15], Gärtner [16], de Acosta [8], de Acosta [9], Kifer [17], Ney & Nummelin [19], Plachky [21], Sievers [24].

1. Notation and the main results. Let $\mathbb{X}$ be a metric space with a metric d(x, y) and the finitely additive Borel field $\mathcal{B}_{f.a.}$. By $\mathcal{P}_{f.a.}(\mathbb{X})$ we denote a complete metric space of all regular finitely additive probability measures on $(\mathbb{X}, \mathcal{B}_{f.a.})$ with the weak topology, $\mathcal{P}(\mathbb{X})$ stands for all countably additive probability measures on $(\mathbb{X}, \mathcal{B})$, where the Borel σ-field $\mathcal{B}$ is generated by $\mathcal{B}_{f.a.}$. By $C_b(\mathbb{X})$ we denote a Banach space of all bounded continuous functions F: $\mathbb{X}\to\mathbf{R}$ with the supremum norm. Through the paper $\{P_\nu\}_{\nu\in\mathcal{J}}$ is a family of probability measures, i. e. $P_\nu\in\mathcal{P}(\mathbb{X})$, $\nu\in\mathcal{J}$; $\mathcal{J}$ is a fixed unbounded (and not necessarily countable) subset of real numbers $\nu\geq1$. To simplify the notation, we write $\{\nu\geq1\}$ instead of $\nu\in\mathcal{J}$.

Following Varadhan [26], we say that $\{P_\nu\}$ satisfies the large deviation principle with a rate function I: $\mathbb{X}\to[0, \infty]$, if the following two conditions are satisfied

(1.1) $$-\inf\{ I(x): x\in A\}\leq \liminf_{\nu\to\infty} 1/\nu \log P_\nu(A)$$

for each open set $A \subset \mathbb{X}$;

(1.2) $$\limsup_{\nu\to\infty} 1/\nu \log P_\nu(A) \leq - \inf\{I(x): x \in A\}$$

for each closed set $A \subset \mathbb{X}$.

We also require $I(.)$ to be lower semicontinuous and with compact level sets $I^{-1}([0, a])$ for each $a \geq 0$. (Then $I(.)$ is determined uniquelly, see e. g. Ellis [15] Theorem II. 3. 2 or Orey [20].)

The following definition is motivated by analogy with weak convergence of measures; the concept occurs explicitly in Deuschel & Stroock [11] and in Lynch & Sethuraman [18] (in the last paper under the name "large deviation tightness"); a number of authors used the same concept less explicitly, see Azencott [1], Baldi [2], de Acosta [8] Theorem 2. 1. (ii), Stroock [25] Theorem 3. 26.

Definition. We say that $\{P_\nu\}$ is $\mathcal{J}$-exponentially tight, if for each $M>0$ there exist a compact set $K \subset \mathbb{X}$ such that

$$\sup_{\nu \in \mathcal{J}} 1/\nu \log P_\nu(K^c) \leq - M.$$

Note that exponential tightness depends both on the topology of $\mathbb{X}$ and on the index set $\mathcal{J}$. Since through this paper $\mathcal{J}$ is fixed, we shall suppress the $\mathcal{J}$-dependence in our terminology. Exponential tightness of probability measures on $\mathbf{R}^d$ is usually verified using exponential moments and the Chebyshev inequality. For probability measures on infinite dimensional spaces, de Acosta [8] theorem 3. 1 is helpful.

Definition. We say that a family $\{P_\nu\}_{\nu\geq 1}$ admits an asymptotic value over a class $\mathcal{F}$ of measurable functions, if

(1.3) $$L(F) = \lim_{\nu\to\infty} 1/\nu \log\left[\int_{\mathbb{X}} \exp\{\nu F(x)\}\, dP_\nu(x)\right]$$

exists and is finite for each function $F \in \mathcal{F}$.

The following theorem shows that each asymptotic value over $C_b(\mathbb{X})$ has a "rate function representation" (1.4). The proof of (1.4) uses only properties of $L(.)$, listed in Lemma L.3.1 below.

Theorem T.1.1. If a family $\{P_\nu\}_{\nu\geq 1}$ admits an asymptotic value $L(.)$ over $C_b(\mathbb{X})$, then there exists a lower semicontinuous function $J: \mathcal{P}_{f.a.}(\mathbb{X}) \to [0, \infty]$ such that

$$(1.4)\qquad \mathcal{L}(F)=\sup\ \{\mu(F)-\mathcal{J}(\mu):\ \mu\in\mathcal{P}_{f.a.}(\mathbb{X})\},$$

and the supremum is attained. Furthermore the following variational expression holds

$$(1.5)\qquad \mathcal{J}(\mu)=\sup\ \{\mu(F)-\mathcal{L}(F):\ F\in C_b(\mathbb{X})\}.$$

Also

$$(1.6)\qquad -\inf\{\mathcal{J}(\mu):\ \mu\in\mathcal{P}_{f.a.},\ \mu(A)=1\}\leq \liminf_{\nu\to\infty}\ 1/\nu \log P_\nu(A)$$

for each open set $A\subset\mathbb{X}$;

$$(1.7)\qquad \limsup_{\nu\to\infty}\ 1/\nu \log P_\nu(A)\leq -\inf\{\mathcal{J}(\mu):\ \mu\in\mathcal{P}_{f.a.},\ \mu(A)=1\}$$

for each closed set $A\subset\mathbb{X}$.

[with the convention $\inf\emptyset=\infty$].

The following variant of T.1.1 gives a converse to Varadhan's [26] theorem 2. 2, and is the basis of our asymptotic value approach to large deviation principles. The proof is simpler than, and independent of, theorem T.1.1; the rate function $I(.)$ need not be convex.

Theorem T.1.2. If a family $\{P_\nu\}_{\nu\geq1}$ admits an asymptotic value $\mathcal{L}(.)$ over $C_b(\mathbb{X})$ and is exponentially tight, then the large deviation principle holds with the rate function $I(x)=\mathcal{J}(\delta_x)$. In particular the following dual variational formulas hold

$$(1.8)\qquad I(x)=\sup\{F(x)-\mathcal{L}(F):\ F\in C_b(\mathbb{X})\};$$

$$(1.9)\qquad \mathcal{L}(F)=\sup\ \{F(x)-I(x):\ x\in\mathbb{X}\}.$$

The next result is our main criterion for the large deviation principle. Its assumptions are rather technical, but additional flexibility is gained in the choice of the family $\mathcal{G}$. Corollaries in section 2 apply this criterion to a subset $\mathcal{G}$ of the concave functions, with (1.10), (1.11), and (1.13) either trivially satisfied or easily checked.

Recall that a family $\mathcal{G}$ of functions $\mathbb{X}\to\mathbf{R}$ separates points of $\mathbb{X}$, if

$\forall x, y\in\mathbb{X}\ \forall a, b\in\mathbf{R}\ \exists\ g\in\mathcal{G}$ such that $g(x)=a$ and $g(y)=b$.

Theorem T.1.3. Let $\{P_\nu\}$ be an exponentially tight family of probability measures. Suppose $\mathcal{G}$ is a subset of the space of all continuous (not necessarily bounded) functions $\mathbb{X}\to\mathbf{R}$, such that the following conditions are satisfied

(1.10) $\mathcal{G}$ separates points of $\mathbb{X}$;

(1.11) $\mathcal{G}$ contains the constant functions and is closed under finite pointwise minima, i. e. if $g_1, \ldots, g_n \in \mathcal{G}$, then $g_1 \wedge \ldots \wedge g_n \in \mathcal{G}$;

(1.12) $\{P_v\}$ admits an asymptotic value over $\mathcal{G}$.

Then $\{P_v\}$ satisfies the large deviation principle.

Moreover, if

(1.13) for each $g \in \mathcal{G}$ there is $0<q<1$ and a measurable function $\mathbf{B}$ such that $g \le q\mathbf{B}$ and

$$\sup_v [\int_{\mathbb{X}} \exp\{v\mathbf{B}(x)\}\, dP_v(x)]^{1/v} < \infty, \tag{1.14}$$

then the rate function $I(.)$ is given by

$$I(x)=\sup\{g(x)-L(g): g \in \mathcal{G}\}. \tag{1.15}$$

The following large deviation principle criterion is a non-trivial consequence of T.1.3.

Theorem T.1.4. Let $\{P_v\}$ be an exponentially tight family of probability measures. Suppose there is a subset $\mathcal{F}$ of the space of all continuous (not necessarily bounded) functions $\mathbb{X} \to \mathbf{R}$ such that the following conditions hold.

(1.16) $\mathcal{F}$ separates points of $\mathbb{X}$;

(1.17) $\mathcal{F}$ is linear (i. e. closed under finite linear combinations);

(1.18) $\{P_v\}$ admits an asymptotic value over $\mathcal{F}$;

(1.19) $\frac{d}{dt} L((1-t)f_1+tf_2)_{|t=0}$ exists for each $f_1, f_2 \in \mathcal{F}$.

Then $\{P_v\}$ satisfies the large deviation principle with a rate function $I(.)$ given by

$$I(x)=\sup\{f(x)-L(f): f \in \mathcal{F}\}. \tag{1.20}$$

Theorem T.1.1 is proved in section 3, theorem T.1.2 is proved in section 4, theorems T.1.3 and T.1.4 are proved in section 6.

2. Corollaries. In this section we list simple but useful consequences of theorems T.1.3 and T.1.4. The following corollary of T.1.3 specifies the family $\mathcal{G}$ and simplifies (1.15) for convex rate functions. The proof is given in section 6.

Corollary C.2.1. Let $\mathbb{V}$ be a locally convex Hausdorff topological linear space with the conjugate space $\mathbb{V}^*$. Suppose $\mathbb{X} \subset \mathbb{V}$ is a metric space in the relative topology, and $\Lambda \subset \mathbb{V}^*$ is a dense linear subspace. Define $\mathcal{G}=\{g: g(x)=\min_i \{\lambda_i(x)+c_i\}, c_i \in \mathbf{R}, \lambda_i \in \Lambda, 1 \le i \le n, n \in \mathbb{N}\}$.

Suppose $\{P_v\}$ is exponentially tight and admits an asymptotic value over $\mathcal{G}$.

Then $\{P_v\}$ satisfies the large deviation principle with a rate function $I(x)$ defined by (1.15).

If in addition $\sup_v 1/v \log[\int_{\mathbb{X}} \exp\{v\lambda(x)\}\, dP_v(x)]<\infty$ for each $\lambda\in\mathbb{V}^*$; $\mathbb{X}\subset\mathbb{V}$ is closed, convex and $I(.)$ defined by (1.15) is convex, then $L(\lambda)$ exists for each $\lambda\in\mathbb{V}^*$ and

(2.1) $$I(x)= \sup\{\lambda(x)-L(\lambda): \lambda\in\mathbb{V}^*\}.$$

Remark R.2.1. In a typical application, exponential tightness is verified with the help of de Acosta [8] theorem 3.1. The fact that $\{P_v\}$ admits an asymptotic value over $\mathcal{G}$ is then verified by showing that $v\to\log[\int_{\mathbb{X}} \exp\{v\lambda(x)\}\, dP_v(x)]$ is close to some super-additive function; the limit then exists by e.g. Dunford & Schwartz [13] VIII.1.4 (c. f. our example 2 for this type of argument).

The following corollary of T.1.4 generalizes to the infinite-dimensional setting Gärtner [16] lemmas 1. 1 and 1. 2; see also Dawson & Gärtner [7] Theorem 3. 4, de Acosta [8] Theorem 2. 1, Ellis [15] Theorem II. 6. 1, Plachky [21], Sievers [24] for related results. The result is especially easy to apply to sequences of probability measures obtained from an i. i. d sequence; see our example 1 below, Baldi [2], Dawson & Gärtner [7] (for an application of C.2.4 below). The corollary is proved in section 6.

Corollary C.2.2. Let $\mathbb{V}$ be a locally convex Hausdorff topological linear space with the conjugate space $\mathbb{V}^*$. Suppose $\mathbb{X}\subset\mathbb{V}$ is a metric space in the relative topology and is closed and convex. Let $\{P_v\}$ be an exponentially tight family of probability measures which admits an asymptotic value $L(\lambda)$ for each bounded continuous functional $\lambda\in\mathbb{V}^*$. Furthermore suppose that $L(\lambda)$ is Gateaux differentiable in each direction $\gamma\in\mathbb{V}^*$ and at each point $\lambda\in\mathbb{V}^*$. Then $\{P_v\}$ satisfies the large deviation principle with a rate function $I(.)$ defined by (2.1).

The assumptions of the next corollary eliminate the need to consider concave functions in (1.3) and (1.15) and avoid any explicit differentiability assumption. Similar result in the more specific context of empirical measures and with $\mathbb{V}^*$ replaced by a dense linear subspace of functionals has been recently presented by Kifer [17] (c.f. our C.2.6, which was added to the preliminary draft of this manuscript after seeing Kifer's paper).

Corollary C.2.3. Let $\mathbb{V}$ be a locally convex Hausdorff topological linear space with the conjugate space $\mathbb{V}^*$. Suppose closed and convex $\mathbb{X}\subset\mathbb{V}$ is a metric space in the relative topology. Let $\{P_V\}$ be an exponentially tight family of probability measures which admits an asymptotic value $L(\lambda)$ over $\lambda\in\mathbb{V}^*$. Also suppose that (2.1) defines a strictly convex function $I(.)$, and that $\sup_{\lambda\in\mathbb{O}} L(\lambda)<\infty$ for some open set $\mathbb{O}\subset\mathbb{V}^*$. Then $\{P_V\}$ satisfies the large deviation principle with the rate function $I(.)$.

Proof. The assumptions guarantee that $L(\lambda)$ is continuous, see e. g. Ekeland & Temam [14], p. 12, Proposition 2.5. To use Corollary C.2.2, it is enough to show that $L(\lambda)$ is Gateaux-differentiable at each point $\lambda\in\mathbb{V}^*$. For a finite dimensional vector space $\mathbb{V}$ the result is stated explicitly in Ruelle [22] p. 252. In general, the proof can be sketched as follows. By strict convexity of $I(.)$, the subgradient $\partial L(\lambda)$ is unique. Indeed, $\mathbf{x}\in\partial L(\lambda)$ if and only if $L(\lambda)+I(\mathbf{x})=\lambda(\mathbf{x})$, see Ekeland & Temam [14], p. 21, Proposition 5. 1. Therefore $\mathbf{x}\in\partial L(\lambda)$ is a unique minimum of the strictly convex function $I(\mathbf{x})—\lambda(\mathbf{x})$. Proposition 5. 2 of Ekeland & Temam [14], p. 23 ends the proof of Gateaux-differentiability of $L(.)$ and the result now follows from Corollary C.2.2.

The following corollary of C.2.2 is essentially a variant of theorem 3. 4 of Dawson & Gärtner [7]. We write it down here to show that it follows from C.2.2, even though the explicit assumption of exponential tightness is absent. Similar variants with no explicit smoothness assumptions can be deduced from C.2.3.

Corollary C.2.4. Let $\mathcal{V}$ be a vector space with a countable Hamel basis. Define $\mathbb{V}$ to be the algebraic dual $\mathcal{V}^a$ with $\sigma(\mathbb{V}, \mathcal{V})$-topology. Let $\mathbb{X}$ be a closed convex subset of $\mathbb{V}$, with the relative topology. Suppose $\{P_V\}$ is a family of probability measures on $(\mathbb{X}, \mathcal{B})$ which admits an asymptotic value $L(\lambda)$ over all bounded continuous functionals $\lambda\in\mathcal{V}$. Furthermore suppose that $L(\lambda)$ is Gateaux-differentiable at each point $\lambda\in\mathcal{V}$. Then $\{P_V\}$ satisfies the large deviation principle with a rate function $I(\mathbf{x})$ defined by (2.1).

Proof. To apply C.2.2, we need only to verify that $\mathbb{X}$ is a metric space (which is actually an inessential assumption made throughout this paper for the sake of simplification) and that $\{P_V\}$ is exponentially tight. The first claim follows trivially from the fact that under our assumptions the topology of $\mathbb{V}$ is metrizable, see Dunford & Schwartz [13] V. 7. 34.

Exponential tightness follows from Chebyshev's inequality and Dunford & Schwartz [13] V. 4. 1 (use the assumption $|\mathcal{L}(\lambda)|<\infty$ for each fixed $\lambda\in\mathcal{V}$).

Remark R.2.2. As in Dawson & Gärtner [7] theorem 3. 4(iii), the assumption that $\mathbb{X}$ is a closed convex subset of $\mathbb{V}$, needed in C.2.1 and consequently in corollaries C.2.2-C.2.4 to get (2.1), can be replaced by what is actually used in the proof, i. e. by $\{x: I(x)<\infty\}\subset\mathbb{X}$.

The next three corollaries are direct applications of C.2.1 and C.2.2 to empirical distributions of a $\mathbb{Z}^d$-indexed random field. In this application $\mathbb{X}=\mathcal{P}(\mathbb{E})$ is a closed and convex subset of the locally convex Hausdorff topological vector space $\mathbb{V}$ of all signed measures on $\mathbb{E}$, with the topology of weak convergence. The space $\mathbb{E}$ will be either a given metric "state space" $\mathbb{F}$, or its product $\mathbb{F}^{\mathbb{Z}^d}$. The natural normalization corresponds to $\mathcal{J}=\{1, 2^d, 3^d, ...\}$; however we shall index empirical; measures by $n\in\mathbb{Z}$ and substitute n^d for ν in all other places.

Let $\mathbb{F}$ be a Polish space. Suppose $\{X_z\}_{z\in\mathbb{Z}^d}$ is an $\mathbb{F}$-valued random field. Define $\mathcal{P}(\mathbb{F})$-valued empirical distributions

$$\nu_n=n^{-d}\Sigma_{k\in\mathcal{C}_n}\delta_{\{X_k\}},$$

where $\mathcal{C}_n=\{z: 1\le z_j\le n\}$.

Define also $\mathcal{P}(\mathbb{F}^{\mathbb{Z}^d})$-valued empirical fields

$$\mu_n=n^{-d}\Sigma_{k\in\mathcal{C}_n}\delta_{\{X_{z+k}\}_{z\in\mathbb{Z}^d}}.$$

We shall say that empirical distributions/fields are exponentially tight, if the induced probability measures on $\mathcal{P}(\mathbb{E})$ are exponentially tight (after re-indexation $\nu=n^d$).

Corollary C.2.5. Let C_o be a dense linear subset of $C_b(\mathbb{F})$. Suppose for each $k\ge1$ and every $F_1, F_2, ..., F_k\in C_o$ there exists

$$\lim_{n\to\infty} n^{-d}\log(E\{\exp(\min_{1\le i\le k}\Sigma_{z\in\mathcal{C}_n}F_i(X_z))\})=\mathcal{L}(F_1, F_2, ..., F_k).$$

If the empirical measures $\{\nu_n\}$ are exponentially tight, then $\{\nu_n\}$ satisfies the large deviation principle with a rate function $I_o(.)$: $\mathcal{P}(\mathbb{F})\to[0,\infty]$ given by

$$I_o(p)=\sup\{p(F_1)\wedge p(F_2)\wedge...\wedge p(F_k)-\mathcal{L}(F_1, ..., F_k): F_1, F_2, ..., F_k\in C_o, k\in\mathbb{N}\}.$$

If in addition $I_o(.)$ is convex, then it can be identified by using k=1 only, i.e.

$$I(p)=\sup\{p(F)-\mathcal{L}(F): F\in C_o\}. \quad (2.2)$$

Proof. This is essentially C.2.1 in the application-adjusted notation. (The integrability condition for (2.1) is here superfluous, since $\mathbf{X}=\mathcal{P}(\mathbb{F})$ is bounded; by continuity, the supremum in (2.2) can be taken over $F\in C_o$ rather than over $F\in C_b(\mathbb{F})$, as would follow from (2.1).)

Remark R.2.3. Convexity of $I_o(.)$ can be verified whenever for each $\mathbf{p}, \mathbf{q}\in\mathcal{P}(\mathbb{F})$ and $F_1, F_2, ..., F_k\in C_o$ one can find $G_1, G_2, ..., G_k, H_1, H_2, ..., H_k\in C_o$ such that simultaneously we have $2\mathcal{L}(F_1, ..., F_k)\geq\mathcal{L}(G_1, ..., G_k)+\mathcal{L}(H_1, ..., H_k)$ and $(p+q)(F_1)\wedge(p+q)(F_2)\wedge...\wedge(p+q)(F_k)=2p(G_1)\wedge p(G_2)\wedge...\wedge p(G_k)=2q(H_1)\wedge q(H_2)\wedge...\wedge p(H_k)$. In practical instances such G_i and H_i are obtained by adding suitable constants to functions F_i.

Corollary C.2.6. Let C_o be a dense linear subset of $C_b(F)$. Suppose for each $k\geq1$ and every $F\in C_o$ there exists

$$\lim_{n\to\infty} n^{-d}\log(E\{\exp(\Sigma_{z\in\mathcal{C}_n} F(X_z))\})=\mathcal{L}(F).$$

Suppose furthermore that $\frac{d}{dt}\mathcal{L}(F+tG)$ exists for each $F, G\in C_o$.

If the empirical measures $\{\nu_n\}$ are exponentially tight on $\mathcal{P}(\mathbb{F})$, then $\{\nu_n\}$ satisfies the large deviation principle with a rate function $I(.)$: $\mathcal{P}(\mathbb{F})\to[0,\infty]$ given by (2.2).

Proof. If $C_o=C_b(\mathbb{F})$, this is C.2.2 in the application adjusted notation. In the general case the large deviation principle is proved exactly as in C.2.2; then (2.2) is established by an argument used in the proof of (2.1), using the additional fact that our $\mathbf{X}$ is bounded in $\mathbb{V}$-norm. The detailed proof is omitted.

Corollary C.2.7 Let C_o be any dense subset of $C_b(\mathbb{F}^{\mathbb{Z}^d})$. Put $Y_z=(X_{z+r})_{r\in\mathbb{Z}^d}\in\mathbb{F}^{\mathbb{Z}^d}$. Suppose for each $k\geq1$ and every $F_1, F_2, ..., F_k\in C_o$ there exists

$$\lim_{n\to\infty} n^{-d}\log(E\{\exp(\min_{1\leq i\leq k}\Sigma_{z\in\mathcal{C}_n} F_i(Y_z))\})=\mathcal{L}(F_1, F_2, ..., F_k).$$

If the empirical fields $\{\mu_n\}$ are exponentially tight, then $\{\mu_n\}$ satisfies the large deviation principle with a rate function $I_o(.)$: $\mathcal{P}(\mathbb{F}^{\mathbb{Z}^d})\to[0,\infty]$ given by

$$I_o(p)=\sup\{p(F_1)\wedge p(F_2)\wedge...\wedge p(F_k)-\mathcal{L}(F_1, ..., F_k): F_1, F_2, ..., F_k\in C_o, k\in\mathbb{N}\}.$$

If in addition $I_o(.)$ is convex, then it can be identified by (2.2).

Proof. This is essentially C.2.1 in the application adjusted notation.

Remark R.2.4. In a typical application, C_o consists of all those bounded continuous functions F: $\mathbb{F}^{\mathbb{Z}^d} \to \mathbf{R}$, which depend on finite number of coordinates only.

The following examples illustrate both the use of the asymptotic value method and the convenience of having several different criteria. We consider sums of random vectors only; similar examples with empirical measures, or "empirical processes" could have been presented equally easily. The first example is the large deviation principle for i. i. d. random vectors, see Donsker & Varadhan [12] theorem 5.3; the second example gives a result that seems to follow via the contraction principle from known large deviation results for the empirical measure of a Markov chain; the third example is synthetic and shows how non-convex rate functions can be handled by the asymptotic value method. The upper bounds in a related to our example 3, but more general setting, have been obtained in de Acosta [8] theorem 5. 1.

Example 1. Suppose $(\mathbb{V}, \|\,.\,\|)$ is a separable Banach space with the conjugate space $\mathbb{V}^*$. Let $\{X_i\}$ be a sequence of $\mathbb{V}$-valued i. i. d. random variables such that $E\{\exp(\alpha\|X_1\|)\}<\infty$ for each $\alpha\in\mathbf{R}$. Then $\{(X_1+\ldots+X_n)/n\}_{n=1,2,\ldots}$ satisfies the large deviation principle with a rate function $I(v)=\sup_{\lambda\in\mathbb{V}^*}\{\lambda(v)-\log E\{\exp(\lambda(X))\}$.

This follows from C.2.2: By the independence assumption $L(\lambda)=\log E\{\exp(\lambda(X))\}$; hence $L(\lambda)$ exists for each $\lambda\in\mathbb{V}^*$ and is Gateaux-differentiable. Exponential tightness follows from theorem 3. 1 of de Acosta [8].

Example 2. Suppose $(\mathbb{V}, \|.\,\|)$ is a separable Banach space with the conjugate space $\mathbb{V}^*$. Let $\{X_n\}$ be a $\mathbb{V}$-valued stationary Markov chain such that $E\{\exp(\alpha\|X\|)\}<\infty$ for each $\alpha\in\mathbf{R}$. Let $\pi(dx)$ be a distribution of X_1. Suppose that there is $C<\infty$ such that 1-step transition probabilities $\Pi(x, dy)$ satisfy

(2.3) $\Pi(x, A)\le C\Pi(y, A)$ for $\pi\otimes\pi$-almost all $x, y\in\mathbb{V}$ and all Borel sets A.

Then $\{(X_1+\ldots+X_n)/n\}_{n=1,2,\ldots}$ satisfies the large deviation principle (with a convex rate function; below we omit the convexity argument).

This follows from C.2.1 applied to $\Lambda=\mathbb{V}^*$. Inequality (2.3) implies $E\{\exp(\sum_{i=1}^{n} q(X_i))\}\leq[CE\{\exp(q(X_1))\}]^n$ for any semi-norm q(.), hence theorem 3. 1 of de Acosta [8] guarantees exponential tightness. We shall verify that $\{(X_1+\ldots+X_n)/n\}$ admits an asymptotic value over the family $\mathcal{G}$, defined in C.2.1. Fix $g(.)\in\mathcal{G}$ and put $M_n=\pi\text{-ess inf } E_x\{\exp(g(X_1+\ldots+X_n))\}$. By the integrability assumption each M_n is finite. Also $M_{n+m}\geq M_n M_n$, since g(.) is concave. This shows that $n^{-1}\log M_n$ has a finite limit, see e. g. Dunford & Schwartz [13] VIII. 1. 4. It remains only to notice that by (2.3) $M_n\leq E\{\exp(g(X_1+\ldots+X_n))\}\leq CM_n$ which shows that $L(g)$ exists.

Example 3. Suppose $\{X_k\}$ is an infinite [0, 1]-valued exchangeable sequence. If a tail σ-field $\bigcap_{n\geq1}\sigma(X_n, X_{n+1}, \ldots)$ is finitely generated, then $\{(X_1+\ldots+X_n)/n\}_{n\geq1}$ satisfies the large deviation principle (with, in general, a non-convex rate function).

This follows from C.2.1 applied to $\mathbb{V}=\Lambda=\mathbf{R}$ and the compact set $\mathbb{X}=[0, 1]$. By the de Finetti theorem, the distribution of $\{X_k\}$ is a mixture of product measures $\otimes_k\pi_\theta(dx_k)$ with a discrete mixing measure $\alpha(d\theta)$. Take a concave continuous function $g\in\mathcal{G}$. Then

$$n^{-1}\log E\{\exp(g(X_1+\ldots+X_n))\}=$$
$$n^{-1}\log\int\exp(g(x_1+\ldots+x_n)\pi_\theta(dx_1)\ldots.\ \pi_\theta(dx_n)\alpha(d\theta)\to\max_\theta L_\theta(g),$$

where the maximum is taken over a finite number of values $\theta\in\text{supp }\alpha$ only, and $L_\theta(.)$ is an asymptotic value corresponding to the product measure $\otimes_k\pi_\theta(dx_k)$, $\theta\in\text{supp }\alpha$.

3. Proof of T.1.1. The following properties of $L(.)$ follow immediately from (1.3) and the proof is omitted[1].

Lemma L.3.1. If a family $\{P_v\}_{v\geq1}$ admits an asymptotic value $L(.)$ over $C_b(\mathbb{X})$, then

(3. 1) $$L((F+G)/2)\leq L(F)/2+L(G)/2$$

for each F, G ∈ $C_b(\mathbb{X})$;

(3. 2) $$\inf_{x\in\mathbb{X}}[F(x)-G(x)]\leq L(F)-L(G)\leq\sup_{x\in\mathbb{X}}[F(x)-G(x)]$$

for each F, G ∈ $C_b(\mathbb{X})$;

(3.3) $L(F+\text{cons})=L(F)+\text{const}$;

(3.4) $L(0)=0$.

---

[1] (3.1) follows from Hölder's inequality.

Denote

$$L_\nu(F) = 1/\nu \log[\int_{\mathbb{X}} \exp\{\nu F(x)\}\ dP_\nu(x)],\ \nu=1, 2, \ldots \tag{3.5}$$

Lemma L.3.2. If $L(g_1), \ldots, L(g_n)$ exist for some measurable functions $g_i(x)$, $1\le i\le n$, then $L(\max\{g_1, \ldots, g_n\})$ exists and

$$L(\max\{g_1, \ldots, g_n\})= \max\{ L(g_1), \ldots, L(g_n)\}.$$

Proof. Since $L_\nu(\max\{g_1, \ldots, g_n\})\ge \max\{ L_\nu(g_1), \ldots, L_\nu(g_n)\}$, we have

$$\liminf L_\nu(\max\{g_1, \ldots, g_n\})\ge \max\{ L(g_1), \ldots, L(g_n)\}.$$

It remains to show that

$$\limsup L_\nu(\max\{g_1, \ldots, g_n\})\le\max\{ L(g_1), \ldots, L(g_n)\}. \tag{3.6}$$

Without losing generality we may assume that

$$\max\{ L(g_1), \ldots, L(g_n)\}=L(g_1).$$

Fix $\varepsilon>0$ and let ν_0 be such that for each $\nu>\nu_0$ and every $1\le i\le n$

$$1/\nu\log\int\exp\{\nu g_i(x)\}dP_\nu(x)\le L(g_i)+\varepsilon.$$

Then $\int\exp\{\nu g_i(x)\}dP_\nu(x)\le e^{\nu L(g_i)+\nu\varepsilon}\le e^{\nu L(g_1)+\nu\varepsilon}$.

Therefore $L_\nu(\max\{g_1, \ldots, g_n\})\le 1/\nu \log\sum_{i=1}^{n} \int\exp\{\nu g_i(x)\}dP_\nu(x)\le$
$1/\nu \log[ne^{\nu L(g_1)+\nu\varepsilon}]\to L(g_1) +\varepsilon$ as $\nu\to\infty$, which proves (3.6).

Note (to be used in the proof of T.1.3). The proof actually shows that

$$\limsup_{\nu\to\infty} L_\nu(\max\{g_1, \ldots, g_n\})=$$
$$\max\{ \limsup_{\nu\to\infty} L_\nu(g_1), \ldots, \limsup_{\nu\to\infty} L_\nu(g_n)\}.$$

Proof of (1.4). Let $\mathcal{J}(.)$ be defined by (1.5) and fix $F_0\in C_b(\mathbb{X})$. By the definition of $\mathcal{J}(.)$, we need to show that

$$L(F_0)= \sup_\mu \inf_F \{\mu(F_0) -\mu(F)+L(F)\}, \tag{3.7}$$

where the supremum is taken over all $\mu\in \mathcal{P}_{f.a.}(\mathbb{X})$ and the infimum is taken over all $F\in C_b(\mathbb{X})$.

Moreover, since by (1.5) $\mathcal{J}(\mu)\ge\mu(F_0)-L(F_0)$ for each $\mu\in \mathcal{P}_{f.a.}(\mathbb{X})$, therefore

$$L(F_0)\ge\sup_\mu \inf_F \{\mu(F_0) -\mu(F)+L(F)\}.$$

Hence to prove (3.7), it remains to show that there is $\sigma\in\mathcal{P}_{f.a.}(\mathbb{X})$ such that

$$L(F_0)\le\sigma(F_0) -\sigma(F)+L(F) \text{ for each } F\in C_b(\mathbb{X}). \tag{3.8}$$

Also, for this σ, the supremum in (1.4) will be attained.

To find σ, define the following sets of functions. Let

$\mathcal{M}=\{F\in C_b(\mathbb{X}):\ \inf_x[F(x)-F_0(x)]>0\}$ and let $\mathcal{N}$ be a set of all finite convex combinations of functions g(x) of the form $g(x)=F(x)+L(F_0)-L(F)$, where $F\in C_b(\mathbb{X})$. It is easily seen that $\mathcal{M}$ and $\mathcal{N}$ are convex; also $\mathcal{M}$ is open and non-empty (e. g. $1+F_0\in\mathcal{M}$). Furthermore $\mathcal{M}$ and $\mathcal{N}$ are disjoint. Indeed, take an arbitrary $\mathcal{N}\ni g=\sum_{k=1}^{n}\alpha_k F_k+L(F_0)-\sum_{k=1}^{n}\alpha_k L(F_k)$.

Then $\inf_x\{g(x)-F_0(x)\}=\inf_x\{\sum_{k=1}^{n}\alpha_k F_k(x)-F_0(x)\}-\sum_{k=1}^{n}\alpha_k L(F_k)+L(F_0)\le$

$$\inf_x\{\sum_{k=1}^{n}\alpha_k F_k(x)-F_0(x)\}-L(\sum_{k=1}^{n}\alpha_k F_k)+L(F_0)\le 0,$$

where the first inequality follows from (3. 1) and the second follows from (3. 2) applied to $F=\sum_{k=1}^{n}\alpha_k F_k(x)$ and $G=F_0$. Therefore $g\notin\mathcal{N}$.

Convex and open $\mathcal{M}$ can be separated from disjoint convex $\mathcal{N}$ by a linear functional, i. e. there is $0\neq f^*\in C_b^*(\mathbb{X})$ such that for some $\alpha\in\mathbf{R}$

(3.9) $$f^*(\mathcal{N})\le\alpha<f^*(\mathcal{M}),$$

see e. g. Ekeland & Temam [14] Ch. 1, section 1. 2, or Dunford & Schwartz [13] V. 2. 8.

Claim: f^* is non-negative.

Indeed, it is easily seen that $F_0(.)$ belongs to $\mathcal{N}$, and, as a limit of $\varepsilon+F_0(x)$ as $\varepsilon\searrow 0$, F_0 belongs also to the closure of $\mathcal{M}$. Therefore by (3.9) we have $\alpha=f^*(F_0)$. To end the proof, take a function F with $\inf_x F(x)>0$. Then $F_1=F+F_0\in\mathcal{M}$ and by (3.9)

$$f^*(F)=f^*(F_1-F_0)=f^*(F_1)-f^*(F_0)>\alpha-f^*(F_0)=0.$$

This ends the proof of the claim.

Without loosing generality, we may assume $f^*(1)=1$; then it is well known, see e. g. Bergström [5] Ch. 2 Section 4 theorem 1, that $f^*(F)=\sigma(F)$ for some $\sigma\in\mathcal{P}_{f.a.}(\mathbb{X})$ (Dunford & Schwartz [13] IV. 6. 2 implies that σ is regular). It remains to check that σ satisfies (3.8). To this end observe that since $F+L(F_0)-L(F)\in\mathcal{N}$, by (3.9) we have

$\sigma(F)+L(F_0)-L(F)\le\alpha=\sigma(F_0)$ for every $F\in C_b(\mathbb{X})$. This ends the proof of (1.4).

Proof of the lower and upper bounds.

For $A\subset\mathbb{X}$ and $\varepsilon>0$ denote $[A]^\varepsilon=\{x: d(x,A)\le\varepsilon\}:=\{x:\inf_{y\in A}d(x,y)\le\varepsilon\}$. Clearly $[A]^\varepsilon$ is closed and $[A]^0$ is the closure of A.

Proposition P.3.1. If a family $\{P_\nu\}$ admits an asymptotic value $\mathcal{L}(.)$ over $C_b(\mathbb{X})$ and $\mathcal{J}(.)$ is defined by (1.5), then for any measurable set $A\subset\mathbb{X}$ we have

(3.10)) $\quad -\inf\{\mathcal{J}(\mu): \mu\in \mathcal{P}_{f.a.}, \mu(\mathrm{int}(A))=1\}\le\liminf_{\nu\to\infty} 1/\nu \log P_\nu(A);$

(3.11) $\quad \limsup_{\nu\to\infty} 1/\nu\log P_\nu(A)\le -\lim_{\varepsilon\searrow 0}\inf\{\mathcal{J}(\mu): \mu\in \mathcal{P}_{f.a.}, \mu([A]^\varepsilon)\ge 1-\varepsilon\}.$

[with the convention $\inf \emptyset=\infty$].

Proof of (3.10). Fix an open set $A\subset\mathbb{X}$ and μ_o, such that $\mu_o(A)=1$. Since μ_o is regular, therefore for each $M\ge 1$ we may choose a closed set $C_M\subset A$ such that $\mu_o(C_M)\ge 1-1/M^2$. Let $F_M: \mathbb{X}\to[-M, 0]$ be a continuous function such that $F_M(x)=-M$ for $x\notin A$, $F_M(x)= 0$ for $x\in C_M$. Then

$$1/\nu \log[\int_{\mathbb{X}} \exp\{\nu F_M(x)\}\, dP_\nu(x)]\le 1/\nu \log[e^{-\nu M}P_\nu(A^c)+P_\nu(A)].$$

Therefore

$$\mathcal{L}_\nu(F_M)\le 1/\nu \log 2+\max\{-M; 1/\nu \log P_\nu(A)\}.$$

Considering separately two cases: $\liminf_{\nu\to\infty} 1/\nu \log P_\nu(A)=-\infty$ and $\liminf_{\nu\to\infty} 1/\nu \log P_\nu(A)>-\infty$, we obtain

$$\liminf_{\nu\to\infty} 1/\nu \log P_\nu(A)\ge\liminf_{M\to\infty} \mathcal{L}(F_M).$$

Indeed, if $\liminf_{\nu\to\infty} 1/\nu \log P_\nu(A)=-\infty$, then $\mathcal{L}_\nu(F_M)\le 1/\nu \log 2 - M$ for all large enough ν, so that $\liminf_{M\to\infty} \mathcal{L}(F_M)=-\infty$. And if $q=\liminf_{\nu\to\infty} 1/\nu \log P_\nu(A)>-\infty$, then $\mathcal{L}_\nu(F_M)\le 1/\nu \log(2)+1/\nu \log P_\nu(A)$ for all large enough ν and for each $M>-q$.

Therefore by (1.4)

$$\liminf_{\nu\to\infty} 1/\nu \log P_\nu(A)\ge\liminf_{M\to\infty}\sup_\mu \{\mu(F_M)-\mathcal{J}(\mu): \mu\in \mathcal{P}_{f.a.} (\mathbb{X})\}\ge$$
$$\liminf_{M\to\infty}\{\mu_o(F_M)-\mathcal{J}(\mu_o)\}.$$

It remains to notice that since $\mu_o(F_M)\ge -M\mu_o(C_M{}^c)\ge -1/M\to 0$, therefore $\liminf_{M\to\infty}\{\mu_o(F_M)-\mathcal{J}(\mu_o)\}\ge -\mathcal{J}(\mu_o)$. Since μ_o is an arbitrary element of $\mathcal{P}_{f.a.}(\mathbb{X})$ such that $\mu_o(A)=1$, this ends the proof of (1.6).

Proof of (3.11). Fix a closed set $A\subset \mathbb{X}$ and $\varepsilon>0$. Let $F_M: \mathbb{X}\to[-M, 0]$ be a continuous function such that $F_M(x)= -M$ for $x\notin [A]^\varepsilon$ and $F_M(x)= 0$ for $x\in A$. Then

$$1/\nu \log[\int_{\mathbb{X}} \exp\{\nu F_M(x)\}\, dP_\nu(x)]\ge 1/\nu \log P_\nu(A).$$

Hence $\limsup_{v\to\infty} 1/v \log P_v(A) \le$

$\inf_M \limsup_{v\to\infty} 1/v \log[\int_{\mathbb{X}} \exp\{vF_M(x)\}\, dP_v(x)]=\inf_M \mathcal{L}(F_M)$.

By (1.4) we obtain $\limsup_{v\to\infty} 1/v \log P_v(A) \le$

$\inf_M\{\sup_\mu \{\mu(F_M)-\mathcal{J}(\mu)\text{: } M>0,\ \mu\in \mathcal{P}_{f.a.}(\mathbb{X}),\ \mu([A]^\varepsilon)\ge 1-\varepsilon \}\vee$

$\sup_\mu \{\mu(F_M)-\mathcal{J}(\mu)\text{: } M>0,\ \mu\in \mathcal{P}_{f.a.}(\mathbb{X}),\ \mu([A]^\varepsilon)<1-\varepsilon \}\} \le$

$\sup_\mu \{ -\mathcal{J}(\mu)\text{: } \mu\in \mathcal{P}_{f.a.}(\mathbb{X}),\ \mu([A]^\varepsilon)\ge 1-\varepsilon \} \vee \inf_M \sup_\mu \{\mu(F_M)\text{: } M>0,\ \mu\in \mathcal{P}_{f.a.}(\mathbb{X}),$
$\mu([A]^\varepsilon)<1-\varepsilon \}$.

Notice that if $\mu([A]^\varepsilon)<1-\varepsilon$, then $\mu(F_M)\le -M\varepsilon\to-\infty$. Hence

$\limsup_{v\to\infty} 1/v \log P_v(A) \le \sup_\mu \{ -\mathcal{J}(\mu)\text{: } \mu\in \mathcal{P}_{f.a.}(\mathbb{X}),\ \mu([A]^\varepsilon)\ge 1-\varepsilon \}=$

$-\inf_\mu \{ -\mathcal{J}(\mu)\text{: } \mu\in \mathcal{P}_{f.a.}(\mathbb{X}),\ \mu([A]^\varepsilon)\ge 1-\varepsilon\}$,

which ends the proof of (3.11).

Proof of (1.7). Since $\mathcal{P}_{f.a.}(\mathbb{X})$ is compact and $\mathcal{J}(.)$ is lower semicontinuous, by the standard subsequence argument we can pass in (3.11) to the limit as $\varepsilon\to 0$.

4 Proof of T.1.2.

Lemma L.4.1. If $\{P_v\}$ is an exponentially tight family of probability measures, then for each $M>0$ there is a compact set $K\subset\mathbb{X}$ with the following property.

If **h** is a measurable function such that $h(x)\le M$ and $\limsup_{v\to\infty} \mathcal{L}_v(h)\ge 0$, then

$$\lim_{v\to\infty}[\mathcal{L}_v(h)- 1/v \log\int_K \exp\{vh(x)\}\, dP_v(x)] = 0.$$

(Recall that $\mathcal{L}_v$ is defined by (3.5).)

Proof. Let K be a compact set such that $\sup_v 1/v \log P_v(K^c)\le -2M$.

Since $\mathcal{L}_v(h)=$

$1/v \log[\int_K \exp\{vh(x)\}\, dP_v(x)+ \int_{K^c} \exp\{vh(x)\}\, dP_v(x)]\le$

$1/v \log 2 +1/v\log([\int_K \exp\{vh(x)\}\, dP_v(x)]\vee \exp(vM)P_v(K^c))$,

therefore

(4.1) $\mathcal{L}_v(h)\le$

$1/v \log 2+ 1/v\log[\int_K \exp\{vh(x)\}\, dP_v(x)]\vee[M+\sup_v 1/v\log P_v(K^c)]\le$

$1/v \log 2+ (-M)\vee 1/v\log[\int_K \exp\{vh(x)\}\, dP_v(x)]$.

Since $M>0$ and $\limsup_{v\to\infty} \mathcal{L}_v(h)\ge 0$, from (4.1) we obtain

$$\limsup_{v\to\infty} \mathcal{L}_v(h) \le \limsup_{v\to\infty} 1/v \log[\int_K \exp\{vh(x)\}\, dP_v(x).$$

This, together with the trivial inequality $1/v \log[\int_K \exp\{vh(x)\} dP_v(x)] \le \mathcal{L}_v(h)$, ends the proof.

Proposition P.4.1. If $\{P_v\}$ admits an asymptotic value $\mathcal{L}(.)$ over $C_b(\mathbb{X})$ and $I(.)$ satisfies (1.9), then (1.1) holds and

(4.2) $$\limsup_{v\to\infty} 1/v \log P_v(A) \le -\lim_{\varepsilon\searrow 0} \inf\{I(x): x\in [A]^{\varepsilon}\}.$$

[with the usual convention $\inf \emptyset = \infty$].

Proof. (4.2) is proved in the same way, as (3.11), except that (1.9) should be used in place of (1.4).

Proof of T.1.2.

As a trivial consequence of (1.8) we get

(4.3) $$\mathcal{L}(F) \ge \sup\ \{F(x)-I(x): x\in\mathbb{X}\}.$$

To prove (1.9) pick a bounded continuous function F. By lemma L.4.1 there is a compact set K such that both $\mathcal{L}(F)$ and $\mathcal{L}(F\chi_K)$ exist and are equal; here χ_K denotes the indicator function of K. By (4.3) we need only to show that

(4.4) $$\limsup_{v\to\infty} \mathcal{L}_v(F\chi_K) \le \sup\ \{F(x)-I(x): x\in K\}$$

for each compact set $K\subset\mathbb{X}$.

To prove (4.4) fix $F_0\in C_b(\mathbb{X})$ and $\varepsilon>0$. Let $s=\sup\ \{F_0(x)-I(x): x\in K\}$. By (1.8) and (3.3), for each $x\in K$, there is $F_x\in C_b(\mathbb{X})$ such that $F_0(x)-F_x(x)<s+\varepsilon$ and $\mathcal{L}(F_x)=0$.

This means that open sets $U_x=\{y\in\mathbb{X}: F_0(y)-F_x(y)<s+\varepsilon\}$ cover K, and we may choose a finite covering $U_{x(1)}, \ldots, U_{x(k)}$. Then

$$F(x) < \max_{1\le i\le k} F_{x(i)}(x)+s+\varepsilon,\ x\in K.$$

By (3. 2) and (3.3)

$$\limsup_{v\to\infty} \mathcal{L}_v(F\chi_K) \le s+\varepsilon+\limsup_{v\to\infty} \mathcal{L}_v(\max_{1\le i\le k} F_{x_i}(x)\chi_K) \le$$

$$s+\varepsilon+\mathcal{L}(\max_{1\le i\le k} F_{x_i}(x))=s+\varepsilon,$$

because Lemma L.3.2 gives $\mathcal{L}(\max_{1\le i\le k} F_{x_i}(x))=0$. This concludes the proof of (1.9).

Since (1.9) has been proved, by P.4.1 we need only to show that $I(.)$ has compact level sets and

(4.5) $$\lim_{\varepsilon\searrow 0}\inf\{I(x): x\in [A]^{\varepsilon}\}=\inf\{I(x): x\in A\},$$

where $I(.)$ is defined by (1.8) and A is a closed set. This is proved by the standard subsequence argument, since, by P.4.2, sets $A\cap I^{-1}[0, \lambda]$ are compact for each $\lambda\geq 0$.

Proposition P.4.2. If $\{P_v\}$ admits an asymptotic value $L(.)$ over $C_b(X)$ and $I(.)$ is defined by (1.8), then the following two conditions are equivalent:

(i) $\{P_v\}$ is exponentially tight;

(ii) $I^{-1}[0, \lambda]$ is a compact set for each $\lambda>0$.

Proof. Since (1.9) was proved, (1.1) follows by P.4.1 and the proof of Stroock [25] Theorem 3. 26 gives the implication (i)$\Rightarrow$(ii). For the implication (ii)$\Rightarrow$(i) see Lynch & Sethuraman [18] lemma 2. 6.

The following proposition complements T.1.1 and P.3.1.

Proposition P.4.3. If a family $\{P_v\}$ of probability measures admits an asymptotic value $L(.)$ over $C_b(X)$ and $I(x)=J(\delta_x)$ is defined by (1.8), then (1.1) holds and

(4.6) $$\limsup_{v\to\infty} 1/v\log P_v(K)\leq -\inf\{I(x): x\in K\}$$

for each compact set $K\subset X$.

Proof. Inequality (1.1) is a trivial consequence of (1.6). To prove (4.6) put F=0 in (4.4) to obtain

$$\limsup_{v\to\infty} 1/v\log P_v(K)=\limsup_{v\to\infty} L_v(F\chi_K)\leq$$
$$\sup\{F(x)-I(x): x\in K\}= -\inf\{I(x): x\in K\}.$$

5. Extension lemmas. The results of this section will be used to identify the rate function.

Lemma L.5.1. Suppose $\{P_v\}$ is tight. If f is a continuous function, then there exist a constant $M>0$ such that

(5.1) $$\lim_{v\to\infty}[L_v(f)-L_v(f\vee(-M))]=0.$$

(Recall that L_v is defined by (3.5).)

Proof. From the tightness assumption and continuity of f it follows that there are $0<\rho<1/2$ and $M>0$ such that $P_v(x: f(x)\geq -M)>\rho$. This implies that

$$\frac{P_v(x: f(x)\geq -M)}{P_v(x: f(x)<-M)}\geq \rho/(1-\rho).$$

Therefore for each $v\geq 1$

(5.2) $$1/\nu \log P_\nu(x: f(x)\geq -M) \geq 1/\nu \log P_\nu(x: f(x)<-M)+1/\nu \log(\rho/(1-\rho)).$$

Define $g(x)=f(x)\vee(-M)$. Since $f\leq g$, therefore $\mathcal{L}_\nu(f)\leq \mathcal{L}_\nu(g)$, which shows that $\limsup_{\nu\to\infty}[\mathcal{L}_\nu(f)-\mathcal{L}_\nu(g)]\leq 0$. To analyze the lower limit, observe that

(5.3) $$\mathcal{L}_\nu(g)=1/\nu \log[e^{-\nu M}P_\nu(x: f(x)<-M)+\int_{f\geq -M}\exp\{\nu f(x)\}\, dP_\nu(x)]\leq$$
$$1/\nu \log 2+\max\{-M+1/\nu \log P_\nu(f(x)<-M);\ 1/\nu \log\int_{f\geq -M}\exp\{\nu f(x)\}\, dP_\nu(x)\}.$$

However, by (5.2) we have

$$1/\nu \log\int_{f\geq -M}\exp\{\nu f(x)\}dP_\nu(x)\geq -M+1/\nu \log P_\nu(f(x)\geq -M)\geq$$
$$-M+1/\nu \log P_\nu(f(x)<-M)+1/\nu \log(\rho/(1-\rho)),$$

and by our choice of ρ we have $\log(\rho/(1-\rho))\leq 0$. Therefore (5.3) implies

$$\mathcal{L}_\nu(g)\leq 1/\nu \log 2-1/\nu \log(\rho/(1-\rho))+1/\nu \log\int_{f\geq -M}\exp\{\nu f(x)\}dP_\nu(x)\leq$$
$$1/\nu \log 2-1/\nu \log(\rho/(1-\rho))+\mathcal{L}_\nu(f).$$

This shows that $\liminf_{\nu\to\infty}[\mathcal{L}_\nu(f)-\mathcal{L}_\nu(g)]\geq 0$ and the lemma is proved.

Lemma L.5.2. Suppose a family $\{P_\nu\}$ of probability measures admits an asymptotic value $\mathcal{L}(.)$ over $C_b(X)$ and is tight. Let $B: X\to R$ be a measurable function such that (1.14) holds. If $f: X\to R$ is a continuous function such that for some $0<q<1$

(5.4) $$f(x)\leq qB(x),$$

then $\mathcal{L}(f)$ exists. Moreover $\mathcal{L}(f)$ is finite and there are constants M_o, N_o such that $\mathcal{L}(f)=\mathcal{L}((M\vee f)\wedge N)$ for each $M\leq M_o$ and each $N\geq N_o$.

Proof. From Lemma L.5.1 it follows that for each continuous function f there is a constant $M>0$ such that if $g(x)=f(x)\vee(-M)$, then both $\mathcal{L}_\nu(f)$ and $\mathcal{L}_\nu(g)$ have the same lim sup and lim inf. Therefore it is enough to show that

$$\liminf_{\nu\to\infty}\mathcal{L}_\nu(g)=\limsup_{\nu\to\infty}\mathcal{L}_\nu(g).$$

Define $H(x)=g(x)\wedge N$, where $N>0$ will be chosen later. Since $g\geq H$ and H is bounded and continuous, therefore $\mathcal{L}(H)$ exists and $\liminf_{\nu\to\infty}\mathcal{L}_\nu(g)\geq\mathcal{L}(H)$.

We shall show that $\limsup_{\nu\to\infty}\mathcal{L}_\nu(g)\leq\mathcal{L}(H)$ for all large enough N. To this end observe that

(5.5) $$\mathcal{L}_\nu(g)\leq 1/\nu \log 2+$$
$$\max\{1/\nu\log\int_{g\geq N}\exp\{\nu g(x)\}dP_\nu(x);\ 1/\nu\log\int_{g<N}\exp\{\nu g(x)\}dP_\nu(x)\}.$$

By the tightness assumption, there are $0<\rho<1/2$ and $N_0>0$ such that

$P_\nu(x: g(x)<N_0)>\rho$. Since $g\geq -M$, this implies that for $N>N_0$

$1/\nu \log\int_{g<N} \exp\{\nu g(x)\}dP_\nu(x)\geq 1/\nu \log[e^{-\nu M}P_\nu(g<N)]\geq -M+1/\nu\log \rho\geq -M+\log \rho$.

By Hölder's inequality we have $\int_{g\geq N} \exp\{\nu g(x)\}dP_\nu(x)\leq$

$(\int \exp\{\nu g(x)/q\}dP_\nu(x))^q(P_\nu(g\geq N))^{1-q}$. Hence using Chebyshev's inequality, (5.4) and (1.14) we obtain

$$\int_{g\geq N} \exp\{\nu g(x)\}\ dP_\nu(x)\leq$$

$$[\int_X \exp\{\nu B(x)\}\ dP_\nu(x)]^q\ [e^{-\nu N/q}\int_X \exp\{\nu g(x)/q\}\ dP_\nu(x)]^{1-q}\leq$$

$$[\int_X \exp\{\nu B(x)\}\ dP_\nu(x)]\ e^{-\nu N(1-q)/q}.$$

Therefore

$$1/\nu \log\int_{g\geq N} \exp\{\nu g(x)\}dP_\nu(x)\leq C-N(1-q)/q,$$

where $C=\log \sup_{\nu\geq 1}[\int_X \exp\{\nu B(x)\}\ dP_\nu(x)]^{1/\nu}$.

In particular $1/\nu \log\int_{g\geq N} \exp\{\nu g(x)\}dP_\nu(x)\leq -M+\log \rho$ for all N large enough. This shows that (5.5) amounts to

$$L_\nu(g)\leq 1/\nu \log 2+ 1/\nu \log\int_{g\leq N} \exp\{\nu g(x)\}\ dP_\nu(x)\leq 1/\nu \log 2+ L_\nu(H),$$

provided that $N>N_0$ and $N>(C+M-\log \rho)q(1-q)^{-1}$. Hence $\limsup_{\nu\to\infty} L_\nu(g)\leq L(H)$.

Proposition P.5.1. Suppose $\mathbb{X}$ is a metric subset of a linear space $\mathbb{V}$ with the dual $\mathbb{V}^*$. If $\{P_\nu\}$ admits an asymptotic value $L(.)$ over $C_b(\mathbb{X})$, $\{P_\nu\}$ is tight and $\sup_\nu[\int_X \exp(\nu\lambda(x))dP_\nu(x)]^{1/\nu}<\infty$ for each $\lambda\in\mathbb{V}^*$, then $L(\lambda)$ exists and is finite, $\lambda\in\mathbb{V}^*$.

Proof. Since $\exp\{|\lambda|\}\leq \exp(\lambda)+\exp(-\lambda)$, therefore $\sup_\nu[\int_X \exp\{\nu 2|\lambda(x)|\}\ dP_\nu(x)]^{1/\nu}<\infty$ for each functional $\lambda\in\mathbb{V}^*$, and the result follows from L.5.2, applied to $f(.)=\lambda(.)$, $q=1/2$ and $B(x)=2|\lambda(x)|$.

The following result extends (1.9) to non-bounded continuous functions, compare de Acosta [9] Lemma 6. 1.

Proposition P.5.2. If for some continuous function f the assumptions of L.5.2 are satisfied and (1.9) holds for all $F\in C_b(\mathbb{X})$, then $L(f)=\sup_x\{f(x)-I(x)\}$.

Indeed, L.5.2 shows that one can find constants M, N such that $\mathcal{L}(f)=\mathcal{L}((M\vee f)\wedge N)$, and the equality is not affected by decreasing M, or increasing N. Take $M<\mathcal{L}(f)$. By (1.9) applied to $\mathcal{L}((M\vee f)\wedge N)$ we get

$\mathcal{L}(f)=\sup_{x:f(x)>M}\{f(x)\wedge N-I(x)\}\vee\sup_{x:f(x)\le M}\{M-I(x)\}$. Since $M<\mathcal{L}(f)$ and $I(.)\geq 0$, this means that $\mathcal{L}(f)=\sup_x\{N\wedge f(x)-I(x)\}$, for each large enough N. Therefore $\mathcal{L}(f)=\sup_N\sup_x\{N\wedge f(x)-I(x)\}=\sup_x\{f(x)-I(x)\}$.

6. Proofs of the Large Deviation Principle Criteria. We begin with the two part proof of Theorem T.1.3: in Part A we show that the large deviation principle holds; in Part B we identify the rate function.

Proof of Theorem T.1.3. Part A (large deviation principle). Fix a bounded continuous function F. By T.1.2, it is enough to prove that $\mathcal{L}(F)$ exists. By adding a constant, see (3.3), without loosing generality we may assume $0\le F\le M$ for some $M\ge 0$. Let $\varepsilon>0$ be fixed. We shall show that

(6.1) $$\limsup_{v\to\infty}\mathcal{L}_v(F)\le\liminf_{v\to\infty}\mathcal{L}_v(F)+2\varepsilon.$$

Since $F\ge 0$, we have $\liminf_{v\to\infty}\mathcal{L}_v(F)\ge 0$. Therefore from Lemma L.4.1 follows that we need only to show that

(6.2) $$\limsup_{v\to\infty}1/v\log\Big[\int_K\exp\{vF(x)\}dP_v(x)\Big]\le$$
$$\liminf_{v\to\infty}1/v\log\int_K\exp\{vF(x)\}dP_v(x)+2\varepsilon,$$

where K is a compact set from the conclusion of Lemma L.4.1.

By the Stone-Weierstrass theorem, see e. g. Schaefer [23] p. 243, there is a finite collection $\{g_i\}_{1\le i\le n}$ of functions in $\mathcal{G}$, such that

(6.3) $$\sup_{x\in K}|F(x)-\max_i g_i(x)|\le\varepsilon.$$

Moreover, since $0\le F(x)\le M$, passing to $g_j\wedge M$ if necessary, we may assume $g_j\le M$ for all j and we can also take $g_0=0$.

From (6.3) we obtain

$$\limsup_{v\to\infty}1/v\log\Big[\int_K\exp\{vF(x)\}dP_v(x)\Big]\le$$
$$\varepsilon+\limsup_{v\to\infty}1/v\log\Big[\int_K\exp\{v[\max_i g_i(x)]\}dP_v(x)\Big].$$

Since by Lemma L.3.2 (see the note following its proof)

$$\limsup_{v\to\infty} 1/v\log[\int_X \exp(v\max_i g_i(x))dP_v(x)] = \max_i L(g_i(x)),$$

therefore

(6.4) $$\limsup_{v\to\infty} 1/v\log[\int_K \exp\{vF(x)\}dP_v(x)] \le \varepsilon + \max_i L(g_i(x)).$$

Using (6.3) once more we get

$$\liminf_{v\to\infty} 1/v\log[\int_K \exp\{vF(x)\}dP_v(x)] \ge$$

$$\liminf_{v\to\infty} 1/v\log[\int_K \exp\{v\max_i g_i(x)\}dP_v(x)] - \varepsilon.$$

Notice that since $\max_i L(g_i(x)) \ge L(g_0) = 0$, by Lemma L.4.1

$$\liminf_{v\to\infty} 1/v\log[\int_K \exp\{v\max_i g_i(x)\}\, dP_v(x)] = \max_i L(g_i(x))$$

and hence

(6.5) $$\liminf_{v\to\infty} 1/v\log[\int_K \exp\{vF(x)\}dP_v(x)] \ge \max_i L(g_i(x)) - \varepsilon.$$

Together (6.5) and (6.4) imply (6.2) and the large deviation principle is proved.

Proof of Theorem T.1.3. Part B (rate function identification).

Let $\vee\mathcal{G} = \{\max\{g_i(.)\}: g_i \in \mathcal{G},\ 1 \le i \le n,\ n \ge 1\}$.

T.1.2 says that the rate function $I_0(.)$ is defined by (1.8). Therefore

(6.6) $$I_0(x) \ge \sup\{h(x) - L(h): h \in \vee\mathcal{G} \cap C_b(X)\}.$$

Claim 1. $I_0(x) \le \sup\{h(x) - L(h): h \in \vee\mathcal{G} \cap C_b(X)\}$.

Indeed, let F be a continuous function, $0 \le F \le M$ and let $\varepsilon > 0$, $x_0 \in X$ be fixed. Let $K = K_M$ be a compact set from the conclusion of Lemma L.4.1, enlarged to ensure $x_0 \in K$. From part I of the proof we see that there is $h \in \vee\mathcal{G} \cap C_b(X)$, $h \ge 0$ such that

$|F(x) - h(x)| < \varepsilon$ at each $x \in K$;

$|L(F) - L(h)| < \varepsilon$.

Therefore

$$F(x) - L(F) \le h(x) - L(h) + 2\varepsilon.$$

Since $\varepsilon > 0$ is arbitrary, this concludes the proof of Claim 1.

Claim 2. $\sup\{h(x) - L(h): h \in \vee\mathcal{G} \cap C_b(X)\} = \sup\{h(x) - L(h): h \in \vee\mathcal{G}\}$

Indeed, inequality "$\le$" is trivial. To prove "$\ge$" fix x_0 and $h = g_1 \vee \ldots \vee g_n \in \vee\mathcal{G}$. From L.5.2. applied to $f := h$, $B(x) := \max B_i(x)$ and $q = \max_i q_i$ [here B_i and q_i come from (1.13), applied to each g_i, $1 \le i \le n$], one can find numbers M, N such that

(6.7) $$L(h) = L((h \vee M) \wedge N).$$

(Lemma L.5.2 can be applied because the large deviation principle holds)

We can also decrease M and increase N so that (6.7) holds and $h(x_0)=(h(x_0)\vee M)\wedge N$. Then

$$h(x_0)-\mathcal{L}(h)=(h(x_0)\vee M)\wedge N-\mathcal{L}((h\vee M)\wedge N)\leq\sup\{h(x_0)-\mathcal{L}(h):\ h\in\vee\mathcal{G}\cap C_b(X)\}.$$

Since **h** and x_0 were arbitrary, this ends the proof of Claim 2.

Claim 3. $\sup\{h(x)-\mathcal{L}(h):\ h\in\vee\mathcal{G}\}=\sup\{g(x)-\mathcal{L}(g):\ g\in\mathcal{G}\}(=I(x))$.

Indeed, inequality "≥" is trivial. To prove "≤" observe that by (1.15) $\mathcal{L}(g)\geq\sup_x\{g(x)-I(x)\}$. Lemma L.3.2 implies that $\mathcal{L}(\max_i g_i)\geq\max_i\sup_x\{g_i(x)-I(x)\}$. Therefore $\sup\{h(x)-\mathcal{L}(h):\ h\in\vee\mathcal{G}\}=\sup_{g_i}\{\max_i g_i(x)-\mathcal{L}(\max_i g_i)\}\leq$

$\sup_{g_i}\inf_y\{\max_i g_i(x)-\max_i g_i(y)+I(y)\}\leq I(x)$, where the last inequality was obtained by taking $y=x$. This proves Claim 3.

Claims 1, 2 and 3 together with (6.6) end the proof of (1.15).

Proof of C.2.1. By the Hahn-Banach theorem, $\mathbb{V}^*$ separates points of $\mathbf{X}$, see e. g. Ekeland & Temam [14] Ch. 1. page 5 Corollary 1. 2. Since Λ is dense and linear, it separates points of $\mathbf{X}$, too, and all the assumptions of Theorem T.1.3 hold, see the proof of Proposition P.5.1 for T.1.3 (1.13). Therefore the large deviation principle holds with a rate function $I_0(.)$ given by (1.15). It remains to show that if $\mathbf{X}$ is a convex closed subset of V and $I_0(.)$ is convex, then $I_0(.)$ is defined by (2.1).

Since for each $\lambda\in\mathbb{V}^*$ the assumptions of L.5.2 are satisfied (take e.g. $q=1/2$, $B=2|\lambda|$), by Proposition P.5.2 the large deviation principle implies $\mathcal{L}(\lambda)=\sup\{\lambda(x)-I_0(x):\ x\in\mathbf{X}\}$. However, since $\mathbf{X}$ is closed and convex, (1.15) gives $I_0(v)=\infty$ at each $v\notin\mathbf{X}$. Therefore $\mathcal{L}(\lambda)=\sup_{v\in\mathbb{V}}\{\lambda(v)-I_0(v)\}$ for every $\lambda\in\mathbb{V}^*$ and the well known result on the bi-conjugate of a convex lower semicontinuous function, see e. g. Ekeland & Temam [14] Ch. 1, gives (2.1).

Proof of T.1.4.

The following lemma is the main step in the proof of T.1.4.

Lemma L.6.1. Fix $n\geq1$. Let measurable functions $f_i(x)$, $1\leq i\leq n$, be such that

(i) $\mathcal{L}(f)$ exists and is a finite number for each linear combination $f=\sum_{i=1}^{n}\alpha_i f_i$;

(ii) $\frac{d}{dt}\mathcal{L}((1-t)f+tg)_{|t=0}$ exists for each pair of convex combinations $f=\sum_{i=1}^{n}\alpha_i f_i$, $g=\sum_{i=1}^{n}\beta_i f_i$.

Then $L(\min\{f_1, ..., f_n\})$ exists and

(6.8) $L(\min\{f_1, ..., f_n\}) = \inf\{L(\alpha_1 f_1 + ... + \alpha_n f_n) : \alpha_i \geq 0, \alpha_1 + ... + \alpha_n = 1\}$.

Proof. Since $(\alpha_1, ..., \alpha_n) \rightarrow L(\alpha_1 f_1 + ... + \alpha_n f_n)$ is continuous (as a convex and bounded function), therefore the infimum on the right hand side of (6.8) is attained. Suppose the infimum is attained at the point $f_0 = \sum_{i=1}^{n} \alpha_i f_i$, where $\alpha_i \geq 0$, i = 1, 2, ..., n, and $\sum_{i=1}^{n} \alpha_i = 1$.

Then $f_1 \wedge ... \wedge f_n = f_0 \wedge f_1 \wedge ... \wedge f_n$ and we need to show that

(6.9) $$L(f_0) = \inf\{L(\alpha_0 f_0 + \alpha_1 f_1 + ... + \alpha_n f_n) : \alpha_i \geq 0, \forall\, i, \alpha_0 + ... + \alpha_n = 1\},$$

implies $L(f_0 \wedge f_1 \wedge ... \wedge f_n) = L(f_0)$.

Notice that since $f_0 \wedge f_1 \wedge ... \wedge f_n \leq f_0$, therefore $\limsup_{V\rightarrow\infty} L_V(f_0 \wedge f_1 \wedge ... \wedge f_n) \leq L(f_0)$, and to prove $L(f_0 \wedge f_1 \wedge ... \wedge f_n) = L(f_0)$ we need only to show that

(6.10) $$\liminf_{V\rightarrow\infty} L_V(f_0 \wedge f_1 \wedge ... \wedge f_n) \geq L(f_0).$$

To end the proof fix $\varepsilon > 0$ and let $\theta = \varepsilon/(1+\varepsilon) \in (0, 1)$. Obviously we have

(6.11) $$f_0 = \theta(f_0 \wedge f_1 \wedge ... \wedge f_n) + (1-\theta)\max_{0\leq k\leq n}\{(1+\varepsilon)f_0 - \varepsilon f_k\}.$$

Using the fact that each $L_V(.)$ is non-decreasing and convex, compare L.3.1, from (6.11) we obtain

(6.12) $$L_V(f_0) \leq \theta L_V(f_0 \wedge f_1 \wedge ... \wedge f_n) + (1-\theta) L_V(\max_{0\leq k\leq n}\{(1+\varepsilon)f_0 - \varepsilon f_k\}).$$

This by Lemma L.3.2 implies

(6.13) $$L(f_0) \leq \theta \liminf_{V\rightarrow\infty} L_V(f_0 \wedge f_1 \wedge ... \wedge f_n) + (1-\theta)\max_{0\leq k\leq n} L((1+\varepsilon)f_0 - \varepsilon f_k).$$

We shall deduce (6.10) from (6.13) by considering separately the following two cases.

Case 1. There is $\varepsilon > 0$ such that $\max_{1\leq k\leq n} L((1+\varepsilon)f_0 - \varepsilon f_k) \leq L(f_0)$.

In this case (6.13) implies $L(f_0) \leq \theta \liminf_V L_V(f_0 \wedge f_1 \wedge ... \wedge f_n) + (1-\theta)L(f_0)$, and (6.10) follows, since $\theta > 0$.

Case 2. $\max_{1\leq k\leq n} L((1+\varepsilon)f_0 - \varepsilon f_k) > L(f_0)$ for each $\varepsilon > 0$.

In this case there is an index k ($1 \leq k \leq n$) such that

$$L((1+\varepsilon_r)f_0 - \varepsilon_r f_k) = \max_{1\leq i\leq n} L((1+\varepsilon_r)f_0 - \varepsilon_r f_i) = \max_{0\leq i\leq n} L((1+\varepsilon_r)f_0 - \varepsilon_r f_i)$$

for the infinite number of values $\varepsilon_r > 0$, r=1, 2, ... and without losing generality we may assume $\varepsilon_r \rightarrow 0$. Then (6.13) implies

$$L(f_0) \leq \theta \liminf_V L_V(f_0 \wedge f_1 \wedge ... \wedge f_n) + (1-\theta) L((1+\varepsilon_r)f_0 - \varepsilon_r f_k).$$

Since $\theta = \varepsilon/(1+\varepsilon)$, this in turn implies that

$$\liminf_{V\to\infty} \mathcal{L}_V(f_0\wedge f_1\wedge\ldots\wedge f_n)-\mathcal{L}(f_0)\geq[\mathcal{L}(f_0)-\mathcal{L}((1+\varepsilon_r)f_0-\varepsilon_r f_k)]/\varepsilon_r.$$

Passing to the limit as $r\to\infty$, and after taking into account the differentiability assumption, we obtain

$$\liminf_{V\to\infty} \mathcal{L}_V(f_0\wedge f_1\wedge\ldots\wedge f_n)-\mathcal{L}(f_0)\geq$$

$$\lim_{r\to\infty}[\mathcal{L}(f_0)-\mathcal{L}((1+\varepsilon_r)f_0-\varepsilon_r f_k)]/\varepsilon_r=\partial/\partial\varepsilon\,\mathcal{L}((1-\varepsilon)f_0+\varepsilon f_k)|_{\varepsilon=0}\geq 0,$$

where the last inequality follows from the assumption that $\mathcal{L}(f_0)\leq\mathcal{L}((1-\varepsilon)f_0+\varepsilon f_k)$ for each $\varepsilon>0$, see (6.9).

Proof of Theorem T.1.4. Without loosing generality we may assume that $\mathfrak{F}$ contains the constant functions, see (3.3). Consider

$$\mathcal{G}=\wedge\mathfrak{F}=\{g: g(x)=f_1(x)\wedge\ldots\wedge f_n(x),\ n\geq1,\ f_i\in\mathfrak{F}\}.$$

It is easy to check that $\wedge f$ satisfies the assumptions of Theorem T.1.3. Indeed, T.1.3. (1.13) holds with $q=2$, $\mathbf{B}=2(f\vee0)$ [condition (1.14) for $\mathbf{B}$ is checked similarly as in the proof of Proposition P.5.1]; T.1.3. (1.10) is assumed for $\mathfrak{F}$ so it holds for $\wedge\mathfrak{F}$, too; T.1.3. (1.12) holds by Lemma L.6.1 ; T.1.3. (1.11) holds by the definition of $\wedge\mathfrak{F}$.

Therefore by Theorem T.1.3. the large deviation principle holds with a rate function

(6.14). $$I_0(x)=\sup\{g(x)-\mathcal{L}(g): g\in\wedge\mathfrak{F}\}.$$

Let $I(.)$ be defined by (1.20). Since the supremum in (1.20) is taken over the smaller set, therefore $I(.)\leq I_0(.)$. To prove the converse inequality, take $g=\min_{1\leq j\leq n} f_j\in\wedge\mathfrak{F}$. By (6.8), there are numbers $\alpha_j\geq0$, $j=1, 2, \ldots, n$, $\sum_{j=1}^n\alpha_j=1$, such that $g(x)-\mathcal{L}(g)=g(x)-\mathcal{L}(\sum_{j=1}^n\alpha_j f_j)$.

Since $\min_j f_j\leq\sum_{j=1}^n\alpha_j f_j$, therefore this implies $g(x)-\mathcal{L}(g)\leq\sum_{j=1}^n\alpha_j f_j(x)-\mathcal{L}(\sum_{j=1}^n\alpha_j f_j)\leq I(x)$.

Since $g\in\wedge\mathfrak{F}$ was arbitrary, this ends the proof.

Proof of C.2.2. We shall show that C.2.1. can be applied. By Lemma L.6.1 $\mathcal{L}(g)$ is defined for each $g=f_1\wedge\ldots\wedge f_n$, where $f_i(x)=\lambda_i(x)+c_i$, $c_i\in\mathbf{R}$, $\lambda_i\in\mathbb{V}^*$, $1\leq i\leq n$, $n\in\mathbb{N}$. To check that (1.15) defines a convex function, we use Lemma L.6.1 again. Indeed, by (6.8) $\mathcal{L}(g)=\mathcal{L}(\sum_{k=1}^n\alpha_k f_k)$ for some $\alpha_k\geq0$, $\sum_{k=1}^n\alpha_k=1$. Since trivially $g(x)\leq\sum_{k=1}^n\alpha_k f_k(x)$, therefore, since each $f(.)$ is linear, we get

$$g(\mu(x))-\mathcal{L}(g)\leq[\sum_{k=1}^n\alpha_k f_k](\mu(x))-\mathcal{L}(\sum_{k=1}^n\alpha_k f_k)=$$

$$\mu([\sum_{k=1}^{n} \alpha_k f_k](x)) - L(\sum_{k=1}^{n} \alpha_k f_k) \le \sup_f\{\mu(f) - L(f)\} \le \mu(I(x))$$

for each (say discrete) measure μ. Since g was arbitrary, this shows $I(\mu(x)) \le \mu(I(x))$, i. e. $I(.)$ is convex.

Acknowledgement. The subject of this research was suggested by conversations with A. Korzeniowski and W. Smolenski. The author also would like to thank A. de Acosta, Y. Kifer, S. Pelikan and W. Smolenski for references and/or preprints.

REFERENCES

[1] R. Azencott, *Grandes deviations et applications*, Lecture Notes in Math. 774 (1980), Springer, Berlin.

[2] P. Baldi, Large Deviations and Stochastic Homogenization, Annali di Matematica Pura et Applicata 151 (1988), pp. 161-177.

[3] M. van den Berg, J. T. Lewis & J. V. Pulé, The Large Deviation Principle and Some Models of an Interacting Boson Gas, Comm. Math. Physics 118 (1988), pp. 61-85.

[5] H. Bergström, *Weak convergence of measures*, Academic Press, New York, 1982.

[6] F. Comets, Large Deviation Estimates for a Conditional Probability Distribution. Application to Random Interaction Gibbs Measures, Probab. Th. Rel. Fields 80 (1989), pp. 407-432.

[7] D. A. Dawson, J. Gärtner, Large Deviations from the McKean-Vlasov Limit for Weakly Interacting Diffusions, Stochastics 20 (1987), pp. 247-308.

[8] A. de Acosta, Upper bounds for Large Deviations of Dependent Random Vectors, Zeitsch. Wahrsch. verw. Gebiete 69 (1985), pp. 551-565.

[9] A. de Acosta, Large Deviations for vector-valued functionals of a Markov chain: lower bounds, Ann. Probab. 16 (1988), pp. 925-960.

[10] A. de Acosta, Large deviations for empirical measures of Markov chains, Case Western Reserve Univ. preprint #88-101 (1988).

[11] J. Deuschel & D. W. Stroock, *Large Deviations*, Academic Press, Boston, 1989.

[12] M. D. Donsker, S. R. S. Varadhan, Asymptotic Evaluation of Certain Markov Process Expectations for Large Time -III, Comm. Pure Appl. Math. 29 (1976), pp. 389-461.

[13] N. Dunford, J. T. Schwartz, *Linear Operators I*, Interscience, New York, 1958.

[14] I. Ekeland, R. Temam, *Convex analysis and variational problems*, North Holland, Amsterdam, 1976.

[15] R. Ellis, Entropy, *Large Deviations and Statistical Mechanics*, Springer-Verlag, New York, 1985.

[16] J. Gärtner, On large deviations from the invariant measure, Theory Probab. Appl. 22 (1977), pp. 24-39.

[17] Y. Kifer, Large Deviations in Dynamical Systems and Stochastic Processes, preprint (1989).

[18] J. Lynch, J. Sethuraman, Large deviations for processes with independent increments, Ann. Probab. 15 (1987), pp. 610-627.

[19] P. Ney, Nummelin, Markov additive process I. Eigenvalue properties and limit theorems, Ann. Probab. 15 (1987), pp. 561-592, Markov additive process II pp. 593-609.

[20] S. Orey, Large Deviations in ergodic theory, In: Seminar in Stochastic Processes 1984, E. Cinclar, K. L. Chung, R. K. Getoor ed., Birhhäuser, Boston, 1985.

[21] D. Plachky, On a theorem by G. L. Sievers, Ann. Math. Stat. 42 (1971), pp. 1442-1443.

[22] D. Ruelle, *Thermodynamic Formalism*, Adison-Wesley, London, 1978.

[23] H. Schaefer, *Topological Vector Spaces*, Springer-Verlag, New York, 1971.

[24] G. L. Sievers, On the probability of large deviations and exact slopes, Ann. Math. Stat. 40 (1969), pp. 1908-1921.

[25] D. W. Stroock, *An introduction to the theory of large deviations*, Springer, New York, 1984.

[26] S. Varadhan, *Large Deviations and Applications*, SIAM, Philadelphia, 1984.

Wlodzimierz Bryc
Department of Mathematics
University of Cincinnati
Cincinnati, OH 45221-0025

Cramér Functional Estimates for Gaussian Measures

V. Goodman* and J. Kuelbs**

1. <u>Introduction</u>. The purpose of this note is to present the probability estimate implicit in Theorem 2.1 of [4], and to interpret it as a refined large deviation upper bound for Gaussian measures. This result was established for use in proving upper bounds for the rate of convergence in functional laws of the iterated logarithm. The estimate is delicate enough to provide the best available upper bounds for this problem in several different settings, so it is possible that it could be of use in other situations as well. Some additional applications are presented in section three.

2. <u>Large deviations and the probability estimate</u>. Let B denote a separable Banach space with norm $\|\cdot\|$ and topological dual B^*. If μ is a mean zero Gaussian measure on B, then it is well known that there is a unique Hilbert space $H_\mu \subseteq B$ such that μ is determined by considering the pair (B, H_μ) as an abstract Wiener space. Lemma 2.1 of [5] and [7] present a construction of H_μ along with various properties of the relationship between H_μ and B. In particular, the set H_μ is the completion of the range of

* Supported in part by NSF Grant DMS-88-07121
** Supported in part by NSF Grant DMS-85-21586

the linear operator $S: B^* \to B$ defined by the Bochner integral

$$(2.1) \qquad Sf = \int_B x f(x) d\mu(x) \qquad (f \in B^*) .$$

The completion is in the norm determined by the inner product

$$(2.2) \qquad (Sf, Sg)_\mu = \int_B f(x) g(x) d\mu(x) ,$$

and for $x \in H_\mu$ we have

$$(2.3) \qquad \|x\| \le \sup_{\|f\|_{B^*} \le 1} \left(\int_B f^2(x) d\mu(x) \right)^{1/2} \|x\|_{H_\mu}$$

where $\|\cdot\|_{H_\mu}$ denotes the inner product norm. Furthermore, the unit ball of H_μ

$$(2.4) \qquad K = \{x \in B : \|x\|_{H_\mu} \le 1\}$$

is a compact subset of B, and when H_μ is infinite dimensional it is well known that $\mu(H_\mu) = 0$ but $\mu(\overline{H}_\mu) = 1$ where $\overline{H}_\mu$ is the B-closure of H_μ. Throughout we will assume that H_μ is infinite dimensional.

In connection with large deviations, recall that if μ is a centered Gaussian measure on B, and X is a B-valued random vector with

$$(2.5) \qquad \mu = \mathcal{L}(X) ,$$

then large deviation results easily imply that

$$\limsup_{\varepsilon\downarrow 0} \varepsilon^2 \log P(\varepsilon X \in A) \le - \inf_{x\in\bar{A}} I(x) , \tag{2.6}$$

where $\bar{A}$ denotes the B-closure of A and $I(\cdot)$ is the rate function

$$I(x) = \begin{cases} \frac{1}{2} \|x\|^2_{H_\mu} & x \in H_\mu \\ +\infty & x \notin H_\mu . \end{cases} \tag{2.7}$$

Some useful references for (2.6) and (2.7) include [1], [2], [6], or [7].

The probability estimate we present can be viewed as a refinement of the inequality (2.6). Since $I(x) \ge 0$ and $I(0) = 0$, it is easy to see from (2.7) and (2.3) that (2.6) contains useful information only for those subsets whose closure does not contain the origin. Hence to state (2.6) in its sharpest form we should identify the largest closed set F such that

$$\inf_{x\in F} I(x) = \inf_{x\in\bar{A}} I(x) . \tag{2.8}$$

Equivalently, we need the smallest open set G containing the origin satisfying

$$\inf_{x\notin G} I(x) = \inf_{x\in\bar{A}} I(x) , \tag{2.9}$$

and writing the inequality (2.6) in terms of such sets we have

$$\limsup_{\varepsilon \downarrow 0} \varepsilon^2 \log P(\varepsilon X \notin G) \le - \inf_{x \notin G} I(x) . \tag{2.10}$$

Because of the option to scale we may assume that G satisfies the normalizing condition

$$\inf_{x \notin G} I(x) = \frac{1}{2} . \tag{2.11}$$

The minimal set satisfying (2.11) is

$$N = \{x : \|x\|_{H_\mu} < 1\} , \tag{2.12}$$

but when H_μ is infinite dimensional N is not open in B (recall K is compact) and $P(\varepsilon X \notin N) = 1$. Hence (2.10) fails if we replace G by the minimizing set N , and we are forced to consider sets G larger than N . The sets we use are starlike neighborhoods of N of the form $N^{\delta E}$ where $\delta > 0$, E is starlike, and

$$N^{\delta E} = \{n + \delta e : n \in N , e \in E\} .$$

The most usual choice for E is βU where $\beta > 0$ and $U = \{x : \|x\| < 1\}$. Since $N^{\delta E} = K^{\delta E}$ when E is open we express our results in terms of the more frequently used set K . However, E need not be open as can be seen from the example in Remark (c) following the statement of Theorem 2.1. In our probability estimate δ will decrease to zero, and in that sense $K^{\delta E}$ is the "minimal set" containing the origin having the right hand side of (2.10) approaching -1/2. Our estimate will also refine (2.10) by including what might be called polynomial terms.

To express this estimate we need some further notation. In particular, we will use the continuous linear operators

$$\Pi_d(x) = \sum_{k=1}^{d} \alpha_k(x)S\alpha_k \quad \text{and} \quad Q_d(x) = x - \Pi_d(x) \qquad (d \tag{2.13}$$

taking B into B. Here $\{\alpha_k : k \geq 1\}$ is a sequence in B^* orthonormal in $L^2(\mu)$, $\{S\alpha_k : k \geq 1\}$ is a CONS in $H_\mu \subseteq B$ defined via (2.1), and when restricted to H_μ, Π_d and Q_d are orthogonal projections onto their ranges. Furthermore, if X is a B-valued centered Gaussian random vector, then $\lim_d \|Q_d(X)\| = 0$ with probability one,

$$E\|Q_d(X)\| \downarrow 0 \quad \text{as} \quad d \nearrow \infty , \tag{2.14}$$

and $\Pi_d(X)$ and $Q_d(X)$ are independent centered Gaussian random vectors.

The estimate is contained in the following theorem and its corollaries.

<u>Theorem 2.1</u>. Let X be a centered Gaussian random vector with values in B. Let $Q_d (d \geq 1)$ be the linear operators of (2.13), K the unit ball of H_μ where $\mu = \mathcal{L}(X)$, and let E denote a starlike subset of B such that for each $r > 0$

$$\lim_{m\to\infty} P(Q_m X \in rE) = 1 . \tag{2.15}$$

For each $\varepsilon > 0$, let $d(\varepsilon)$ be the positive integer satisfying

$$(2.16) \qquad d(\varepsilon) = \inf\{m \geq 1 : P(Q_m X \in (m\varepsilon\log \varepsilon^{-1})E) \geq \frac{1}{2}\},$$

and set

$$(2.17) \qquad \delta(\varepsilon) = d(\varepsilon)\varepsilon^2\log \varepsilon^{-1} .$$

Then, for all $\varepsilon > 0$ sufficiently small

$$(2.18) \qquad P(\varepsilon X \notin K + \delta(\varepsilon)E) \leq \exp\{-\frac{1}{2}\varepsilon^{-2} + d(\varepsilon)\log((d(\varepsilon)\varepsilon^2)^{-1})\}.$$

Remark: (a) The proof will show that the dimension function $d(\varepsilon)$ can be chosen such that $d(\varepsilon) = o((\varepsilon\log \varepsilon^{-1})^{-1})$ and $\delta(\varepsilon) = o(\varepsilon)$ as $\varepsilon\downarrow 0$. As a result, the right hand expression in (2.18) is of the form

$$\exp\{-\frac{1}{2}\varepsilon^{-2} + o(\varepsilon^{-1})\} .$$

Furthermore, we will use estimates generated by (2.18) for sequences of the form $\varepsilon_n = (\log n)^{-1/2}$ which show that the probabilities in (2.18) with $\varepsilon = \varepsilon_n$ are summable. The remark following Theorem 3.1 mentions further refinements.

(b) The inequality (2.18) should be compared with (2.10) with $G = K + \delta(\varepsilon)E$ even if G is not open. In particular, the reader should note that if $E \subseteq \beta U$ where $\beta > 0$ and U is the open unit ball of B, then

$$\liminf_{\varepsilon\downarrow 0}\ \inf_{x\notin K+\delta(\varepsilon)E} I(x) = \frac{1}{2}\,. \tag{2.19}$$

To verify (2.19) first observe that for each $\varepsilon > 0$,

$$\inf_{x\notin K+\delta(\varepsilon)E} I(x) \geq \frac{1}{2}\,.$$

Hence we need only check the quantity in (2.19) is less than or equal to $\frac{1}{2}$. For this, simply choose a sequence $\{x_n\} \subseteq H_\mu$ such that $x_n \notin K$ and for some $y \in K$, $\lim_n \|x_n - y\|_{H_\mu} = 0$. Since $x_n \notin K$ and K is compact there exists a $\delta > 0$ such that $x_n \notin K^\delta$. Hence, for each $n \geq 1$ and $E \subseteq \beta U$

$$\liminf_{\varepsilon\downarrow 0}\ \inf_{x\notin K+\delta(\varepsilon)E} I(x) \leq I(x_n)\,,$$

and since $\lim_n I(x_n) = I(y) = \frac{1}{2}$ we have (2.19). Of course, we used $\lim_{\varepsilon\downarrow 0} \delta(\varepsilon) = 0$ in the above, but this will be proved in Lemma 2.1 below.

(c) Examples of starlike sets which satisfy (2.15) obviously include $E = \beta U$, $\beta > 0$ and U the open unit ball of B, by applying (2.14). For another class of examples, set $B = C[0,1]$, let μ be Wiener measure, and define, for $0 < \alpha < \frac{1}{2}$,

$$\|f\|_\alpha = \sup_{\substack{0\leq s,t\leq 1\\ s\neq t}} \frac{|f(s) - f(t)|}{|s - t|^\alpha}\,. \tag{2.20}$$

Since the paths of Brownian motion are $\mathrm{Lip}(\alpha)$ for $0 < \alpha < \frac{1}{2}$ with probability one, the set

(2.21) $$E = \{f : \|f\|_\alpha < \infty \text{ and } \sup_{0\le t\le 1} |f(t)| < \beta\}$$

is starlike and (2.15) follows for every $r > 0$ since (2.14) holds. A useful consequence of Theorem 2.1 is the following corollary.

Corollary 2.1. Let $\tilde{d}(\varepsilon)$ be the dimension function defined by

(2.22) $$\tilde{d}(\varepsilon) = \inf\{m \ge 1 : E\|Q_m X\| \le \frac{m}{2} \Gamma \varepsilon \log \varepsilon^{-1}\},$$

where

(2.23) $$\Gamma = \sup_{x\in K} \|x\|,$$

and

(2.24) $$\tilde{\delta}(\varepsilon) = \tilde{d}(\varepsilon)\varepsilon^2 \log \varepsilon^{-1}.$$

Then, for all $\varepsilon > 0$ sufficiently small

(2.25) $$P(\varepsilon X \notin K + 7\tilde{\delta}(\varepsilon)\Gamma U) \le \varepsilon^{\tilde{d}(\varepsilon)} e^{-1/2\varepsilon^{-2}}.$$

Remark. The term $\varepsilon^{\tilde{d}(\varepsilon)}$ gives an estimate of what might be called the polynomial terms in the upper bound.

Proof of Theorem 2.1. First we prove a lemma.

Lemma 2.1. If E is starlike and for each $r > 0$ (2.15) holds, then the dimension function $d(\varepsilon)$ defined by (2.16) is such that

(2.26) $$d(\varepsilon) = o((\varepsilon\log \varepsilon^{-1})^{-1}) \quad \text{as} \quad \varepsilon\downarrow 0$$

and the quantity $\delta(\varepsilon)$ in (2.17) satisfies

(2.27) $$\delta(\varepsilon) = o(\varepsilon) \quad \text{as} \quad \varepsilon\downarrow 0 .$$

Proof. Since H_μ is assumed infinite dimensional with $\mu = \mathcal{L}(X)$, we have

$$\lim_{\varepsilon\downarrow 0} d(\varepsilon) = \infty .$$

Choose $\delta > 0$ arbitrarily. Then there exists m_0 such that $m \geq m_0$ implies

$$P(Q_m X \in \delta E) \geq \frac{3}{4} .$$

Hence for all $m \geq m_0$ we have

$$P(Q_m X \in m(\varepsilon\log \varepsilon^{-1})E) \geq \frac{3}{4} .$$

if $m(\varepsilon\log \varepsilon^{-1}) \geq \delta$ as E is starlike. Hence $d(\varepsilon) \leq \dfrac{\delta}{\varepsilon\log \varepsilon^{-1}} + 1$ for all $d(\varepsilon) \geq m_0$. Since $\delta > 0$ is arbitrary, (2.26) holds, and (2.27) is obvious from (2.17) and (2.26). Thus the lemma is proved.

Writing $\varepsilon X = \varepsilon\Pi_{d(\varepsilon)}X + \varepsilon Q_{d(\varepsilon)}X$, we have

$$(2.28)\qquad P(\varepsilon X \notin K + \delta(\varepsilon)E) \le P(\varepsilon\Pi_{d(\varepsilon)}X \notin K)$$

$$+ P(\varepsilon\Pi_{d(\varepsilon)}X \in K\ ,\ \varepsilon X \notin K + \delta(\varepsilon)E)\ .$$

Now we observe that the last probability in (2.28) is dominated by

$$(2.29)\qquad P(\varepsilon\|\Pi_{d(\varepsilon)}X\|_{H_\mu} \le 1\ ,\ \varepsilon Q_{d(\varepsilon)}X \notin \delta(\varepsilon)E$$
$$+ (1 - \varepsilon^2\|\Pi_{d(\varepsilon)}X\|^2_{H_\mu})^{1/2}Q_{d(\varepsilon)}K).$$

To verify this assume that $\varepsilon\Pi_{d(\varepsilon)}(x) \in K$ and assume

$$\varepsilon Q_{d(\varepsilon)}(x) = \delta(\varepsilon)e + (1 - \varepsilon^2\|\Pi_{d(\varepsilon)}(x)\|^2_{H_\mu})^{1/2}Q_{d(\varepsilon)}(k)$$

where $e \in E$, $k \in K$. Then

$$\varepsilon x = \tilde{k} + \delta(\varepsilon)e$$

where

$$\tilde{k} = \varepsilon\Pi_{d(\varepsilon)}(x) + (1 - \varepsilon^2\|\Pi_{d(\varepsilon)}(x)\|^2_{H_\mu})^{1/2}Q_{d(\varepsilon)}(k).$$

Now

$$\|\tilde{k}\|^2_{H_\mu} = \varepsilon^2\|\Pi_{d(\varepsilon)}x\|^2_{H_\mu} + (1 - \varepsilon^2\|\Pi_{d(\varepsilon)}(x)\|^2_{H_\mu})\|Q_{d(\varepsilon)}(k)\|^2_{H_\mu} \le 1$$

since $\|Q_{d(\varepsilon)}(k)\|_{H_\mu} \le 1$ and $\Pi_{d(\varepsilon)}$ and $Q_{d(\varepsilon)}$ are orthogonal on H_μ.

Hence

$$\varepsilon x \in K + \delta(\varepsilon)E ,$$

and (2.29) dominates $P(\varepsilon\Pi_{d(\varepsilon)}(X) \in K , \varepsilon X \notin K + \delta(\varepsilon)E)$.

Since $d(\varepsilon)$ satisfies (2.16) we have

$$P(Q_{d(\varepsilon)}X \in d(\varepsilon)\ \varepsilon(\log \varepsilon^{-1})E) \geq \frac{1}{2} ,$$

and hence by Borell's inequality [3] with $\delta(\varepsilon) = d(\varepsilon)\ \varepsilon^2 \log \varepsilon^{-1}$

$$(2.30) \qquad P(\varepsilon Q_{d(\varepsilon)}X \notin \delta(\varepsilon)E + (1 - \varepsilon^2 a^2)^{1/2} Q_{d(\varepsilon)}K) \leq 1 - \Phi((1 - \varepsilon^2 a^2)^{1/2})$$

where $\Phi(t)$ is the standard normal distribution function.

Using the independence of $\Pi_{d(\varepsilon)}X$ and $Q_{d(\varepsilon)}X$ we have from (2.29) and (2.30) that

$$(2.31) \qquad P(\varepsilon\Pi_{d(\varepsilon)}(X) \in K , \varepsilon X \notin K + \delta(\varepsilon)E)$$

$$\leq E\left[I(\varepsilon\|\Pi_{d(\varepsilon)}X\|_{H_\mu} \leq 1)(1 - \Phi((1 - \varepsilon^2\|\Pi_{d(\varepsilon)}X\|^2_{H_\mu})^{1/2}))\right] ,$$

and combining (2.28) and (2.31) we get

$$(2.32) \qquad P(\varepsilon X \notin K + \delta(\varepsilon)E) \leq P(\varepsilon\|\Pi_{d(\varepsilon)+1}(X)\|_{H_\mu} > 1) .$$

Now $P(\|\Pi_d(X)\|^2_{H_\mu} > \lambda^2) = P(g_1^2 + \ldots + g_d^2 > \lambda^2)$ where $g_1, g_2, \ldots$ are i.i.d. $N(0,1)$, and by Markov's inequality for $|t| < \frac{1}{2}$

$$P(g_1^2 + \dots + g_d^2 > \lambda^2) \le \exp\{-\lambda^2 t\} \prod_{j=1}^{d} E(e^{tg_j^2})$$

$$(2.33) \qquad = \exp\{-\lambda^2 t - \frac{d}{2}\log(1 - 2t)\} .$$

Setting $1 - 2t = \varepsilon^2 d(\varepsilon)$ in (2.33), and recalling $\varepsilon^2 d(\varepsilon) \to 0$ as $\varepsilon \to 0$, we have for $\lambda = \varepsilon^{-1}$, $t = (1 - \varepsilon^2 d(\varepsilon))/2$, and all $\varepsilon > 0$ sufficiently small

$$(2.34) \qquad P(\varepsilon X \notin K + \delta(\varepsilon)E) \le \exp\{-\frac{1}{2}\varepsilon^{-2} + \left(\frac{d(\varepsilon)}{2}\right)(1 - \log(\varepsilon^2 d(\varepsilon)))\}$$

$$\le \exp\{-\frac{1}{2}\varepsilon^{-2} + d(\varepsilon)\log((\varepsilon^2 d(\varepsilon))^{-1})\}$$

since $\lim_{\varepsilon\downarrow 0} d(\varepsilon) = \infty$. Thus the theorem is proved.

Proof of Corollary 2.1. The first step of the proof is the following elementary lemma.

Lemma 2.2. Let $E = \Gamma U$ where Γ is given by (2.23). If $d(\varepsilon)$, $\tilde{d}(\varepsilon)$, $\tilde{\delta}(\varepsilon)$ are defined by (2.16), (2.22), and (2.24), respectively, then

$$(2.35) \qquad \tilde{d}(\varepsilon) \ge d(\varepsilon) ,$$

$$(2.36) \qquad \tilde{d}(\varepsilon) = o((\varepsilon\log \varepsilon^{-1})^{-1}) \quad \text{as} \quad \varepsilon\downarrow 0 ,$$

$$(2.37) \qquad \tilde{\delta}(\varepsilon) = o(\varepsilon) \quad \text{as} \quad \varepsilon\downarrow 0 , \text{ and}$$

$$(2.38) \qquad \tilde{d}(\varepsilon') \sim \tilde{d}(\varepsilon) \quad \text{if} \quad \varepsilon' \sim \varepsilon \quad \text{as} \quad \varepsilon\downarrow 0 .$$

Proof. Since U is the open unit ball of B ,

$$P(Q_mX \notin m(\varepsilon\log \varepsilon^{-1})\Gamma U) \le E\|Q_mX\|/(m\ \Gamma\ \varepsilon\log \varepsilon^{-1}) . \tag{2.39}$$

Hence for $m = \tilde{d}(\varepsilon)$ as in (2.22) we have

$$P(Q_mX \notin \tilde{d}(\varepsilon)(\varepsilon\log \varepsilon^{-1})\Gamma U) \le \frac{1}{2} .$$

Thus $\tilde{d}(\varepsilon) \ge d(\varepsilon)$ and (2.35) holds. Furthermore, (2.14) implies the right hand side of (2.39) is less than or equal to $\frac{1}{2}$ for some choice of $m = o(2/(\Gamma\ \varepsilon\log \varepsilon^{-1}))$. Thus the dimension function $\tilde{d}(\varepsilon)$ of (2.22) satisfies (2.36), and (2.37) holds.

Now $\varepsilon' < \varepsilon$ implies $\tilde{d}(\varepsilon') \ge \tilde{d}(\varepsilon)$ for small ε . This follows since $x\log x^{-1}$ decreases at $x = 0$ for $x > 0$ and $E\|Q_mX\|$ decreases as m increases. Hence fix $\delta > 0$ and choose $\varepsilon > \varepsilon' > \frac{\varepsilon}{1+\delta}$. Then, for $\varepsilon > 0$ sufficiently small

$$\tilde{d}(\varepsilon') = \inf\left\{m \ge 1 : E\|Q_mX\| \le \frac{m}{2}\ \Gamma\varepsilon'\ \log \frac{1}{\varepsilon'}\right\}$$

$$\le \inf\left\{m \ge 1 : E\|Q_mX\| \le \frac{m}{2}\ \Gamma\ \left(\frac{\varepsilon}{1+\delta}\right)\log\ \left(\frac{1+\delta}{\varepsilon}\right)\right\} \tag{2.40}$$

$$\le \inf\{m \ge 1 : E\|Q_mX\| \le \frac{m}{2}\ \Gamma\ (\varepsilon\log \varepsilon^{-1})(1 - 2\delta)\} .$$

The last inequality in (2.40) holds since, for $\delta > 0$,

$$\left(\frac{\varepsilon}{1+\delta}\right)\ \log\ \left(\frac{1+\delta}{\varepsilon}\right) \ge (\varepsilon\log \varepsilon^{-1})(1 - 2\delta)$$

provided $\varepsilon > 0$ is sufficiently small. Now

$$\inf\{m \geq 1 : E\|Q_m X\| \leq \frac{m}{2} \Gamma (\varepsilon \log \varepsilon^{-1})(1 - 2\delta)\}$$

$$\leq \inf\{m \geq 1 : E\|Q_m X\| \leq \frac{\tilde{d}(\varepsilon)}{2}\Gamma(\varepsilon \log \varepsilon^{-1}) \frac{m}{\tilde{d}(\varepsilon)}(1 - 2\delta)\}$$

$$\leq \inf\{m \geq 1 : m(1 - 2\delta)/\tilde{d}(\varepsilon) \geq 1\}$$

$$\text{since } E\|Q_{\tilde{d}(\varepsilon)} X\| \leq \frac{\tilde{d}(\varepsilon)}{2} \Gamma(\varepsilon \log \varepsilon^{-1})$$

$$\leq \frac{\tilde{d}(\varepsilon)}{1-2\delta} + 1 ,$$

and hence

$$\tilde{d}(\varepsilon') \leq \frac{\tilde{d}(\varepsilon)}{1-2\delta} + 1 .$$

Since $\delta > 0$ is arbitrary, this implies $\tilde{d}(\varepsilon') \sim \tilde{d}(\varepsilon)$ if $\varepsilon \sim \varepsilon'$. Thus (2.38) holds and the lemma is proved.

To prove (2.25), recall $\Gamma = \sup_{x \in K} \|x\|$ with $U = \{x : \|x\| < 1\}$, and hence

$$P(\varepsilon X \notin K + 7\tilde{\delta}(\varepsilon) \Gamma U) \leq P(\varepsilon X \notin (1 + 4\tilde{\delta}(\varepsilon))K + 2\tilde{\delta}(\varepsilon)\Gamma U)$$

$$= P(\left(\frac{\varepsilon}{1+4\tilde{\delta}(\varepsilon)}\right)X \notin K + \frac{2\tilde{\delta}(\varepsilon)}{1+4\tilde{\delta}(\varepsilon)} \Gamma U)$$

(2.42)

$$\leq P(\left(\frac{\varepsilon}{1+4\tilde{\delta}(\varepsilon)}\right)X \notin K + \tilde{\delta}\left(\frac{\varepsilon}{1+4\tilde{\delta}(\varepsilon)}\right) \Gamma U)$$

for $\varepsilon > 0$ sufficiently small. To see the last inequality, recall $\tilde{\delta}(\varepsilon) = \tilde{d}(\varepsilon)\varepsilon^2 \log \varepsilon^{-1}$ and $\tilde{d}(\varepsilon) \sim \tilde{d}(\varepsilon')$ if $\varepsilon' \sim \varepsilon$ as $\downarrow 0$ by (2.38). Letting $\varepsilon' = \varepsilon/(1 + 4\tilde{\delta}(\varepsilon))$, (2.42) implies

$$P(\varepsilon X \notin K + 7\tilde{\delta}(\varepsilon)\Gamma U) \le P(\varepsilon' X \notin K + \tilde{\delta}(\varepsilon')\Gamma U)$$

(2.43)

$$\le P(\varepsilon' X \notin K + \delta(\varepsilon')\Gamma U)$$

since

$$\tilde{\delta}(\varepsilon') = \tilde{d}(\varepsilon')(\varepsilon')^2 \log(1/\varepsilon')$$

$$\ge d(\varepsilon')(\varepsilon')^2 \log(1/\varepsilon')$$

$$= \delta(\varepsilon') .$$

Applying Theorem 2.1 to (2.43) we get for $\varepsilon > 0$ and $\varepsilon' = \dfrac{\varepsilon}{1+4\tilde{\delta}(\varepsilon)}$ sufficiently small that

$$P(\varepsilon X \notin K + 7\tilde{\delta}(\varepsilon)\Gamma U) \le \exp\{-\frac{1}{2}(\varepsilon')^{-2} + d(\varepsilon')\log((d(\varepsilon')(\varepsilon')^2)^{-1})\}$$

$$\le \exp\{-\frac{1}{2}\varepsilon^{-2} - 4\tilde{\delta}(\varepsilon)\varepsilon^{-2} - 8\tilde{\delta}^2(\varepsilon)^{-2} + \frac{3}{2}\tilde{d}(\varepsilon)\log(\varepsilon^{-2})\}$$

by (2.35) and (2.38) as $\varepsilon \sim \varepsilon'$ with $d(\varepsilon') \longrightarrow \infty$.

$$\le \exp\{-\frac{1}{2}\varepsilon^{-2} - 4\tilde{d}(\varepsilon)\log(\varepsilon^{-1}) + 3\tilde{d}(\varepsilon)\log(\varepsilon^{-1})\}$$

$$= \exp\{-\frac{1}{2}\varepsilon^{-2} - \tilde{d}(\varepsilon)\log(\varepsilon^{-1})\}$$

$$= \varepsilon^{\tilde{d}(\varepsilon)} e^{-1/2\varepsilon^{-2}} .$$

Thus (2.25) holds and Corollary 2.1 holds.

For some applications it is useful to have the following variation of Corollary 2.1. It's proof follows that of Corollary 2.1, and hence is not included.

Corollary 2.2. Let $\rho(\varepsilon)$ be any integer valued function satisfying

$$\rho(\varepsilon) \geq \inf\{m \geq 1 : E\|Q_m X\| \geq \frac{m}{2} \Gamma\, \varepsilon \log \varepsilon^{-1}\}, \tag{2.44}$$

where Γ is as in (2.23), and such that

$$\rho(\varepsilon) = O((\varepsilon \log \varepsilon^{-1})^{-1}) \quad \text{as} \quad \varepsilon \downarrow 0, \text{ and} \tag{2.45}$$

$$\rho(\varepsilon) \sim \rho(\varepsilon') \quad \text{if} \quad \varepsilon \sim \varepsilon' \quad \text{as} \quad \varepsilon \downarrow 0. \tag{2.46}$$

If

$$\gamma(\varepsilon) = \rho(\varepsilon)\varepsilon^2 \log(\varepsilon^{-1}), \tag{2.47}$$

then $\gamma(\varepsilon) = O(\varepsilon)$ and

$$P(\varepsilon X \notin K + 7\gamma(\varepsilon)\Gamma U) \leq \varepsilon^{\rho(\varepsilon)} e^{-1/2\varepsilon^{-2}}. \tag{2.48}$$

3. Some applications. Let X be a centered Gaussian random vector with K as above, and assume $X_1, X_2, \ldots$ are i.i.d. copies of X. If $\{\varepsilon_n\}$ is a sequence of strictly positive numbers such that

$\varepsilon_n \longrightarrow 0$, we define

(3.1) $$\mathcal{N} = \mathcal{N}(\{\varepsilon_n\}) = \#\{n \geq 1 : X_n/(2Ln)^{1/2} \notin K + \varepsilon_n U\}$$

and

(3.2) $$\mathcal{L} = \mathcal{L}(\{\varepsilon_n\}) = \sup\{n \geq 1 : X_n/(2Ln)^{1/2} \notin K + \varepsilon_n U\} .$$

In (3.1) and (3.2), $Lx = \max(1, \log_e x)$, and as before $U = \{x : \|x\| < 1\}$. Here we investigate integrability properties of $\mathcal{N}$ and $\mathcal{L}$. The paper [8] contains some related results, but does not consider $\varepsilon_n \longrightarrow 0$.

<u>Theorem 3.1</u>. Let X be a centered Gaussian random vector, and assume

(3.3) $$\varepsilon_n = \gamma/(Ln)^{1/2} \qquad (\gamma > 0,\ n \geq 1) .$$

Then the following hold.

(A) If $\mathcal{N}$ is defined by (3.1), then for all $\gamma > 0$ and $t < 1$

(3.4) $$E(\exp\{t\mathcal{N}\, L\, \mathcal{N}\}) < \infty .$$

(B) If $\mathcal{L}$ is defined by (3.2), then for all $\gamma > 0$, $r > 0$

(3.5) $$E((\log \mathcal{L})^r) < \infty .$$

<u>Remark</u>. Theorem 3.1 holds for arbitrary centered Gaussian X

provided $\{\varepsilon_n\}$ satisfies (3.3). For various special X it is possible to choose $\{\varepsilon_n\}$ converging to zero faster than what is prescribed in (3.3), and still have the sequence of probabilities

$$p_n = P(X/(2Ln)^{1/2} \notin K + \varepsilon_n U)$$

satisfying

$$\sum_{n=1}^{\infty} p_n < \infty$$

The proof of Theorem 3.1 will show that for such X, both (3.4) and (3.5) hold. The results in [4] show that Brownian motion, the Brownian sheet, and various self-similar Gaussian processes are all of this type.

Proof. The first step of the proof is to show

$$\sum_{n=1}^{\infty} p_n < \infty , \tag{3.6}$$

where

$$p_n = P(X_n/(2Ln)^{1/2} \notin K + \varepsilon_n U) . \tag{3.7}$$

Since $X, X_1, X_2, \ldots$ are i.i.d. X_n in (3.6) can be replaced by X.

Given $\{\varepsilon_n\}$ as in (3.3), fix $\gamma > 0$ and choose

$$\rho(\varepsilon) = \frac{\gamma}{7\Gamma}(\varepsilon \log \varepsilon^{-1})^{-1} \tag{3.8}$$

in Corollary 2.2. This is possible since $\tilde{d}(\varepsilon)$ of Corollary 2.1 satisfies (2.36). Then

$$\gamma(\varepsilon) = \frac{\gamma}{7\Gamma}\,\varepsilon \ , \tag{3.9}$$

and

$$P(\varepsilon X \notin K + 7\gamma(\varepsilon)\Gamma U) \le \exp\{-\tfrac{1}{2}\,\varepsilon^{-2} + \rho(\varepsilon)\log\,\varepsilon\} \ . \tag{3.10}$$

By combining (3.8), (3.9), and (3.10), we get

$$P(\varepsilon X \notin K + \gamma\varepsilon U) \le \exp\{-\tfrac{1}{2}\,\varepsilon^{-2} - \frac{\gamma}{7\Gamma}\varepsilon^{-1}\} \ .$$

Hence, setting $\varepsilon = (2Ln)^{-1/2}$, we have $\gamma\varepsilon \le \varepsilon_n$ and for all sufficiently large n

$$P(X/(2Ln)^{1/2} \notin K + \varepsilon_n U) \le \exp\{-Ln - \frac{\gamma}{7\Gamma}\,(Ln)^{1/2}\} \ . \tag{3.11}$$

Thus (3.6) holds.

To prove (A), let

$$f_n = I(X_n/(2Ln)^{1/2} \notin K + \varepsilon_n U) \ . \tag{3.12}$$

Then $f_1, f_2, \ldots$ are independent Bernoulli random variables with

$$\mathcal{N} = \sum_{n=1}^{\infty} f_n \ . \tag{3.13}$$

Since each $f_n \ge 0$, and (3.6) holds with $p_n = P(f_n = 1)$, we have for $t > 0$

$$E(\exp\{t\mathcal{N}\}) = \prod_{n=1}^{\infty} E(\exp\{tf_n\})$$

$$= \prod_{n=1}^{\infty} (1 + p_n(e^t - 1))$$

$$\le \prod_{n=1}^{\infty} \exp\{p_n(e^t - 1)\}$$

$$= \exp\{\Lambda(e^t - 1)\}$$

where $0 < \Lambda = \sum_{n=1}^{\infty} p_n < \infty$. Thus by Markov's inequality, for each $t > 0$

$$P(\mathcal{N} > \lambda) \le \exp\{-\lambda t\} E(\exp\{t\mathcal{N}\})$$

$$\le \exp\{-\lambda t + \Lambda(e^t - 1)\} .$$

Minimizing this upper bound with respect to t, we find t must satisfy $t = \log(\lambda/\Lambda)$. Now $\Lambda > 0$, so for every $\delta > 0$

$$P(\mathcal{N} > \lambda) \le \exp\{-\lambda(\log \lambda - \log \Lambda) + \lambda - \Lambda\}$$

$$\le \exp\{-(1 - \delta)\lambda \log \lambda\}$$

provided λ is sufficiently large, say $\lambda \ge \lambda(\delta)$. Since $Lx = \max(1, \log_e, x)$ we have

$$E(\exp\{t\mathcal{N}\ L\ \mathcal{N}\}) \le e^{te} + \int_e^{\infty} \exp\{t\ \lambda\ \log\ \lambda\}dF(\lambda)$$

where $F(\lambda) = P(\mathcal{N} \le \lambda)$, and integrating by parts we obtain for each $t < 1$ that

$$E(\exp\{t\mathcal{N}\ L\ \mathcal{N}\}) \le 2e^{te} + \int_e^{\infty} P(\mathcal{N} > \lambda)\ \exp\{t\ \lambda\ \log\ \lambda\}(t(1 + \log\ \lambda))d\lambda$$
$$< \infty\ .$$

The last inequality holds since $t < 1$ and thus there exists a $\delta > 0$ such that $t < 1 - \delta < 1$ and for all $\lambda \ge \lambda(\delta)$

$$P(\mathcal{N} > \lambda)\ \exp\{t\ \lambda\ \log\ \lambda\} \le \exp\{-((1 - \delta) - t)\lambda\ \log\ \lambda\}\ .$$

Thus part (A) holds.

To prove part (B) we first observe that

$$E((\log\ \mathcal{L})^r) = \sum_{n=1}^{\infty} (\log\ k)^r P(\mathcal{L} = k)$$

$$= \sum_{k=1}^{\infty} (\log\ k)^r p_k \prod_{j=k+1}^{\infty} (1 - p_j)\ .$$

Since $\sum_{n=1}^{\infty} p_n < \infty$ we have

$$\lim_{k} \prod_{j=k+1}^{\infty} (1 - p_j) = 1\ .$$

Hence,

$$E((\log \mathcal{L})^r) < \infty$$

if

$$\sum_{n=1}^{\infty} (\log k)^r p_k < \infty . \tag{3.14}$$

From (3.11) we have (3.14), and hence the theorem is proved.

REFERENCES

[1] de Acosta, A. (1985), On large deviations of sums of independent random vectors, Lecture Notes in Mathematics 1153, Springer-Verlag, Berlin and New York, 1-14.

[2] Azencott, R. (1980), Grandes deviations et applications, Lecture Notes in Mathematics 774, Springer-Verlag, Berlin and New York, 1-249.

[3] Borell, C. (1975), The Brunn-Minkowski inequality in Gauss space, Invent. Math. 30, 207-216.

[4] Goodman, V. and Kuelbs, J. (1989), Rates of Clustering for some Gaussian self-similar processes, preprint.

[5] Goodman, V., Kuelbs, J., and Zinn, J. (1981), Some results on the LIL in Banach spaces with applications to weighted empirical processes, Ann. Probab. 9, 713-742.742.

[6] Jain, N. (1985), An introduction to large deviations, Lecture Notes in Mathematics 1153, Springer-Verlag, Berlin and New York, 273-296.

[7] Kuelbs, J. (1976), A strong convergence theorem for Banach space valued random variables, Ann. Probab. 4, 744-771.

[8] Lai, T.L. (1976), On r-quick convergence and a conjecture of Strassen, Ann. of Probab. 4, 612-627.

[9] Stroock, D. (1984), An Introduction to the Theory of Large Deviations, Springer-Verlag, New York.

Victor Goodman
Department of Mathematics
Indiana University
Bloomington, Indiana 47405

James Kuelbs
Department of Mathematics
University of Wisconsin
Madison, Wisconsin 53706

Asymptotic Behaviour of the Laplace Transform of Weighted Occupation Times of Random Walks and Applications

Bruno Rémillard[1,2]

1. Introduction. In this paper we study limits for quantities of the form

$$n^{-\alpha}\log E_{x_n}\left(\exp\left\{n^{-\beta}\sum_{k:1}^{n} V(S_k)\right\}\right) \tag{1.1}$$

where $(S_k)_{k\geq 0}$ is a simple symmetric random walk on $\mathbb{Z}$, E_x denotes the expectation with respect to the random walk starting at $x \in \mathbb{Z}$, $V \in \mathcal{V} = \{V : \mathbb{Z} \to \mathbb{R};$ V has finite support$\}$, $\alpha > 0$, $\beta \geq 0$ and x_n is an appropriate sequence of integers.

This problem was posed to the author by Ted Cox and Rick Durrett who were studying large deviations for weighted occupation time of an infinite system of independent random walks.

Before stating our main results, let us remark that the limiting behavior of $n^{-\beta}\sum_{k:1}^{n} V(S_k)$ depends on whether or not $\overline{V} = \sum_x V(x) \neq 0$. It is well-known that when $\overline{V} \neq 0$, $n^{-1/2}\sum_{k:1}^{n} V(S_k)$ converges in law to $2^{-1/2}\,\overline{V}\,X_{1/2}$ where the law of $X_{1/2}$ is the Mittag-Leffler distribution of index 1/2; on the other hand, when $\overline{V} = 0$, $n^{-1/4}\sum_{k:1}^{n} V(S_k)$ converges in law to a non-degenerate random variable.

---

[1] Part of this work was done wile the author was visiting Cornell University under a Natural Sciences and Engineering Research Council of Canada Postdocotral Fellowship.

[2] Supported in part by the Natural Sciences and Engineering Research Council of Canada.

AMS 1980 *subject classifications.* Primary 60F10; secondary 60J15, 60J65.

Keywords and phrases. Large deviations, occupation time, random walk, infinite particle systems.

Therefore one can easily guess that the behavior of (1.1) is quite different when $V \in V_0 = \{V \in V ; \overline{V} = 0\}$ or when $V \in V_+ = \{V \in V ; \overline{V} > 0\}$. This is indeed true and in Section 2, we prove the

THEOREM 1.1 Suppose $V \in V_0 \cup V_+$ and let $W_n = \sum_{k:1}^{n} V(S_k)$.

a) If $V \in V_0$, $\beta \in (0, 1/4)$ and $x_n/n^{1-2\beta} \to x$, then for any $\lambda \in \mathbb{R}$

$$\lim_{n\to\infty} n^{4\beta-1} \log E_{xn}\left(e^{\lambda n^{-\beta} W_n}\right) = \left(\frac{\lambda^4 B_V^2}{2} - |x|\lambda^2 B_V\right)^+$$

where $B_V = \frac{1}{2}\sum_x (G(x) + G(x-1))^2$, $G(x) = \sum_{y \le x} V(y)$;

b) If $V \in V_+$, $\beta \in (0, 1/2)$ and $x_n/n^{1-\beta} \to x$ then

$$\lim_{n\to\infty} n^{2\beta-1} \log E_{xn}\left(e^{\lambda n^{-\beta} W_n}\right) = \begin{cases} 0 & \text{if } \lambda \le 0 \\ \left(\frac{\lambda^2 (\overline{V})^2}{2} - |x|\lambda \overline{V}\right)^+ & \text{if } \lambda \ge 0 \end{cases}$$

Section 3 is devoted to applications to infinite systems of independent random walks. To describe the results, let a system of independent particles performing independent random walks on $\mathbb{Z}$ be given and let $\xi_n(x)$ = number of particles at site x at time n.

In Cox and Durrett (1989) the authors proved among other things, that if the initial configuration $\xi_0(\cdot)$ satisfies

(*) $\xi_0(\cdot)$ is non random and $\lim_{n\to\infty} (2n)^{-1} \sum_{|x| \le n} \xi_0(x) = 1$

then

(1.2) $$\lim_{n\to\infty} n^{6\beta-2} \log E\left(\exp\left\{\lambda n^{-\beta}\sum_{k:0}^{n}\left(\xi_k(0)-\xi_k(1)\right)\right\}\right) = \frac{\lambda^6}{4}, \quad \beta\in(0,1/4).$$

Adapting the proof of (1.2) to our sett ng we can prove the

THEOREM 1.2 Suppose $\xi_0(\cdot)$ satisfies (*) and let $L_n(V) = \sum_x \sum_{k:0}^{n} V(x)\xi_k(x)$, $V\in \mathcal{V}$ be given.

a) If $V\in \mathcal{V}_0$ and $\beta\in(0,1/4)$

$$\lim_{n\to\infty} n^{6\beta-2}\log E\left(e^{n^{-\beta}L_n(V)}\right) = (1/4)\,B_V^3$$

b) If $V\in \mathcal{V}$ and $\beta\in(0,1/2)$

$$\lim_{n\to\infty} n^{3\beta-2}\log E\left(e^{n^{-\beta}L_n(V)}\right) = (1/4)\,((\overline{V})^+)^3$$

Next let $D_n(x) = n^{-1}\sum_{k:0}^{n}\xi_k(x)$. Considering $D_n(\cdot) = (D_n(x))_{x\in\mathbb{Z}}$ as a random variable with values in $\Lambda = \{\lambda : \mathbb{Z}\to\mathbb{R}_+\}$ (with the product topology), Lee (1989) proved that if $(\xi_0(x))_{x\in\mathbb{Z}}$ are i.i.d Poisson random variables with mean 1, then $(\Lambda, D_n(\cdot), n^{1/2})$ is a large deviation system with action functional $I(\lambda) = \sup_{V\in\mathcal{V}}\left(\left(\sum_x \lambda(x)V(x)\right) - H(V)\right)$, i.e. $\lim_{n\to\infty}\sup n^{-1/2}\log P\left(D_n\in S\right) \le -\inf_{\lambda\in S} I(\lambda)$ for every closed set $S\subset\Lambda$, and $\lim_{n\to\infty}\inf n^{-1/2}\log P\left(D_n\in S\right)\ge -\inf_{\lambda\in S} I(\lambda)$ for every open set $S\subset\Lambda$.

Moreover $H(V) = \overline{V}\int_0^1 f_{1/2}((t/2)^{1/2}\overline{V})dt$ where $f_{1/2}(\cdot)$ is the moment generating function of $X_{1/2}$. Therefore, if we define $\psi(x) = x\int_0^1 f_{1/2}((t/2)^{1/2}x)dt$, then one can easily prove that

$$I(\lambda) = \begin{cases} J(c) = \sup_{x\ge 0}(cx-\psi(x)) & \text{if } \lambda(x)\equiv c\ge 0 \\ +\infty & \text{otherwise} \end{cases}.$$

The function $\psi(\cdot)$ is the same function as in Cox and Griffeath (1984).

Here, as a corollary to Theorem 1.2, we will prove the following result:

THEOREM 1.3 Suppose $\xi_0(\cdot)$ satisfies (*) and $\alpha \in (0,1)$. Then $(\Lambda, n^{-\alpha}D_n, n^{(3\alpha+1)/2})$ is a large deviation system with action functional $I(\cdot)$ given by

$$I(\lambda) = \begin{cases} 4(c/3)^{3/2} & \text{if} \quad \lambda(x) \equiv c \geq 0 \\ +\infty & \text{otherwise} \end{cases}$$

Finally in Section 4 we prove the analogues of Theorem 1.1 for continuous time random walks and for Brownian motions.

2. Proof of Theorem 1.1 Let ℓ_2 be the Hilbert space of all $f : \mathbb{Z} \to \mathbb{R}$ such that $\|f\|^2 = \sum_x f^2(x) < \infty$, with the scalar product $(\cdot,\cdot)$ defined by $(f,g) = \sum_x f(x)g(x)$, $f,g \in \ell_2$.

Throughout this section $V \in \mathcal{V}$ is fixed and $V(x) = 0$ if $|x| > h \geq 0$. We will now define some operators on ℓ_2. First set $\Pi f(x) = E_x(f(S_1)) = 1/2(f(x+1) + f(x-1))$, $f \in \ell_2$. It is easy to check that Π is a bounded self-adjoint operator on ℓ_2 and $(\Pi f, f) \leq \|f\|^2$ $\forall f \in \ell_2$. Next define $T_\lambda f(x) = e^{\frac{\lambda}{2}V(x)}\Pi\left(e^{\frac{\lambda}{2}V} f\right)(x)$, $\lambda \in \mathbb{R}$, $f \in \ell_2$. Then $\{T_\lambda, \lambda \in \mathbb{R}\}$ are also bounded self-adjoint operators on ℓ_2 and $(T_\lambda)^n f(x) = E_x\left(\exp\left\{\lambda W_n + \lambda/2(V(x) - V(S_n))\right\} f(S_n)\right)$ $\left(\text{recall that } W_n = \sum_{k:1}^{n} V(S_k)\right)$.

Proposition 2.1 Define $c(\lambda) = \log\left(\sup_{\substack{f \in \ell_2 \\ \|f\|=1}} (T_\lambda f, f)\right)$.

a) c is well defined, $c(\lambda) \geq 0$, $c(0) = 0$ and $c(\cdot)$ is convex;

b) $\sup_{|x| \leq n} E_x\left(e^{\lambda W_n}\right) \leq 5\, n \exp(2\,|\lambda|\,|V|_\infty + nc(\lambda))$, where $|V|_\infty = \sup_x |V(x)|$.

Proof: a) Define $f_n(x) = (2n + \cdot)^{-1/2} 1_{\{|x| \le n\}}$. Then $||f_n|| = 1$ and

$(T_\lambda f_n, f_n) = (2n+1)^{-1} \sum_{x:-n+1}^{n} \exp(\lambda/2(V(x) + V(x-1))) \ge$

$(2n/2n+1)\exp\left(\lambda/4n \sum_{x:-n+1}^{n} (V(x) + V(x-1))\right) = (2n/2n+1)e^{\lambda\bar{V}/2n}$ whenever $n \ge h+1$.

Hence $\sup_{\substack{f \in \ell_2 \\ ||f||=1}} (T_\lambda f, f) \ge \sup_{n \ge 1} \left(2n/2n+1\ e^{\lambda\bar{V}/2n}\right) \ge 1$, proving that c is well defined and

$c(\lambda) \ge 0$. Next $0 \le c(0) = \log\left(\sup_{\substack{f \in \ell_2 \\ ||f||=1}} (\Pi f, f)\right) \le 0$. Finally $\log (T_\lambda f, f)$ is convex in λ

for every $f \in \ell_2$ such that $f \ge 0$, $f \equiv 0$. Since $c(\lambda) =$

$\log\left(\sup_{\substack{f \in \ell_2 \\ ||f||=1}} (T_\lambda f, f)\right) = \sup_{\substack{f \in \ell_2, f \ge 0 \\ ||f||=1}} \log(T_\lambda f, f)$, $c(\cdot)$ is also convex.

b) Set $f_n(x) = 1_{\{|x| \le 2n\}}$. Then $T_\lambda^n f_n(x) = E_x\left(e^{\lambda W n + \frac{\lambda}{2}V(x) - \frac{\lambda}{2}V(S_n)} 1_{\{|S_n| \le 2n\}}\right) \ge$

$E_x\left(e^{\lambda W_n}\right)e^{-|\lambda|\,|V|_\infty}$ for $|x| \le n$. Therefore $(T_\lambda^n f_n, f_n) = \sum_{|y| \le 2n} T_\lambda^n f_n(y) \ge$

$e^{-|\lambda|\,|V|_\infty} \sup_{|x| \le n} Ex\left(e^{\lambda W_n}\right)$. By the spectral theorem, $(T_\lambda^n f_n, f_n) \le e^{nc(\lambda)} \| f_n \|^2$ if n is

odd. On the other hand, if n is even, $E_x\left(e^{\lambda W_n}\right) \le e^{|\lambda|\,|V|_\infty} E_x\left(e^{\lambda W_{n-1}}\right) \le$

$e^{2|\lambda|\,|V|_\infty + (n-1)c(\lambda)} ||f_n||^2$. Since $||f_n||^2 = 4n + 1 \le 5n$, we obtain

$\sup_{|x| \le n} E_x\left(e^{\lambda W_n}\right) \le 5\,n\,e^{2\,|\lambda|\,|V|_\infty + nc(\lambda)}$, $\forall\, \lambda \in \mathbb{R}$, proving b). ∎

Before stating the next proposition, we need to introduce some notations and definitions.

Let $\theta(\lambda)$ be such that $\log \operatorname{ch} \theta(\lambda) = c(\lambda)$, where $\operatorname{ch} x = (e^x + e^{-x})/2$, $\operatorname{sh} x = (e^x - e^{-x})/2$. Next, for each $(\lambda, \theta) \in \mathbb{R} \times \mathbb{R}_+$, let $f_{\lambda,\theta}(\cdot)$ be the unique solution of the following system:

$$\begin{cases} f(-h-2) = 1 \\ f(-h-1) = e^{\theta} \\ T_\lambda f(x) = (\operatorname{ch} \theta)\, f(x),\ x \geq -h-1 \end{cases}$$

Finally set $\Phi(\lambda, \theta) = e^{\theta} f_{\lambda,\theta}(h+2) - f_{\lambda,\theta}(h+1)$.

PROPOSITION 2.2 a) If $\theta > 0$ and $\Phi(\lambda, \theta) = 0$ for some $\lambda \in \mathbb{R}$, then

$$F_{\lambda,\theta}(x) = \begin{cases} e^{\theta x}, & x \leq -h-2 \\ e^{-\theta(h+2)} f_{\lambda,\theta}(x), & x \geq -h-2 \end{cases}$$

is such that $F_{\lambda,\theta} \in \ell_2$, $T_\lambda F_{\lambda,\theta}(x) = (\operatorname{ch}\theta)\, F_{\lambda,\theta}(x)$ $\forall\, x \in \mathbb{Z}$ i.e. $F_{\lambda,\theta}$ is an eigenvector for T_λ, and $F_{\lambda,\theta}(x) = e^{-\theta x} f_{\lambda,\theta}(h+2)$, $x \geq h+1$.

b) $\theta(\lambda) = \sup\{\theta \geq 0 \,;\, \Phi(\lambda,\theta) = 0\}$ (where the sup over the empty set is defined to be equal to zero).

Proof: a) Suppose that $\theta > 0$ and $\Phi(\lambda,\theta) = 0$ for some λ. Then $F_{\lambda,\theta}$ is well defined. It follows from the definition of $f_{\lambda,\theta}$ that $f_{\lambda,\theta}(x) = e^{-\theta(x-(h+2))} f_{\lambda,\theta}(h+2)$, $x \geq h+1$. Hence $F_{\lambda,\theta}(x) = e^{-\theta x} f_{\lambda,\theta}(h+2)$, $x \geq h+1$. So $F \in \ell_2$ and it is easy to see that $T_\lambda F_{\lambda,\theta}(x) = \operatorname{ch}\theta\, F_{\lambda,\theta}(x)$, $\forall x \in \mathbb{Z}$.

b) Suppose that $\sup\{\theta \geq 0 \,;\, \Phi(\lambda,\theta) = 0\} = \theta_0 > 0$. Since $\Phi(\lambda,\theta)$ is jointly continuous in (λ,θ), $\Phi(\lambda,\theta_0) = 0$ and $e^{c(\lambda)} = \operatorname{ch}\theta(\lambda) \geq (T_\lambda F_{\lambda,\theta_0}, F_{\lambda,\theta_0})/\| F_{\lambda,\theta_0} \|^2 = \operatorname{ch}\theta_0$ which proves that $\sup\{\theta \geq 0 \,;\, \Phi(\lambda,\theta) = 0\} \leq \theta(\lambda)$. To prove the reverse inequality, suppose that $\theta(\lambda) > 0$. It follows that for large n

$$1 < \sup_{\substack{f \in \ell_2,\ f(x)=0,\ |x|>n \\ \|f\|=0}} (T_\lambda f, f) \ := \mathrm{ch}\theta_n(\lambda) \quad \text{and } \theta_n(\lambda) \uparrow \theta(\lambda).$$

From the definition of $\theta_n(\lambda)$, we see that there exists $f_n \in \ell_2$, $f_n(x) = 0$, $|x| > n$, $\| f_n \| = 1$ such that $T_\lambda f_n(x) = \mathrm{ch}\theta_n(\lambda)\, f_n(x)$, $|x| \le n$. Therefore

$$\begin{pmatrix} f_n(x+1) \\ f_n(x) \end{pmatrix} = A_{\lambda,\theta_n(\lambda)}(x) \begin{pmatrix} f_n(x) \\ f_n(x-1) \end{pmatrix}, \quad |x| \le n, \text{ where}$$

$$A_{\lambda,\theta}(x) = \begin{pmatrix} 2\mathrm{ch}\theta \exp\{-\lambda/2(V(x) + V(x+1)\} & -\exp\{\lambda/2(V(x-1) - V(x+1)\} \\ 1 & 0 \end{pmatrix}$$

Since $V(x) = 0$ for $|x| > h$, $A_{\lambda,\theta}(x) = A_{\lambda,\theta} = \begin{pmatrix} 2\mathrm{ch}\theta & -1 \\ 1 & 0 \end{pmatrix}$, if $|x| \ge h + 2$.

Since $f_n(\pm(n+1)) = 0$, we get

$$\text{(2.1)} \qquad \left\langle A_{\lambda,\theta_n(\lambda)}(n) \cdots A_{\lambda,\theta_r(\lambda)}(-n), \begin{pmatrix} 1 \\ 0 \end{pmatrix}, \begin{pmatrix} 1 \\ 0 \end{pmatrix} \right\rangle = 0$$

where $\langle \cdot\, , \cdot \rangle$ is the usual scalar product on $\mathbb{R}^2$. Taking into account that $A_{\lambda,\theta}(x)$ is constant for $|x| \ge h + 2$, (2.1) becomes

$$\text{(2.2)} \qquad \left\langle A^{n-h-1}_{\lambda,\theta_n} A_{\lambda,\theta_n}(h+1) \cdots A_{\lambda,\theta_n}(-h-1)\, A^{n-h-1}_{\lambda,\theta_n} \begin{pmatrix} 1 \\ 0 \end{pmatrix}, \begin{pmatrix} 1 \\ 0 \end{pmatrix} \right\rangle = 0$$

Now $A^k_{\lambda,\theta} = \begin{pmatrix} \mathrm{sh}\theta(k+1)/\mathrm{sh}\theta & -\mathrm{sh}\theta k/\mathrm{sh}\theta \\ \mathrm{sh}\theta k/\mathrm{sh}\theta & -\mathrm{sh}\theta(k-1)/\mathrm{sh}\theta \end{pmatrix}$ and

$\lim_{n\to\infty} e^{-\theta_n(\lambda)(n-h-1)} A^{n-h-1}_{\lambda,\theta_n(\lambda)} = \begin{pmatrix} e^{\theta(\lambda)} & -1 \\ 1 & -e^{-\theta(\lambda)} \end{pmatrix}$, $A_{\lambda,\theta(x)}$ being jointly continuous in (λ,θ) and $\theta_n(\lambda)$ converging to $\theta(\lambda)$. Next, multiplying both sides of (2.2) by $e^{-2\theta_n(\lambda)(n-h-1)}$ and letting n go to infinity we obtain

$$\left\langle \begin{pmatrix} e^{\theta(\lambda)} & -1 \\ 1 & -e^{-\theta(\lambda)} \end{pmatrix} A_{\lambda,\theta(\lambda)}(h+1) \cdots A_{\lambda,\theta(\lambda)}(-h-1) \begin{pmatrix} e^{\theta(\lambda)} & -1 \\ 1 & -e^{-\theta(\lambda)} \end{pmatrix} \begin{pmatrix} 1 \\ 0 \end{pmatrix}, \begin{pmatrix} 1 \\ 0 \end{pmatrix} \right\rangle = 0 \tag{2.3}$$

Finally (2.3) is equivalent to

$$\left\langle A_{\lambda,\theta(\lambda)}(h+1) \cdots A_{\lambda,\theta(\lambda)}(-h-1) \begin{pmatrix} e^{\theta(\lambda)} \\ 1 \end{pmatrix}, \begin{pmatrix} e^{\theta(\lambda)} \\ -1 \end{pmatrix} \right\rangle = 0 \tag{2.4}$$

But $A_{\lambda,\theta(\lambda)}(x) \cdots A_{\lambda,\theta(\lambda)}(-h-1) \begin{pmatrix} e^{\theta(\lambda)} \\ 1 \end{pmatrix} = \begin{pmatrix} f_{\lambda,\theta(\lambda)}(x+1) \\ f_{\lambda,\theta(\lambda)}(x) \end{pmatrix}$, $x \ge -h-1$.

Hence (2.4) can be written in the following form:

$$e^{\theta(\lambda)} f_{\lambda,\theta(\lambda)}(h+2) - f_{\lambda,\theta(\lambda)}(h+1) = \Phi(\lambda,\theta(\lambda)) = 0 \tag{2.5}$$

This completes the proof of b). ∎

The next lemma is the final ingredient needed to prove Theorem 1.1.

LEMMA 2.3

a) Suppose that $V \in \mathcal{V}_+$. Then

i) $c(\lambda) > 0 \quad \forall \lambda > 0$;

ii) $\lim_{\lambda \downarrow 0} \theta(\lambda)/\lambda = \overline{V}$;

iii) $\exists\, \varepsilon > 0$ such that $\theta(\lambda) = 0 \quad \forall \lambda \in [-\varepsilon, 0]$;

iv) $\lim_{\varepsilon\downarrow 0} c(\lambda\varepsilon)/\varepsilon^2 = \begin{cases} \lambda^2(\bar{V})^2/2 & , \lambda \geq 0 \\ 0 & , \lambda \leq 0 \end{cases}$.

b) Suppose that $V \in V_0$, $V \not\equiv 0$. Then

i) $c(\lambda) > 0 \quad \forall\lambda \neq 0$;

ii) $\lim_{\lambda\to 0} \theta(\lambda)/\lambda^2 = B_V$;

iii) $\lim_{\varepsilon\downarrow 0} c(\lambda\varepsilon)/\varepsilon^4 = \lambda^4 B_V^2 / 2$.

Proof: Suppose that $V \in V$ and let $g_\theta(\cdot)$ be such that $\Pi g_\theta(x) = (\text{ch}\theta) g_\theta(x)$, $x \geq 0$, $\theta \geq 0$. Set $\hat{V}(x) = V(x) + V(x-1)$. Then for any $b \geq -h-1$, the following hold:

(2.6)
$$e^{\frac{\lambda}{2}\hat{V}(b+1)}\left(f_{\lambda,\theta}(b+1)g_\theta(0) - f_{\lambda,\theta}(b)g_\theta(-1)\right) = \left(e^{\theta}\, g_\theta(b+h+2) - g_\theta(b+h+1)\right)$$
$$- \sum_{x:-h-1}^{b} f_{\lambda,\theta}(x)\left(\left(e^{\lambda/2\,\hat{V}(x)} - 1\right)g_\theta(b+1-x) + \left(e^{\lambda/2\,\hat{V}(x+1)} - 1\right)g_\theta(b-1-x)\right)$$

Remark: (2.6) follows from the definitions of $f_{\lambda,\theta}$ and g_θ, and from the following "integration by part" formula:

$$2\sum_{x:a}^{b} \left(g(x)\Pi f(x) - f(x)\Pi g(x)\right) =$$
$$g(b)f(b+1) - g(b+1)f(b) + g(a)f(a-1) - g(a-1)f(a) \quad , \quad a \leq b, \ f,g \in \ell_2.$$

If we apply (2.6) with $b = h + 1$ and $g_\theta(x) = \dfrac{e^\theta \text{sh}\theta(x+1) - \text{sh}\theta x}{\text{sh}\theta}$ we obtain

(2.7)
$$\Phi(\lambda,\theta) - \left(e^\theta g_\theta(2h+3) - g_\theta(2h+2)\right) =$$
$$- \sum_{|x|\leq h+1} f_{\lambda,\theta}(x)\left(\left(e^{\frac{\lambda}{2}\hat{V}(x)} - 1\right)g_\theta(h+2-x) + \left(e^{\frac{\lambda}{2}\hat{V}(x+1)} - 1\right)g_\theta(h-x)\right)$$

Let us remark that $g_\theta(x) - 1 \sim \theta(x+1)$ $\left(\text{i.e. } \lim\limits_{\theta \downarrow 0} \frac{g_\theta(x) - 1}{\theta} = x + 1\right)$, and $e^\theta g_\theta(2h+3) - g_\theta(2h+2) \sim 2\theta$. Since $f_{\lambda,\theta}(x)$ is jointly continuous in (λ,θ), it follows from (2.7) that

$$\lim_{\lambda \downarrow 0} \Phi(\lambda, \lambda a)/\lambda = 2(a - \overline{V}) \ , \quad a > 0$$

If $V \in V_+$, we see that there exists $n_0 \in \mathbb{N}$ such that

$$\Phi\left(\tfrac{1}{n}, \overline{V}/2n\right) < 0 < \Phi\left(\tfrac{1}{n}, \tfrac{3\overline{V}}{2n}\right) \quad \forall n > n_0$$

Hence $\theta(1/n) > 0 \ \forall n > n_0$ which is equivalent to $c(1/n) > 0 \ \forall n > n_0$. By proposition 2.1a) $c(\cdot)$ is convex so $c(\lambda) > 0 \ \forall \lambda > 0$ proving a.i).
Suppose now that $\theta(\lambda) > 0 \ \forall \lambda \neq 0$ (we already know that $\theta(\lambda) > 0 \ \forall \lambda > 0$). It follows from (2.7) that

$$\text{(2.8)} \quad e^{\theta(\lambda)} g_{\theta(\lambda)}(2h+3) - g_{\theta(\lambda)}(2h+2) = \sum_{|x| \le h+1} f_{\lambda,\theta(\lambda)}(x)\left(\left(e^{\lambda/2\,\widehat{V}(x)} - 1\right)g_{\theta(\lambda)}(h+2-x) + \left(e^{\lambda/2\,\widehat{V}(x+1)} - 1\right)g_{\theta(\lambda)}(h - x)\right)$$

We know that the ℓ.h.s. of (2.8) is $\sim 2\theta(\lambda)$ and the r.h.s. is $\sim 2\lambda\overline{V}$ as $|\lambda| \to 0$. Therefore $\lim\limits_{\lambda \downarrow 0} \theta(\lambda)/\lambda = \overline{V}$ and $\lim\limits_{\lambda \downarrow 0} \frac{\theta(-\lambda)}{-\lambda} = \overline{V}$. The first equality proves a.ii) and the second equality cannot hold because $\theta(-\lambda) > 0$ and $\overline{V} > 0$. Hence $\theta(-\varepsilon) = 0$ for some $\varepsilon > 0$; $c(\cdot)$ being convex we get $\theta(\cdot) = 0$ on $[-\varepsilon, 0]$ proving a.iii). Finally a.iv) is clearly a consequence of a.ii) and a.iii).

We will now prove b). Suppose that $V \in V_0$ and set $\hat{g}_\theta(x) = \operatorname{sh}\theta x/\operatorname{sh}\theta$, $x \ge -1$. If we apply (2.6) with $\hat{g}_\theta(\cdot)$, we get for $|b| \le h + 1$

$$e^{\lambda/2\,\widehat{V}(b+1)}\left(f_{\lambda,\theta}(b)-1\right)+\left(e^{\lambda/2\,\widehat{V}(b+1)}-1\right)=$$

$$\left(e^{\theta}\widehat{g}_{\theta}(b+h+2)-\widehat{g}_{\theta}(b+h+1)-1\right) \tag{2.9}$$

$$-\sum_{x:-h-1}^{b} f_{\lambda,\theta}(x)\left(\left(e^{\lambda/2\,\widehat{V}(x)}-1\right)\widehat{g}_{\theta}(b+1-x)+\left(e^{\lambda/2\,\widehat{V}(x+1)}-1\right)\widehat{g}_{\theta}(b-1-x)\right)$$

If we choose $\theta = a(\lambda) > 0$ in such a way that $\lim_{\lambda\to 0} a(\lambda)/\lambda = 0$ we obtain from (2.9)

$$\lim_{\lambda\to 0}\frac{f_{\lambda,a(\lambda)}(b)-1}{\lambda}=\zeta(b)=\frac{1}{2}\widehat{G}(b)-\sum_{x\le b}\widehat{G}(x),\text{ where }\widehat{G}(b)=\sum_{x\le b}\widehat{V}(x). \tag{2.10}$$

Next, we transform (2.7) into

$$\Phi(\lambda,\theta)=e^{\theta}g_{\theta}(2h+3)-g_{\theta}(2h+2)$$

$$-\sum_{|x|\le h+1}\left(f_{\lambda,\theta}(x)-1\right)\left(\left(e^{\lambda/2\,\widehat{V}(x)}-1\right)g_{\theta}(h+2-x)+\left(e^{\lambda/2\,\widehat{V}(x+1)}-1\right)g_{\theta}(h-x)\right)$$

$$-\sum_{|x|\le h+1}\left(e^{\lambda/2\,\widehat{V}(x)}-1-\frac{\lambda}{2}\widehat{V}(x)\right)g_{\theta}(h+2-x) \tag{2.11}$$

$$-\sum_{|x|\le h+1}\left(e^{\lambda/2\,\widehat{V}(x+1)}-1-\frac{\lambda}{2}\widehat{V}(x+1)\right)g_{\theta}(h-x)$$

$$-\frac{\lambda}{2}\sum_{|x|\le h+1}\left(\widehat{V}(x)(g_{\theta}(h+2-x)-1)+\widehat{V}(x+1)(g_{\theta}(h-x)-1)\right).$$

Using (2.10) and (2.11) we get for any $c > 0$

$$\lim_{\lambda\to 0}\Phi(\lambda,\lambda^2 c)/\lambda^2 =$$

$$1/2\sum_{|x|\le h+1}\zeta(x)\,(\widehat{V}(x)+\widehat{V}(x+1))+1/8\sum_{|x|\le h+1}(\widehat{V}^2(x)+\widehat{V}^2(x+1)) \tag{2.12}$$

Now $1/2\sum_{|x|\le h}\zeta(x)\,(\widehat{V}(x)+\widehat{V}(x+1)) = 1/4\sum_{x}\widehat{G}(x)\,(\widehat{G}(x+1)-\widehat{G}(x-1))$

$$- 1/2 \sum_{x} \sum_{b \le x} \widehat{G}(b) \left(\widehat{V}(x+1) + \widehat{V}(x)\right) =$$

$$0 - 1/2 \sum_{b} \widehat{G}(b) \left(\sum_{x \ge b} \widehat{V}(x) + \sum_{x \ge b+1} \widehat{V}(x) \right) = 1/2 \sum_{b} \widehat{G}(b) \left(\widehat{G}(b) + \widehat{G}(b-1) \right).$$

Therefore (2.12) is equivalent to

$$\text{(2.13)} \qquad \lim_{\lambda \to 0} \Phi(\lambda, \lambda^2 c) \big/ \lambda^2 = 2(c - B_V).$$

Since $B_V > 0$, we get from (2.13) that $\theta(\pm 1/n) > 0$ $\forall n > n_0$ for some $n_0 \in \mathbb{N}$. Hence $c(\lambda) > 0$ $\forall \lambda \neq 0$, proving b.i). Next, if we apply (2.8) we get $\lim_{\lambda \to 0} \theta(\lambda) \big/ \lambda = 0$ so $\lim_{\lambda \to 0} \dfrac{f_{\lambda, \theta(\lambda)}(x) - 1}{\lambda} = \zeta(x)$ and from (2.11) with $\theta = \theta(\lambda)$ we get

$$\text{(2.14)} \qquad A_0(\lambda) = e^{\theta(\lambda)} g_{\theta(\lambda)}(2h+3) - g_{\theta(\lambda)}(2h+2) = A_1(\lambda) + A_2(\lambda)$$

where $A_0(\lambda) \sim 2\,\theta(\lambda)$, $A_1(\lambda) \sim 2\,\lambda^2 B_V$ and $A_2(\lambda) \sim \lambda\,\theta(\lambda) k$ for some constant k.

Therefore, b.ii) follows from (2.14) and

$\lim_{\varepsilon \downarrow 0} \varepsilon^{-4} c(\lambda\varepsilon) = \lim_{\varepsilon \downarrow 0} \varepsilon^{-4} \log \operatorname{ch} \theta(\lambda\varepsilon) = \lambda^4 B_V^2 / 2$, which completes the proof of b). ■

We are now in a position to prove Theorem 1.1.

Proof: a) Suppose that $V \in \mathcal{V}_0$, $\beta \in (0, 1/4)$ and $x_n/n^{1-2\beta} \to x$. We first prove

$$\text{(2.15)} \qquad \lim_{n \to \infty} \sup n^{4\beta-1} \log E_{x_n}\left(e^{\lambda n^{-\beta} W_n}\right) \le \left(\lambda^4 B_V^2 / 2 - \lambda^2 B_V |x|\right)^+$$

If $x = 0$ then $|x_n| \le n$ for n large and it follows from Proposition 2.1.b) that $n^{4\beta-1} \log E_{x_n}\left(e^{\lambda n^{-\beta} W_n}\right) \le n^{4\beta-1}\left(\log 5n + 2|\lambda|\, |V|_\infty n^{-\beta}\right) + n^{4\beta}\, c\left(\lambda/n^\beta\right)$. Therefore

(2.15) follows from the last inequality and Lemma 2.3.b.iii). Next suppose that $x \neq 0$; then $|x_n| > h$ for large n and using the same trick as in Cox and Durrett (1989), we get $E_{x_n}\left(e^{\lambda n^{-\beta} W_n}\right) \leq 1 + e^{|\lambda| |V|_\infty n^{-\beta}} \sum_{k:1}^{n} P_{x_n}(T_h = k) E_{\alpha h}\left(e^{\lambda n^{-\beta} W_{n-k}}\right)$, where $T_h = \inf\{k \geq 0 ; |S_k| = h\}$, and $\alpha = \text{sign}(x)$.

From the inequality $\log \mathrm{ch}\lambda \leq \frac{\lambda^2}{2}$ we obtain

$$P_0(S_m \geq a) \leq \inf_{\lambda > 0} \left(e^{-\lambda a} E_0\left(e^{\lambda S_m}\right)\right) = \inf_{\lambda > 0} \left(e^{-\lambda a}(\mathrm{ch}\lambda)^m\right) \leq e^{\frac{-a^2}{2m}}, \ a > 0.$$

Hence $P_{x_n}(T_h = k) \leq P_0(S_k \geq |x_n| - h) \leq e^{-(|x_n| - h)^2/2k}$.

Using Proposition 2.1.b) and the preceeding estimates we obtain

$$n^{4\beta-1} \log E_{x_n}\left(e^{\lambda n^{-\beta} W_n}\right) \leq$$

(2.16) $$n^{4\beta-1}\log 2 + \max\left(0, n^{4\beta-1}\left(\log 5n^2 + 3|\lambda| \, |V|_\infty n^{-\beta}\right) + a_n\right), \text{ where}$$

$$a_n = \sup_{t \in (0,1]} (1 - t)\left(n^{4\beta} c(\lambda / n^\beta)\right) - 1/2t \left(\frac{|x_n| - h}{n^{1-2\beta}}\right)^2$$

Using Lemma 2.3.b.iii), it is easy to see that

(2.17) $$\lim_{n \to \infty} a_n = \sup_{t \in (0,1]} \left((1 - t) \lambda^4 B_V^2 / 2 - x^2 / 2t\right) \leq \left(\lambda^4 B_V^2 / 2 - |x| \lambda^2 B_V\right)^+.$$

Using (2.16) and (2.17) we obtain (2.15) for any $|x| > 0$; hence (2.15) holds true.

We will now prove the lower bound

(2.18) $$\liminf_{n \to \infty} n^{4\beta-1} \log E_{x_n}\left(e^{\lambda n^{-\beta} W_n}\right) \geq \left(\lambda^4 B_V^2 / 2 - |x| \lambda^2 B_V\right)^+$$

To prove (2.18) we will use the eigenfunction $F_\lambda = F_{\lambda,\theta(\lambda)}$ defined in Proposition 2.2. Since $T^n_{\lambda n^{-\beta}} F_{\lambda n^{-\beta}}(x) = e^{nc(\lambda/n^\beta)} F_{\lambda n^{-\beta}}(x)$, $\forall x \in \mathbb{Z}$, and $\lim_{n\to\infty} f_{\lambda,\theta(\lambda n^{-\beta})}(x) = 1\ \forall x$,

we have: $\lim_{n\to\infty} \sup_x F_{\lambda n^{-\beta}}(x) = 1$ and $\lim_{n\to\infty} n^{4\beta-1}\log F_{\lambda n^{-\beta}}(x_n) = -|x|\,\lambda^2 B_V$ by Proposition 2.2 and Lemma 2.3.

Next $E_{x_n}\left(e^{\lambda n^{-\beta}W_n}\right) \geq e^{-n^{-\beta}|\lambda|\,|V|_\infty} T^n_{\lambda n^{-\beta}} F_{\lambda n^{-\beta}}(x_n)/\sup_x F_{\lambda n^{-\beta}}(x)$. Therefore

$$\liminf_{n\to\infty} n^{4\beta-1}\log E_{x_n}\left(e^{\lambda n^{-\beta}W_n}\right) \geq \lambda^4 B_V^2/2 - |x|\,\lambda^2 B_V.$$

To complete the proof of (2.18) we only need to prove that

$\liminf_{n\to\infty} n^{4\beta-1}\log E_{x_n}\left(e^{\lambda n^{-\beta}W_n}\right) \geq 0$ for $|x| > \lambda^2 B_V/2$ and this follows from

$E_{x_n}\left(e^{\lambda n^{-\beta}W_n}\right) \geq P_{x_n}(T_h > n) \geq P_0\left(\max_{0\leq k\leq n} S_k \leq |x_n| - h\right)$ which converges to 1 because $|x_n|/n^{1/2} \to +\infty$. (2.15) and (2.18) together yield a).

The proof of b) being similar, it is left to the reader. ■

Remarks: (1) For the case $\beta = 0$, one can easily prove that if $V \in \mathcal{V}$ and $x_n/n \to 0$ then

$$\lim_{n\to\infty} n^{-1}\log E_{x_n}\left(e^{\lambda W_n}\right) = c(\lambda).$$

(2) Suppose $V \in \mathcal{V}_+$ and $\beta \in (0, 1/2)$. We already know that $\lim_{n\to\infty} P_0(W_n/n^{1/2} \leq x) = P\left(2^{-1/2}\bar{V}X_{1/2} \leq x\right)$, so one can ask if the following is true:

$$\lim_{n\to\infty} n^{2\beta-1}\log E\left(\exp\left\{\lambda 2^{\frac{1}{2}}\bar{V}\, n^{(1-2\beta)/2} X_{\frac{1}{2}}\right\}\right) = \begin{cases} \frac{\lambda^2}{2}(\bar{V})^2 & \text{if } \lambda \geq 0 \\ 0 & \text{if } \lambda \leq 0 \end{cases}.$$

The answer is yes because

$$E\left(e^{\lambda X_{1/2}}\right) = f_{\frac{1}{2}}(\lambda) = \pi^{-1/2} e^{\lambda^2} \int_{-2\lambda}^{\infty} e^{-x^2/4}\, dx \,,\ \lambda \in \mathbb{R}.$$

In view of Remark (2) one can also ask if

(2.19) $$\lim_{n \to \infty} n^{4\beta-1} \log E\left(e^{\lambda n^{(1-4\beta)/4} W}\right) = \lambda^4 B_V^2/2,$$

where W is a random variable satisfying $\lim_{n \to \infty} P_0 (W_n/n^{1/4} \leq x) = P(W \leq x)$.

As proved in Kesten (1962), we know that such a W is given by $W = \sigma 2^{-\frac{1}{4}}(2h+1)^{1/2}(X_{1/2})^{1/2}Z$, where Z and $X_{1/2}$ are independent, Z being Gaussian with mean 0 and variance 1, and $\sigma > 0$ is such that

$\lim_{n \to \infty} P_0\left((V(Y_1) + \ldots + V(Y_n))/\sigma n^{1/2} \leq x\right) = P(Z \leq x)$, where $(Y_k)_{k \geq 0}$ is a Markov chain on [-h,h] associated with the operator

$$Sf(x) = \begin{cases} \Pi f(x) & \text{if } |x| < h \\ (f(h) + f(h-1)/2 & \text{if } x = h \\ (f(-h) + f(-h+1))/2 & \text{if } x = -h \end{cases} \quad , \ f \in \ell_2.$$

It follows that $\lim_{n \to \infty} n^{4\beta-1} \log E\left(e^{\lambda n^{(1-4\beta)/4} W}\right) =$

$$\lim_{n \to \infty} n^{4\beta-1} \log f_{\frac{1}{2}}\left(2^{-3/2} \lambda^2 \sigma^2 (2h+1)\, n^{(1-4\beta)/2}\right) = \frac{\lambda^4 \sigma^2 (2h+1)^2}{8}.$$

Therefore (2.19) holds iff

(2.20) $$\sigma^2 = (2/2h+1)\, B_V$$

(2.20) will follow if we prove

$$(2.21) \qquad \lim_{n\to\infty} E_x\left(\exp\left\{\lambda n^{-1/2} \sum_{k:1}^{n} V(Y_k)\right\}\right) = \exp\left\{\frac{\lambda^2 B_V}{2h+1}\right\}.$$

The proof of (2.21) is similar to the proof of Lemma 2.3. We can show that there exists a convex function $\tilde{c}(\lambda) \geq 0$ such that $\lim_{\lambda\to 0} \tilde{c}(\lambda)/\lambda^2 = (1/2h+1)B_V$ and

$$a_n e^{n\tilde{c}(\lambda n^{-1/2})} \leq E_x\left(\exp\left\{\lambda n^{-1/2} \sum_{k:1}^{n} V(Y_k)\right\}\right) \leq b_n e^{n\tilde{c}(\lambda n^{-1/2})}$$

where $a_n \to 1$ and $b_n \to 1$. The details are left to the reader.

(3) (2.20) provides an easier way to compute σ^2 that the one given in Kesten (1962).

3. Applications to systems of independent random walks. As mentioned in Section 1, Theorem 1.2.a) is proved in Cox and Durrett (1989) with $V(x) = \begin{cases} 1 & \text{if } x=0 \\ -1 & \text{if } x=1 \\ 0 & \text{otherwise} \end{cases}$. Repeating the same arguments, the proof of our theorem is easy (see Cox and Durrett (1989) for the details).

Proof of Theorem 1.2: a) Suppose that $V \in \mathcal{V}_0$ and $\beta \in (0, 1/4)$. Let $L_n(V) = \sum_x \sum_{k:0}^{n} \xi_k(x)V(x)$. If $\xi_0(\cdot)$ satisfies (*), then $\log E\left(e^{n^{-\beta}L_n(V)}\right) = \sum_x \xi_0(x)\log E_x\left(e^{n^{-\beta}(V(x)+W_n)}\right) = n^{-\beta}L_0(V) + \sum_x \xi_0(x)\log E_x\left(e^{n^{-\beta}W_n}\right)$, where $W_n = \sum_{k:1}^{n} V(S_k)$. Since $V(x) = 0$ for $|x| > h$, $\sum_x \xi_0(x)\log E_x\left(e^{n^{-\beta}W_n}\right) = \sum_{|x|\leq n+h} \xi_0(x)\log E_x\left(e^{n^{-\beta}W_n}\right)$.

As in Cox and Durrett (1989) one can show that for some $M > (1/2)B_V$,

$$\lim_{n\to\infty} n^{6\beta-2} \sum_{Mn^{1-2\beta} < |x| \le n+h} \xi_0(x)\log E_x\left(e^{n^{-\beta}W_n}\right) = 0.$$

Next we define $\mu_n(A) = n^{2\beta-1} \sum_{x\in n^{1-2\beta}A} \xi_0(x)$, $A \subset [-M,M]$ and $\phi_n(y) = n^{4\beta-1}\log E_{[yn^{1-2\beta}]}\left(e^{n^{-\beta}W_n}\right)$, where $[x]$ stands for the integer part of x. By Theorem 1.1, $\phi_n(y_n) \to \phi(y) = \left(B_V^2/2 - |y|\,B_V\right)^+$ boundedly as $y_n \to y$ in $[-M,M]$. It follows that $\lim_{n\to\infty} n^{6\beta-2}\log E\left(e^{n^{-\beta}L_n(V)}\right) = \lim_{n\to\infty}\int_{[-M,M]} \phi_n(y)\mu_n(dy) = \int_{-M}^{M} \phi(y)dy = (1/4)B_V^3$ proving a).

The proof of b) being similar, it is omitted. ∎

We will now prove Theorem 1.3..

Proof of theorem 1.3: It follows from Theorem 1.2 that for any $V \in \mathcal{V}$ and $\alpha \in (0,1)$:

$$\text{(3.1)} \qquad \lim_{n\to\infty} n^{-(1+3\alpha)/2}\log E\left(\exp\left\{n^{(1+3\alpha)/2}\left(\langle D_n,V\rangle/n^{\alpha}\right)\right\}\right) = (1/4)((\overline{V})^+)^3,$$

where $\langle D_n,V\rangle = \sum_x V(x)\,D_n(x)$ and $\xi_0(\cdot)$ satisfies (*).

Keeping this in mind, let $\mathcal{F}$ be the class of finite sets $F \subset \mathbb{Z}$, let $P_F : \Lambda \to \Lambda_F$, $\Lambda_F = \mathbb{R}_+^F$, be the canonical projection and let $\mathcal{V}_F = \{V \in \mathcal{V};\ V(x) = 0 \text{ on } F^c\}$, $F \in \mathcal{F}$.

From well-known results on large deviations, (3.1) entails that for every $F \in \mathcal{F}$, $\alpha \in (0,1)$,

(3.2) $(\Lambda_F, P_F(D_n/n^{\alpha}), n^{(1+3\alpha)/2})$ is a large deviation system with action functional I_F given by

$$I_F(\lambda) = \sup_{V \in \mathcal{V}_F} \left(\langle \lambda, V\rangle - (1/4)\left((\overline{V})^+\right)^3\right) = \begin{cases} 4(c/3)^{3/2} & \text{if } \lambda(x) \equiv c > 0 \\ +\infty & \text{otherwise} \end{cases},$$

$\lambda \in \Lambda_F$.

Since Λ is the projective limit of the spaces Λ_F, $F \in \mathcal{F}$, we can apply Theorem 3.3 of Dawson and Gärtner (1987) to get from (3.2):

(3.3) $(\Lambda, D_n/n^{\alpha}, n^{(1+3\alpha)/2})$ is a large deviation system with action functional $I(\cdot)$ given by
$$I(\lambda) = \sup_{F \in \mathcal{F}} I_F(P_F(\lambda)) = \begin{cases} 4(c/3)^{3/2} & \text{if } \lambda(x) \equiv c \geq 0 \\ +\infty & \text{otherwise} \end{cases}, \lambda \in \Lambda.$$

The proof is complete.

Remarks: (1) We can prove that when $\xi_0(\cdot)$ satisfies (*), $V \in \mathcal{V}_0$ and $\beta = 1/4$ (resp. $V \in \mathcal{V}$ and $\beta = 1/2$),

$$\lim_{n \to \infty} n^{-1/2} \log E\left(e^{n^{-\beta} L_n(V)}\right) = \tag{3.4}$$
$$\int_{-\infty}^{\infty} \log\left(1 + \left(\int_0^1 \frac{|x| e^{-x^2/2t}}{(2\pi t^3)^{1/2}} \left(f_{1/2}\left(2^{-\frac{1}{2}} A_V (1-t)^{1/2}\right) - 1\right) dt\right)\right) dx$$

where $A_V = B_V$ (resp. $\overline{V}$).

(2) When $(\xi_0(x))_{x \in \mathbb{Z}}$ are i.i.d. Poisson with mean 1, we obtain for $V \in \mathcal{V}_0$ and $\beta = 1/4$ (resp. $V \in \mathcal{V}$ and $\beta = 1/2$)

$$\lim_{n\to\infty} n^{-1/2}\log E\left(e^{n^{-\beta}L_n(V)}\right) = \int_{-\infty}^{\infty}\left(\int_0^1 \frac{|x\, e^{-x^2/2t}}{(2\pi t^3)^{1/2}}\left(f_{1/2}\left(2^{-\frac{1}{2}}A_V(1-t)^{1/2}\right)-1\right)dt\right)dx =$$

(3.5) $$(2/\pi)^{1/2}\int_0^1\left(f_{1/2}\left(2^{-\frac{1}{2}}A_V(1-t)^{1/2}\right)-1\right)\Big/ t^{1/2}\, dt = \psi(A_V) =$$

$$A_V\int_0^1 f_{1/2}\left(2^{-\frac{1}{2}}A_V t^{1/2}\right)dt, \text{ where } A_V = B_V(\text{resp. } \overline{V}).$$

The details of the proofs of (3.4) and (3.5) can be found in Cox and Durrett (1989).

4. Extension of the results to continuous time random walk and Brownian motion. With the some minor modifications the analogues of Theorem 1.1 hold for continuous time random walk (Theorem 4.1) and Brownian motion (Theorem 4.2).

We begin with the continuous time random walk $(S_t)_{t\geq 0}$. Let E_x denotes the expectation with respect to $(S_t)_{t\geq 0}$, where $S_0 = x$, and let $W_t = \int_0^t V(S_u)d_u$, $t \geq 0$, $V \in \mathcal{V}$ fixed.

Theorem 4.1 a) If $V \in \mathcal{V}_0$, $\beta \in (0, 1/4)$ and $x_t / t^{1-2\beta} \to x$, then

$$\lim_{t\to\infty} t^{4\beta-1}\log E_{x_t}\left(e^{\lambda t^{-\beta}W_t}\right) = \left(\lambda^4 D_V^2/2 - |x|\,\lambda^2 D_V\right)^+, \text{ where } D_V = 2\sum_x G^2(x), \text{ and}$$

$G(x) = \sum_{y\leq x} V(y)$.

b) If $V \in \mathcal{V}_+$, $\beta \in (0, 1/2)$ and $x_t / t^{1-\beta} \to x$, then

$$\lim_{t\to\infty} t^{2\beta-1}\log E_{x_t}\left(e^{\lambda t^{-\beta}W_t}\right) = \begin{cases}\left(\lambda^2(\overline{V})^2/2 - |x|\lambda\overline{V}\right)^+ & \text{if } \lambda \geq 0\\ 0 & \text{if } \lambda \leq 0\end{cases}$$

Proof: The proof is similar to the proof of Theorem 1.1.

From the spectral theorem, for any $f \in \ell_2$, $\sum_x f(x)\, E_x\left(e^{\lambda W_t} f(S_t)\right) \le \|f\|_2\, e^{t c(\lambda)}$,

where $c(\lambda) = \sup_{\substack{f \in \ell_2 \\ \|f\| = 1}} \left((\Pi - I)f, f\right) + \lambda(Vf, f)$.

It can be shown that $c(0) = 0$, $c(\lambda) \ge 0\ \forall\, \lambda \in \mathbb{R}$ and $c(\cdot)$ is convex; moreover $c(\lambda) = \sup\{\operatorname{ch}\theta - 1;\ \theta \ge 0,\ \Phi(\lambda, \theta) = 0\}$, where $\Phi(\lambda, \theta) = e^{\theta} F_{\lambda,\theta}(h+1) - F_{\lambda,\theta}(h)$ and

$F_{\lambda,\theta}(\cdot)$ is the unique function satisfying $\begin{cases} F_{\lambda,\theta}(x) = e^{\theta x}, & x \le -h \\ (\Pi + \lambda V) F_{\lambda,\theta}(x) = (\operatorname{ch}\theta) F_{\lambda,\theta}(x), & x \ge -h \end{cases}$

Next $E_x\left(e^{\lambda t^{-\beta} W_t}\right) \ge \left(F_{\lambda,\theta(\lambda t^{-\beta})}(x) \Big/ \sup_y F_{\lambda,\theta(\lambda t^{-\beta})}(y)\right) e^{tc(\lambda t^{-\beta})}$

and $\sup_{|x| \le Mt} \left(e^{\lambda t^{-\beta} W_t}\right) \le e^{|\lambda| t^{1-\beta} |V|_\infty} P_0(|S_t| \ge Mt) + (2Mt + 1) e^{tc(\lambda t^{-\beta})}$.

Since $\limsup_{t \to \infty} \frac{1}{t} \log P_0(|S_t| \ge Mt) \le -\frac{M^2}{2}$, we see that for t large enough $E_x\left(e^{\lambda t^{-\beta} W_t}\right) \le 1 + 3Mt\, e^{tc(\lambda t^{-\beta})}$. The rest of the proof is quite similar and even easier that the proof of Theorem 1.1. For example, the "integration by part" formula yields

$$\begin{aligned} (4.2) \qquad -2\lambda \sum_{x:-h}^{b} F_{\lambda,\theta}(x) g_\theta(b - x) V(x) &= g_\theta(0) F_{\lambda,\theta}(b + 1) \\ &- g_\theta(-1) F_{\lambda,\theta}(b) - e^{-\theta(h+1)}\left(e^{\theta} g_\theta(b+h+1) - g_\theta(b+h)\right), \end{aligned}$$

where $g_\theta(\cdot)$ is such that $\Pi g_\theta(x) = \operatorname{ch}\theta\, g_\theta(x)$, $x \ge 0$.

For appropriate choices of $g_\theta(\cdot)$, we can prove that when $V \in \mathcal{V}_+$, $\theta(\lambda) = 0$ on $[-\varepsilon, 0]$ for some $\varepsilon > 0$ and $\lim_{\lambda \downarrow 0} \theta(\lambda)/\lambda = \overline{V}$; when $V \in \mathcal{V}_0$, $\lim_{\lambda \to 0} \theta(\lambda)/\lambda^2 = D_V$ and $D_V = \left((2h+1)/2\right) \rho^2$, where ρ is such that

$$\lim_{t\to\infty} E_0\left(\exp\left\{i\lambda W_t \Big/ (\rho t^{1/2} 2^{-1/2}(2h+1))^{1/2}\right\}\right) = f_{\frac{1}{2}}\left(-\frac{\lambda^2}{2}\right). \quad \blacksquare$$

The next theorem is the analogue of Theorem 1.1 for Brownian motion.

Let $\mathcal{V}$ be the set of all $V : \mathbb{R} \to \mathbb{R}$ such that V is measurable, bounded and $V(x) = 0 \ \forall |x| > h$ for some $h > 0$. Further let E_x denotes expectation with respect to the Brownian motion $(B_t)_{t \geq 0}$ starting at x.

As before $\mathcal{V}_+ = \{V \in \mathcal{V} ; \overline{V} = \int V(x)dx > 0\}$ and $\mathcal{V}_0 = \{V \in \mathcal{V} ; \overline{V} = 0\}$.

Theorem 4.2. Suppose $V \in \mathcal{V}$ and let $W_t = \int_0^t V(B_u)du$.

a) If $V \in \mathcal{V}_0$, $\beta \in (0, 1/4)$ and $x_t / t^{1-2\beta} \to x$, then $\lim_{t\to\infty} t^{4\beta-1}\log E_{x_t}\left(e^{\lambda t^{-\beta} W_t}\right) =$

$\left(\lambda^4 K_V^2 / 2 - |x|\, \lambda^2 K_V\right)^+$, where $K_V = 2\int_{-\infty}^{\infty} G^2(x)dx$ and $G(x) = \int_{-\infty}^{x} V(y)dy$. Moreover

$$\lim_{t\to\infty} E_x\left(e^{i\lambda W_t/(2^{1/2}t^{1/2}K_V)^{1/2}}\right) = f_{\frac{1}{2}}\left(-\frac{\lambda^2}{2}\right).$$

b) If $V \in \mathcal{V}_+$, $\beta \in (0, 1/2)$ and $x_t / t^{1-\beta} \to x$, then

$$\lim_{t\to\infty} t^{2\beta-1}\log E_{x_t}\left(e^{\lambda t^{-\beta} W_t}\right) = \begin{cases} (\lambda^2(\overline{V})^2/2 - |x|\lambda\overline{V})^+, & \lambda \geq 0 \\ 0, & \lambda \leq 0 \end{cases}.$$

Proof: Define $c(\lambda) = \sup\limits_{\substack{f \in C_0^\infty \\ \|f\| = 1}} \lambda(Vf, f) - I(f)$, where C_0^∞ is the space of infinitely differentiable functions $f : \mathbb{R} \to \mathbb{R}$ with compact support, $(f,g) = \int_{-\infty}^{\infty} f(x)g(x)dx$ is the usual scalar product on the Hilbert space $H = L^2(dx)$, $\|f\|^2 = (f, f)$, $f \in H$, and

$I(f) = \frac{1}{2}\int_{-\infty}^{\infty} (\dot{f}(x))^2 dx$, where $\dot{f}$ stands for the derivative of f.

The reason why we introduce $c(\cdot)$ will become clear. For any $f, g \in H$, $\int f(x)\ E_x\left(e^{\lambda W_t} g(B_t)\right) dx \le \|f\| \cdot \|g\|\ e^{tc(\lambda)}$. Therefore $E_x\left(e^{\lambda t^{-\beta} W_t}\right) \le e^{|\lambda|\,|V|_\infty t^{1-\beta}}\, P_x(\|B_t\| \ge m_t) + E_x\left(e^{\lambda t^{-\beta} W_t}\, 1_{\{|B_t| \le m_t\}}\right)$ where m_t is a sequence to be chosen later, and $|V|_\infty = \sup_{x \in \mathbb{R}} |V(x)|$. Set $p(x) = (2\pi)^{-1/2} e^{-x^2/2}$. Then $\|p\| = (4\pi)^{-1/4}$ and by the Markov property of $(B_t)_{t \ge 0}$,

$$
\begin{gathered}
E_x\left(e^{\lambda t^{-\beta} W_t}\, 1_{\{|B_t| \le m_t\}}\right) \le \\
e^{|\lambda|\, t^{-\beta} |V|_\infty} \int_{-\infty}^{\infty} p(y - x)\, E_y\left(e^{\lambda t^{-\beta} W_{t-1}}\, 1_{\{|B_{t-1}| \le m_t\}}\right) dy \le \\
(2m_t)^{1/2} (4\pi)^{-1/4}\, e^{|\lambda|\, t^{-\beta} |V|_\infty + (t-1)\, c(\lambda t^{-\beta})} \le \\
(m_t)^{1/2}\, e^{|\lambda|\, t^{-\beta} |V|_\infty + t\, c(\lambda / t^{\beta})}
\end{gathered}
$$

since $c(\cdot) \ge 0$, $c(\cdot)$ is convex and $c(0) = 0$.

Choosing $m_t = 2tM$, $M > 2(|\lambda|\ |V|_\infty)^{1/2}$, we get for large t:

$$
\sup_{|x| \le Mt} e^{|\lambda|\, t^{-\beta} |V|_\infty} P_x(B_t| \ge 2M_t) \le 2e^{-\frac{tM^2}{4}}.
$$

Hence $\sup_{|x| \le tM} E_x\left(e^{\lambda t^{-\beta} W_t}\right) \le 2e^{-\frac{tM^2}{4}} + (2Mt)^{1/2} \exp\left\{ |\lambda|\, t^{-\beta}\, |V|_\infty + tc(\lambda / t^{\beta}) \right\}$. As before, when $c(\lambda) > 0$, there exists F_λ such that

$$
F_\lambda(x) = \begin{cases} e^{x(2c(\lambda))^{1/2}} & , \ x \le -h \\ e^{-h(2c(\lambda))^{1/2}} f_\lambda(x) & , \ |x| \le h \\ e^{-x(2c(\lambda))^{1/2}} f_\lambda(h) & , \ x \ge h \end{cases} \quad \text{and } E_x\left(e^{\lambda W_t} F_\lambda(B_t)\right) = F_\lambda(x) e^{tc(\lambda)},
$$

where $f_\lambda = f_{\lambda, c(\lambda)}$ and $f_{\lambda,\theta}$ is the unique solution of $f_{\lambda,\theta}(-h) = 1$, $\dot{f}_{\lambda,\theta}(-h) = (2\theta)^{1/2}$, $1/2\, \ddot{f}_{\lambda,\theta}(x) + \lambda\, V(x)\, f_{\lambda,\theta}(x) = \theta\, f_{\lambda,\theta}(x)$ almost surely, $x > -h$.

Moreover $c(\lambda) = \sup\{\theta \ge 0;\ \Phi(\lambda,\theta) = \dot{f}_{\lambda,\theta}(h) + (2\theta)^{1/2} f_{\lambda,\theta}(h) = 0\}$. Integrating by part $\int_{-h}^{x} \ddot{f}_{\lambda,\theta}(y)\, g_\theta(x - y)\, dy$, where $g_\theta(x) = a \operatorname{ch} x(2\theta)^{1/2} + (2\theta)^{-1/2}\, b \operatorname{sh} x(2\theta)^{1/2}$, $\theta > 0$, we obtain

$$\dot{f}_{\lambda,\theta}(x)a + f_{\lambda,\theta}(x)b + 2\lambda \int_{-h}^{x} f_{\lambda,\theta}(y)\, g_\theta(x-y)\, V(y)dy = \tag{4.2}$$
$$\left(a(2\theta)^{1/2} + b\right) \exp\left\{(x+h)\,(2\theta)^{1/2}\right\},\ x \geq -h.$$

With proper choices of a and b, one can prove that if $V \in \mathcal{V}_0$ and $V \not\equiv 0$, $c(\lambda) > 0$ $\forall\lambda \neq 0$ and $\lim_{\varepsilon \downarrow 0} c(\lambda\varepsilon)/\varepsilon^4 = \lambda^4 K_V^2/2$; when $V \in \mathcal{V}_+$, $c(\lambda) > 0$ $\forall\lambda > 0$, $c(0) = 0$ on $[-\varepsilon,0]$ for some $\varepsilon > 0$ and $\lim_{\varepsilon \downarrow 0} c(\varepsilon)/\varepsilon^2 = \frac{1}{2}(\overline{V})^2$. The rest of the proof goes as in Theorem 1.1, except for weak convergence, where the result follows from Papanicolaou, Stroock and Varadhan (1977). ■

Acknowledgments: The author is indebted to Ted Cox and Rick Durrett for helpful discussions.

REFERENCES

[1] Cox, J.T. and Durrett, R.(1989) Large deviations for independent random walks. To appear in Prob. Theory. Rel. Fields.

[2] Cox, J.T. and Griffeath, D.(1984) Large deviations for Poisson systems of independent random walks. Z. Wahrsch. verw. Gebiete **66** 543-558.

[3] Dawson, D.A. and Gärtner, J.(1987) Large deviations from the McKean-Vlasov limit for weakly interacting diffusions. Stochastics **20** 247-308.

[4] Kesten, H.(1962) Occupation times for Markov and semi-Markov chains. Trans. AMS **103** 82-112.

[5] Lee, T.Y.(1989) Large deviations for systems of noninteracting recurrent particles. Ann. Probab. **17** No. 1 46-57.

[6] Papanicolaou, G.C., Stroock, D.W. and Varadhan, S.R.S.(1977) Martingale approach to some limit theorems. Papers from the Duke Turbulence Conference. Duke Univ. Math. Ser. Vol. III, Duke Univ., Durham, N.C.

Bruno Rémillard
Département de mathématiques et d'informatique
Université du Québec à Trois-Rivières
Case postale 500
Trois-Rivières, Qué. Canada
G9A 5H7

Progress in Probability

Edited by:

Progress in Probability includes all aspects of probability theory and stochastic processes, as well as their connections with and applications to other areas such as mathematical statistics and statistical physics. Each volume presents an in-depth look at a specific subject, concentrating on recent research developments. Some volumes are research monographs, while others will consist of collections of papers on a particular topic.

Proposals should be sent directly to the series editors or to Birkhäuser Boston, 675 Massachusetts Avenue, Suite 601, Cambridge, MA 02139.

1 ÇINLAR/CHUNG/GETOOR. Seminar on Stochastic Processes, 1981
2 KESTEN. Percolation Theory for Mathematicians
3 ASMUSSEN/HERING. Branching Processes
4 CHUNG/WILLIAMS. Introduction to Stochastic Integration
5 ÇINLAR/CHUNG/GETOOR. Seminar on Stochastic Processes, 1982
6 BLOOMFIELD/STEIGER. Least Absolute Deviation
7 ÇINLAR/CHUNG/GETOOR. Seminar on Stochastic Processes, 1983
8 BOUGEROL/LACROIX. Products of Random Matrices with Application to Schrödinger Operator
9 ÇINLAR/CHUNG/GETOOR. Seminar on Stochastic Processes, 1984
10 KIFER. Ergodic Theory of Random Transformations
11 EBERLEIN/TAQQU. Dependence in Probability and Statistics
12 ÇINLAR/CHUNG/GETOOR. Seminar on Stochastic Processes, 1985
13 ÇINLAR/CHUNG/GETOOR/GLOVER. Seminar on Stochastic Processes, 1986
14 DEVROYE. A Course in Density Estimation
15 ÇINLAR/CHUNG/GETOOR/GLOVER. Seminar on Stochastic Processes, 1987
16 KIFER. Random Perturbations of Dynamical Systems
17 ÇINLAR/CHUNG/GETOOR/GLOVER. Seminar on Stochastic Processes, 1988
18 ÇINLAR/CHUNG/GETOOR/FITZSIMMONS/WILLIAMS. Seminar on Stochastic Processes, 1989
19 ALEXANDER/WATKINS. Spatial Stochastic Processes: A Festschrift in Honor of Ted Harris on His 70th Birthday
20 HAAGERUP/HOFFMANN-JØRGENSEN/NIELSEN. Probability in Banach Spaces 6: Proceedings of the Sixth International Conference, Sandbjerg, Denmark 1986
21 EBERLEIN/KUELBS/MARCUS. Probability in Banach Spaces 7: Proceedings of the Seventh International Conference
22 PINSKY. Diffusion Processes and Related Problems in Analysis, Volume I: Diffusions in Analysis and Geometry